Extreme-Temperature and Harsh-Environment Electronics

Physics, technology and applications

Extreme-Temperature and Harsh-Environment Electronics

Physics, technology and applications

Vinod Kumar Khanna
CSIR-Central Electronics Engineering Research Institute Pilani-333031 (Rajasthan), India

IOP Publishing, Bristol, UK

ISBN 978-0-7503-1155-7 (ebook)
ISBN 978-0-7503-1156-4 (print)
ISBN 978-0-7503-1157-1 (mobi)

DOI 10.1088/978-0-7503-1155-7

Version: 20170301

IOP Expanding Physics
ISSN 2053-2563 (online)
ISSN 2054-7315 (print)

British Library Cataloguing-in-Publication Data: A catalogue record for this book is available from the British Library.

Published by IOP Publishing, wholly owned by The Institute of Physics, London

IOP Publishing, Temple Circus, Temple Way, Bristol, BS1 6HG, UK

US Office: IOP Publishing, Inc., 190 North Independence Mall West, Suite 601, Philadelphia, PA 19106, USA

Cover image: Computer illustration of NASA's Juno spacecraft over Jupiter's pole. Credit: Mark Garlick / Science Photo Library.

To my late father Shri Amarnath Khanna
For his earnest endeavours to shape my educational career.
To my beloved mother Smt. Pushpa Khanna
For her love and blessings to guide me on the path of Life.
To my daughter Aloka
For bringing joy and happiness in the family.
To my wife Amita
For her unflinching and unfailing support.

Contents

Preface

Customarily electronic devices and circuits are required to operate at room temperature. This is more a matter of convenience and convention than optimization. On lowering the temperature, the performance of electronic devices is improved two-fold or by several orders of magnitude. This upgradation is observed in various forms, e.g. increased speed of digital systems, a better signal-to-noise ratio and greater bandwidth for analog systems, improved sensitivity for sensors, greater precision and range for measuring instruments, and overall deceleration of the ageing process of materials. However, low temperature is not always beneficial, e.g. the current gain and breakdown voltage of bipolar transistors are degraded with decreasing temperature. Broadly speaking, low-temperature electronics (LTE) has two offshoots: semiconductor-based electronics and superconductor-based electronics. Semiconductor-based electronics pertains to electronics that can be made to operate at any temperature from room temperature, or even much higher, down to the lowest cryogenic temperatures, 1 K or below. On the other hand, present-day systems based on superconductivity are confined to operation at low cryogenic temperatures, below about 10 K, which has been a serious impediment to their widespread use. The advent of high-temperature superconductors and associated systems appears to hold some promise.

The other side of the story is high-temperature electronics (HTE). 'High temperature' is any temperature above 125 °C. This cut-off point is frequently specified as the upper limit at which standard commercial silicon devices are supposed to function properly. However, tests on standard commercial components indicate that even temperatures up to 150 °C may be applied to selected silicon components. Certain niche applications have to operate beyond the melting point of many materials used in present-day industrial electronics. Examples are monitors and down-hole well-drilling tools for energy exploration. Aircraft and turbine engine controls also have to withstand high temperatures.

The scope of extreme temperature electronics (ETE) encompasses operation in temperature ranges outside the traditional commercial, industrial or military ranges, i.e. −55 °C/−65 °C to +125 °C. There are three categories of ETE, known as HTE referring to temperatures over +125 °C, LTE for temperatures below −55 °C/−65 °C, and cryogenic temperature electronics, below −150 °C.

Apart from the extreme temperature applications discussed above, mention may be made of chemically corrosive conditions. Excessively high humidity causes corrosion in electronic devices. Low humidity favors the building up of static electricity. Atmospheric corrosion, an electrochemical process that occurs on metals covered by a thin film of water and ions, is often responsible for damage to electrical and electronic components leading to premature failures even in indoor ambient conditions.

Ionizing radiation consisting of electromagnetic waves such as x-rays and γ-rays, and particulate radiation, i.e. protons, electrons, neutrons, etc, causes malfunctions and failures in electronic components and circuits. The extent of damage depends on the type of radiation, its intensity or energy, the time of exposure and hence the dose, in addition to the distance between the radiation source and the electronic equipment.

Unconventional electronics is subdivided into different areas, e.g. electronics capable of operating at high temperatures for deep well and geothermal logging, light weight ground and air vehicles, and space exploration; electronics benefitting from low-temperature operation as well as electronics able to withstand low temperatures for infrared systems, satellite communications and medical equipment, and newer opportunities in wireless and mobile communications, computers, and measurement and scientific equipment; electronics capable of countering the detrimental effects of humid and chemically corrosive environments for use in tropical climates and industries such as pulp and paper processing, oil and petroleum refining, mining, foundry, chemicals, etc; and radiation environment electronics for the space, medical and nuclear power industries.

This book has three objectives:

(i) to explore the beneficial/harmful impacts of extreme temperatures on electronic devices and circuits, as well as to enquire into the complexities introduced by harsh conditions such as damp, dirty and radiation-filled environments;
(ii) to describe the techniques adopted to utilize the advantages of these unconventional situations; and
(iii) to present the remedial measures taken to counteract and deal with these unfavorable circumstances.

This book is written for graduate and research students in electrical and electronic engineering. It will serve as a useful supplement to microelectronics course material, treating this specialized discipline with breadth and depth. The book answers several questions that come to mind when one starts thinking and imagining beyond a normal electronics course. The book is also written to fulfill the needs of electronic device and process design engineers, as well as circuit and system developers engaged in this fast-moving field, covering both fundamentals and applications. Scientists and professors engaged in this field will also find it useful as a comprehensive guide to the state-of-the-art electronic technologies for hostile environments.

Vinod Kumar Khanna
CSIR-CEERI, Pilani

Acknowledgements

Above all, I must thank Almighty God for giving me the vigor and wisdom to complete this work. I am grateful to all my colleagues and Director, CSIR-CEERI, Pilani for constant encouragement in my efforts. I wish to thank my editor and editorial assistant for rejuvenating my interest and boosting my confidence from time to time. Their kind co-operation and support led to timely completion of the work. I owe a profound debt of gratitude to all authors/editors of research papers, magazine articles and web pages, on whose work this book is based. Most of these excellent works are cited in the bibliographies at the end of each chapter. However, if anyone has inadvertently escaped mention, the same may be forgiven. Last but not least, I am grateful to my family for providing the serene atmosphere and relieving me of many domestic responsibilities so that I could devote more time to book writing. To all the above, I express my sincere thanks.

Vinod Kumar Khanna
CSIR-CEERI, Pilani

About this book

Extreme-Temperature and Harsh-Environment Electronics: Physics, technology and applications presents a unified perspective combining the impact of extremely high and exceedingly low temperatures on the operation of semiconductor electronics, in addition to the influence of hostile environments such as high-humidity conditions, as well as surroundings contaminated by chemical vapors, nuclear radiations or those disturbed by mechanical shocks and vibrations. Incorporating the preliminary background material and thus laying down foundations for easy understanding, the progress in mainstream silicon, silicon-on-insulator and gallium arsenide electronics is sketched. Contemporary wide bandgap semiconductor technologies such as silicon carbide, gallium nitride and diamond electronics are explored. After a brief treatment of superconductivity, concepts of superconductive electronics are introduced. Progress in Josephson junctions, SQUIDs and RFSQ logic circuits is highlighted. The state of the art in high-temperature superconductor-based power delivery is surveyed. Succeeding chapters look at various protection schemes that have been devised to shield electronic circuits and equipment from adverse ambient conditions. These conditions range from the presence of high moisture concentrations in the atmosphere, to showers of high-energy particles such as the alpha particles, protons and nuclei of heavy elements that flood the atmosphere from outer space. A particularly attractive area is the dampening of vibration effects to protect electronics from the quivering disturbances or jerks that are always present near large machines or during accidental falls of electronic equipment.

In this book, a lucid description of this vast panorama of topics is reinforced by an elegant mathematical treatment. Broad in scope, this comprehensive treatise provides a coherent, well-organized and amply illustrated exposition of the subject which will be immensely useful for graduate and research students, professional engineers and researchers engaged in this frontier of technology. Its three-pronged approach encompassing physical aspects, technological breakthroughs and application examples will make reading interesting and enjoyable.

Author biography

Vinod Kumar Khanna

Vinod Kumar Khanna, born on 25 November 1952 in Lucknow, Uttar Pradesh, India, is an Emeritus Scientist at the CSIR-Central Electronics Engineering Research Institute, Pilani, Rajasthan, India, and Emeritus Professor, AcSIR (Academy of Scientific and Innovative Research), India. He is former Chief Scientist and Head of the MEMS and Microsensors Group, CSIR-CEERI, Pilani, and Professor, AcSIR. During his service tenure of more than 34 years at CSIR-CEERI, starting in April 1980, he has worked on various research and development projects on power semiconductor devices (high-current and high-voltage rectifiers, high-voltage TV deflection transistors, power Darlington transistors, fast switching thyristors, power DMOSFETs and IGBTs), PIN diode neutron dosimeters and PMOSFET gamma ray dosimeters, ion-sensitive field-effect transistors (ISFETs), microheater-embedded gas sensors, capacitive MEMS ultrasonic transducers (CMUTs) and other MEMS devices. His research interests have covered micro- and nanosensors and power semiconductor devices. From 1977 to 1979, he was a research assistant in the Physics Department, Lucknow University.

Dr Khanna's deputations abroad include Technische Universität Darmstadt, Germany, 1999; Kurt-Schwabe-Institut für Mess- und Sensortechnik e.V., Meinsberg, Germany, 2008; Institute of Chemical Physics, Novosibirsk, Russia, 2009; and Fondazione Bruno Kessler, Povo, Trento, Italy, 2011. He also participated in and presented research papers at the IEEE-IAS Annual Meeting, Denver, Colorado, USA, 1986.

Dr Khanna received his MSc degree in physics with specialization in electronics from the University of Lucknow in 1975, and his PhD degree in physics from Kurukshetra University, Kurukshetra, Haryana, in 1988 for his work on a thin-film aluminum oxide humidity sensor. A fellow of the Institution of Electronics and Telecommunication Engineers (IETE), India, he is a life member of the Indian Physics Association (IPA), Semiconductor Society (India) (SSI), and Indo-French Technical Association.

Dr Khanna has published ten previous books, six chapters in edited books, and 185 research papers in national/international journals and conference proceedings; he holds two US and two Indian patents.

Abbreviations, acronyms, chemical symbols and mathematical notation

A	ampere
AC	alternating current
Ag	silver (argentum)
Ag_2O	silver oxide
Ag_2S	silver sulfide
AH	absolute humidity
Al	aluminum
ALD	atomic layer deposition
AlGaAs	aluminum gallium arsenide
AlGaN	aluminum gallium nitride
AlN	aluminum nitride
Al_2O_3	aluminum oxide
As_2O_3	arsenic trioxide
As_2O_5	arsenic pentoxide
Au	gold (aurum)
$BaCO_3$	barium carbonate
BCS	Bardeen–Cooper–Schrieffer (theory of superconductivity)
Be	beryllium
BGJFET	buried-grid JFET
B_2H_6	diborane
B_2O_3	boron trioxide
BiCMOS	bipolar CMOS
BJT	bipolar junction transistor
BN	boron nitride
BOX	buried oxide
BPF	bandpass filter
BSCCO	bismuth–strontium–calcium–copper oxide
BSG	borosilicate glass
C	carbon
°C	degrees centigrade
CeO_2	cerium (IV) oxide
CH_4	methane
$(CH_3)_3B$	trimethyl boron
CHFET	complementary heterojunction FET
$C_7H_7N_3$	tolytriazole
$C_{12}H_{24}N_2O_2$	dicyclohexylammonium nitrite
$C_6H_8O_7$	citric acid
$C_6H_{10}O_4$	adipic acid
CMOS	complementary metal–oxide–semiconductor (FET)
COB	chip-on-board (assembly)

CO-OP	controlled over pressure (process)
$CoSi_2$	cobalt silicide
CQT	cascaded quadruplet trisection coupling structure (filter geometry)
Cr	chromium
CRT	cathode ray tube
CTE	coefficient of thermal expansion
CVD	chemical vapor deposition
Cu	copper
CuO	copper oxide
Cu_2S	copper sulfide
CZ	Czochralski (single-crystal silicon)
dB	decibel
DC	direct current
DD	displacement damage
DI-BSCCO	dynamically innovative-BSCCO
DICE	dual interlocked storage cell
DICHAN	dicyclohexylammonium nitrite
2DEG	two-dimensional electron gas
2DHG	two-dimensional hole gas
DGVTJFET	dual-gate vertical channel trench JFET
DMOSFET	double-diffused MOSFET
DMVTJFET	depletion-mode VTJFET
DRAM	dynamic random access memory
e-beam	electron beam
EG	electronic grade (polysilicon)
EHP	electron hole pair
EIA	Electronics Industries Association
EL Ni	electroless nickel
EMI	electromagnetic interference
EMVTJFET	enhancement-mode VTJFET
ESD	electrostatic discharge
ETO	ethylene oxide
eV	electron volt
fcc	face-centered cubic
FCL	fault current limiter
FET	field-effect transistor
FIRST	Far Infra-Red Space Telescope
Ga	gallium
GaAs	gallium arsenide
GaN	gallium nitride
Ga_2O_3	gallium trioxide
Ge	germanium
GHz	gigahertz
GTO	gate turn-off (thyristor)
Gy	gray
h	hour
HASL	hot air solder leveling
HBT	heterojunction bipolar transistor
HEMT	high electron mobility transistor
HFET	heterojunction FET

HfO_2	hafnium oxide
Hg	mercury (hydrargyrum)
HPHT	high-pressure and high-temperature
HTE	high-temperature electronics
HTS	high-temperature superconductor
HVAC	high-voltage alternating current
Hz	hertz
IBAD	ion beam-assisted deposition
IC	integrated circuit
IF	intermediate frequency
i-GaN	intrinsic gallium nitride
IGBT	insulated gate bipolar transistor
IM Au	immersion gold
InAlN	indium aluminum nitride
InAs	indium arsenide
InGaAs	indium gallium arsenide
InP	indium phosphide
InSb	indium antimonide
ISOPHOT	Infrared Space Observatory Photometer
JFET	junction FET
JJ	Josephson junction
JTE	junction termination extension
K	kelvin (scale of temperature)
km	kilometer
keV	kilo-electronvolt
kV	kilovolt
$LaAlO_3$	lanthanum aluminate
LaB_6	lanthanum hexaboride
LAO	lanthanum aluminate
LCD	liquid crystal display
LCJFET	lateral channel JFET
LEC	liquid encapsulated Czochralski
LED	light-emitting diode
Li	lithium
LO	local oscillator
LPCVD	low-pressure CVD
LPF	low-pass filter
LTE	low-temperature electronics
m	meter
mA	milli-ampere
MAG	maximum available gain
MEA	more electric aircraft
MEMS	microelectromechanical systems
MESFET	metal–semiconductor FET
meV	milli-electronvolt
Mg	magnesium
MG	metallurgical grade (polysilicon)
MgO	magnesium oxide
MHz	megahertz
MISFET	metal–insulator–semiconductor FET

mK	millikelvin
mm	millimeter
MMIC	monolithic microwave IC
$Mn(NO_3)_2$	manganese nitrate
MnO_2	manganese dioxide
mΩ	milli-ohm
MOCVD	metal–organic CVD
MOSFET	metal–oxide–semiconductor FET
$MoSi_2$	molybdenum disilicide
MPCVD	microwave plasma-enhanced CVD
MPS	merged PiN/Schottky (diode)
mS	millisiemen
MTTF	mean-time-to-failure
MV	megavolt
NaOH	sodium hydroxide
Nb	neobium
Ni	nickel
NiCr	nichrome
NMOS	n-channel MOSFET
NP0	negative positive zero
NO_2	nitrogen dioxide gas
ns	nanosecond
OCVD	open circuit voltage decay
OP-AMP	operational amplifier
OSP	organic solderability preservative
Pb	lead (plumbum)
pBN	pyrolytic boron nitride
PCB	printed circuit board
PCBA	printed circuit board assembly
PECVD	plasma-enhanced CVD
PH_3	phosphine
PLD	pulsed laser deposition
PMOS	p-channel MOSFET
$POCl_3$	phosphorous oxychloride
ps	picosecond
PSG	phosphosilicate glass
Pt	platinum
PTFE	polytetrafluoroethylene
QUAD	quadruple
rad	radiation absorbed dose
RAM	random access memory
ReBCO	rare-earth barium copper oxide
RF	radio frequency
RFSQ	rapid flux single quantum
RH	relative humidity
RHBD	radiation hardening by design
RHBP	radiation hardening by process
RoHS	restriction of hazardous substances
RPCVD	reduced pressure CVD
R–S	reset-set (flip flop)

R_2SiO	structural unit of silicone where R is an organic group
RTA	rapid thermal annealing
RTV	room-temperature vulcanization
Ru	ruthenium
s	second
S	sulfur, siemen
SBD	Schottky barrier diode
Se	selenium
SEB	single event burnout
SEE	single event effect
SEFI	single event functional interrupt
SEGR	single event gate rupture
SEI	Sumitomo Electric Industries Ltd
SEJFET	static expansion channel JFET
SEL	single event latchup
SES	single event snapback
SET	single event transient
SEU	single event upset
SFCL	superconducting FCL
Si	silicon
SIAFET	static induction-injected accumulated FET
SiC	silicon carbide
SiH_4	silane
$SiHCl_3$	trichlorosilane
SiH_2Cl_2	dichlorosilane
SiGe	silicon–germanium (alloy)
$Si_{1-x}Ge_x$	silicon–germanium (alloy) where x is the mole fraction of germanium in the alloy with a value from 0 to 1
SMD	surface mount device
Si_3N_4	silicon nitride
$Si_xO_yN_z$	silicon oxynitride
SiO_2	silicon dioxide
$Si(OC_2H_5)_4$	tetraethylorthosilicate (TEOS)
SISO	spiral in/spiral out (resonator)
Sn	tin (stannum)
SOI	silicon-on-insulator
SQUID	superconducting quantum interference device
SRH	Shockley–Read–Hall (recombination)
SS	subthreshold swing
T	tesla
TaN	tantalum nitride
Ta_2O_5	tantalum pentoxide
$TaSi_2$	tantalum disilicide
TBCCO	thallium–barium–calcium–copper oxide
TC	temperature coefficient
TCR	TC of resistance
TCS	trichlorosilane
Te	tellurium
TEOS	tetraethylorthosilicate
TID	total ionizing dose

Ti	titanium
TiN	titanium nitride
TiO_2	titanium dioxide
$TiSi_2$	titanium disilicide
TMB	trimethylboron
TMR	triple modular redundancy
TTL	transistor–transistor logic
UHVCVD	ultra-high vacuum CVD
UMOSFET	U-shaped MOSFET
UV	ultraviolet
VCI	volatile corrosion inhibitor (coating)
V_2O_5	vanadium pentoxide
VTJFET	vertical trench JFET
W	tungsten, watt
Wb	weber
WN_x	tungsten nitride
WSi_2	tungsten disilicide
$YBa_2Cu_3O_7$	YBCO
YBCO	yttrium barium cuprate
YIG	yttrium iron garnet
Y_2O_3	yttrium oxide
YSZ	yttria stabilized zirconia
Zn	zinc
ZrO_2	zirconium oxide
ZTC	zero temperature coefficient biasing point of a MOSFET

Roman alphabet symbols

a	depth of the active region in MESFET
A	parameter in the model of Quay *et al*, area
$\mathbf{A}$	magnetic vector potential
A^*	effective Richardson constant
A, B, C	parameters of the Bludau *et al* model
A_V	voltage gain
b	parameter in Arora–Hauser–Roulston equation, parameter in Chynoweth equations
$B(0)$	magnetic induction at the surface
$B(x)$	magnetic induction along the x-direction
BV_{CBO}	collector–base breakdown voltage of a bipolar transistor with emitter open
BV_{CEO}	collector–emitter breakdown voltage of a bipolar transistor with base open
BV_{DSS}	drain–source breakdown voltage of a MOSFET with gate shorted to source
C_1, C_2, C_3	parameters in the model of threshold ionization energy
c	velocity of light, damping coefficient (vibration theory)
C	capacitance
C_{ds}, C_{DS}	drain–source capacitance
C_{gd}	gate–drain capacitance
C_{iss}	intrinsic capacitance
C_{rss}	reverse transfer capacitance
C_{ox}	oxide capacitance per unit area
d	diameter, thickness, length, deformation
D	diffusion coefficient of carrier, diffusion coefficient of dopant
$d\mathbf{l}$	linear element
D_{nB}	diffusion constant of electrons in the p-base
$D_{nB}(x)$	position-dependent diffusion coefficient of electrons in the base
$\widetilde{\boldsymbol{D}}_{nB}$	position-averaged diffusion coefficient across the base profile
D_{pB}	diffusion constant of holes in the base
D_{pE}	diffusion constant of holes in the n^+-emitter
$d\mathbf{s}$	areal element
e	electronic charge
E	electric field
E_C	energy of conduction band edge
E_g, E_G	energy bandgap
E_{gB0}	silicon bandgap at zero doping of the base layer (in an HBT)
$\Delta E_{gB,A}$	bandgap narrowing of base layer due to acceptor impurity doping effect (in an HBT)
$\Delta E_{gBGe}(x = 0)$	bandgap offset of base layer at $x = 0$ (in an HBT)
$\Delta E_{gB,Ge}(x = W_B)$	bandgap offset of base layer at $x = W_B$ (in an HBT)
$E_{gB}(x)$	position-dependent energy bandgap of SiGe base layer (in an HBT)
E_F	Fermi energy level
E_V	energy of valence band edge
E_{fn}	quasi-Fermi level of electrons
E_{fp}	quasi-Fermi level of holes
E_g	energy bandgap
$E_g(0)$	energy bandgap at 0 K

$\langle E_{\mathrm{p}} \rangle$	average energy loss due to phonon scattering
F	force, free energy
F_0	peak value of the force waveform
f_{max}	maximum frequency of oscillation
f_{n}	natural frequency
f_{T}	transition frequency (unity-gain frequency)
g	acceleration due to gravity
g_{d}	output conductance
g_{m}	transconductance of MOSFET
g_{mb}	body-effect conductance
g_{m0}	transconductance value (maximum) at $V_{\mathrm{GS}} = 0$
g_{ms}	transconductance of MOSFET in saturated condition
h	Planck's constant
$\hbar$	reduced Planck's constant = $h/2\pi$
H_{C}	critical magnetic field of a superconductor
h_{FE}	current gain of a bipolar transistor in a common-emitter connection
i, I	current
I_{b}	biasing current (superconductor, Josephson junction, SQUID)
I_{B}	base current of a bipolar transistor
I_{C}	collector current of a bipolar transistor, critical current (superconductor)
I_{CBO}	collector–base reverse current of a bipolar transistor with emitter open
I_{CCH}	current drawn from the supply during logic high output state
I_{CEO}	collector–emitter reverse current of a bipolar transistor with base open
I_{d}	current in SBD
I_{D0}	reverse saturation current (leakage current) of SBD
I_{DS}	drain–source current of a MOSFET
I_{DSS}	drain–source leakage current of a MOSFET with the gate shorted to the source, the saturated drain current of a MESFET at $V_{\mathrm{GS}} = 0$
I_{E}	emitter current of a bipolar transistor
I_{F}	forward current
I_{fc}	full saturation current
I_{IL}	input low current
I_{nB}	electron current from the n^+-emitter to the p-base
I_{OFF}	off-state current
I_{ON}	on-state current
I_{p}	persistent current (superconductor)
I_{pE}	hole current from the p-base to the n^+-emitter of a bipolar transistor
I_{R}	reverse current
I_{S}	saturation current, screening current (superconductor)
J, j	current density
J_{B}	base current density
J_{C}, j_{C}	collector current density of a bipolar transistor, critical current density of a superconductor
J_{n}	Bessel functions of the first kind
k	stiffness, spring constant
K, K_1, K_2	constants
k_{B}	Boltzmann constant
K_{FE}	damage coefficient related to current gain
K_τ	damage coefficient associated with carrier lifetime
L	diffusion length of carrier, channel length of an FET, inductance
L_{nB}	diffusion length of electrons in the base

L_{pE}	diffusion length of holes in the emitter
M	number of equivalent valleys in the conduction band of SiC, collector multiplication factor
m	index in MOSFET drain current equation in Shoucair's analysis
m_n^*, m_p^*	effective masses of electrons and holes for density-of-state calculations
m_0	rest mass of electron = 9.11×10^{-31} kg
N	dopant concentration
N_C, N_V	effective densities of states in the conduction and valence bands of a semiconductor
$N_{C,SiGe}(x)$	position-dependent effective density of states in the conduction band of SiGe
$N_{V,SiGe}(x)$	position-dependent effective density of states in the valence band of SiGe
$\widetilde{N}$	position-averaged ratio of effective densities of states in SiGe and Si, across the base profile
N_A	total acceptor impurity concentration
N_A^-	ionized acceptor impurity concentration
N_{AB}	acceptor concentration in the p-base layer
N_{crit}	critical impurity density
N_D	total donor impurity concentration
N_D^+	ionized donor impurity concentration
N_{DE}	doping concentration in the n-emitter layer
$N_{D(g)}$	doping concentration of polysilicon gate
n_i	intrinsic carrier concentration (of a semiconductor)
n	electron concentration, ideality factor of SBD
N_I	density of charged impurities, trap density
$n_{iB}(x)$	position-dependent intrinsic carrier concentration in the base
n_{pB}	number of electrons in the p-base
$\mathbf{p}$	canonical momentum of a classical particle
p	hole concentration
$p_B(x)$	position-dependent hole concentration in the base varying with position x
p_{H_2O}	partial pressure of water vapor present in air
$p_{H_2O}^*$	equilibrium vapor pressure of water vapor
p_{nE}	number of holes in the n^+-emitter
p_0	TC of threshold voltage
q	electronic charge
q_0	parameter in Shoucair's threshold voltage equation (the value of V_{Th} at 0 K, as found by extrapolation)
P_0	amplitude of the transmitted force
q	electronic charge
$Q_a(T)$	conducting channel charge
Q_f	fixed charge in the silicon dioxide
$R_{CHANNEL}$	resistance of channel region of a MOSFET
R_{DRIFT}	resistance of drift region of a MOSFET
R_D^{Si}	on-resistance of Si diode
R_D^{SiC}	on-resistance of SiC diode
$R_{DS(ON)}$	drain–source on-resistance of a MOSFET
R_n	external resistor shunting a JJ
R_s, R_S	series resistance of the source

S, s	cross-section, area
t	time
T	temperature in the Kelvin scale, periodic time of clock signal, transmissibility
T_L	lattice temperature
t_{ox}	oxide thickness
U_1, U_2	states of lowest energy on the opposite sides of the tunnel barrier
v	velocity of electron
v, V	voltage
V_0	amplitude of sinusoidal voltage
V_a	early voltage
V_{AC}	AC voltage
v_{BE}	base–emitter voltage
v_{BC}	base–collector voltage
V_{bi}	built-in potential
V_{CE}	collector–emitter voltage, voltage conversion efficiency
v_{CES}	collector–emitter voltage in saturation mode
V_d	voltage across SBD
V_{DC}	DC voltage
V_{DD}	drain supply
V_{DS}	drain–source voltage
V_D^{Si}	forward voltage of Si diode
V_D^{SiC}	forward voltage of SiC diode
V_{FB}	flatband voltage
V_{GS}	gate–source voltage
V_{GS}(ZTC)	V_{GS} at ZTC point
v_{nB}	velocity of electrons at the emitter end of the base
V_{NMH}	high-level noise margin
V_{NML}	low-level noise margin
V_{OH}	output high voltage
V_{OL}	output low voltage
V_{peak}	peak voltage
V_{po}	pinch-off voltage of MESFET
v_{pE}	velocity of holes at the base end of the emitter
V_R	reverse bias
V_{SB}	substrate bias
V_{SS}	source supply
V_{sub}, $V_{substrate}$	substrate voltage
V_{Th}	threshold voltage of a MOSFET
$V_{Thermal}$	thermal voltage
v_{sat}	saturation velocity of carrier
v_n^{sat}	saturation velocity of electron
v_p^{sat}	saturation velocity of hole
w	depletion layer width
W, W_B	base width of a bipolar transistor, channel width of an FET
W_D	depletion region thickness at the drain
W_I	threshold ionization energy
W_S	depletion region thickness at the source

Greek/other symbols

α	fitting parameter in the Varshini equation, index of temperature for mobility variation, ionization coefficient, current gain of a bipolar transistor in the common-base configuration, an empirical constant determining the saturation voltage of the drain current of a MESFET
α_{it}	coefficient for interface states
α_n	ionization coefficient of an electron
α_{ot}	coefficient for oxide-trapped charges
α_p	ionization coefficient of a hole
α_T	base transport factor
β	fitting parameter in the Varshini equation, current gain of a bipolar transistor in the common-emitter configuration, transconductance parameter of a MESFET containing the electron mobility μ_n
β_{max}	maximum current gain
β_{NPN}	current gain of n–p–n transistor
β_{PNP}	current gain of p–n–p transistor
β_{Si}	current gain of Si BJT
$\beta_{Si/SiGe}(T)$	current gain of Si/SiGe HBT
γ	emitter injection efficiency of a bipolar transistor, the effective threshold voltage displacement with V_{DS} for a MESFET
δ	phase difference
ΔE_A	activation energy of acceptor impurity
ΔE_D	activation energy of donor impurity
ΔE_g	energy bandgap difference between the emitter and base semiconductor materials
$\Delta\xi_g$, $\Delta\xi_{gBE}$	decrease in bandgap of the emitter relative to base
$(\Delta E_g)_{Si/SiGe}$	difference between energy bandgaps of the Si emitter and Si_xGe_{1-x} base
ΔE_{gE}	bandgap narrowing of the emitter
ΔE_{gB}	bandgap narrowing of the base
ΔE_V	valence band offset
ΔN_{it}	interface trap density
ΔN_{ot}	oxide trapped charge density
Δt	time interval
$\Delta\Phi$	phase difference between junctions A, B in a SQUID
ε	dielectric constant
ε_0	permittivity of free space
ε_{ox}	relative permittivity of silicon dioxide
ε_s	relative permittivity of silicon
η	emission coefficient or ideality factor in diode equation
η_a,η_b	empirical constants
θ	gauge-invariant phase difference, contact angle at solid–liquid interface
$\theta(\mathbf{r})$	phase of the wave function
κ	TC of threshold voltage
λ	mean free path of carrier, channel length modulation parameter of MESFET
λ or λ_L	London penetration length/depth
λ_0	high-energy low-temperature asymptotic phonon mean free path
μ	mobility of carrier
μ_0	vacuum permeability

μ_i	initial mobility
μ_I	ionized impurity scattering-limited mobility
μ_f	final mobility
μ_L	lattice scattering-limited mobility
μ_n	mobility of an electron
μ_{nB}	mobility of electrons in the base
μ_p	mobility of a hole
μ_{pE}	mobility of holes in the emitter
ξ	coherence length (for Cooper pairs)
ξ_g	bandgap in the emitter
Ξ	an integer
ρ	resistivity of a substance
ρ_1,ρ_2	densities of the Cooper pairs
ρ_{ox}	charge density in the oxide
$\rho(\mathbf{r})$	Cooper pair concentration in the superconductor where **r** is the position vector
σ	conductivity
σ_v	an empirical constant
τ	carrier lifetime, time constant
τ_B	base transit time
τ_E	emitter transit time
τ_F	forward transit time, final lifetime
τ_{HL}	high-level lifetime
τ_I	initial lifetime
τ_{LL}	low-level lifetime
τ_p	lifetime of holes
τ_{pE}	lifetime of holes in the emitter
ϕ	fluence
ϕ	phase difference between the wavefunctions ψ_1 and ψ_2; $\phi = \phi_2 - \phi_1$
ϕ_1, ϕ_2	phases of the wavefunctions of Cooper pairs
ϕ_0	phase difference when $V = 0$
Φ	magnetic flux threading a loop
Φ_{A1}, Φ_{A2}	phases adjacent to the junction A in a SQUID
Φ_{B1}, Φ_{B2}	phases adjacent to the junction B in a SQUID
Φ_A, Φ_B	phase differences across the JJs A and B in a SQUID
ϕ_B	Schottky barrier height
Φ_{bn}	Schottky barrier height of metal–semiconductor interface
Φ_0	magnetic flux quantum (fluxon)
$\Phi_{External}$	external magnetic flux
ϕ_F	bulk potential
ϕ_{ms}	metal–semiconductor work function
$\boldsymbol{\psi}$	wavefunction
ψ_1 and ψ_2	wavefunctions of Cooper pairs
$\psi(\mathbf{r})$	wavefunction of Cooper pair electron where **r** is the position vector
ω	angular frequency
ω_n	natural angular frequency
Ω	ohm
Ω_s	excitation frequency
∇	nabla (differential operator)

Chapter 1

Introduction and overview

1.1 Reasons for moving away from normal practices in electronics

Extreme temperature and harsh environment electronics begins where routine conventional electronics designed for operation in room-temperature friendly environments ends. It breaks away from the traditional treatment of electronics to cover aspects which may, at times, appear less friendly and more antagonistic to electronic circuit operation, although there are exceptions to this rule, as we shall see below. Necessity is the mother of invention. The rationale for taking excursions from routine electronics is that in many applications it becomes imperative to build electronic systems that have to perform satisfactorily and reliably for long durations in unfavorable circumstances (Werner and Fahmer 2001). Such conditions may prevail when we dig deep into the Earth, when we move out into space, when electronic equipment is placed near nuclear reactors and particle accelerators, when the operation of heavy machinery creates vibrations in buildings, when equipment has to withstand high humidity, rainy and stormy weather, to name just a few of the aggressive situations one can contemplate. So, the need for deviation from conventional electronics is primarily driven by the increasing demands from the users and customers who work in such hostile conditions (Johnson *et al* 2004).

But there is a secondary reason as well. This reason originates from the realization that many physical phenomena, such as superconductivity, take place only at temperatures which are far below room temperature. To apply such phenomena for human use, the temperature has to be deliberately decreased close to absolute zero, or at least to the vaporization temperature of liquid nitrogen. Here, the basic operational principle of electronic devices and circuits imposes the requirement of breaking the norms of working at normal temperatures. This applies not only from the viewpoint of superconductivity phenomena; many electrical parameters of semiconductor devices show improvement as the temperature is decreased. Thus one should not consider that deviations from the norm will always lead to an aggressive and incompatible situation. It may be a fortunate situation as well.

doi:10.1088/978-0-7503-1155-7ch1

In fact, the properties of semiconductors change over a very wide range as one increases the temperature from −273 K to 1000 K. This variation of properties of semiconductors is visible in the form of changes in the electrical behavior of the electronic devices fabricated from them. Some electrical parameters tend to improve at low or high temperatures, while others show deterioration. The comprehensive study of these behavioral trends helps scientists take advantage of the changes that are beneficial in the utilization of electronic circuits when we are operating outside the recommended range of temperatures.

In response to both types of situation portrayed above, one stemming from application-specific requirements and the other from phenomenological needs, one has to move away from regular practice and deal with challenging situations to meet one's aims.

1.2 Organization of the book

This book is subdivided into 20 chapters. Chapter 2 elaborates on the motivations for departure from treading the established path of electronics. In this chapter, the reader will come across many situations and applications that show the deficiencies of conventional electronics in solving problems. These examples also illustrate the need to shift away from the normal course to benefit from the advantages of utilizing phenomena that take place only under special conditions.

The remainder of the book (chapters 3–20) is grouped into two parts (see figure 1.1). Part I, consisting of chapters 3–14, deals with extreme-temperature electronics (ETE). It examines countering the harmful effects of very high temperatures as well as utilizing the beneficial effects of very low temperatures, as in superconductive electronics, and utilizing the characteristics of semiconductor devices which improve with a fall in temperature, e.g. the leakage currents. Thus temperature effects manifest themselves as both a curse and a boon for electronics exhibiting hostile as well as friendly behavior. Part II, comprising chapters 15–20, covers only those effects which are detrimental to electronic circuit operation, such as: highly humid climatic conditions; corrosive environments inside or in the neighborhood of chemical factories; in radiation-contaminated areas, such as near nuclear power stations, x-ray or gamma-ray equipment in hospitals; and in vibrating buildings amidst the busy and heavy traffic of metropolitan cities or where heavy machinery is being used, making the surroundings noisy and shaky.

1.3 Temperature effects

1.3.1 Silicon-based electronics

Chapter 3 explains how the properties of semiconductors change as a function of temperature. Since variation of these properties is reflected in the thermal behavior of devices, a thorough understanding of this chapter lays a firm foundation for understanding the contents of ensuing chapters. The upper temperature limit of a semiconductor material is fixed by its bandgap energy. Device operation is governed by the concentration of free carriers in a pure semiconductor, known as the intrinsic carrier concentration. The intrinsic carrier concentration is an exponential function

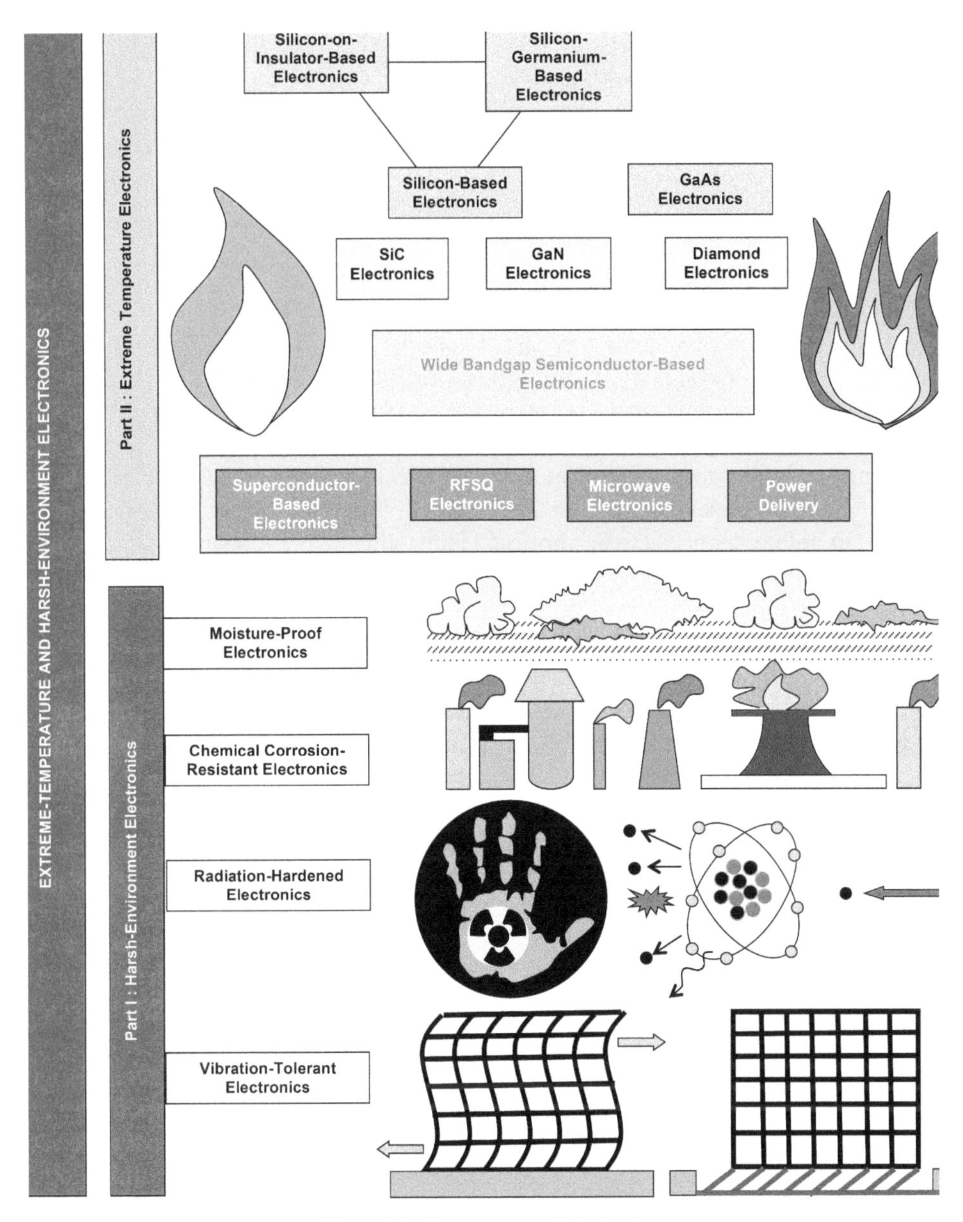

Figure 1.1. Organization of the book.

of temperature. An increase in temperature augments the energy of electrons in the valence band of a semiconductor material. At a particular temperature called the intrinsic temperature, the thermal energy of electrons exceeds the bandgap energy of the semiconductor. Then the electrons are promoted from the valence band to the conduction band. The number of thermally generated carriers becomes equal to the number of free carriers due to impurity doping, either n- or p-type. Then there are no longer any distinguishable n- or p-type regions. The p–n junction reduces to a resistor and its function is paralyzed.

On the opposite side, as the temperature of a semiconductor decreases towards absolute zero, the ionization of impurity atoms to release free carriers ceases. Then there are no, or very few, carriers available for conduction. In this carrier freeze-out regime as well, the normal operation of a semiconductor device stops. In fact, a semiconductor is an insulator at zero kelvin.

Apart from the liberation of electrons from their bonds rendering possible their availability for conduction of electrical current, other noteworthy phenomena include the scattering of carriers by lattice atoms and impurity ions, which affect the ease of movement of carriers through the crystal lattice, i.e. the carrier mobility. Mobility is a strong function of temperature. As lattice atoms vibrate with greater amplitudes with rising temperature, the electrons undergo more collisions on their paths and mobility decreases. Impurity scattering limited mobility varies in the opposite way, because the increased vibration of impurity ions makes them less effective in influencing electron motion. The temperature dependence of carrier mobilities is also affected by whether a semiconductor is non-degenerately or degenerately doped.

From chapter 4 onwards up to chapter 9 we move towards discussion, assessment of the capabilities and appraisal of the critical issues concerning electronic devices and circuits made from semiconductor materials of progressively increasing energy bandgap, and hence intrinsic temperature. In this sequence, the starting material of interest is silicon, which has been the favorite of electronic engineers for a very long time and has reigned as the king of electronic materials. Silicon electronics forms the contents of chapters 4 and 5. In chapter 4, bipolar silicon devices are addressed. Chapter 5 focuses on MOS silicon devices. Silicon electronics has two forms: bulk silicon and silicon-on-insulator (SOI) technologies. The objectives of these chapters are to present simple analytical formulae for the temperature coefficients (TCs) of silicon bipolar and MOS devices. These derivations help in appreciating the degradation or upgradation in electrical characteristics of bipolar/MOS silicon devices and circuits subjected to constantly rising temperatures.

The forward voltage drop across a p–n diode or Schottky diode decreases with temperature, the current gain of a bipolar transistor increases with temperature and the breakdown voltage of a diode increases with temperature. In almost all circuit applications, the leakage currents of p–n junctions should be kept infinitesimally small with respect to the signal currents. These leakage currents increase exponentially with temperature. In complementary metal–oxide semiconductor (CMOS) structures, junction leakage occurs at the junctions between source/substrate and drain/substrate due to the minority carrier diffusion current near the depletion region, together with electron–hole pair (EHP) generation inside the depletion region. Gate leakage increases with thinning of the gate oxide as MOSFETs are scaled towards smaller dimensions. The threshold voltage decreases linearly with rise in temperature and hot carrier effects become less pronounced with increasing temperature. High-cost SOI technology considerably helps in obviating the leakage current issues in bulk silicon devices at high temperatures.

Silicon and germanium have been two key materials forming the backbone of electronics. Silicon–germanium, an alloy of silicon with germanium, combines the best properties of both materials. Silicon/silicon–germanium heterojunction bipolar transistors (HBTs) overcome the problem of a rapid fall in the current gain and switching frequency of silicon bipolar temperatures at cryogenic temperatures. Chapter 6 treats the mathematical theory of the HBT and shows how this device serves as a replacement for bipolar transistors, showing much better performance at these temperatures.

Silicon was the first-generation material of the twentieth century. It played a leading role in ushering in the microelectronics revolution. Time and again, it has appeared that silicon electronics has hit a wall, prompting the search for alternative materials. Silicon was followed by gallium arsenide in the second generation of semiconductors. Together with silicon, gallium arsenide set off the information technology and wireless revolution around the turn of the twenty-first century. Chapter 7 deals with gallium arsenide, which is superior to silicon in the fabrication of ultra-fast radio-frequency (RF) devices, and is also suitable for making optoelectronic devices such as LEDs and laser diodes. In contrast to silicon technology, where the primary devices are bipolar transistors and MOSFETs, GaAs relies on MESFETs and HBTs.

1.3.2 Wide bandgap semiconductors

The wide bandgap semiconductors, silicon carbide and gallium nitride, belong to the third-generation semiconductors. They heralded the optoelectronics and HTE era at the beginning of the twenty-first century. The capability of silicon carbide and gallium nitride chips to operate at higher temperatures, voltages and frequencies promoted the research interest in these materials. Approximately three times the energy used in the case of silicon is needed to transport an electron from the conduction to valence band in SiC and GaN, which makes both ideal candidates for realizing high-temperature devices with high breakdown strength. Naturally, the power electronic modules built from these materials are significantly more energy efficient.

The manufacturers of silicon carbide wafers have been able to minimize the defect density considerably as well as increase the size of the wafers. Technological breakthroughs in SiC materials and process techniques have led to the realization of several devices such as p–n diodes, Schottky barrier diodes (SBDs), JFETs, bipolar transistors, etc. SiC JFETs are very attractive for HTE, but SiC MOSFETs still need a lot of improvement. In SiC, the interface state density is high and the carrier mobility is very low. Scrupulous efforts are being made in the development of SiC thyristors, IGBTs and gate turn-off thyristors (GTOs). Chapter 8 provides glimpses into achievements in silicon carbide electronics.

Gallium nitride is an excellent material for the fabrication of light-emitting diodes and power transistors. Advancing from gallium nitride wafers on sapphire substrates to free-standing gallium nitride wafers with low dislocation density is likely to boost confidence in GaN technology. GaN MESFETs and HEMTs have been developed and tested at high temperatures. Chapter 9 takes a look at some of the GaN devices.

After silicon carbide and gallium nitride, the superior properties of diamond make it suitable to catapult another major upheaval in microelectronics. Diamond has been crowned as the ultimate semiconductor material due to its spectacular combination of properties, including high thermal conductivity and radiation hardness. Diamond electronics is still immature because the full capabilities of diamond have so far evaded utilization. The ability to synthesize diamond from the vapor phase to produce large-area films has generated a lot of interest in diamond-based devices, and the scenario seems to be gradually changing as single-crystal electronic grade (EG) diamond is becoming commercially available. Chapter 10 deals with the synthesis, processing and characteristics of diamond films and related devices.

1.3.3 Passive components and packaging

The aforementioned few chapters concentrate on wide-bandgap semiconductors and the active devices fabricated from them, the aim being to fabricate semiconductor devices that are able to withstand successively higher temperatures. Apart from semiconductors, the other important materials used in electronic devices are the metals used for contact electrodes. No electronic circuit can be fabricated without passive components. Furthermore, the devices have to be packaged within safe enclosures to protect them from mechanical damage, as well as atmospheric and weather effects. The electronic circuit will fail at elevated temperatures if the active or passive components, metallization or packaging are not up to the mark. From these considerations, chapter 11 diverts attention from semiconductors towards resistors, capacitors, metal interconnections and packaging. Carbon resistors show good thermal stability. Diamond resistors are fabricated from CVD diamond on an aluminum nitride substrate. Teflon capacitors can be used up to 200 °C and mica capacitors up to 260 °C. Diamond capacitors are based on a dielectric film of diamond with Au contacts. Low dielectric loss and constant capacitance are observed up to 450 °C.

In addition to good adhesion with the underlying silicon, the metallization must be able to withstand thermal cycling and must not decompose or undergo chemical reactions at high temperatures. Films of refractory metals and refractory metal silicides formed by chemical vapor deposition (CVD) serve as a useful metallization for high temperatures. This method provides selective deposition over silicon regions, leaving aside oxides and insulating areas. Electrolytic Cr–Ni–Au metallization is a robust scheme for wire bonding applications exposed to high temperatures (250 °C).

During die attachment, care must be taken to match the coefficients of thermal expansion of the die, die-attach and substrate. This avoids any mechanical stressing or fracturing of the die during thermal cycling. Die-attach materials proven for room-temperature operation cannot be used owing to their low glass transition temperature.

For the reliability of wire bonds at high temperatures, the metals used for the wire and the metallization of bonding pad must be mutually compatible. Poor compatibility of metals involved in wire bonding leads to two types of problem. Either an intermetallic compound is formed at the interface between the metals causing brittleness of the bond and breakability, or voids are formed at the interface by diffusion, an effect called the Kirkendall effect, thereby weakening the bond. Inopportunely, the common Au–Al combination between the Au wire and Al metallization pad is prey to such phenomena. These arguments entice us to use the same metal for the wire bond and the bond pad.

For high-temperature operation, hermetic ceramic packages are far better than plastic packages. They also serve as moisture and contamination impenetrable barriers. Limiting the ingress of moisture and dirt prevents corrosion. Regrettably ceramic packages are larger, heavier and costlier than plastic packages, which cannot be used past 150 °C. High melting point solders with a melting point >250 °C must be used.

1.3.4 Superconductivity

The following three chapters (chapters 12–14) are concerned with superconductivity, both low-temperature and high-temperature. Superconducting films exhibit low resistance even at frequencies of approximately a few hundred GHz, paving the way for their utilization in magnetometry using superconducting quantum interference devices (SQUIDs), microwave filters, transmission lines, etc. Rapid flux single quantum (RFSQ) logic electronics uses Josephson junctions (JJs) to perform logic operations based on the quantization of magnetic flux. HTS-based power transmission uses cables comprising hundreds of strands of HTS wire with a cryogenic cooling system to maintain the required low temperature. In dense urban localities, power substations often reach capacity limits. HTS systems bind these substations together circumventing expensive transformer upgradation. HTS power delivery is used to load pockets in high-demand metropolitan areas with a saturated grid.

1.4 Harsh environment effects

1.4.1 Humidity and corrosion effects

Temperature is an important parameter, and has been the focus of the attention of electronic engineers because by increasing/decreasing the temperature, both advantageous and disadvantageous results can be produced on electronic equipment. In addition to temperature, humidity is another vital parameter. Humidity-related failures and remedial schemes are discussed in chapters 15 and 16. Through moisture condensation, water droplets accumulate on the surfaces of semiconductor devices, producing ionic current flow. The effects of humidity depend on the materials used, the dimensions of the components and their layout. Humidity accelerates the corrosion rate. Corrosion is the deterioration of the materials constituting the electronic devices and circuits under the influence of reactive gases in the environment. The principal reason for the vulnerability of electronic products to corrosion is the large variety of metals and alloys used in the electronics industry.

A few of these are aluminum, copper and its alloys, silver and its alloys, tin and its alloys, titanium, chromium, nickel, gold, platinum, palladium, tungsten, etc. Apart from humidity, corrosion depends on inorganic and organic contaminants, atmospheric pollutants, salt spray, noxious gases, and residues from soldering and other assembly processes. Corrosion increases the contact resistances between joints and leakage currents between wires and decay products. Corrosion by water, dust and gases causes short circuits and produces ugly surfaces. It can initiate cross-talks. As a consequence of the reduced spacing between components in the wake of miniaturization of circuits, a material loss of a few pictograms due to corrosion is adequate to spawn a fault. Therefore, the impact of corrosion becomes more magnified in damaging electronic circuits. Chapter 17 suggests ways to mitigate corrosion problems.

1.4.2 Radiation effects

Next to temperature and humidity comes the radiation from space as well as terrestrial sources. These types of radiation have a negative influence on electronic circuits located on Earth, those placed in orbiting satellites or used in long-distance space flights. These defects range from performance degradation to complete loss of functionality. Due to these radiation-induced failures, satellite lifetimes are shortened and space missions are disrupted. Radiation effects can be one or more of the following types: single event effects (SEEs), displacement damage (DD) and total ionizing dose (TID) in the form of a cumulative effect over a long time span. Radiation-hardened circuits are modified versions of non-hardened equivalents, incorporating revised designs for fault-tolerance and software approaches to deal with disturbances, along with suitable manufacturing process amendments. Bipolar ICs are more radiation-hard than CMOS circuits. Radiation-hard circuits are fabricated in SOI or silicon-on-sapphire substrates, instead of the common bulk silicon wafers, because of the higher leakage currents produced by radiation in junction isolated devices. Chapter 18 describes the detrimental effects of radiation exposure on electronic circuits. Feasible ways to thwart the disturbances and damage produced by radiation in electronic circuits are suggested in chapter 19.

1.4.3 Vibration and mechanical shock effects

Last but not the least, the damaging effects of vibrations, impacts, kicks, drops and shocks, as caused by acceleration/deceleration and impulsive forces on electronic circuitry cannot be ignored. The concluding chapter 20 describes the common techniques that must be adopted for protection from vibrations.

1.5 Discussion and conclusions

The temperature effects on electronic circuits must be carefully understood, taking due consideration of their pros and cons. Pernicious effects must be suppressed. Beneficial effects must be gainfully exploited to execute utilitarian functions, as an enhancement to system functionality. The effects of humidity, corrosion, radiation and vibration need to be addressed according to the specific application. They have to be dealt with on a case-by-case basis. Some ideas of this chapter are succinctly described in the following poem:

Safeguarding electronics

Extreme temperatures are sometimes hostile, sometimes cozy;
Sometimes thorny, sometimes rosy;
Away from room temperature, some device parameters downgrade,
Other parameters upgrade.
But high humidities and aggressive chemicals
are always detrimental.
Vibrations too are harmful.
So, please be careful
To make electronic design fault-tolerant,
And package environment-resistant.
Wide bandgap semiconductors are promising,
Superconductive electronics is amazing.
To ensure success of mission,
Choose electronic design, materials and fabrication
According to application;
Provide proper surface passivation
to protect from moisture invasion.
Prevent chemical corrosion.
Take precautions for radiation
And cushion electronic product against vibration.
Keep in touch with new developments and innovations.

Review exercises

1.1. Contemplate two situations, one negative or unfavorable situation and one positive or favorable situation, which urge electronic engineers to develop circuits that can work in non-conventional conditions.
1.2. What property of a semiconductor material fixes the upper permissible limit of temperature up to which an electronic device made from it may be operated?
1.3. Why is operation of a semiconductor device not possible at a temperature exceeding its intrinsic temperature?
1.4. Give some examples illustrating the beneficial effects of low temperatures for electronic device operation.
1.5. Explicate, with reference to carrier mobility, the outcomes of scattering of electrons by lattice atoms and impurity ions. Elucidate the influence of temperature on mobility variations caused by the two types of scattering.
1.6. Cite three examples showing the variation of the most important electrical parameters of semiconductor devices with temperature.

1.7. Does temperature affect the leakage current of a p–n junction? If so, why does it happen? Describe the various types of leakage currents that a CMOS structure is susceptible to.
1.8. In what manner is a silicon–germanium heterojunction transistor superior to a silicon bipolar transistor for operation at cryogenic temperatures?
1.9. Mention two application areas where gallium arsenide devices find wide usage.
1.10. In what ways can silicon carbide and gallium nitride outperform silicon and gallium arsenide devices?
1.11. Is it possible to synthesize diamond artificially?
1.12. Give examples of resistors and capacitors developed for high-temperature applications.
1.13. What kind of contact metallization can be used in HTE?
1.14. What precautions are necessary for selecting appropriate die-attach materials to be used at high temperatures?
1.15. What problems occur due to poor compatibility of the wire metal and metallization pad?
1.16. What are the advantages and disadvantages of hermetic ceramic packages?
1.17. Give three applications of superconductivity.
1.18. What is the physical basis of RFSQ logic?
1.19. Why is HTS-based power delivery beneficial in upgrading dense, congested urban power network?
1.20. Why are electronic products prone to corrosion effects? How do humid environments aggravate corrosion effects? How does corrosion impair the performance of electronic devices?
1.21. What are the three types of radiation effects on electronic circuits? What are the approaches followed for countering these effects?
1.22. Which type of ICs are more radiation-hard: bipolar or CMOS?
1.23. Which type of silicon wafers are used for the fabrication of radiation-hard circuits: bulk silicon wafers or SOI wafers?
1.24. Mention one adversary of electronic circuits, apart from temperature, humidity, corrosion and radiation.

References

Johnson R W, Evans J L, Jacobsen P, Thompson J R and Christopher M 2004 The changing automotive environment: high-temperature electronics *IEEE Trans. Electron. Packag. Manuf.* **27** 164–76

Werner M R and Fahmer W R 2001 Review on materials, microsensors, systems and devices for high-temperature and harsh-environment applications *IEEE Trans. Ind. Electron.* **48** 249–57

Chapter 2

Operating electronics beyond conventional limits

Degradation effects on the performance characteristics of semiconductor devices at elevated temperatures are described. The reason why the thermal management approach is insufficient to deal with the menace of high temperatures is explained as are the beneficial and disadvantageous effects of low temperatures on device operation. Many interesting areas for the application of HTE are highlighted, notably in the automobile, aerospace and well-logging industries. The effects of humidity, radiation and vibration on device behavior are described. The terminology of low-temperature, high-temperature, extreme-temperature and harsh environment electronics is elaborated. A useful strategy to overcome the problems faced consists in addressing all of the issues, starting from the chip design through process steps up to the packaging stage.

2.1 Life-threatening temperature imbalances on Earth and other planets

Very hot temperatures are uncomfortable for human beings. Very cold temperatures are no less excruciating. At times, such temperatures can prove perilous to human health. For those working in a prickly hot environment, the most serious concern is posed by heat stroke, causing loss of consciousness or fainting (syncope). During heat stroke, the temperature-regulating mechanism of the body malfunctions. Relatively less serious is heat exhaustion due to overheating. This leads to heavy sweating accompanied by a fast pulse rate.

In very cold temperatures, the most serious danger is posed by hypothermia. It is a precarious overcooling of the body. In this condition, the body loses heat more speedily than it produces heat. The body temperature may fall below 35 °C. Another serious effect of exposure to severe cold is frostbite. This results in injury to the skin and underlying tissues of the extremities of the body, such as the fingers, toes, nose and ear lobes, at temperatures near or below the freezing point of water.

doi:10.1088/978-0-7503-1155-7ch2

It is interesting to know the temperatures of some of the hottest and coldest places on Earth. Table 2.1 gives an idea of some such places, both hot and cold (Tavanaei 2013). One can easily imagine how demanding it is to live and survive in these places. The very thought of these temperatures makes us tremble and chill with fear. But temperatures on other planets of the solar system are even more extreme (Williams 2014, Howell 2014). Depending on their respective distances from the Sun, the surface temperatures of the planets vary from >400 °C on Mercury and Venus to < −200 °C on the distant planets (figure 2.1). One can imagine the conditions on these planets from table 2.2. These temperatures may be considered as one of the reasons for the non-existence of living organisms on these planets.

2.2 Temperature disproportions for electronics

What is true about humans holds also for electronics. But in electronics, the scales of intolerable temperatures are different. In electronics, the traditional range of operating temperatures lies between −55 °C/−65 °C at the lower extremity to +125 °C on the higher side. Temperatures falling outside these classical boundaries are known as extreme temperatures. ETE is a branch of electronics that deals with the properties and operation of electronic materials, devices, circuits and systems under the severe thermal conditions lying above or below the limits specified above (Kirschman 2012). It has two sub-branches. (i) Electronics addressing operation above +125 °C to as high as electronics can be made to operate. This sub-branch is known as HTE. (ii) Electronics concerned with operation below −55 °C/−65 °C to −273 °C or 0 K. This sub-branch is called LTE.

2.3 High-temperature electronics

The first query that can be raised is the following: why should one worry about abnormal situations, which may constitute a small segment of the electronics

Table 2.1. The six hottest and six coldest places on earth.

Sl. No.	Hottest places		Coldest places	
	Place	Temperature (°C)	Place	Temperature (°C)
1.	Lut Desert in Iran	+70.7	Vostok Station, Antarctica	−89.2
2.	Death Valley, California, North America	+56.7	Oymyakon, Russia	−71.2
3.	Al'Aziziyah, Northwest Libya, Africa	+57.8	Verkhoyansk, Russia	−69.8
4.	Ghudamis, Libya, Africa	+55	North Ice, Greenland	−66
5.	Kebili, Tunisia, Africa	+55	Snag, Yukon, Canada	−63
6.	Timbuktu, Mali, West Africa	+54.5	Prospect Creek, Alaska, USA	−62

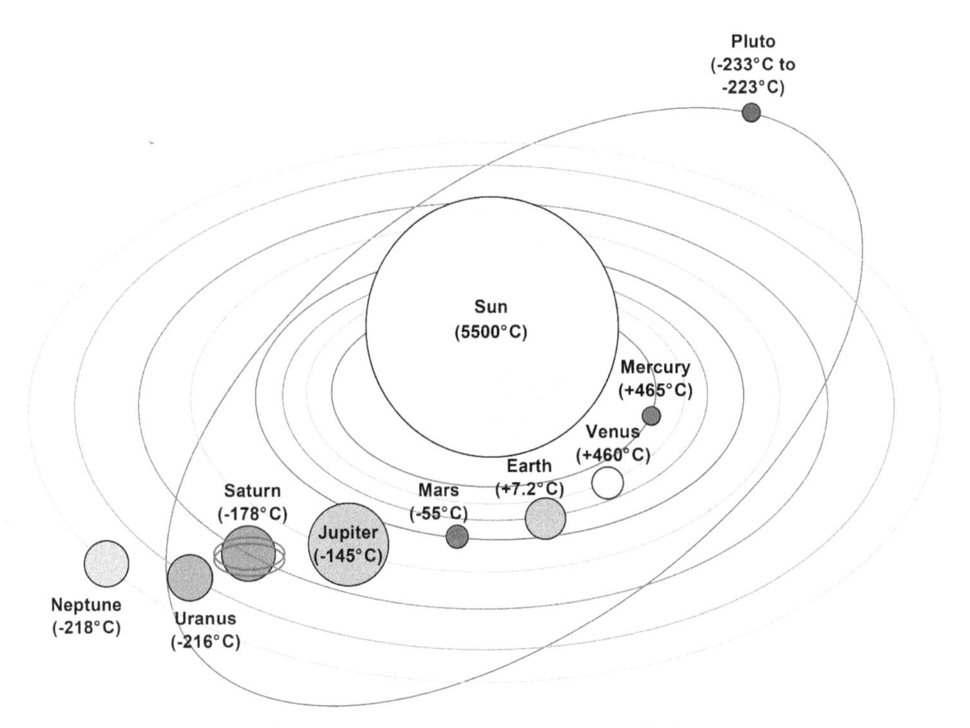

Figure 2.1. Illustrating the wide temperature variation in the solar system.

Table 2.2. Average surface temperatures of planets.

Sl. No.	Planet	Average distance from Sun (km)	Temperature (°C)
1	Mercury	5.7×10^7	+465 (side exposed to Sun); −184 (dark side)
2	Venus	1.08×10^8	+460
3	Earth	1.50×10^8	+7.2 varying from +70.7 (the deserts of Iran) to −89.2 (Antarctica)
4	Mars	2.28×10^8	−55, ranging between as high as +20 at the equator during midday, to as low as −153 at the poles
5	Jupiter	7.79×10^8	−145
6	Saturn	1.43×10^9	−178
7	Uranus	2.88×10^9	−216
8	Neptune	4.50×10^9	−218
9	Pluto	5.91×10^9	−233 to −223

industry? This remark is valid because electronics is generally required to operate under conditions favorable to human life. The answer is that this segment, even though small, is vital strategically. It comprises several important sectors in which HTE has firmly strengthened its roots. One of these sectors is the automotive industry, i.e. motor vehicles. Another sector is the aerospace industry, dealing with travel in the Earth's atmosphere or in the region above it. The final sector is the well-logging industry, which records the properties of subsurface rocks and fluids by drilling a bore hole into the Earth's interior. In each of these sectors, specific problems are encountered. These issues cannot be solved without the help of HTE.

The second question that arises is: what will happen if traditional electronics is used in combination with active or passive cooling when designing electronics intended to function outside the normal temperature ranges? Yes, customarily engineers have followed this approach. Thermal management systems serve as useful alternatives to high-temperature devices. But these systems add undesired weight and volume to the total system. Additional overhead is introduced in the form of longer wires, extra connectors and/or cooling systems. As a result, the power-to-volume and power-to-weight ratios are reduced. These can negatively affect the advantages endowed by electronics with respect to the overall system operation. Thermal management systems also increase the potential for failure. By removing the heat sink and long interconnects, savings in the overall mass and volume of the power electronic modules can be achieved to the extent of one order of magnitude.

In certain circumstances, the use of cooling systems is impossible, principally when compactness is desired. In other applications, it is more appealing to operate electronics at high temperatures with the intent of increasing either the reliability of the system or to reducing the total expenditure. When objectives of this nature become primary, multiple challenges are faced in building the electronic systems: the design techniques are dramatically altered; the choice of semiconductor material is of paramount significance; the selection of other materials such as those used in metallization becomes a decisive determinant of performance; packaging materials and technologies differ from routine practice; and new qualification criteria and methodologies must be developed.

Needless to say, in the absence of HTE the cost of managing and monitoring hot environments becomes astronomically high. The direct costs escalate due to the increased system complexity and the specialized cabling requirements. Indirect costs also mount up through the increase in the weight of the electronic system due to cooling paraphernalia. The systems operated with these additional cooling jackets are also less reliable than the electronic systems which confront the intricacies of high temperatures *ab initio* and throughout the system (McCluskey *et al* 1997). Examples of applications demanding HTE are given in the following subsections (Delatte 2010). See also figure 2.2.

2.3.1 The automotive industry

The ignition circuit used in a car is shown in figure 2.3. It consists of a battery, a switching transistor, a step-up transformer and a spark plug. A voltage ~20–50 kV is produced across the gap in the spark plug when the current flowing through the power transistor stops, such that the magnetic field around the primary coil collapses, inducing high voltage in the secondary winding.

Figure 2.4 shows how a car engine is electronically controlled using various types of sensors, namely the throttle position sensor, fuel pressure sensor, air flow sensor, knock sensor, temperature sensor and oxygen sensors. The electronics provide an

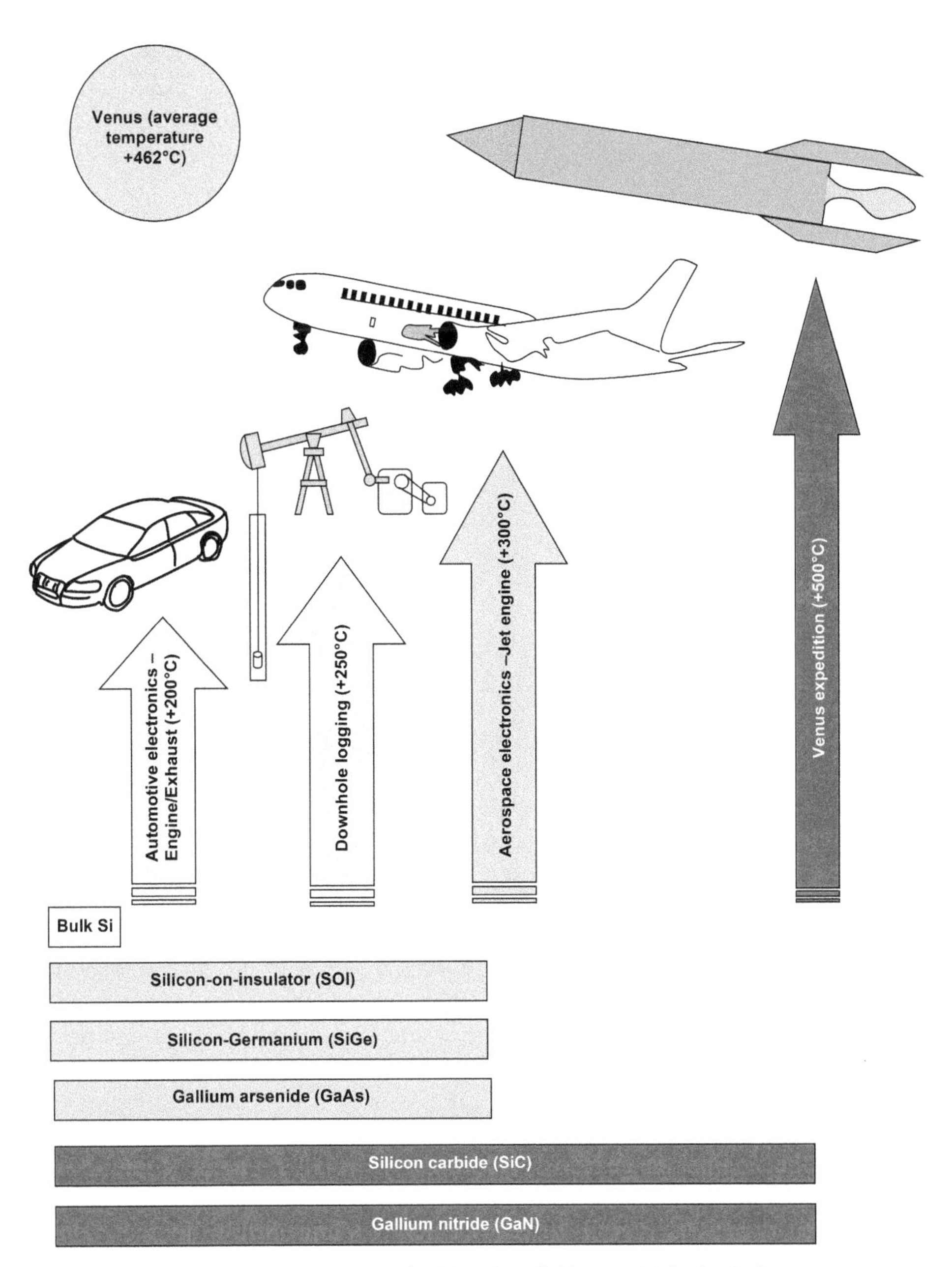

Figure 2.2. Applications of HTE and available materials/technologies.

optimum timing cycle of ignition for all speeds and widely different loading conditions of the engine. Compared to a mechanical ignition circuit, electronic ignition provides superior starting and smooth running. It minimizes fuel consumption and thus causes reduced atmospheric pollution. In particular at very low and very high speeds, it provides a far-better performance. Servicing costs are also reduced.

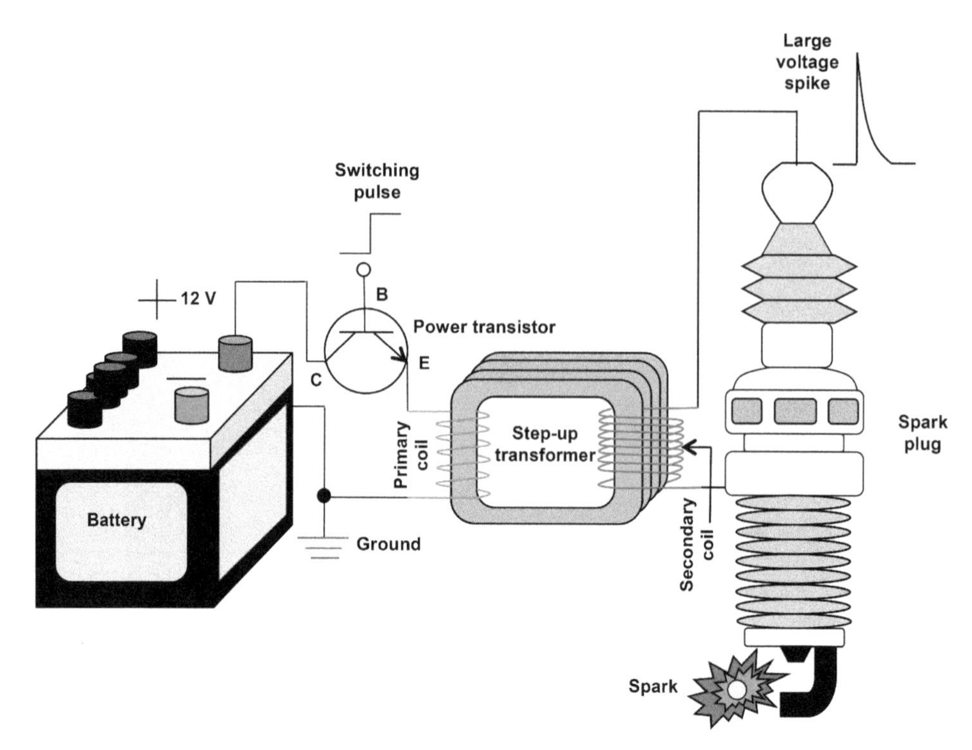

Figure 2.3. Car ignition circuit.

In the automotive industry, engineers designing the body of the motor vehicles are continuously persuaded to augment the space inside the cabin, and thereby the capacity to carry more passengers and load. Cabin space can be increased by squeezing the engine and power module into smaller cubicles. This in turn necessitates the reduction of the size of electronic systems and wiring. Therefore, the electronic control systems must be mounted as proximate as possible to the engine. They may be placed in locations such as the gearbox. They may also be placed in transmission components. Consequently, a close-packed system is obtained. This kind of system can work satisfactorily if the electronic boards and components in the system can perform accurately at the approximate temperatures reached in the engine or gearbox. Additionally, smaller engine compartments impose stricter margins on heat sink sizes. Hence, vehicle aerodynamics is modified. This modified dynamics is able to provide less cooling airflow to the radiator. An increase in temperature under the hood invariably follows. In all the above circumstances, the capabilities provided by HTE help to protect the electronic system from adverse thermal effects.

Another aspect of the advancement of electronics in the automotive industry relates to a recent paradigm shift. There is a strong desire to move away from purely mechanical and hydraulic systems to electromechanical or mechatronics schemes. This migration is essential to improve reliability. At the same time maintenance costs are reduced. But this requires the placement of sensors, signal conditioning

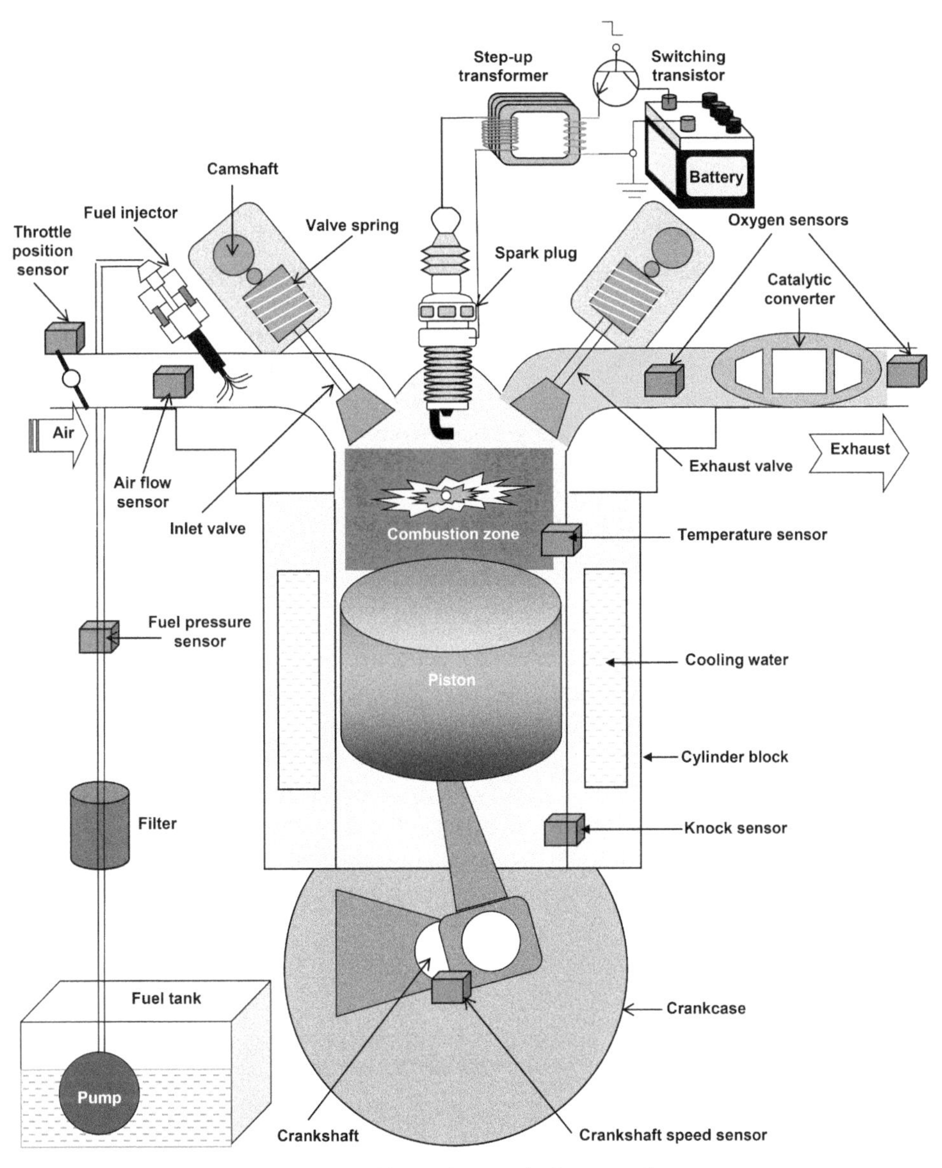

Figure 2.4. Electronic control of a car engine.

and control electronics adjoining heat sources which produce a high-temperature environment. The maximum temperature and exposure time vary according to the type of vehicle and the location of the electronics on the vehicle. In some vehicles, a higher degree of integration of electrical and mechanical systems is achieved, such as collocation of the transmission and transmission controller. This leads to simplification of the manufacturing, testing and maintenance of automotive subsystems, but is associated with an increase in temperature. Here also, HTE is resorted to.

Remarkable developments have taken place in the hybrid automotive industry. These developments have generated an enormous need for different power electronic modules. The desired modules are direct current (DC)/DC converters and DC/alternating current (AC) inverters capable of working at elevated temperatures. On average, the junction temperatures for ICs are 10 °C to 15 °C higher than the ambient temperature. For power devices, they are around 25 °C above the ambient temperature. Hence, the electronics used in the automotive industry, particularly those placed close to the engine, must be able to work at temperatures above 150 °C + 25 °C = 175 °C.

In electric cars and those based on a hybrid approach, power electronics is used for motor drives by integration of power converters and smart power devices into the drive train. This requirement increases the temperature requirements for automotive-qualified semiconductor devices from 150 °C to a 200 °C peak value (Huque *et al* 2008). This again is an application area of HTE.

With increase in the availability and reduction in price of HTE, its applications in automotive systems will continue to grow.

2.3.2 The aerospace industry

The aerospace industry entails civil and military aviation, and space flights and missions. Figures 2.5–2.7 illustrate different types of aircraft engines. Piston-engine powered aircraft obtain their thrust from a propeller driven by the engine. Turbopropeller aircraft obtain their thrust from a propeller driven by a gas turbine, while turbofan aircraft are similar but obtain their thrust from the internal fan driven by the turbine. In both the turbopropeller and turbofan aircrafts, a small amount of thrust is also obtained from the hot exhaust gases. The propeller/fan sucks in air, which is passed through a compressor and sent to combustor chamber

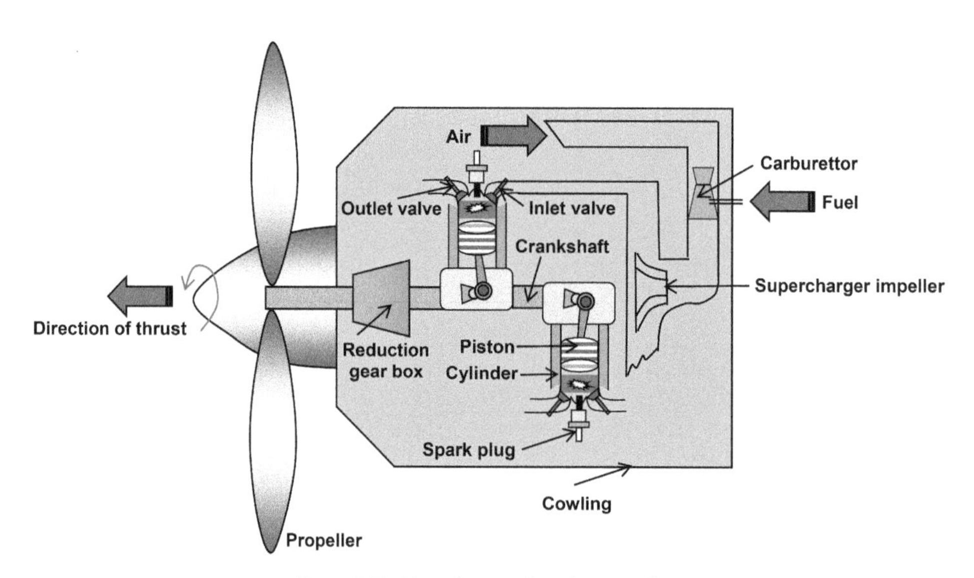

Figure 2.5. Aircraft propeller piston engine.

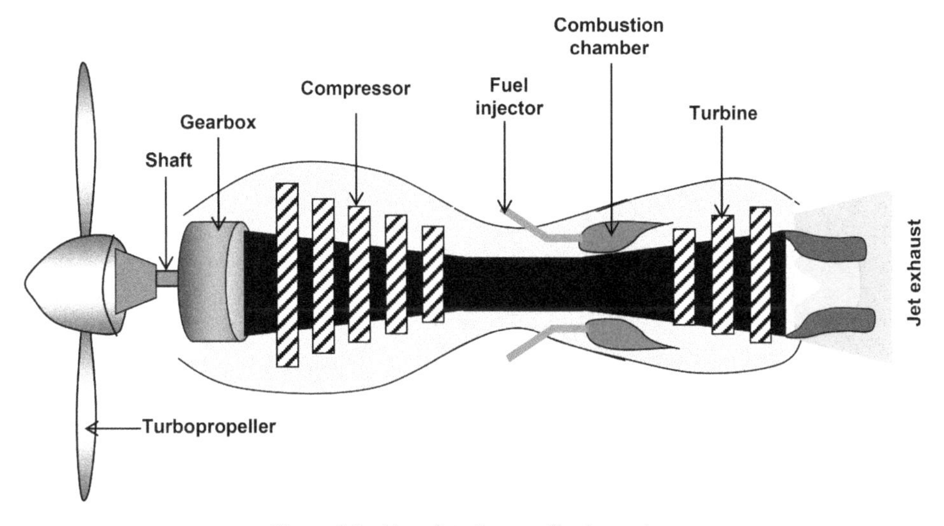

Figure 2.6. Aircraft turbopropeller jet engine.

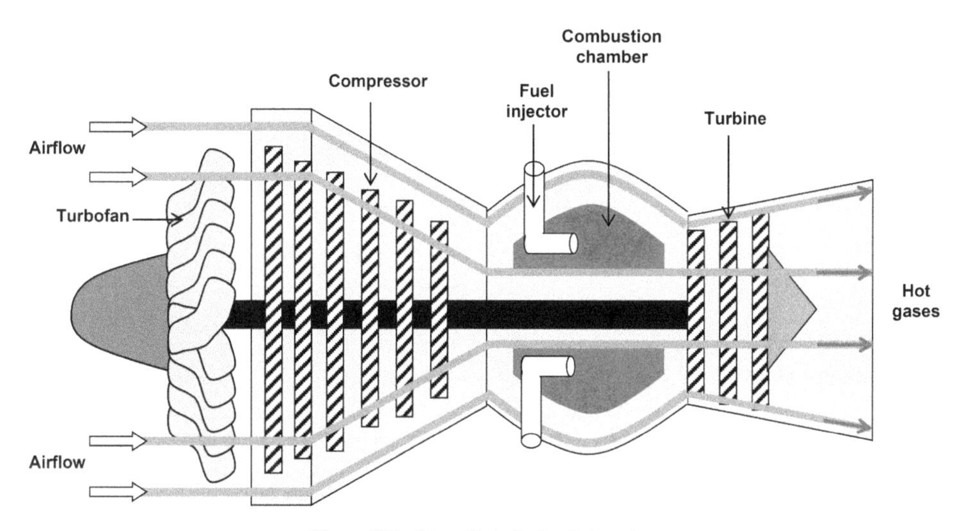

Figure 2.7. Aircraft turbofan jet engine.

where it is ignited. While part of the air passes through the engine core where combustion occurs, the remaining air, known as the bypass air, moves around the core through a duct, producing additional thrust. The hot air moves the turbine blades which are coupled to the propeller/fan and help in its rotation, sucking more air. A rocket engine is shown in figure 2.8. It forms its exhaust jet using only the fuel stored inside. It also has an oxygen tank because it has to travel in space where an oxygen supply is not available. In all the above engines, whether aircraft or rockets, the fuel ignition is electronically monitored.

The ignition system of an aircraft is shown in figure 2.9. The dual magnetos used in an aircraft engine provide redundancy. In the case that one magneto fails, the other

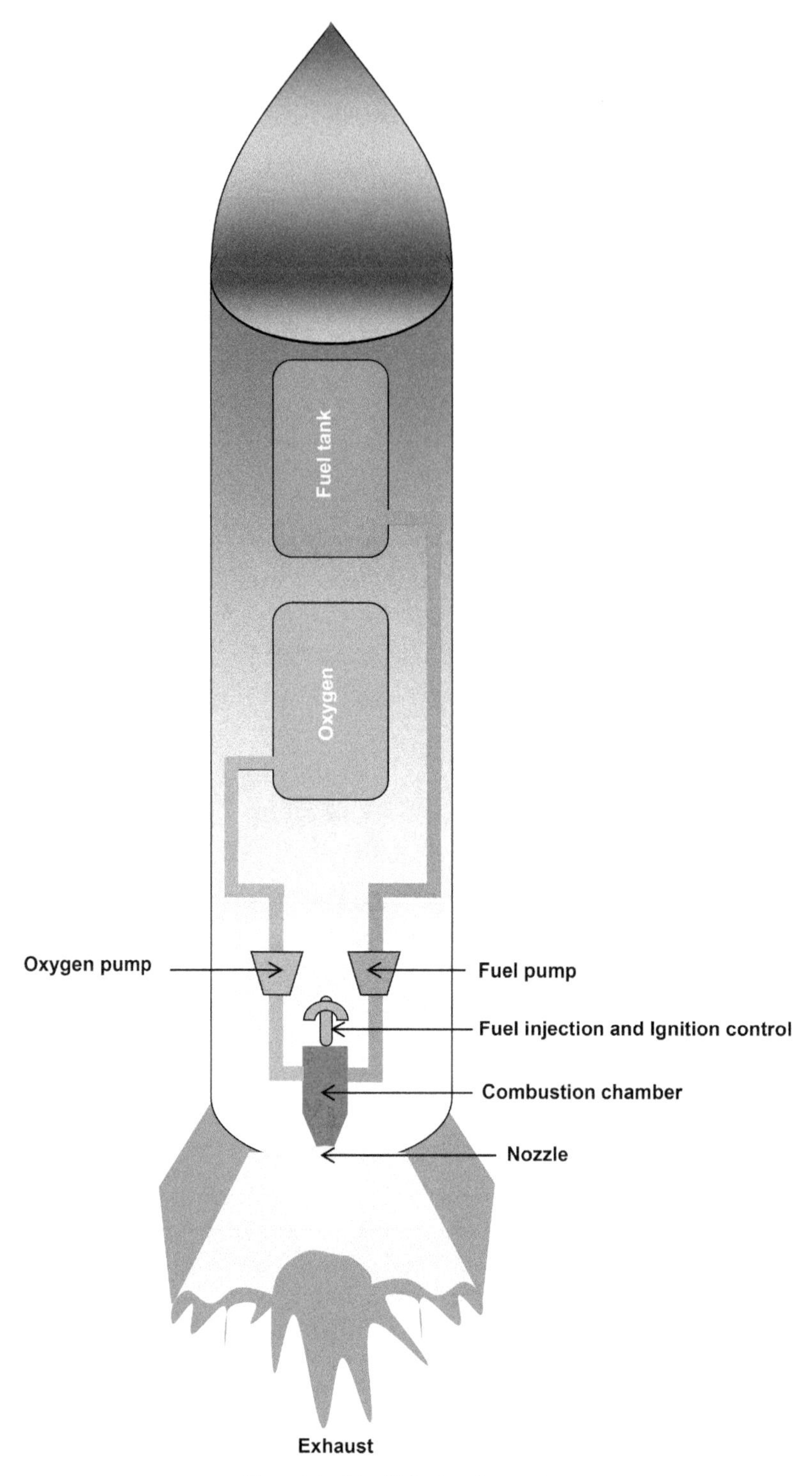

Figure 2.8. Liquid fuel rocket engine.

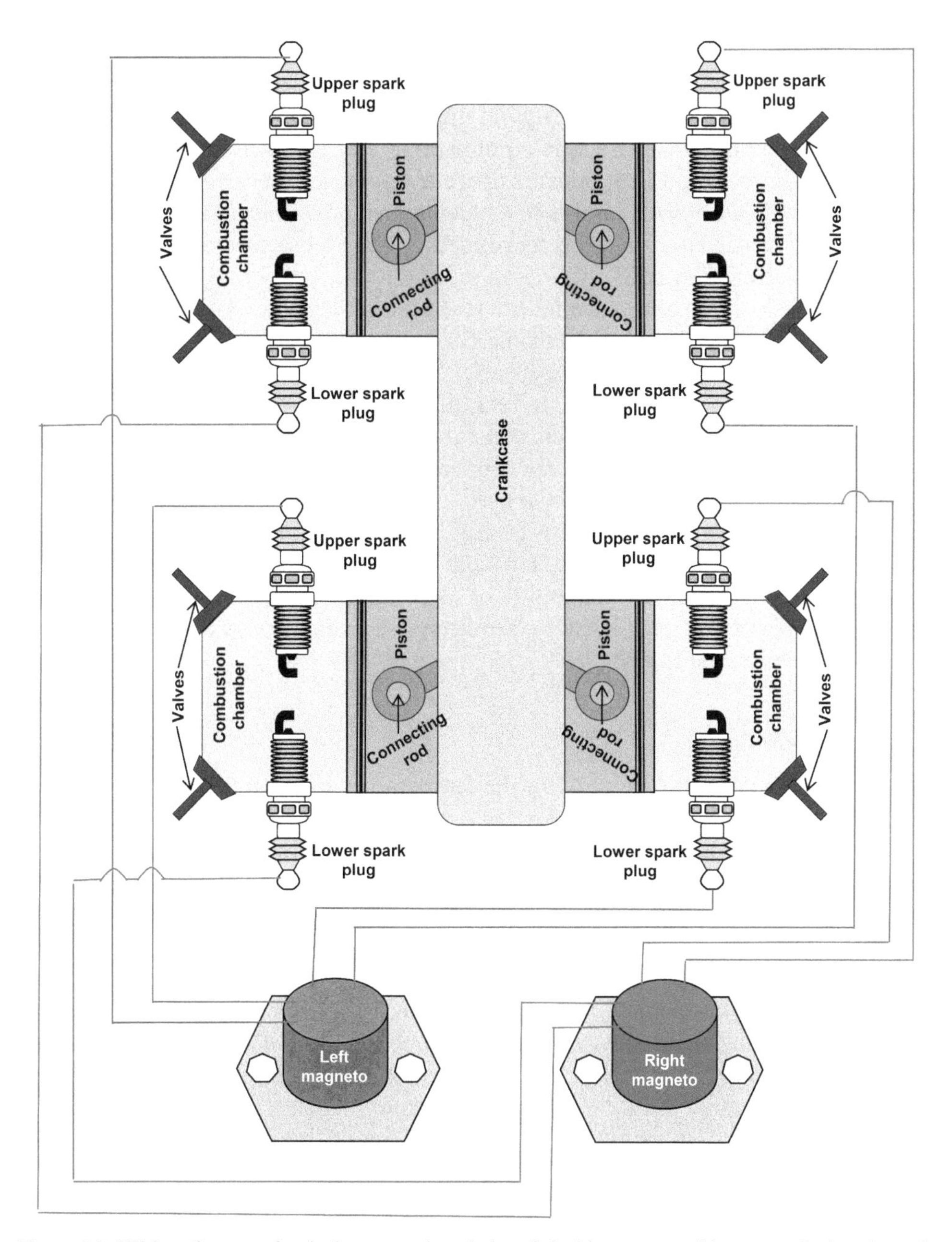

Figure 2.9. Wiring diagram of a dual magneto-based aircraft ignition system with two spark plugs in each cylinder for redundancy.

magneto serves as a backup. In addition, the ignition efficiency is improved by ensuring thorough combustion of the air–fuel mixture from both sides towards the center.

Currently the aerospace industry is following in the footsteps of the automotive industry, in so far as changeover to newer electromechanical technologies is concerned. Like the automotive industry, there is a propensity for substitution of outdated hydraulic actuators with lighter and more cost-effective electrical equivalents. Engine components and braking systems are the focus of attention. In the lubrication subsystems of turbine engines, mechanical pumps are being replaced by electrical oil pumps. Consequently, motor-drive electronics is moving into close vicinity with lubricants. The lubricants operate at temperatures over 200 °C. Thus in the aerospace industry, HTE assists in the deployment of electronic systems closer to heat sources, which are operating beyond the usual temperature boundaries laid down for electronic devices.

Furthermore, in the aerospace industry, a growing tendency is observed towards the inclusion of more electrical/electronic systems inside the aircraft. There is a clear-cut trend to make the aircraft more electric. This initiative of 'more electric aircraft' (MEA) partly seeks to eliminate traditional centralized engine controllers. They are replaced with distributed control systems. Centralized control requires large, heavy wire connections. These connections contain hundreds of conductors. Multiple connector interfaces are also used. Moving to a distributed control scheme brings the engine controls closer to the engine. The complexity of the interconnections is reduced by a factor of ten. Through these efforts, tons of aircraft weight is set aside. As a result, the reliability of the system is enhanced, but all these alterations are possible only if the electronic systems can bear up against the higher temperatures near the engine.

2.3.3 Space missions

In satellite applications (figure 2.10), the temperature of electronic assemblies may be controlled.

Space missions from the Earth to the Moon (figure 2.11) must take into consideration the wide temperature fluctuations that need to be tolerated on board the space flight and at the lunar surface. The temperature on the Moon plunges down to −153 °C during the lunar night and rises up to 107 °C during the lunar day. This happens because the Moon has no atmosphere, unlike the Earth. On Earth, the atmosphere acts as a blanket, and traps the heat received from the Sun making its escape slower. During the day, sunlight passes through the atmosphere and warms up the soil. At night, the energy is emitted by the soil as infrared radiation, but it cannot escape through the atmosphere easily; hence the planet warms up. Nights are colder than days but unlike the Moon, the night temperature does not fall to a very low level. To deal with the dramatic range in temperature on the Moon, spacesuits are heavily insulated with layers of fabric and are also covered with reflective outer layers. In addition, they have internal heaters and cooling systems. They use liquid heat exchange pumps to remove excess heat.

In deep-space missions (figure 2.12) such as planetary expeditions to Mercury, Venus, Mars or Jupiter, the ability to operate at very high temperatures up to 300 °C is a mandatory requirement, e.g. for Mercury and Venus.

Electronics can also be made to function by regulating the temperature within acceptable levels. However, in comparison to temperature regulation, the greater reliability and stable performance of high-temperature components provides more

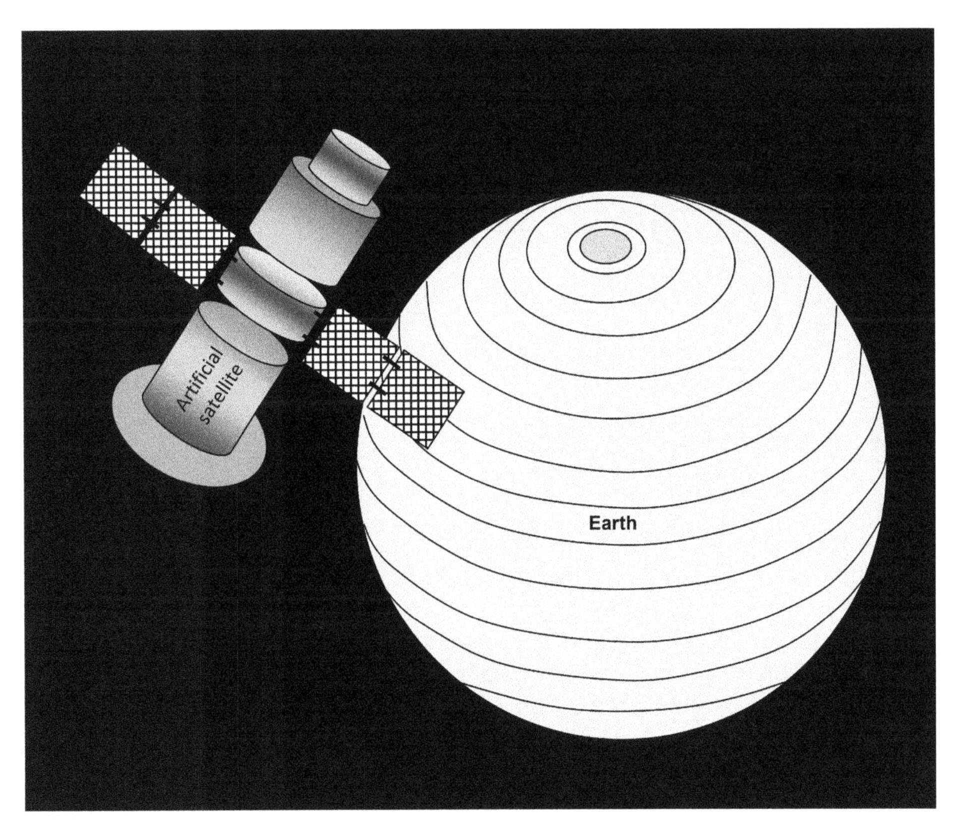

Figure 2.10. An artificial satellite orbiting the Earth.

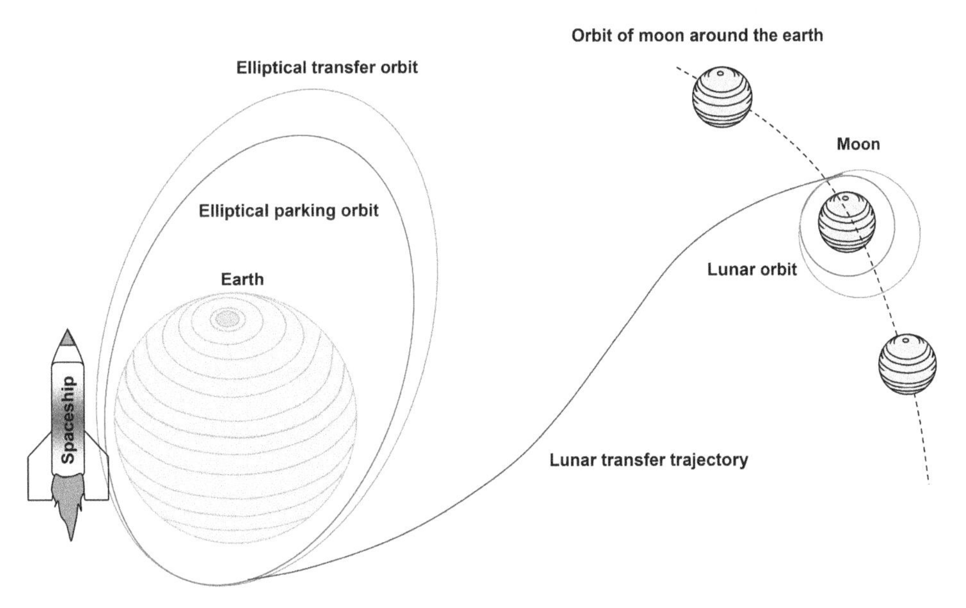

Figure 2.11. A space mission from Earth to the Moon.

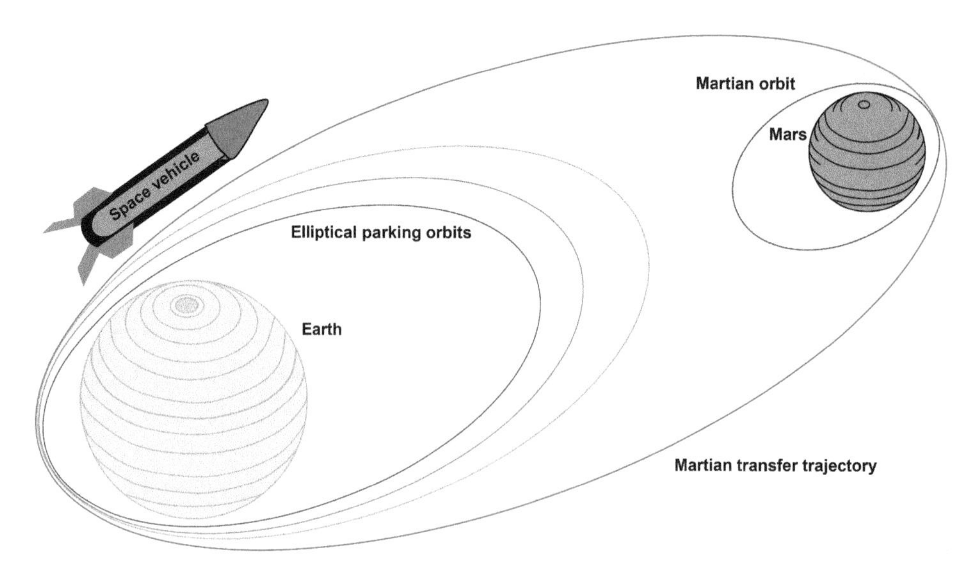

Figure 2.12. A space mission from Earth to Mars.

robust systems. It helps to remove the financial burden of temperature regulation and cooling. Thus HTE becomes a necessity in space expeditions.

At the opposite extreme temperature condition, space systems are also exposed to extremely cold conditions, e.g., for Mars and Jupiter, which falls under the domain of LTE.

2.3.4 Oil well logging equipment

Petrochemical exploration companies are striving to exploit natural resources lying underground at depths of several kilometers beneath the surface of the Earth. Well logging or bore hole logging is the practice of chronicling a detailed record of the geological rock strata, using comparable physical characteristics, when penetrated by a bore hole. In this application, the electronics are required to operate at a temperature which varies with the depth of the well below ground level. A crude estimate of this temperature is obtained from the geothermal gradient. This gradient is the increase in temperature per unit kilometer increase in depth into the Earth's interior. Near the surface and away from the tectonic plate boundaries, this gradient is 25 °C km^{-1}–30 °C km^{-1}. The tectonic plates are made up of the layers of the Earth's rocky crust, the uppermost envelop, together called the lithosphere.

In the past, explorations have been terminated at temperatures around 150 °C to 175 °C. At present, such easily accessible natural resources have been exhausted. Moreover, technological advancements have encouraged deeper drilling operations. Terrestrial regions of higher thermal gradients must also be explored. In these regions, temperatures exceeding 200 °C at pressures >25 psi are encountered in the deepest wells, more than 5 km deep. In such hostile environments, cooling techniques are neither viable nor effective.

The use of HTE in the exploration sector is multifaceted (figure 2.13). The parameters measured vary widely. The resistivity of the rocks indicates their

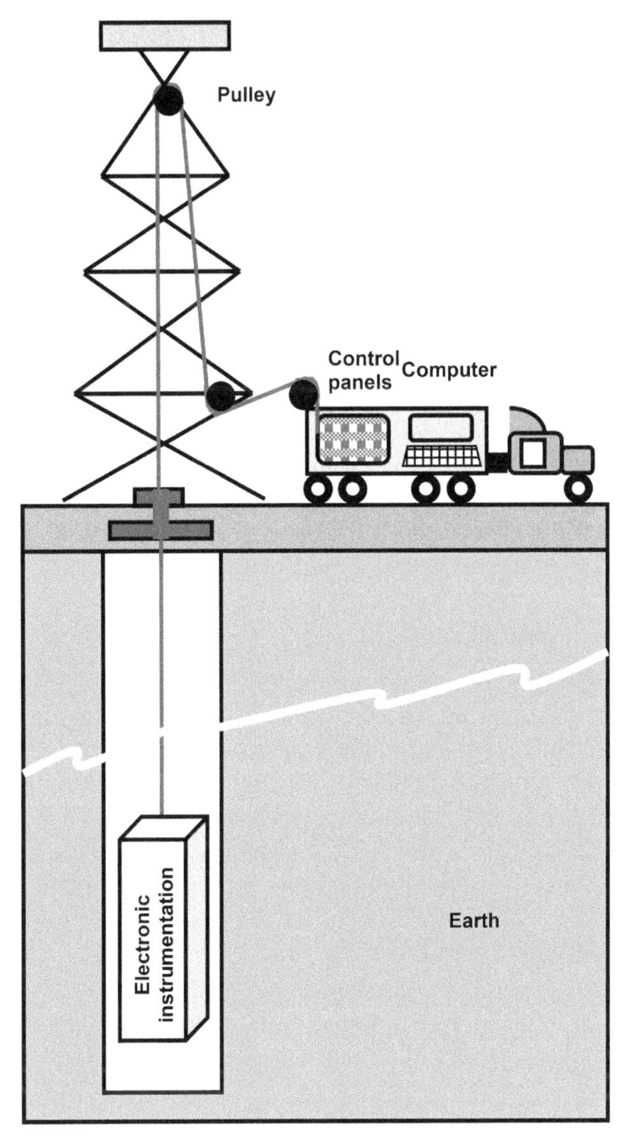

Figure 2.13. Data logging by suspending the electronic measuring unit inside the bore well and pulling it out.

electrical properties. Radioactive decay describes the type of radioactivity in the material. Similarly, acoustic travel time, magnetic resonance and other properties are measured to ascertain the general characteristics of the geological formation. Apart from these, porosity, permeability and water/hydrocarbon saturation properties are determined. Gross physical characteristics such as the mineral composition, color, texture, grain size, etc, of a rock come under its lithology. The data thus gathered allow the geologist to determine the kinds of rock in the formation, the types of fluids present, their location, and the extractability of adequate amounts of hydrocarbons from fluid-bearing zones. Accuracy is essential for all measurement and data-acquisition systems, because data taken from deep bores play a decisive role in the analysis of formations leading to identification of deposits such as oil or gas. From the data accumulated, the most promising localities to commence extraction are agreed upon and extraction is performed only at these sites. Because large monetary investments are at stake and rely on the data collected, the systems must operate accurately at these exceptionally high temperatures, emphasizing the importance of HTE.

The role of HTE does not end with well exploration but extends much further. It is also required post-exploration, in the equipment used to complete the bore and in the extraction of resources. During the completion and extraction phases, electronic systems monitor the pressure, temperature and vibrations. They regulate multiphase flow by actively controlling valves. To cater to these needs, a complete chain of high-performance components is put into service. The reliability of the system is of the utmost priority. The cost of downtime due to equipment failure is high. An electronic assembly failure on a drill string operating several kilometers underground can take more than a day to restore, a Herculean task. If replacement is necessary, enormous expenditure may be incurred.

2.3.5 Industrial and medical systems

These systems present special requirements for electronic modules. To fulfil industrial and medical needs, these systems must offer the specified exactness, constancy and permanence at high temperatures in the range of 175 °C. In industrial process control, electronic control and processing modules must be mounted close to sensors and actuators. Such mounting prevents noise from entering the system through long wiring. Medical electronic systems need HTE to, for example, take advantage of the benefits of high-temperature monitoring of sterilization systems.

2.4 Low-temperature electronics

LTE is referred to by different names such as cold electronics, cryogenic electronics or cryoelectronics (Gutiérrez *et al* 2001). It involves the operation of electronic materials, devices, circuits and systems at temperatures significantly below the conventional range, which conventionally ends at −55 °C (Kirschman 1990). Electronic devices and systems have been operated at temperatures down to within a degree of absolute zero (0 K = −273 °C). Because of the variety of materials, devices and effects, no single temperature can be given as the boundary between the conventional and the low-temperature ranges. However, the cryogenic range is considered to start from about 100 K (−173 °C) and move towards the colder side.

In general, three main regions are differentiated in discussions on low-temperature operation (Gutiérrez *et al* 2001): (i) the liquid-nitrogen (77 K) range, (ii) the liquid-helium (4.2 K) range and (iii) the deep cryogenic temperatures descending down to the millikelvin range. The first region may lead to more or less commercial applications. The second region is mainly used for cold electronics associated with space missions such as the Infrared Space Observatory Photometer (ISOPHOT) and Far Infra-Red Space Telescope (FIRST). Extremely low temperatures are found in the world of astrophysical applications, such as bolometers. The lower the operating temperature, the greater is the transference of activities from potential industrial applications to research-oriented fields.

The motivation to study LTE lies in the following three reasons (Clark *et al* 1992). First and foremost, it is clear that temperature has a profound impact on several important properties of materials. Notable properties are the drift velocity of carriers and conductance of substances. In ICs, noise margins are influenced. Keeping the effect of temperature on such properties under surveillance can provide useful insights into the behavior of electronic circuits at room temperature and liquid-nitrogen temperature. This could potentially lead to the ability to counteract intrinsic problems, such as reliability, through methods that would have otherwise remained unknown.

The second reason why LTE must be investigated is to explore the areas in which a low-temperature ambient is essential and wherein room-temperature electronics cannot penetrate. In this area falls electronics which cannot operate at room temperature, but become feasible only at low temperatures. This area is exemplified by two well-known phenomena, namely superconductance and the Josephson effect. They occur only at significantly low (77 K) temperatures. Furthermore, transistor characteristics, such as the threshold voltage of MOSFETs, do not scale properly with temperature. Thus, understanding the effects of low temperatures is upgraded from a simple option to a mandatory requirement.

Lastly, the performance of existing CMOS technology can be further improved if transformed into LTE, i.e. operated under low-temperature conditions as opposed to room temperature (Peeples *et al* 2000). Low-temperature conditions have also eliminated hazardous problems such as latchup. They have significantly benefited applications such as dynamic random access memory (DRAM) (Henkels *et al* 1989). The drawback, however, is the enormous cost necessary to maintain the temperature at a sufficiently low magnitude so that the desired phenomena can take place. Figure 2.14 illustrates how LTE can be utilized as a boon to electronics, helping in significant improvement in the performance of existing technologies and also affording the development of new technologies, notably by exploiting superconductivity.

For the operation of silicon devices there is a minimum temperature, called the freeze-out temperature, at which the thermal energy inside the semiconductor is too small and is inadequate to activate the donor or acceptor impurity atoms, depending on whether the material is n-type or p-type. Freeze-out is the condition when dopants are no longer ionized and the semiconductor behaves as an insulator. The freeze-out temperature depends on the semiconductor doping concentration. At an elevated level of doping, which is much higher than that used in standard silicon technology, one can operate devices at temperatures close to 0 K. On the other side, consumer electronic devices are clearly not aimed at operation below 233 K. However, exceptional, specific applications exist. The main snags with cryogenic design arise from the carrier freeze-out phenomena and changes in carrier mobilities that occur near absolute zero. These effects must be fully understood to correctly model devices. They must be incorporated accurately into electronic design software.

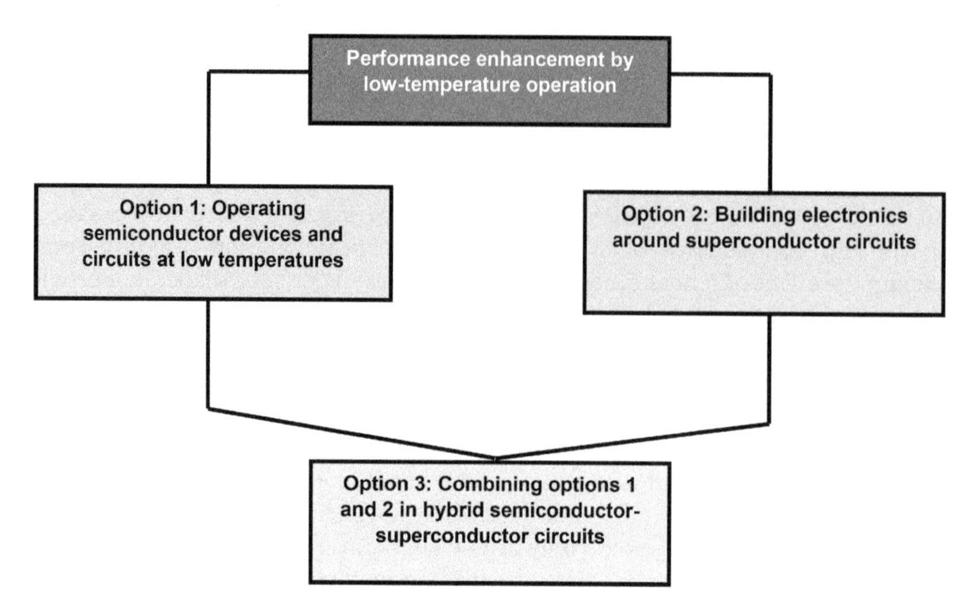

Figure 2.14. Cold electronics using semiconductors and superconductors.

2.5 The scope of extreme-temperature and harsh-environment electronics

A harsh environment is described as any location or surroundings in which survival is arduous or outside the bounds of possibility. For a human being, a harsh or inhospitable environment is an unwelcoming set of conditions that can cause harm to the body over a period of time. The concept is extendable to electronics. An electrical circuit is easily spoiled and ceases to function if it is exposed to excessively high temperatures and fast temperature cycling. But temperature is not the only uncongenial variable which is to be blamed for failure.

In addition to thermal effects, there are other factors which may be more detrimental to electronics than temperature. Systems can be destroyed if submerged in water. Prolonged exposure to a high-humidity ambient may also damage the circuit. Subjecting a circuit to ingress of particulate matter is no less harmful. Electrostatic discharge (ESD) effects are known to cause dielectric breakdown of MOS devices if suitable precautions are not taken. Electromagnetic interference (EMI) upsets the normal functioning of circuits. Apart from these factors, vibrations, physical shocks and collisions can impair the functionality of devices and circuits. Therefore, harsh environments include high-temperature atmospheres as a sub-component. But low temperatures may not necessarily fall in this domain. A few low-temperature phenomena improve device characteristics, a few others degrade them. Nevertheless, they always need elaborate arrangements to provide the cooling which make the systems very bulky. Therefore, in this book temperature effects are treated separately from harsh environment effects, such as those of humidity, radiation and shocks. Thermal effects can be both disadvantageous and advantageous.

2.5.1 High-temperature operation: a serious vulnerability

Compared to the low-temperature end, the high-temperature end causes more problems in electronics. Frequently, the high-temperature restriction is not determined by the inherent limitations of the semiconductor materials, but by the properties of the materials used for interconnection between devices and packaging of finished chips.

2.5.2 Upgradation/degradation of performance by cooling

Contrary to human beings, very low temperatures may not be always punitive to electronics. Instead, they may often lead to better performance of devices and circuits. In general, as temperature decreases, the electrical characteristics of a component can undergo a gradual or an abrupt change. It may even stop operating completely. At lower temperatures, field-effect transistors (FETs) exhibit increased gain and speed, and unwanted leakage current is decreased. A MOSFET or CMOS device can operate below the liquid-nitrogen temperature (−196 °C). Its performance is upgraded. On the other hand, a silicon bipolar transistor discontinues functioning at about −150 °C. Essentially, it offers such low gain that it becomes unworkable. Band-structure changes are responsible for these malfunctions, not the freeze-out phenomenon. Freeze-out does not occur in silicon until about −230 °C (Titus 2012).

With falling temperature, parasitic resistances and capacitances in interconnections decrease. Heat transfer improves and the noise level decreases. An important low-temperature application includes pacing up digital electronics and another application is noise reduction in microwave pre-amplification.

2.5.3 Corrosion: humidity and climatic effects

When the atmospheric relative humidity (RH) is very high, >80%, a lot of water vapor condenses on the metal surfaces of semiconductor devices. Optimal performance of computers is achieved by maintaining the humidity levels tightly between 45% and 55% RH limits. The reliability of computers is also assured at these humidities. The recommendation is for warning alerts to be triggered when humidity falls below 40% RH or rises above 60% RH. Critical alerts may be sounded at humidity levels less than 30% RH and above 70% RH (Grundy 2005).

Condensation of moisture molecules in the superincumbent air on metallic surfaces leads to the formation of a thin film of water on these surfaces. Application of an electric field to the metal results in the elution of metal ions, which is the process of stripping of ions as if by washing with a solvent. The neighboring conductor pulls the eluted metal ions. In this way, current can flow between adjoining conductors through the film formed by eluted metal ions, short-circuiting the device and thereby causing its failure. Often, the metal film resembles a dendrite of a human nerve cell with threadlike extensions. At the time of short circuit, it disappears because of its brittleness. Device failure induced by migration of metal ions is a cumbersome process, which is not reproducible after its occurrence (Apiste Corporation 2015).

Corrosion induced by climatic effects increases the contact resistances of joints. Leakage currents between wires are enhanced. It also causes materials to decay. The surfaces of electronic devices acquire an ugly appearance. Different types of operational faults are observed (Hienonen and Lahtinen 2000).

2.5.4 Deleterious effects of nuclear and electro-magnetic radiations on electronic systems

We dwell in a world which is filled with radiations of different kinds. The radiation content in our environment varies dramatically from place to place. It is also variable with respect to time (Boscherini *et al* 2003). While civilians reside in a comparatively safe environment, military and space systems are confronted with consistent threats of assorted types of radiation of high intensities. Many times, military operations have to withstand man-made nuclear radiation. Catastrophic effects are produced in electronic systems exposed to such hazardous radiation-contaminated environments. The impact of this radiation on electronic systems may be apparent in various forms. It may lead to momentary cessation of services or to erasure of chip memory, or ultimately to death of the device by burning out. As radiation effects are primarily issues faced by military and space engineers, manufacturers of commercial consumer electronics goods hardly ever pay any attention to these problems. As a result, consumer electronic systems have no built-in features safeguarding them against the injurious effects of radiation. They often fail prematurely in such environments.

Considering the types of radiation encountered by electronic systems, there are two broad categories, namely particulate type radiation and electromagnetic waves or photons. Particulate radiation consists of charged and neutral particles, such as electrons or beta particles, protons or hydrogen nuclei, alpha particles or helium nuclei, ions resulting from fission reactions called fission fragments, neutrons and heavy ions. Photonic radiation comprises gamma and x-rays. Radiation damage is the general term describing the detrimental consequences of radiations of all types.

Radiation can be distinguished into two other types, those containing charged particles (α-particles, β-rays, protons) and those of a neutral nature (neutrons and γ-rays). Fundamentally, exposure to the two types of radiation produces different kinds of effects on electronic devices and systems. These effects are subdivided into three broad categories.

(i) *Single event effects (SEEs).* Here passage of a single, highly ionizing particle alters the operational state of a semiconductor device in a single event, e.g. it may change the logic state of a transistor.

(ii) *Total ionizing dose (TID) effects.* These are long-term effects which take place on prolonged exposure to radiation. They cause build-up of charge at the boundaries or interfacial regions between different layers in semi-conductor devices.

(iii) *Displacement damage (DD) effects.* Collisions take place between the irradiating particles and lattice atoms. These collisions displace the lattice atoms from their positions, resulting in the formation of Schottky or

Frenkel defects in the crystal lattice. The crystal lattice is permanently damaged. Secondary particles liberated as a result of the collisions may cause further displacement of atoms, activating a cascading of collisions. Several defects in the form of vacancies and interstitials can be found along the tracks of the incoming particles. Such defects may occur in clusters at the ends of the tracks.

2.5.5 Vibration and shock effects

When first considering the harsh environmental conditions to be encountered by an electronic product, a design engineer is likely to overlook the effects of vibrations and shocks. However, vibration and shock effects represent a major cause of failure in many systems. During the lifecycle of a product, it may have to sustain such effects at many stages. Conceivably, it may experience shocks during shipping or transportation when the cargo is emplaned or deplaned, or carried in a vehicle for delivery to the customer. During everyday use, it may sometime fall from its support. If it is too fragile, it will stop working. In many applications, ranging from automobile, train and aerospace systems, to oil drilling apparatus/hardware, power houses and power generating plants, the product has to face vibrational stresses of moderate or severe magnitudes (Askew 2015).

Protection from random vibrations is provided by mounting the electronic systems on resilient or elastic supports. The trade-off between the damping and stiffness of supporting platforms is based on the dynamic responses of the resulting electronic enclosures subject to the maximum rattle space provided in a given application. Thus the design rule for resilient mounts is to optimize the isolation between vibration and the electronic system within the limits of the maximum deflections permitted. By pursuing this approach, reliable vibration isolators can be built in view of the intensity/amplitude of vibrations experienced (Veprik 2003).

2.6 Discussion and conclusions

It is evident that electronic systems designed to function under normal operational conditions cannot be expected to perform well under extreme-temperature and harsh environmental conditions, of which a few representative examples were presented in this chapter. The deleterious effects on electronic devices can occur due to the following reasons: (i) operation at very high temperatures for which the device has not been fabricated; (ii) operation at very low temperatures for which the device was not planned; (iii) exposure to very wet conditions; (iv) exposure to radiations of different varieties; and (v) subjecting the device to vibrations and shocks. In this list, low temperatures are known to influence device operation in both directions, i.e. in upgradation as well as degradation. High temperatures too can be beneficial on some occasions. Therefore, temperature effects are considered as a separate subject which can be harmful or beneficial. The rest of the subject is treated under harsh-environment effects. The field of electronic devices and systems intended to operate in these abnormal circumstances is therefore a subject in itself, and this subject constitutes the focal theme of this book. Various interesting facets, issues and challenges faced by engineers and scientists in overcoming the problems will be highlighted and their solutions will be addressed as the reader progresses through the text. The utilization of the helpful effects of low-temperature operation will also be explained.

As the reader browses through chapters of this book, it will be evident that necessary precautionary measures need to be taken right at the outset, i.e. from product conceptualization for a particular application. The design of the product must take into consideration all adverse situations that the electronic device is likely to be confronted with during its lifetime. This practice should be adhered to during planning its fabrication and packaging steps, right up to system assembly (Watson and Castro 2012).

Review exercises

2.1. Explain the following terms: (i) sun stroke, (ii) hypothermia and (iii) frostbite.
2.2. Which is the hottest place on the Earth's surface? What is its temperature?
2.3. Which is the coldest place on the Earth's surface? What is its temperature?
2.4. Name the hottest and coldest planets in the Solar System. What are their temperatures?
2.5. Explain the following terms: (i) HTE, (ii) LTE, (iii) ETE and (iv) harsh-environment electronics.
2.6. Why is using traditional electronics combined with active or passive cooling inadequate to meet the challenges posed by high temperatures?
2.7. Give three reasons explaining the need for HTE in the automotive industry.
2.8. Explain the significance of HTE in aerospace industries.
2.9. Why does the temperature on Earth not fall to very low magnitudes during the night? Explain the role of HTE in space missions.
2.10. What is the value of the thermal gradient per kilometer below the Earth's surface? Explain the need for developing high-temperature electronic devices for deep oil well exploration.
2.11. Give one example of the use of HTE in medicine.
2.12. Write three alternative names for LTE.
2.13. Mention the three major regions of low-temperature operation of semiconductor devices. Indicate the targeted applications of each region.
2.14. Name two properties of semiconductors which are benefitted by low-temperature operation.
2.15. Name two phenomena which take place only at low temperatures.
2.16. What is meant by carrier freeze-out? What are the minimum temperatures up to which traditional MOS and bipolar devices can operate? Does reduction of the gain of a bipolar transistor at low temperatures result from carrier freeze-out or some other phenomenon?

2.17. Between what limits of humidity is the performance of computers optimized? What happens when moisture condenses on the metal surfaces of semiconductor devices and a voltage is applied to them?
2.18. What are the two main classes of radiation which affect the performance of semiconductor devices?
2.19. Name and discuss the three types of effects that radiation produces in semiconductor devices.
2.20. How is a semiconductor device protected from the effects of vibrations and shocks?

References

Apiste Corporation 2015 Effect of humidity on electronic devices *Apiste Corporation* www.apiste-global.com/enc/technology_enc/detail/id=1263

Askew D 2015 Vibration protection of electronic components in harsh environments *Mouser Electronics Inc.* www.mouser.in/applications/harsh-environment-vibration-protection/

Boscherini M, Adriani O, Bongi M, Bonechi L, Castellini G, D'Alessandro R and Gabbanini A *et al* 2003 Radiation damage of electronic components in space environment *Nucl. Instrum. Methods Phys. Res.* A **514** 112–6

Clark W F, El-Kareh B, Pires R G, Titcomb S L and Anderson R L 1992 Low temperature CMOS—a brief review *IEEE Trans. Components Hybrids Manuf. Technol.* **15** 397–403

Delatte P 2010 Designing high-temp electronics for auto and other apps *EE Times* www.eetimes.com/document.asp?doc_id=1272966&

Grundy R 2005 Recommended data centre temperature and humidity: preventing costly downtime caused by environment conditions *AVTECH News* www.avtech.com/About/Articles/AVT/NA/All/-/DD-NN-AN-TN/Recommended_Computer_Room_Temperature_Humidity.htm

Gutiérrez D E A, Deen M J and Claeys C (ed) 2001 *Low Temperature Electronics: Physics, Devices, Circuits and Applications* (New York: Academic), 964 pages

Henkels W H, Lu N C C, Hwang W, Rajeevakumar T V, Franch R L, Jenkins K A, Bucelot T J, Heidel D F and Immediato M J 1989 A 12-ns low-temperature DRAM *IEEE Trans. Electron. Devices* **36** 1414–22

Hienonen R and Lahtinen R 2000 *Corrosion and Climatic Effects in Electronics* (Espoo: Technical Research Centre of Finland), 420 pages

Howell E 2014 How far are the planets from the Sun? *Universe Today* www.universetoday.com/15462/how-far-are-the-planets-from-the-sun/

Huque M A, Islam S K, Blalock B J, Su C, Vijayaraghavan R and Tolbert L M 2008 Silicon-on-insulator based high-temperature electronics for automotive applications *IEEE International Symposium on Industrial Electronics* (*Cambridge, 30 June–2 July*) (Piscataway, NJ: IEEE) pp 2538–43

Kirschman R K 1990 Low-temperature electronics *IEEE Circuits Devices* **6** 12–24

Kirschman R K 2012 *Extreme-Temperature Electronics, Tutorials* www.extremetemperatureelectronics.com

McCluskey F P, Grzybowski R and Podlesak T (ed) 1997 *High Temperature Electronics* (New York: CRC Press), 337 pages

Peeples J W, Little W, Schmidt R and Nisenoff M 2000 Low temperature electronics workshop *Sixteenth Annual Semiconductor Thermal Measurement and Management Symposium* (*San Jose, CA, 21–23 March*) (Piscataway, NJ: IEEE) pp 107–8

Tavanaei G 2013 The five coldest and hottest places on Earth *Epoch Times* www.theepochtimes.com/n3/101262-the-five-coldest-and-hottest-places-on-earth/

Titus J 2012 Design electronics for cold environments *ECN Mag.* www.ecnmag.com/articles/2012/12/design-electronics-cold-environments

Veprik A M 2003 Vibration protection of critical components of electronic equipment in harsh environmental conditions *J. Sound Vib.* **259** 161–75

Watson J and Castro G 2012 High-temperature electronics pose design and reliability challenges *Analogue Dialogue* **46–04** 1–7

Williams M 2014 What is the average surface temperature of the planets in our solar system? *Universe Today* www.universetoday.com/35664/temperature-of-the-planets/

Part I

Extreme-temperature electronics

IOP Publishing

Extreme-Temperature and Harsh-Environment Electronics
Physics, technology and applications
Vinod Kumar Khanna

Chapter 3

Temperature effects on semiconductors

The impact of temperature on the important properties of semiconducting materials used for electronic devices and circuit fabrication is examined, with a focus on silicon. The properties considered are the energy bandgap (the Varshini and Blaudau *et al* models), intrinsic carrier concentration and saturation velocity of carriers (the Quay model, and Ali-Omar and Reggiani model). Various mobility equations are discussed, e.g. the Arora–Hauser–Roulston equation, Klaassen equations and those in the MINIMOS model. The differences between uncompensated and compensated semiconductors regarding the temperature dependence of the mobility and carrier concentration are described. The ionization regimes of semiconductors are also described, namely the carrier freeze-out regime, extrinsic or saturation regime, and intrinsic regime. The conceptual development in this chapter paves the way for the temperature-related discussions in forthcoming chapters.

3.1 Introduction

A proper understanding of the operation of semiconductor devices at very low as well as at very high temperatures can only be obtained on the basis of a general comprehension of the physical properties of semiconductor materials, together with a correct perception of their variation under extreme thermal conditions. These properties often change drastically compared to the more familiar room-temperature behavior. The present chapter investigates how the properties of a semiconductor change with temperature. Throughout, silicon is taken as the focal material, but the treatment of silicon helps us to interpret the behavior of other materials, taking into consideration the relevant differences between material properties.

3.2 The energy bandgap

The energy bandgap is a fundamental property of a semiconductor which determines the electrical characteristics of the devices fabricated from it. For practical purposes, such as the simulation of semiconductor components, the variation of the

doi:10.1088/978-0-7503-1155-7ch3

bandgap value with the temperature and doping concentration, must be accurately known to the device designer (Stefanakis and Zekentes 2014).

The energy bandgap of semiconductors always diminishes when its temperature is increased (Van Zeghbroeck 2011). This behavior can be appreciated if one considers that the interatomic spacing becomes larger when the amplitude of the atomic vibrations increases. This increase in interatomic spacing is caused by the enhancement in thermal energy by intensification of thermal motion at higher temperatures. The effect is quantified by the linear expansion coefficient of a material. With an increase of interatomic spacing, the potential seen by the electrons in the material decreases. It is this decreased potential which is responsible for the reduction of the energy bandgap. A direct modulation of the interatomic distance also brings about bandgap changes. By applying high compressive stress the bandgap increases, whereas on applying a tensile stress it decreases.

A semi-empirical relationship for the variation of the energy gap (E_g) of semiconductors with temperature (T) was proposed by Varshini (1967):

$$E_g = E_g(0) - \frac{\alpha T^2}{T + \beta}, \tag{3.1}$$

where $E_g(0)$ (eV), α (eV K^{-1}) and β (K) are the fitting coefficients of the model. The symbol $E_g(0)$ represents the bandgap of the material at 0 K, and T is the temperature in the Kelvin scale. Values of the parameters α and β for Si, GaAs and 4H-SiC are compiled in table 3.1. The equation has been found to satisfactorily represent the experimental data for diamond, Ge, Si, 6H-SiC, GaAs, InP and InAs.

Applying the Varshini model, variation of the energy bandgaps of common semiconductors is presented in table 3.2.

Table 3.1. Varshini fitting parameters.

Material	$E_g(0)$ (eV)	α (eV K^{-1})	β (K)	Reference
Germanium	0.7437	4.774×10^{-4}	235	Sze (1981)
Silicon	1.1695	4.73×10^{-4}	636	Singh (1993), Ioffe Institute (2015b)
Gallium arsenide	1.521	5.58×10^{-4}	220	Wilkinson and Adams (1993)
4H-SiC	3.285	3.3×10^{-4}	240	Stefanakis and Zekentes (2014)

Table 3.2. Bandgaps at different temperatures according to the Varshini model.

Material	4.2 K	77.2 K	300 K	600 K
Germanium	0.74366	0.7346	0.66339	0.53787
Silicon	1.169486	1.16555	1.12402	1.031733
Gallium arsenide	1.520956	1.50981	1.42442	1.276024
4H-SiC	3.284976	3.2788	3.23	3.143571

Table 3.3. Bandgaps of silicon at different temperatures according to the Bludau *et al* model.

Temperature	4.2 K	77.2 K	100 K	200 K	300 K
Bandgap	1.170	1.167	1.165	1.148	1.124

For Si, a precise assessment of the bandgap energy E_g between 2 K and 300 K (Bludau *et al* 1974) led to the approximate formula as follows:

$$E_g(T) = A + BT + CT^2. \tag{3.2}$$

The values of parameters A, B and C are provided for two temperature zones. In the temperature range $0 < T \leqslant 190$ K,

$$A = 1.170 \text{ eV}, \quad B = 1.059 \times 10^{-5} \text{ eV K}^{-1} \quad \text{and} \quad C = -6.05 \times 10^{-7} \text{ eV K}^{-2},$$

while for the temperature interval $150 \leqslant T \leqslant 300$ K,

$$A = 1.1785 \text{ eV}, \quad B = -9.025 \times 10^{-5} \text{ eV K}^{-1} \quad \text{and} \quad C = -3.05 \times 10^{-7} \text{ eV K}^{-2}.$$

Table 3.3 shows the changes in the Si bandgap with temperature on the basis of the Bludau *et al* model.

3.3 Intrinsic carrier concentration

The intrinsic carrier concentration of a semiconductor is a key material property which occurs every now and then in the formulae describing the operations of various devices. This basic parameter exhibits a strong dependence on temperature. Further, the nature of variation is vastly different for different semiconductors, as shown in figure 3.1.

Under thermal equilibrium conditions, the intrinsic carrier concentration in a semiconductor is given by

$$n_i = \sqrt{N_C N_V} \exp\left(-\frac{E_g}{2k_B T}\right), \tag{3.3}$$

where N_C, N_V are the effective densities of states in the conduction and valence bands, respectively. N_C, N_V are expressed as

$$N_C = 2\left(\frac{2\pi m_n^* k_B T}{h^2}\right)^{1.5} \tag{3.4}$$

$$N_V = 2\left(\frac{2\pi m_p^* k_B T}{h^2}\right)^{1.5}, \tag{3.5}$$

where m_n^*, m_p^* are the effective masses of electrons and holes for density of states calculations, k_B is the Boltzmann constant ($= 1.381 \times 10^{-23}$ m^2 kg s^{-2} K^{-1}), h is

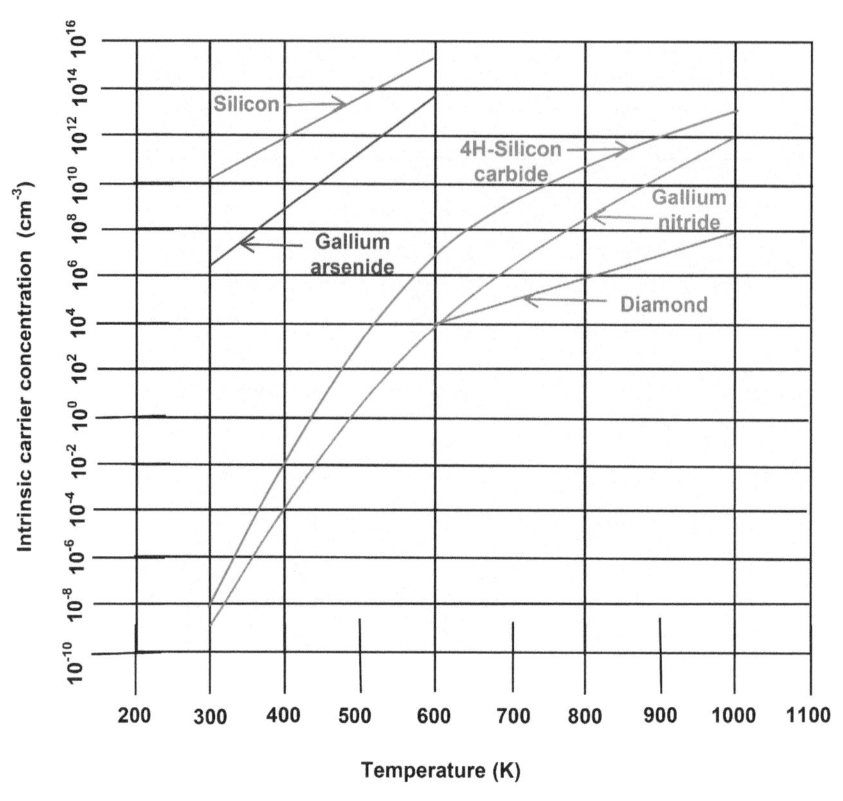

Figure 3.1. Change in the intrinsic carrier concentration of semiconductors on raising the temperature above room temperature (300 K).

Table 3.4. Effective densities of states (Van Zeghbroeck 1997).

Material	m_n^*/m_0	m_p^*/m_0	N_C (cm^{-3})	N_V (cm^{-3})
Germanium	0.56	0.29	$2.023 \times 10^{15} T^{1.5}$	$7.54 \times 10^{14} T^{1.5}$
Silicon	1.08	0.81	$5.42 \times 10^{15} T^{1.5}$	$3.52 \times 10^{15} T^{1.5}$
Gallium arsenide	0.067	0.47	$8.37 \times 10^{13} T^{1.5}$	$1.56 \times 10^{15} T^{1.5}$

m_0 = free electron rest mass = 9.11×10^{-31} kg

Planck's constant (= 6.626×10^{-34} m^2 kg s^{-1}). The calculated values of N_C, N_V for different semiconductors are tabulated in table 3.4.

In the above calculations, the masses m_n^*, m_p^* are presupposed to be constant with respect to temperature. But in reality, this assumption is not true. Careful review and correlation of experimental data on density-of-states effective masses m_n^*, m_p^*, and on the intrinsic concentration in silicon have revealed that the effective masses are, in fact, temperature- and energy-dependent (Barber 1967). Based on the measured temperature dependence of the energy gap, the obvious temperature variation of both the hole and electron effective masses was approximated in a first-order evaluation. Substitution of these temperature-dependent effective masses into the theoretical expression for intrinsic carrier concentration yielded a close agreement with reported measurements of n_i, confined within the bounds of experimental error. To perform a more rigorous calculation including the temperature-dependent effective masses of carriers as well as the energy bandgap of the semiconductor, we write

$$n_i = 2\left(\frac{2\pi k_B}{h^2}\right)^{1.5}\left(m_n^* m_p^*\right)^{0.75} T^{1.5} \exp\left(-\frac{E_g}{2k_B T}\right) \tag{3.6}$$

$$\begin{aligned}
&= 2\left\{\frac{2 \times 3.14 \times 1.381 \times 10^{-23}}{(6.626 \times 10^{-34})^2}\right\}^{1.5}\left(\frac{m_n^* m_p^*}{m_0^2} \times m_0^2\right)^{0.75} T^{1.5} \exp\left(-\frac{E_g}{2k_B T}\right) \\
&= 2\left(\frac{8.67268 \times 10^{-23}}{4.3903876 \times 10^{-67}}\right)^{1.5} \times (m_0^2)^{0.75} \times \left(\frac{m_n^* m_p^*}{m_0^2}\right)^{0.75} T^{1.5} \exp\left(-\frac{E_g}{2k_B T}\right) \\
&= 5.55272 \times 10^{66} \times m_0^{1.5} \times \left(\frac{m_n^* m_p^*}{m_0^2}\right)^{0.75} T^{1.5} \exp\left(-\frac{E_g}{2k_B T}\right) \\
&= 5.55272 \times 10^{66} \times (9.11 \times 10^{-31})^{1.5} \times \left(\frac{m_n^* m_p^*}{m_0^2}\right)^{0.75} T^{1.5} \exp\left(-\frac{E_g}{2k_B T}\right) \\
&= 4.8281787 \times 10^{21} \times \left(\frac{m_n^* m_p^*}{m_0^2}\right)^{0.75} T^{1.5} \exp\left(-\frac{E_g}{2k_B T}\right) \mathrm{m}^{-3} \\
&= 4.83 \times 10^{15} \times \left(\frac{m_n^* m_p^*}{m_0^2}\right)^{0.75} T^{1.5} \exp\left(-\frac{E_g}{2k_B T}\right) \mathrm{cm}^{-3}.
\end{aligned} \tag{3.7}$$

For silicon, the equations describing the temperature-dependent effective masses m_n^* and m_p^* in terms of free electron rest mass m_0 are (Caiafa *et al* 2003):

$$m_n^* = (-1.084 \times 10^{-9} T^3 + 7.580 \times 10^{-7} T^2 + 2.862 \times 10^{-4} T + 1.057) m_0 \tag{3.8}$$

$$m_p^* = (1.872 \times 10^{-11} T^4 - 1.969 \times 10^{-8} T^3 + 5.857 \times 10^{-6} T^2 + 2.712 \times 10^{-4} T + 0.584) m_0. \tag{3.9}$$

The temperature dependence of the bandgap energy is included through the equations of Bludau *et al* (1974) as follows:

$$E_g = 1.17 + 1.059 \times 10^{-5} T - 6.05 \times 10^{-7} T^2 \text{ for } T \leqslant 190\,\mathrm{K} \tag{3.10}$$

$$E_g = 1.1785 - 9.025 \times 10^{-5} T - 3.05 \times 10^{-7} T^2 \text{ for } 300\,\mathrm{K} \geqslant T \geqslant 190\,\mathrm{K}. \tag{3.11}$$

At $T = 4.2$ K,

$$\begin{aligned} m_n^* &= \{-1.084 \times 10^{-9}(4.2)^3 + 7.580 \times 10^{-7}(4.2)^2 \\ &\quad + 2.862 \times 10^{-4}(4.2) + 1.057\}m_0 \\ &= \{-8.031 \times 10^{-8} + 1.337 \times 10^{-5} + 1.202 \times 10^{-3} + 1.057\}m_0 \\ &= 1.058m_0 \end{aligned} \tag{3.12}$$

$$\begin{aligned} m_p^* &= \{1.872 \times 10^{-11}(4.2)^4 - 1.969 \times 10^{-8}(4.2)^3 \\ &\quad + 5.857 \times 10^{-6}(4.2)^2 + 2.712 \times 10^{-4}(4.2) + 0.584\}m_0 \\ &= \{5.825 \times 10^{-9} - 1.459 \times 10^{-6} + 1.033 \times 10^{-4} \\ &\quad + 1.139 \times 10^{-3} + 0.584\}m_0 = 0.585m_0 \end{aligned} \tag{3.13}$$

$$\begin{aligned} E_g &= 1.17 + 1.059 \times 10^{-5}(4.2) - 6.05 \times 10^{-7}(4.2)^2 \\ &= 1.17 + 4.4478 \times 10^{-5} - 1.067 \times 10^{-5} \\ &= 1.17\text{eV} \end{aligned} \tag{3.14}$$

$$\begin{aligned} \therefore n_i &= 4.83 \times 10^{15} \times \left(\frac{1.058m_0 \times 0.585m_0}{m_0^2}\right)^{0.75} (4.2)^{1.5} \\ &\quad \exp\left(-\frac{1.17}{2 \times 8.617 \times 10^{-5} \times 4.2}\right) \text{cm}^{-3} \\ &= 4.83 \times 10^{15} \times 0.6978 \times 8.6074 \times 1.0087 \times 10^{-702}\ \text{cm}^{-3} \\ &= 4.83 \times 10^{15} \times 6.0585 \times 10^{-702}\ \text{cm}^{-3} = 2.926 \times 10^{-686}\ \text{cm}^{-3}. \end{aligned} \tag{3.15}$$

Similar calculations are performed for $T = 77.2$ K and $T = 300$ K. Above 300 K, bandgaps are calculated by Varshini's formula, e.g. at $T = 600$ K, the computational procedure is given below:

$$\begin{aligned} m_n^* &= \{-1.084 \times 10^{-9}(600)^3 + 7.580 \times 10^{-7}(600)^2 + 2.862 \times 10^{-4}(600) + 1.057\}m_0 \\ &= \{-0.234144 + 0.27288 + 0.17172 + 1.057\}m_0 = 1.267456m_0 \end{aligned} \tag{3.16}$$

$$\begin{aligned} m_p^* &= \{1.872 \times 10^{-11}(600)^4 - 1.969 \times 10^{-8}(600)^3 + 5.857 \times 10^{-6}(600)^2 \\ &\quad + 2.712 \times 10^{-4}(600) + 0.584\}m_0 \\ &= (2.426112 - 4.25304 + 2.10852 + 0.16272 + 0.584)m_0 = 1.028312m_0 \end{aligned} \tag{3.17}$$

Table 3.5. Intrinsic carrier concentrations in silicon at different temperatures. m_0 = free electron rest mass = 9.11×10^{-31} kg.

Temperature (K) /Parameter	4.2	77.2	300	600	800 K	900 K	1000 K
m_n^*/m_0	1.058	1.0831	1.1818	1.2675	1.216	1.1383	1.017
m_p^*/m_0	0.585	0.6315	0.81249	1.0283	2.1359	3.5	5.74
E_g (eV)	1.17	1.167	1.124	1.0317	0.9587	0.92	0.88
n_i (cm^{-3})	2.93×10^{-686}	1.954×10^{-20}	8.81×10^9	4.02×10^{15}	2.135×10^{17}	9.76×10^{17}	3.47×10^{18}

$$E_g = E_g(0) - \frac{\alpha T^2}{T + \beta} = 1.1695 - \frac{4.73 \times 10^{-4}(600)^2}{600 + 636} \tag{3.18}$$

$$= 1.1695 - 0.13776699 = 1.03173 \text{eV}$$

$$\therefore n_i = 4.8281787 \times 10^{15} \times \left(\frac{1.267456 m_0 \times 1.028312 m_0}{m_0^2}\right)^{0.75} (600)^{1.5}$$

$$\exp\left(-\frac{1.03173}{2 \times 8.617 \times 10^{-5} \times 600}\right) \text{cm}^{-3} \tag{3.19}$$

$$= 4.83 \times 10^{15} \times 1.2198 \times 14696.93846 \times 4.64256 \times 10^{-5} \text{cm}^{-3}$$

$$= 4.0199 \times 10^{15} \text{cm}^{-3}.$$

This calculation procedure is followed up to 1000 K. Table 3.5 lists the computed values of n_i.

Figure 3.2 presents the calculated change in the intrinsic carrier concentration of silicon with an increase in temperature.

3.4 Carrier saturation velocity

The proportionality relationship between the average velocity of carriers and the applied electric field is violated at high electric fields. At these values of electric field, the carrier velocity attains a maximum value for both electrons and holes.

Quay *et al* put forward a simple and precise temperature-dependent model for the saturation velocity as a function of temperature in semiconductors (Quay *et al* 2000)

$$v_{sat}(T_L) = \frac{v_{sat}(300)}{(1 - A) + A(T_L/300)}. \tag{3.20}$$

This model is a two-parameter model. The first parameter $v_{sat}(300)$ is the saturation velocity at the lattice temperature T_L = 300 K. The second parameter A is the TC portraying the temperature dependence of the different materials complying with the model. In case of silicon, $v_{sat}(300) = 1.02 \times 10^7$ cm s^{-1}, $A = 0.74$ for electrons; and $v_{sat}(300) = 0.72 \times 10^7$ cm s^{-1}, $A = 0.37$ for holes. The model was demonstrated for

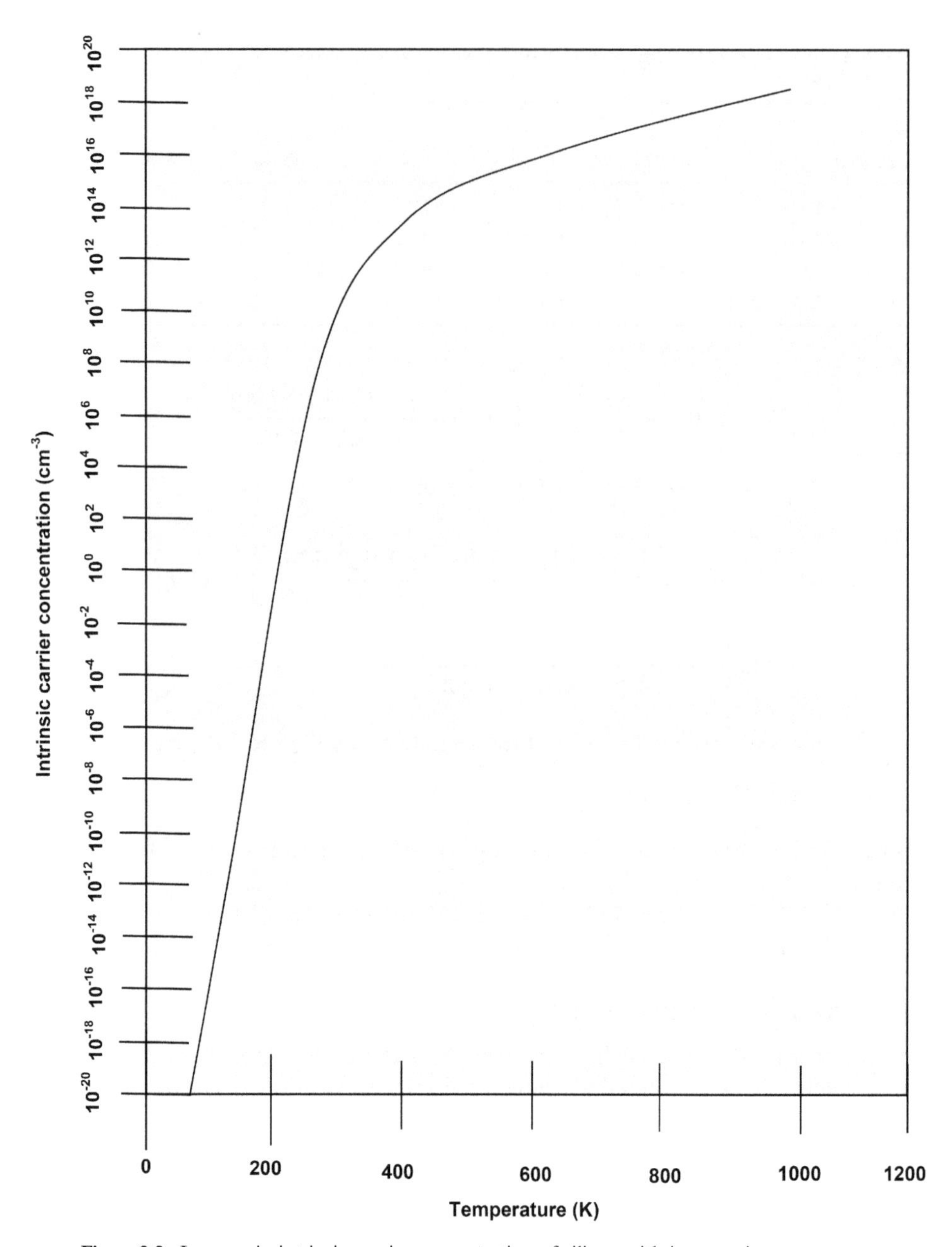

Figure 3.2. Increase in intrinsic carrier concentration of silicon with increase in temperature.

the v_{sat} data of electrons in silicon in the temperature range 0 K to 500 K by Jacoboni *et al* (1977). However, the model applies in general to a large number of technologically relevant semiconductor materials from 200 K to 500 K. Table 3.6 presents the calculated values from the model of Quay *et al* (2000) for silicon.

Table 3.6. Electron and hole saturation velocities in silicon from the model of Quay *et al* (2000).

Temperature (K)	4.2	77.2	300	500
$v_{sat}(T_L)$, cm s^{-1} for electrons	3.77×10^7	2.265×10^7	1.02×10^7	6.83×10^6
$v_{sat}(T_L)$, cm s^{-1} for holes	1.13×10^7	9.928×10^6	7.2×10^6	5.775×10^6

Table 3.7. Electron and hole saturation velocities in silicon from the Ali-Omar and Reggiani model (1987).

Temperature (K)	4.2	77.2	300	500
$v_{sat}(T_L)$, cm s^{-1} for electrons	1.45×10^7	1.424×10^7	9.995×10^6	7.948×10^6
$v_{sat}(T_L)$, cm s^{-1} for holes	9.05×10^6	9.047×10^6	7.982×10^6	6.735×10^6

After Ali-Omar and Reggiani (1987),

$$v_n^{sat} = 1.45 \times 10^7 \sqrt{\tanh\left(\frac{155\text{K}}{T}\right)} \text{ cm s}^{-1} \tag{3.21}$$

$$v_p^{sat} = 9.05 \times 10^6 \sqrt{\tanh\left(\frac{312\text{K}}{T}\right)} \text{ cm s}^{-1}. \tag{3.22}$$

The values obtained from this model are listed in table 3.7.

3.5 Electrical conductivity of semiconductors

The conductivity of a semiconductor is obtained by adding the contributions from the electron and hole populations as

$$\sigma = q(\mu_n n + \mu_p p), \tag{3.23}$$

where q is the charge of the electron, n and p stand for the densities of electrons and holes, and μ_n and μ_p refer to the mobilities of the electrons and holes, respectively. In a doped semiconductor under equilibrium conditions, the number of majority carriers greatly exceeds that of minority carriers. Then the above equation reduces to a single term involving the majority carrier.

The main point to be noted is that the conductivity of a semiconductor is determined by two factors, namely, the concentration of charge carriers moving freely and at disposal to transmit current, and the mobility or freedom of movement of these carriers. Mobility determines the extent to which the free carriers are affected by electric fields. It is defined as the average drift velocity acquired by the carriers in unit electric field strength. In a semiconductor, both the carrier concentration and mobility are temperature-dependent. Thus, it is important to view the conductivity as a function of temperature. This assertion is expressed as

$$\sigma = q\left\{\mu_n(T)n(T) + \mu_p(T)p(T)\right\}. \tag{3.24}$$

To understand how the conductivity of a semiconductor changes with temperature and what its value will be at low temperatures relative to room temperature, it is necessary to know the influence of temperature on carrier concentrations and mobilities.

3.6 Free carrier concentration in semiconductors

Discrete power semiconductor devices as well as ICs are designed and fabricated to operate between designated temperature limits. These temperature limits are specified by manufacturers. The standard practice is that the device/circuit designer selects the doping level or levels. It is generally assumed that the dopants are approximately 100% ionized. It is taken for granted that the dopant atoms are completely exhausted through release of free carriers from them. Also, it is postulated that the operating temperature is neither too high nor too low relative to room temperature. If this assumption regarding temperature ceases to hold, a profound impact on the values of a plethora of device parameters may be noticed. The depletion width or threshold voltage of an FET may drastically change.

Exponential temperature dependence dominates the temperature variation of intrinsic carrier concentration $n_i(T)$. In order to determine the total carrier concentration, space-charge neutrality must be taken into account. Hence,

$$n(T) = N_D^+(T) - N_A^-(T) + \frac{n_i^2(T)}{n(T)} \tag{3.25}$$

$$p(T) = N_A^-(T) - N_D^+(T) + \frac{n_i^2(T)}{p(T)}, \tag{3.26}$$

where $N_D^+(T)$ is the ionized donor concentration and $N_A^-(T)$ is the ionized acceptor concentration, which are clearly different from the number of neutral donor and acceptor atoms, N_D, N_A. Both $N_D^+(T)$ and $N_A^-(T)$ are temperature-dependent.

With temperature falling to very low values in the vicinity of 0 K (large $1/T$), n_i becomes infinitesimally small. Then, in an intrinsic material, the number of EHPs plummets to negligible proportions. The donor electrons are bound to the corresponding electron donating atoms. The holes are also fastened to the acceptor atoms concerned. When the temperature is too low, the percentage ionization of the dopant or dopants is appreciably less than 100%. This turnaround of the percentage of ionization of the dopants as opposed to their near 100% room-temperature values is called freeze-out.

3.7 Incomplete ionization and carrier freeze-out

Impurity freeze-out is modeled using Fermi–Dirac statistics and the degeneracy factors associated with the conduction and valence energy bands. The ionized donor and acceptor concentrations $N_D{}^+$, $N_A{}^-$ are expressed in terms of the total donor and acceptor concentrations N_D, N_A as (Cole and Johnson 1989, Silvaco 2000)

$$\frac{N_D^+}{N_D} = \left[1 + g_{CD} \exp\left\{\left(\frac{\Delta E_D}{k_B T}\right) + \left(\frac{E_{fn} - E_C}{k_B T}\right)\right\}\right]^{-1} \tag{3.27}$$

$$\frac{N_A^-}{N_A} = \left[1 + g_{VD} \exp\left\{\left(\frac{\Delta E_A}{k_B T}\right) + \left(-\frac{E_{fp} - E_V}{k_B T}\right)\right\}\right]^{-1}, \tag{3.28}$$

where g_{CD}, g_{VD} are the degeneracy factors for conduction and valence bands, respectively, with the usually assumed values $g_{CD} = 2$ and $g_{VD} = 4$; E_{fn}, E_{fp} are the electron and hole quasi-Fermi energy levels; $(E_{fn} - E_C)$ is the position of the electron quasi-Fermi level relative to the conduction band edge for phosphorous-doped silicon; $(E_{fp} - E_V)$ is the position of the hole quasi-Fermi level relative to the valence band edge for boron-doped silicon; E_C, E_V are the energies of the conduction and valence band edges; $\Delta E_D = E_C - E_D$ is the activation energy for the phosphorous impurity in silicon (= 0.045 eV); and $\Delta E_A = E_A - E_V$ is the activation energy for the boron impurity in silicon (= 0.045 eV).

The temperature dependence of the activation energies of shallow donor and acceptor levels is not considered significant (Jonscher 1964).

At 4.2 K,

$$\frac{\Delta E_D}{k_B T} = \frac{0.045}{8.617 \times 10^{-5} \times 4.2} = 124.3389. \tag{3.29}$$

For $n = 1 \times 10^{15}$ cm^{-3}, the electron Fermi level E_{fn} is related to the electron concentration n as

$$\left(\frac{E_{fn} - E_C}{k_B T}\right)_{4.2\,K} = \left\{\ln\left(\frac{n}{N_C}\right)\right\}_{4.2\,K} = \ln\left\{\frac{1 \times 10^{15}}{N_C}\right\}, \tag{3.30}$$

where N_C is the effective density of states in the conduction band at 4.2 K:

$$\begin{aligned}
(N_C)_{4.2\,K} &= 2\left(\frac{2\pi k_B T}{h^2}\right)^{1.5} \left\{\left(m_n^*\right)_{4.2\,K}\right\}^{1.5} \\
&= 2\left\{\frac{2 \times 3.14 \times 1.381 \times 10^{-23} \times 4.2}{(6.626 \times 10^{-34})^2}\right\}^{1.5} \times (1.058 \times 9.11 \times 10^{-31})^{1.5} \\
&= 2\left(\frac{2 \times 3.14 \times 1.381 \times 10^{-23} \times 4.2}{4.3904 \times 10^{-67}}\right)^{1.5} \times (1.058 \times 9.11 \times 10^{-31})^{1.5} \\
&= 2\left(\frac{3.6425 \times 10^{-22}}{4.3904 \times 10^{-67}}\right)^{1.5} \times (1.058 \times 9.11 \times 10^{-31})^{1.5} \\
&= 2\{8.2965 \times 10^{44}\}^{1.5} \times (9.638 \times 10^{-31})^{1.5} \\
&= 2 \times 2.3897 \times 10^{67} \times 9.46 \times 10^{-46} \\
&= 4.521 \times 10^{22}\ \text{m}^{-3} = 4.521 \times 10^{16}\ \text{cm}^{-3}
\end{aligned} \tag{3.31}$$

$$\therefore \left(\frac{E_{fn} - E_C}{k_B T}\right)_{4.2\,K} = \ln\left(\frac{1 \times 10^{15}}{4.521 \times 10^{16}}\right) = \ln(2.2119 \times 10^{-2}) = -3.8113. \quad (3.32)$$

Therefore,

$$\frac{N_D^+}{N_D} = [1 + 2\exp\{(124.3389) + (-3.8113)\}]^{-1} = [1 + 2\exp(120.5276)]^{-1}$$
$$= [1 + 2 \times 2.21 \times 10^{52}]^{-1} = (4.42 \times 10^{52})^{-1} = 2.262 \times 10^{-53}. \quad (3.33)$$

At $T = 40$ K

$$m_n^* = \{-1.084 \times 10^{-9}(40)^3 + 7.580 \times 10^{-7}(40)^2 + 2.862 \times 10^{-4}(40) + 1.057\}m_0$$
$$= \{-0.000069376 + 0.0012128 + 0.011448 + 1.057\}m_0 = 1.06959m_0 \quad (3.34)$$

$$\frac{\Delta E_D}{k_B T} = \frac{0.045}{8.617 \times 10^{-5} \times 40} = 13.0556. \quad (3.35)$$

For $n = 1 \times 10^{15}$ cm^{-3}, the quasi-Fermi level E_{fn} is related to the electron concentration n as

$$\left(\frac{E_{fn} - E_C}{k_B T}\right)_{40\,K} = \left\{\ln\left(\frac{n}{N_C}\right)\right\}_{40\,K} = \ln\left\{\frac{1 \times 10^{15}}{N_C}\right\}, \quad (3.36)$$

where N_C is the effective density of states in the conduction band at 40 K:

$$(N_C)_{40\,K} = 2\left(\frac{2\pi k_B T}{h^2}\right)^{1.5}\left\{(m_n^*)_{40\,K}\right\}^{1.5}$$
$$= 2\left\{\frac{2 \times 3.14 \times 1.381 \times 10^{-23} \times 40}{(6.626 \times 10^{-34})^2}\right\}^{1.5} \times (1.06959 \times 9.11 \times 10^{-31})^{1.5}$$
$$= 2\left(\frac{2 \times 3.14 \times 1.381 \times 10^{-23} \times 40}{4.390388 \times 10^{-67}}\right)^{1.5} \times (1.06959 \times 9.11 \times 10^{-31})^{1.5} \quad (3.37)$$
$$= 2\left(\frac{3.4691 \times 10^{-21}}{4.390388 \times 10^{-67}}\right)^{1.5} \times (1.06959 \times 9.11 \times 10^{-31})^{1.5}$$
$$= 2\{7.9016 \times 10^{45}\}^{1.5} \times (9.743965 \times 10^{-31})^{1.5}$$
$$= 2 \times 7.0238 \times 10^{68} \times 9.61842 \times 10^{-46}$$
$$= 1.35 \times 10^{24}\ \text{m}^{-3} = 1.35 \times 10^{18}\ \text{cm}^{-3}$$

$$\therefore \left(\frac{E_{fn} - E_C}{k_B T}\right)_{4.2\,K} = \ln\left(\frac{1 \times 10^{15}}{1.35 \times 10^{18}}\right) = \ln(7.4 \times 10^{-4}) = -7.20886. \quad (3.38)$$

Table 3.8. Fractional ionization at different temperatures.

Temperature (K)	4.2	40	77.2	300	600
N_D^+/N_D	2.262×10^{-53}	1.443×10^{-3}	0.58	0.99964	0.999978
N_D^+ (cm^{-3})	2.262×10^{-38}	1.443×10^{-12}	5.8×10^{14}	9.9964×10^{14}	9.99978×10^{14}

Therefore,

$$\frac{N_D^+}{N_D} = [1 + 2\exp\{(13.0556) + (-7.20886)\}]^{-1} = [1 + 2\exp(5.84674)]^{-1}$$
$$= [1 + 2 \times 346.1]^{-1} = (693.2)^{-1} = 1.443 \times 10^{-3}. \quad (3.39)$$

Similarly, N_D^+/N_D calculations are performed at 77.2 K, 300 K and 600 K. These values are given in table 3.8.

3.8 Different ionization regimes

For a semiconductor, three definite regimes of operation are observed in relation to temperature, as shown in figure 3.2. These regimes are distinctly discernible in figure 3.3. The following sections will bring out the salient features of these regimes.

3.8.1 At temperatures $T < 100$ K: carrier freeze-out or incomplete ionization regime

At temperatures which are not too low but still low enough, e.g. <100 K (= −173 °C), the thermal energy within a semiconductor is lacking in intensity to cause activation of the full donor and acceptor impurity content. As a result, the carrier concentration is not equal to the concentration of dopant atoms. This region of operation below 100 K, in which the thermal energy within the silicon is inadequate to fully ionize the impurity atoms, is known as the freeze-out regime. The statement about the availability of insufficient energy to ionize all impurity atoms is only partially true. In the quest for the truth, one must delve deeper into the actual situation. Two cases must be clearly distinguished.

Case I: intrinsic and low-doped or non-degenerate semiconductors. For a semiconductor in thermal equilibrium, the occupation of all electronic levels, both free and localized, is determined by the Fermi–Dirac distribution function $f(E)$ symbolizing the probability that an electronic state at energy E is occupied by an electron

$$f(E) = \left\{\exp\left(\frac{E - E_F}{k_B T}\right) + 1\right\}^{-1}, \quad (3.40)$$

where k_B is the Boltzmann constant (= 8.617×10^{-5} eV K^{-1}) and E_F is a parameter known as the Fermi energy, which is the energy at which the probability of occupation by an electron is exactly one-half. The function $\{1 - f(E)\}$ is the

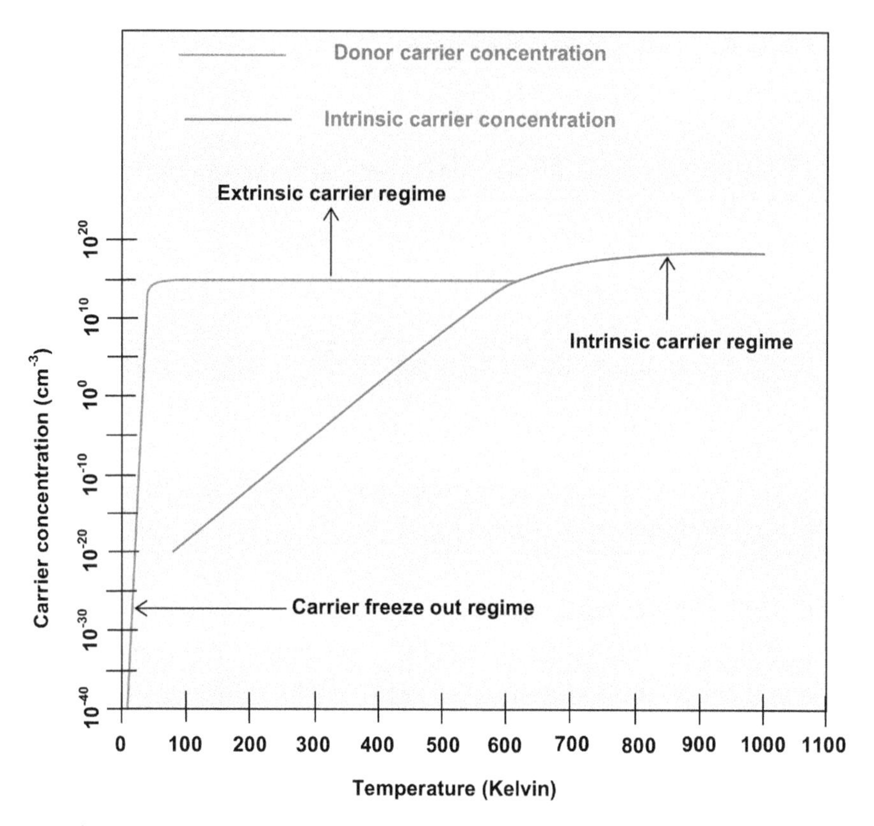

Figure 3.3. Diagram showing the appearance of the three operational regimes in an n-type semiconductor with $N_D = 1 \times 10^{15}$ cm^{-3} with an increase of temperature: carrier freeze-out, extrinsic and intrinsic carrier concentration regimes.

probability that an electron is not found at energy E, i.e. the probability of finding a hole there.

For silicon, the range of doping considered here extends from 1.45×10^{10} cm^{-3} to 1×10^{18} cm^{-3}. At $T = 0$ K, when $E < E_F$, i.e. for energy levels lying below E_F, $(E - E_F)$ is a negative quantity and hence,

$$f(E) = \{\exp(-\infty) + 1\}^{-1} = \{0 + 1\}^{-1} = 1 \tag{3.41}$$

meaning that all the states below the Fermi level are filled; hence all the electrons are present in the valence band. In other words, there are no holes available for conduction.

At $T = 0$ K, when $E > E_F$, i.e. for energy levels lying above E_F, $(E - E_F)$ is a positive quantity, so that

$$f(E) = \{\exp(+\infty) + 1\}^{-1} = \{\infty + 1\}^{-1} = 0, \tag{3.42}$$

i.e. all the states above the Fermi level are empty so that there are no free electrons in the conduction band.

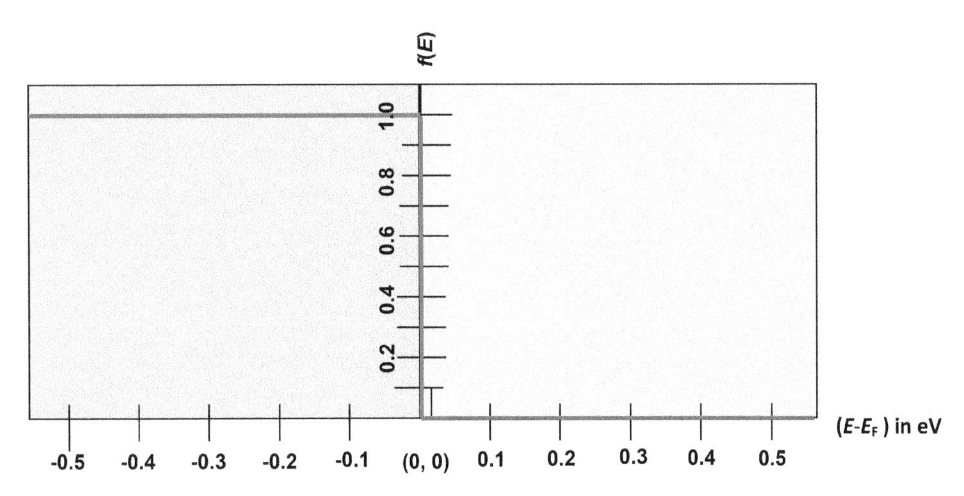

Figure 3.4. Plot of Fermi–Dirac distribution function $f(E)$ with respect to $(E - E_F)$ at $T = 0$ K.

At $T = 0$ and at all temperatures, when $E = E_F$, i.e.

$$f(E) = \{\exp(0) + 1\}^{-1} = \{1 + 1\}^{-1} = 0.5. \tag{3.43}$$

Thus the graph of $f(E)$ against $(E - E_F)$ is a step function in which, for $(E - E_F) < 0$, $f(E) = 1$, but when $(E - E_F) > 0$, $f(E) = 0$. The probability of finding an electron with energy equal to the Fermi energy in a semiconductor is ½ at any temperature including absolute zero. The graph of $f(E)$ versus $(E - E_F)$ is symmetrical around the Fermi level E_F. This graph is sketched in figure 3.4.

Let us now plot a graph between $f(E)$ and $(E - E_F)$ at the temperature $T = 10$ K (figure 3.5). We note that for energy $E - E_F = -3k_BT$, the exponential term in the equation for function $f(E)$ becomes <0.05. Hence the Fermi–Dirac distribution increases to ~1. Also, for energy $E - E_F = 3k_BT$, the exponential term in the equation for function $f(E)$ becomes >20. Hence the Fermi–Dirac distribution decays to ~0. Therefore, the $f(E)$ values will be determined for three negative values $E - E_F = -k_BT, -2k_BT, -3k_BT$ and three positive values of $E - E_F = k_BT, 2k_BT, 3k_BT$.

For negative values of $E - E_F$, it is found that when $E - E_F = -k_BT, -2k_BT, -3k_BT = -8.617 \times 10^{-4}, -1.723 \times 10^{-3}, -2.585 \times 10^{-3}$ eV, $f(E) = 0.731, 0.881, 0.9526$. Thus as $E - E_F$ decreases from $-k_BT$ to $-3k_BT$, $f(E)$ increases from 0.731 to 0.9526. Similarly for positive values of $E - E_F = k_BT, 2k_BT, 3k_BT = 8.617 \times 10^{-4}$, 1.723×10^{-3} eV, 2.585×10^{-3} eV, $f(E) = 0.269, 0.119, 0.0474$, respectively. Thus as $E - E_F$ increases from k_BT to $3k_BT$, $f(E)$ decreases from 0.269 to 0.0474. Hence, the transition from $f(E) = 1$, corresponding to nearly full occupation of levels by electrons, to $f(E) = 0$, corresponding to almost empty levels, occurs in the range of energies $\pm 3k_BT = \pm 0.0026$ eV. For the very narrow range of energy levels enclosed within ± 0.0026 eV on either side of E_F, there is a very small, although finite, probability that some states are vacant on the valence band side and some states are filled on the conduction band side.

The gist of the discussion is that at temperatures near absolute zero, the transition from $f(E) = 1$ to $f(E) = 0$ being very rapid, the range of energies in which the

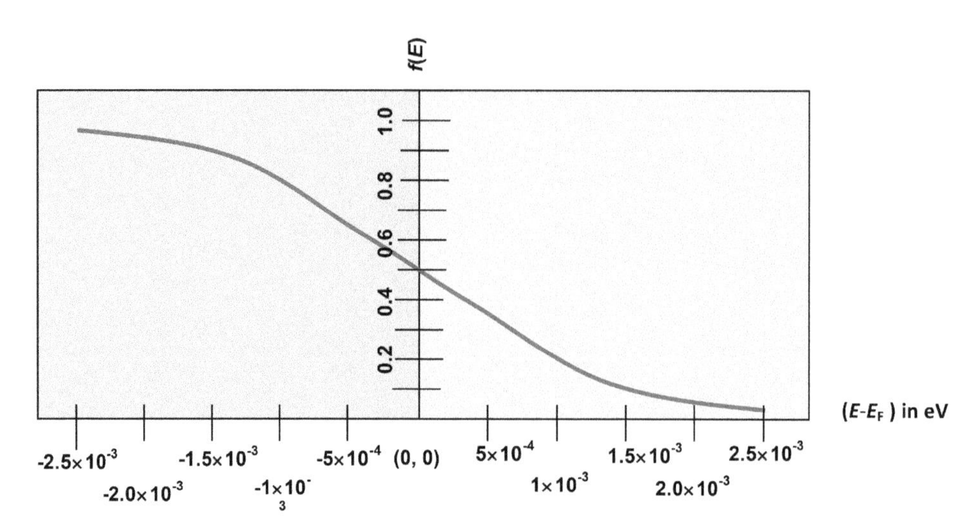

Figure 3.5. Plot of Fermi–Dirac distribution function $f(E)$ with respect to $(E - E_F)$ at $T = 10$ K.

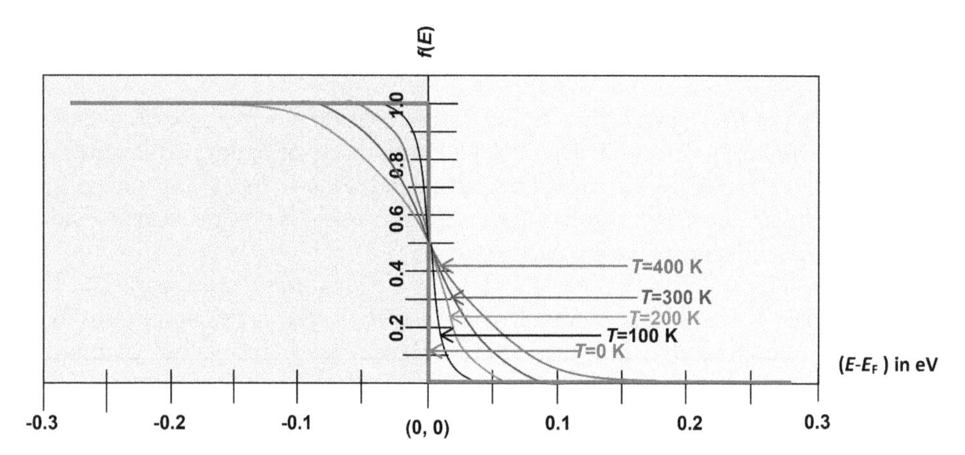

Figure 3.6. Graphs of the Fermi–Dirac distribution function with respect to energy difference $(E - E_F)$ from Fermi level E_F at different temperatures.

probability of free carriers is non-zero is very small, so the number of free carriers in the semiconductor is insignificant. Inasmuch as the number of carriers is negligibly small due to the cooling effect, it appears as if the carriers have been completely frozen out. The freeze-out effect is a serious problem hampering the operation of semiconductor devices at cryogenic temperatures. Figure 3.6 showing the $f(E)$ versus $(E - E_F)$ plots in the temperature range 0 K to 400 K corroborates the above assertion.

Case II: degenerate semiconductors. In a low-doped semiconductor, the Fermi level is inside the bandgap. For an n-type semiconductor, it is above the intrinsic Fermi level E_i and closer to the edge of the conduction band. For a p-type semiconductor it is below the intrinsic Fermi level E_i and closer to the valence band edge. Depending on whether the semiconductor is doped with donor or acceptor impurities, the higher the doping concentration is, the closer the Fermi level is to the edge of the conduction band or valence band. A non-degenerate semiconductor is defined as a semiconductor for which the Fermi energy is at least $3k_BT$ energy units away from either band edge.

As the doping level is continuously increased, at a particular stage, the Fermi level moves within an allowed band. It moves either inside the conduction or valence band. The semiconductor is then said to be degenerate. In this condition, the carrier densities no longer obey classical statistics. Thus a degenerate semiconductor is defined as a heavily doped semiconductor. In such a semiconductor, the Fermi level lies either in the conduction band or in the valence band, causing the material to behave as a metal. However, a degenerate semiconductor still has far fewer charge carriers than a true metal. So, its behavior is in many ways intermediary between a semiconductor and a metal.

At high impurity concentrations, the individual impurity atoms become such close neighbors that their doping levels merge into an impurity band. In other words, the onset of impurity interaction causes the broadening of the discrete energy levels of isolated impurities into an impurity band, with the resulting reduction of the effective activation energy below the value appropriate to high-purity material. This reduces somewhat the tendency to freeze out carriers at any given temperature, and also produces some conduction in the broadened impurity band.

At a sufficiently high density of impurities, to the increasing broadening of the impurity band is added a shift downwards of the conduction band or upwards of the valence band, which is due to carrier–carrier interaction at high carrier densities. At this stage there is no longer any clear-cut distinction between the localized levels and the free bands; the activation energy becomes zero and the carrier density ceases to be a function of temperature. The important inference and verdict from the point of view of device operation is that there is no freeze-out of carriers even at the lowest temperatures; therefore the conductivity remains high. So, the statement about carrier freeze-out is true for a non-degenerate semiconductor. It is not true for a degenerate semiconductor.

3.8.2 At temperatures $T \sim 100$ K, and within $100\text{ K} < T < 500$ K: extrinsic or saturation regime

A progressive increase in ionization takes place as the temperature is raised. At a temperature around 100 K, a large fraction of the donor atoms has undergone ionization. At this point, the carrier concentration becomes a function of doping. It can be stated that at temperatures between 100 K and 500 K, i.e. from −173 °C to 227 °C, plentiful thermal energy resides within the silicon crystal. This thermal energy can ionize the impurity atoms. The region of operation where available dopants have been ionized and free carriers are liberated is known as the extrinsic regime or saturation region. It is the region where

$$N_D^+(T) = N_D \tag{3.44}$$

$$N_A^-(T) = N_A \tag{3.45}$$

$$n_i(T) \ll |N_D - N_A|. \tag{3.46}$$

Recalling the discussion regarding the $f(E)$ with $E - E_F$ plot at 0 K and 10 K, now increasing the temperature to 100 K, the range of energies ($E - E_F$) for transition from $f(E) = 1$ to $f(E) = 0$ broadens to $\pm 10 \times 0.0026$ eV = ± 0.026 eV and more carriers become available. With a further increase in temperature to 300 K, the range spreads more. At 300 K, its width is $\pm 30 \times 0.0026 = \pm 0.078$ eV. Since the acceptor level energy $E_a = \Delta E_A$ for boron in silicon is 0.045 eV while donor level energy $E_d = \Delta E_D$ for phosphorous in silicon is also 0.045 eV, almost all the impurity atoms are ionized, contributing carriers from the donor/acceptor states. The effect of incomplete ionization of dopants is often neglected in simulations of silicon devices, as it is considered to be non-meaningful at room temperature.

3.8.3 At temperatures $T > 500$ K: intrinsic regime

If the temperature is too high, the thermal generation effect causes the majority carrier concentration to become excessively higher than the dopant concentration in what is called the intrinsic temperature regime. As the temperature increases beyond 550 K (= 277 °C), the intrinsic carrier concentration approaches and then exceeds the impurity concentration. At these high temperatures, the thermally generated intrinsic carriers outnumber the dopant-produced carriers, and the silicon returns to intrinsic-type behavior. In the intrinsic regime, the majority carrier concentration is nearly equal to the intrinsic concentration, n_i.

$$n_i(T) > |N_D - N_A|. \tag{3.47}$$

In this intrinsic region, the carrier concentration increases with temperature. The exhaustion regime lies between these two extremes, intrinsic and freeze-out (Pieper and Michael 2005).

At 600 K, the range of energies $E - E_F$ for $f(E) = 1$ to $f(E) = 0$ increases to $\pm 60 \times 0.0026 = \pm 0.156$ eV. As the temperature increases towards 600 K, electrons are dislodged from silicon atoms by thermal excitation, creating EHPs, and a large concentration of free carriers is built up, much greater than that due to impurity doping.

Figure 3.7 provides an explanation of the carrier concentration changes in a semiconductor from 0 K to 600 K in terms of the thermally induced EHP generation from the perspective of the energy band model.

3.8.4 Proportionality to bandgap at $T \geqslant 400$ K

One particularly interesting case occurs at high temperatures (above 400 K = 127 °C or higher) when mobility is dominated by lattice scattering ($\mu \propto T^{-3/2}$). In such cases, the conductivity can easily be shown to vary with temperature as

$$\sigma \propto \exp\left(-\frac{E_g}{2k_B T}\right). \tag{3.48}$$

In this case, conductivity depends only on the semiconductor bandgap and the temperature, as in an intrinsic semiconductor.

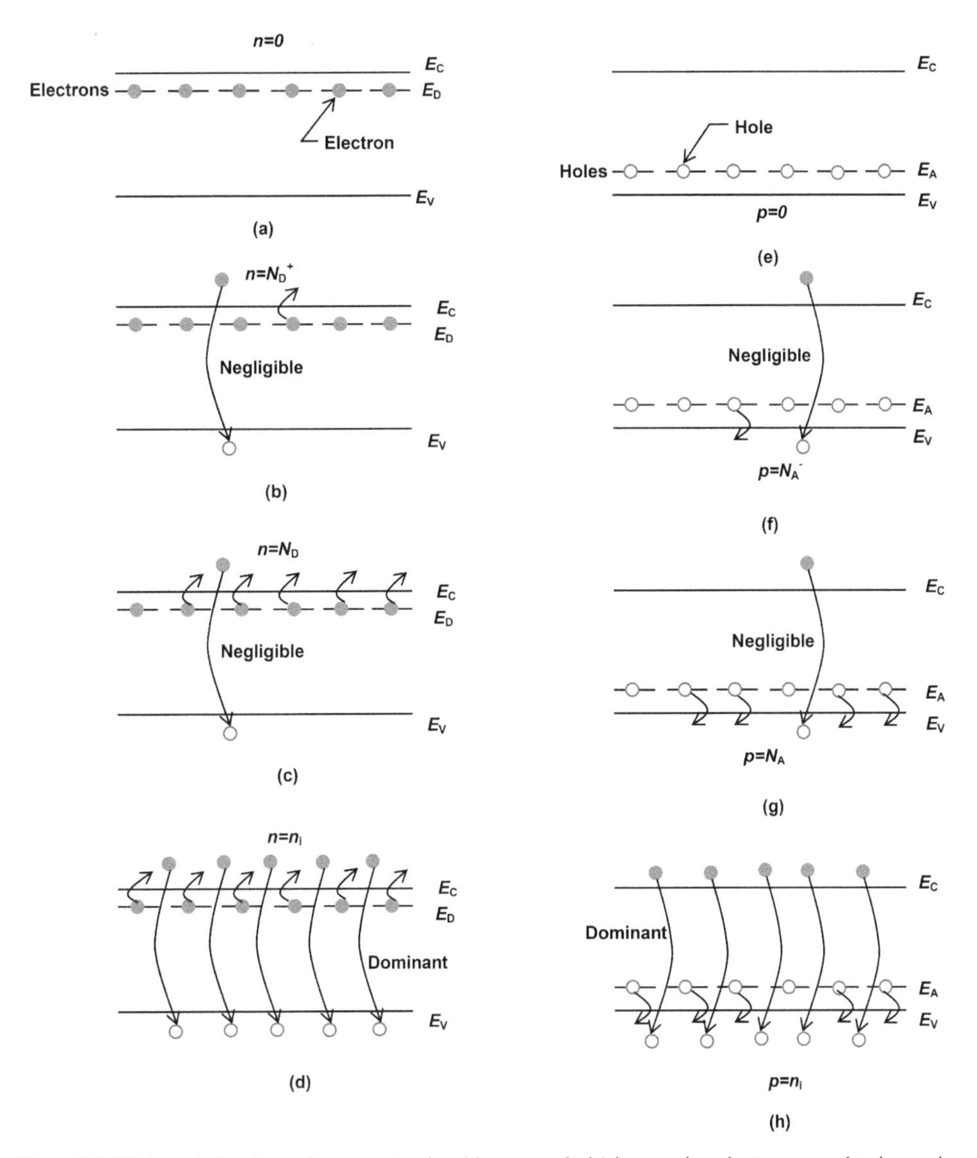

Figure 3.7. Wide variation in carrier concentration (electron or hole) in a semiconductor exposed to increasing temperature with reference to (a), (b), (c), (d) energy band diagrams for an n-type semiconductor at 0 K, low temperature <100 K, moderate temperature ~300 K and high temperature ~600 K, respectively; (e), (f), (g), (h) corresponding band diagrams for a p-type semiconductor at above temperatures. n, p = free electron and hole concentrations; N_D^+, N_A^- = ionized donor and acceptor concentrations; N_D, N_A = total donor and acceptor concentrations; n_i = intrinsic carrier concentration.

3.9 Mobilities of charge carriers in semiconductors

Carrier mobility is a phenomenological parameter whose value critically controls the performance characteristics of numerous semiconductor devices. As examples of these devices, mention may be made of diodes, and bipolar and FETs. Principally, the mobility of electrons and holes in a semiconductor is decided by two disparate types of scattering mechanisms. These mechanisms are: lattice scattering and impurity scattering (Van Zeghbroeck 2011).

3.9.1 Scattering by lattice waves

In scattering by lattice waves, the absorption or emission of either acoustical or optical phonons occurs. The density of phonons in a solid is known to increase with temperature. As a consequence, the scattering time due to this mechanism will decrease with temperature, as will the mobility. Theoretical calculations indicate that *acoustic* phonon interaction dominates the mobility in non-polar semiconductors, e.g. silicon and germanium. The natural expectation is that it is proportional to $T^{-3/2}$. On the other hand, the mobility due to *optical* phonon scattering only is expected to be proportional to T^{-2}. The effect of temperature on lattice mobility μ_L is conveyed by the equation

$$\mu_L \propto T^{-\alpha}, \tag{3.49}$$

where $1.5 < \alpha < 2.5$. The lower limit takes account of acoustic phonon scattering whereas the upper limit takes account of optical phonon scattering.

3.9.2 Scattering by ionized impurities

The extent of scattering due to electrostatic forces between the carrier and the ionized impurity depends on two factors: (i) the number of impurity ions and (ii) the interaction time between the carriers and the impurity ions. Larger impurity concentrations cause more scattering of carriers and thereby lead to a lower mobility. The interaction time is directly linked to the relative velocity of the carrier and the impurity. So, it is related to the thermal velocity of the carriers. As this thermal velocity increases with the ambient temperature, the interaction time decreases with an increase in temperature. Therefore, the amount of scattering decreases. The result is a mobility increase with temperature. To first order, the mobility due to impurity scattering varies directly as $T^{3/2}/N_I$. The denominator N_I represents the density of charged impurities, regardless of sign. Hence, we can write

$$\mu_I \propto T^{3/2}/N_I. \tag{3.50}$$

The temperature dependences of mobility at low temperatures due to ionized impurity scattering and that at high temperatures caused by lattice scattering are shown in figure 3.8.

3.9.3 Mobility in uncompensated and compensated semiconductors

There are two cases of interest: uncompensated and compensated semiconductors. In uncompensated material, the density of charged impurities is equal to the density of ionized shallow donors or acceptors, which decreases as the carrier density goes down with temperature. Shallow impurities are impurities which require little energy, typically around the thermal energy at room temperature or less, to ionize.

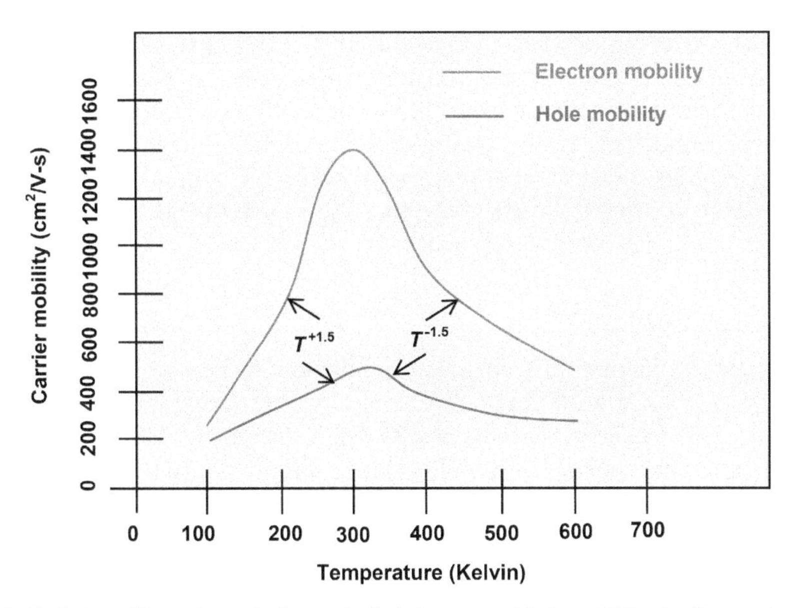

Figure 3.8. Variation of impurity scattering-controlled electron and hole mobilities in silicon at low temperatures and lattice-scattering determined electron and hole mobilities at high temperatures in accordance with the $T^{1.5}$ and $T^{-1.5}$ laws, respectively.

Because of the decrease in density of charge centers with falling temperature, the effect on mobility due to impurity scattering becomes temperature-dependent in an uncompensated semiconductor.

Suppose a semiconductor contains both types of impurities: shallow donors and shallow acceptors. Also, let the concentrations of donors and acceptors be equal. Such a semiconductor is called a compensated semiconductor. It is so-named because in this semiconductor, equal amounts of donors and acceptors compensate each other, yielding no free carriers. The simultaneous presence of shallow donors and shallow acceptors in a semiconductor causes the electrons ejected from the donor to be captured by the acceptor state. This ionizes the acceptor without producing a free electron or hole. The resulting density of charged centers in compensated material, which contains equal quantities of both shallow donors and shallow acceptors, is equal to double the minority compensating density ($2N_A$ in an n-type material). Due to the large number of charged centers present, the scattering increases and the mobility in a compensated material is less than in an uncompensated material.

However, in this type of compensated material, the density of charged centers is virtually independent of temperature because approximately equal numbers of donor and acceptor atoms are oppositely ionized at a given temperature, cancelling the effects of each other. The natural outcome is that the effect of impurity scattering on mobility does not depend on temperature. It is virtually temperature-independent.

Thus it is evident that the temperature variation of mobility due to impurity scattering depends on whether it is compensated or uncompensated. The effect of temperature changes accordingly if a semiconductor belongs to the former or latter category. This effect is very sensitive to the extent of compensation. The degree or extent of compensation determines the material polarity, i.e. whether it is n-type or p-type. The ionized impurity concentration is the difference between the donor and acceptor concentrations. If the donor concentration is larger than the acceptor concentration, the material is n-type with positively ionized impurities. If the acceptor concentration is larger than the donor concentration, the material is p-type with negatively ionized impurities.

3.9.4 Resultant mobility

The observed carrier mobility in a semiconductor is the combined effect of the lattice-scattering limited mobility and the impurity-scattering limited mobility. The temperature dependence of mobility due to lattice scattering is approximately $T^{-\alpha}$. The temperature dependence of mobility due to impurity scattering is $T^{+3/2}$. Hence, carrier mobility is found to follow the trend of increasing with diminishing temperature if lattice scattering plays a dominant role, but decreasing with falling temperature if impurity scattering overrides lattice scattering. At lower temperatures, carrier movements become feebler and slower. Thus carriers have more time to interact with charged impurities. Consequently, as the temperature is lowered, impurity scattering increases. In consequence, the mobility decreases. This is just the reverse of the effect of lattice scattering.

As temperature goes down, the upward trend of mobility due to the ionized impurity scattering mechanism is counterbalanced by the downward trend of mobility due to lattice scattering. The resultant mobility μ is the harmonic mean of the components due to the two mechanisms μ_L, μ_I

$$\mu = \frac{\mu_L \mu_I}{\mu_L + \mu_I}, \tag{3.51}$$

i.e. it is effectively determined by the smaller of the two mobilities.

Case I: pure materials. At cryogenic temperatures in reasonably pure materials, the ionized impurity scattering is normally dominant, which is responsible for the decrease in mobility below 100 K.

Case II: degenerate semiconductors. The situation described above changes considerably in several respects in semiconductor materials containing sufficiently high densities of impurities to give rise to interaction between neighboring impurity centers. As a very rough estimate, the critical impurity density N_{crit} for shallow 'hydrogenic' centers corresponds to a mean inter-impurity distance equal to the radius of the Bohr orbit of the ground state a_1. Thus, $N_{crit} = a_1^{-3}$ depends on the effective mass and the dielectric constant K of the material or, what amounts to the same in this simple model, on the impurity activation energy. As to order of magnitude, the critical density is 10^{18} cm^{-3} in silicon, 10^{16} cm^{-3} in germanium and 10^{13} cm^{-3} in n-type InSb (Jonscher 1964).

The mobility is affected to a large extent by degeneracy. In the first place, the ionized impurity scattering depends on the effective energy of the carriers contributing to conduction and in degenerate systems this is the temperature-independent Fermi energy. Second, if the free carriers are present in reasonably high densities, their own screening action reduces the range of action of ionized centers. As a result, one can observe relatively high mobilities down to very low temperatures, even in highly impure materials.

3.10 Equations for mobility variation with temperature

3.10.1 Arora–Hauser–Roulston equation

Arora *et al* proposed the following empirical equations for electron/hole mobility in silicon up to a dopant concentration N of 10^{20} cm^{-3} for the temperature range $T = 200$ K to 500 K (Arora *et al* 1982):

$$\mu_n(N, T) = 88\left(\frac{T}{300}\right)^{-0.57} + \frac{7.4 \times 10^8 T^{-2.33}}{1 + 0.88\left\{\frac{N}{1.26 \times 10^{17}(T/300)^{2.4}}\right\}(T/300)^{-0.146}} \tag{3.52}$$

$$\mu_p(N, T) = 54.3\left(\frac{T}{300}\right)^{-0.57} + \frac{1.36 \times 10^8 T^{-2.23}}{1 + 0.88\left\{\frac{N}{2.35 \times 10^{17}(T/300)^{2.4}}\right\}(T/300)^{-0.146}}. \tag{3.53}$$

The above equations were derived by them on the basis of experimental data and the amended Brooks–Herring theory of mobility. They found that the electron mobility data in the range 200 K to 500 K fitted satisfactorily with the lattice scattering limited mobility μ_{Ln} formula

$$\mu_{Ln} = 8.56 \times 10^8 T^{-2.33} \tag{3.54}$$

and ionized impurity scattering μ_{In} formula

$$\mu_{In} = (7.3 \times 10^{17} T^{1.5})/[N_I\{\ln(b + 1) - b/(b + 1)\}], \tag{3.55}$$

where N_I is the number of ionized impurity atoms

$$b = \frac{1.52 \times 10^{15} T^2}{n\{2 - (n/N)\}} \tag{3.56}$$

n is the electron concentration per cm^3. The two mobilities μ_{Ln}, μ_{In} were combined together by the mixed scattering formula to arrive at equation (3.52).

Similarly, the hole mobility data for the 150 K to 400 K range could be fitted to

$$\mu_{Lp} = 1.58 \times 10^8 T^{-2.23} \tag{3.57}$$

$$\mu_{\mathrm{Ip}} = (5.6 \times 10^{17} T^{1.5})/[N_{\mathrm{I}}\{\ln(b+1) - b/(b+1)\}], \tag{3.58}$$

$$b = \frac{2.5 \times 10^{15} T^2}{p\{2 - (p/N)\}} \tag{3.59}$$

where p is the hole concentration per cm^3. As before, the two mobilities μ_{Lp}, μ_{Ip} were combined together by the mixed scattering formula to obtain equation (3.53).

3.10.2 Klaassen equations

A physics-based analytical model was presented by Klaassen expressing the temperature-dependent mobility of charge carriers in terms of donor, acceptor, as well as electron and hole concentrations (Klaassen 1992a).

For lattice scattering,

$$\mu_{\mathrm{i,L}} = \mu_{\mathrm{max}} \left(\frac{300}{T} \right)^{\theta_{\mathrm{i}}} \tag{3.60}$$

where $\theta_{\mathrm{n}} = 2.285$ for electrons, $\theta_{\mathrm{p}} = 2.247$ for holes and μ_{max} values are given with impurity scattering for different dopants.

For majority impurity scattering of electrons/holes,

$$\mu_{\mathrm{i,I}}(N_{\mathrm{I}}, c) = \mu_{\mathrm{i,N}} \left(\frac{N_{\mathrm{ref,1}}}{N_{\mathrm{I}}} \right)^{\alpha_1} + \mu_{\mathrm{i,c}} \left(\frac{c}{N_{\mathrm{I}}} \right), \tag{3.61}$$

where (i, I) → (n, D) or (p, A) (n for electron, p for hole, D for donor and A for acceptor), c is the carrier concentration and

$$\mu_{\mathrm{i,N}} = \frac{\mu_{\mathrm{max}}^2}{\mu_{\mathrm{max}} - \mu_{\mathrm{min}}} \left(\frac{T}{300} \right)^{3\alpha_1 - 1.5} \tag{3.62}$$

$$\mu_{\mathrm{i,c}} = \frac{\mu_{\mathrm{max}} \mu_{\mathrm{min}}}{\mu_{\mathrm{max}} - \mu_{\mathrm{min}}} \left(\frac{300}{T} \right)^{0.5}. \tag{3.63}$$

For phosphorous, $\mu_{\mathrm{max}} = 1414.0$ cm^2 V^{-1} s^{-1}, $\mu_{\mathrm{min}} = 68.5$ cm^2 V^{-1} s^{-1}, $N_{\mathrm{ref,1}} = 9.2 \times 10^{16}$ cm^{-3}, $\alpha_1 = 0.711$. For boron, $\mu_{\mathrm{max}} = 470.5$ cm^2 V^{-1} s^{-1}, $\mu_{\mathrm{min}} = 44.9$ cm^2 V^{-1} s^{-1}, $N_{\mathrm{ref,1}} = 2.23 \times 10^{17}$ cm^{-3}, $\alpha_1 = 0.719$.

3.10.3 MINIMOS mobility model

As in other models, the lattice mobility $\mu_{\mathrm{n,p}}^{\mathrm{L}}$ is modeled as a simple power law. In the MINIMOS 4 mobility model, the equations for mobilities of electrons and holes are (Selberherr *et al* 1990):

$$\mu_{\mathrm{n}}^{\mathrm{L}} = 1430(T/300\mathrm{K})^{-2}\ \mathrm{cm^2/V-s} \tag{3.64}$$

$$\mu_{\rm p}^{\rm L} = 460(T/300\ {\rm K})^{-2.18}\ {\rm cm^2/V-s}. \tag{3.65}$$

The ionized impurity mobility $\mu_{\rm n,p}^{\rm LI}$ is modeled by using the Caughey–Thomas equation

$$\mu_{\rm n,p}^{\rm LI} = \mu_{\rm n,p}^{\rm min} + \frac{\mu_{\rm n,p}^{\rm L} - \mu_{\rm n,p}^{\rm min}}{1 + \left(N_{\rm I}/N_{\rm n,p}^{\rm ref}\right)^{\alpha_{\rm n,p}}} \tag{3.66}$$

where

$$\mu_{\rm n}^{\rm min} = 80(T/300\ {\rm K})^{-0.45}\ {\rm cm^2/V-s} \quad {\rm for}\ T \geqslant 200\ {\rm K} \tag{3.67}$$

$$\mu_{\rm n}^{\rm min} = 80(200\ {\rm K}/300\ {\rm K})^{-0.45}(T/200\ {\rm K})^{-0.15}\ {\rm cm^2/V-s} \quad {\rm for}\ T < 200\ {\rm K} \tag{3.68}$$

$$\mu_{\rm p}^{\rm min} = 45(T/300\ {\rm K})^{-0.45}\ {\rm cm^2/V-s\ for}\ T \geqslant 200\ {\rm K} \tag{3.69}$$

$$\mu_{\rm p}^{\rm min} = 45(200\ {\rm K}/300\ {\rm K})^{-0.45}(T/200\ {\rm K})^{-0.15}\ {\rm cm^2/V-s} \quad {\rm for}\ T < 200\ {\rm K} \tag{3.70}$$

$$N_{\rm n}^{\rm ref} = 1.12 \times 10^{17}\left(\frac{T}{300}\right)^{3.2} \tag{3.71}$$

$$N_{\rm p}^{\rm ref} = 2.23 \times 10^{17}\left(\frac{T}{300}\right)^{3.2} \tag{3.72}$$

$$\alpha_{\rm n,p} = 0.72\left(\frac{T}{300}\right)^{0.065}. \tag{3.73}$$

3.11 Mobility in MOSFET inversion layers at low temperatures

The mobility of electrons in the conducting channel formed under the surface of the device monotonically increases as temperature falls to 5 K (Hairapetian *et al* 1989). Considering the electric field dependence of mobility, this increase of mobility takes place at all values of the electric field, as evidenced from measurements at all gate voltages. This increase in the inversion layer mobility of electrons is greatly beneficial to device operation at low temperatures. In NMOS devices, at an electric field of 3.5×10^5 V cm^{-1}, the experimentally measured electron mobility in the inversion layer was found to increase from approximately 450 cm^2 Vs^{-1} at 293 K to 2800 cm^2 Vs^{-1} at 77 K, 4700 cm^2 Vs^{-1} at 25 K and 5800 cm^2 Vs^{-1} at 5 K. In a PMOS device, at an electric field of 2.0×10^5 V cm^{-1}, the hole mobility was 160 cm^2 Vs^{-1} at 293 K, but increased to 400 cm^2 Vs^{-1} at 100 K, 620 cm^2 Vs^{-1} at 25 K and 680 cm^2 Vs^{-1} at 5 K.

Both electron and hole mobilities increased with decreasing temperature when the respective NMOS or PMOS devices were operating under strong inversion. But a difference was noticed between the behavior of electron and hole mobilities, when the device was operating near the threshold voltage instead of the strong inversion regime. For operation near the threshold voltage, the PMOS device exhibited a peak value of hole mobility around 50 K. The hole mobility started to decrease at temperatures <50 K. However, the NMOS device did not show such a peak. The mobility continued to increase up to 5 K.

3.12 Carrier lifetime

The minority-carrier lifetime is defined as the average time during which an excess minority carrier, electron or hole, undergoes recombination (Park 2004). It is determined by three recombination mechanisms: (i) band-to-band radiative recombination, (ii) trap-assisted (or Shockley–Read–Hall (SRH)) recombination and (iii) Auger recombination. Mechanisms (i) and (ii) involve only two carriers, but mechanism (iii) is a three-carrier process in which the energy liberated during electron–hole recombination is imparted to a third carrier, either an electron or a hole. It is not released in the form of heat or light. Amongst these three mechanisms, band-to-band recombination has a relatively inconsequential effect in silicon because the radiative lifetime in silicon is extremely large. The remaining two mechanisms (ii) and (iii) govern the electrical properties of silicon devices. SRH recombination is correlated with the concentration of deep energy-level metallic impurities contaminating the silicon wafer and also the lattice defects created in the silicon. Lifetime degradation occurs during crystal growth as well as in the device fabrication steps, mainly through impurities in processing chemicals, gases, quartzware and containers, and also through thermal shocks in high-temperature diffusion and oxidation furnaces. The lifetime killing deteriorates the device characteristics. Although Auger lifetime is not related to impurity density, it varies inversely with the carrier concentration.

When a semiconductor device is in the operating condition, the minority carrier density in the active device layer is the decisive factor affecting the value of recombination lifetime in silicon. Two interesting situations shall be considered. At low minority carrier density, also called low-level injection, the non-equilibrium minority-carrier concentration is smaller than the majority-carrier concentration at thermal equilibrium. Then the SRH recombination mechanism is preponderant and so relevant for carrier lifetime in silicon. At high carrier density, also known as high-level injection, the non-equilibrium minority-carrier concentration is larger than the equilibrium majority carrier concentration. In this case, the Auger recombination mechanism predominates and hence influences the recombination lifetime.

It is found that the efficiency of recombination centers is high at lower temperatures (Hudgins *et al* 1994). Evidence for this statement comes from the decrease in the effective carrier lifetime at these temperatures. It is reported that for recombination centers that are poised at shallow locations in the forbidden energy gap, the capture cross section increases very rapidly with falling temperature (Jonscher

1964). Interpretation for this steep rise in capture cross-section has been provided in terms of the popular 'giant trap' model. In this model, the carriers are not captured directly into the ground state of the capturing center. They are first trapped in one of the higher excited states. Afterwards, they cascade down to the ground state. The cascading resembles a waterfall or a series of waterfalls. The excited states provide a much larger capture cross section than would be expected from the ground state. Extensive and colossal cross-sections in the range of 10^{-15} cm^2 to 10^{-13} cm^2 have been observed for a wide variety of traps in Si and Ge. Some of these traps involve binding energies several times larger than the Debye energy (Lax 1959).

Recombination is an inelastic process. In this process, a major chunk of the binding energy is carried away by acoustic phonons (Park *et al* 1988). The impurity ionization energy E_I is much greater than the energy of an acoustic phonon, i.e., $E_I \gg k_B T$. Naturally, direct recombination into the ground state involves several phonons. The multiphonon procedure has equivalently a very small cross section. Therefore, the controlling activity is captured into highly excited states. The capturing is ensued by a cascade of single phonon emissions and absorptions. These happen as the carrier slowly diffuses into the ground state.

The giant trap model applies, however, only to the shallow 'hydrogenic' impurities. It is not clear how the theory can be extended to deeper levels. Regrettably, these are the levels which are normally associated with recombination.

Focusing attention on the temperature range between room temperature and about 90 K (Ichimura *et al* 1998), a very frail temperature dependence is observed in the carrier lifetime of as-polished wafers. In contrast, the carrier lifetime decreases sharply with decreasing temperature for oxidized wafers. In all specimens, an initial hasty decay occurs in the photoconductivity decay curves followed by a sluggish component in the photoconductivity decay curves at temperatures below 150 K. From numerical simulations, the inference was drawn that the decrease in the carrier lifetime with decreasing temperature was caused by recombination through shallow recombination centers. These centers had an energy level within 0.15 eV from the band edge. Moreover, the slow component was due to minority carrier traps with a small majority carrier capture cross-section.

The low-level lifetime is of no interest at low temperatures, since injected densities are invariably high in comparison with equilibrium densities (Jonscher 1964). The temperature dependence of high-level lifetime is given by a simple power law (Palmer *et al* 2003)

$$\tau_{HL} = 5 \times 10^{-7}\left(\frac{T}{300}\right)^{1.5} \tag{3.74}$$

At $T = 4.2$ K,

$$\tau_{HL} = 5 \times 10^{-7}\left(\frac{4.2}{300}\right)^{1.5} = 8.28 \times 10^{-10}\,\text{s}. \tag{3.75}$$

Similarly, τ_{HL} values are calculated at 77 K, 300 K and 600 K. These are tabulated in table 3.9.

Table 3.9. High-level carrier lifetimes at different temperatures.

Temperature (K)	4.2	77.2	300	600
Carrier lifetime (s)	8.2×10^{-10}	6.53×10^{-8}	5×10^{-7}	1.41×10^{-6}

A simple temperature exponent of 1.5 does not address the competing mechanisms of SRH recombination (carrier lifetime increases as temperature increases) and Auger recombination (carrier lifetime decreases as temperature increases) under the conditions of medium to large carrier concentrations. For rigorous modeling applications, the effects of intrinsic lifetime, SRH recombination, and Auger processes must be included, assuming high-level injection and quasi-neutrality (electron concentration $n \approx$ hole concentration p) (Klaassen 1992b).

3.13 Wider bandgap semiconductors than silicon

3.13.1 Gallium arsenide

The energy gap of GaAs obeys the formula (Ioffe Institute 2015a)

$$E_g(T) = 1.519 - 5.405 \times 10^{-4} T^2/(T + 204)\,\text{eV} \quad \text{for } 0 < T < 1000\,\text{K}. \tag{3.76}$$

Intrinsic carrier concentration in GaAs follows the relation (Madelung *et al* 2002)

$$n_i(T) = 1.05 \times 10^{16} T^{3/2} \exp\{-1.604/(2k_B T)\} \quad \text{for } 33 < T < 475\,\text{K}. \tag{3.77}$$

3.13.2 Silicon carbide

For 4H-SiC, the energy gap is (Ioffe Institute 2015d)

$$E_g(T) = E_g(0) - 6.5 \times 10^{-4} T^2/(T + 1300), \tag{3.78}$$

where $E_g(0) = 3.23$ eV is the bandgap at 300 K.

For 4H-SiC in the intrinsic carrier concentration formula, the effective density of states in the conduction band is

$$\begin{aligned} N_c(T) &\approx 4.82 \times 10^{15} M (m_c/m_0)^{1.5} T^{1.5}\,(\text{cm}^{-3}) \approx 4.82 \times 10^{15} (m_{cd}/m_0)^{1.5} T^{1.5} \\ &= 3.25 \times 10^{15} T^{1.5}\,(\text{cm}^{-3}), \end{aligned} \tag{3.79}$$

where $M = 3$ denotes the number of equivalent valleys in the conduction band. The symbol $m_c = 0.37 m_0$ represents the effective mass of the density of states in one valley of the conduction band and $m_{cd} = 0.77\, m_0$ is the effective mass of the density of states.

The effective density of states in the valence band conforms to the equation

$$N_v(T) = 4.8 \times 10^{15} T^{1.5}\,(\text{cm}^{-3}). \tag{3.80}$$

3.13.3 Gallium nitride

The energy gap–temperature relationship for gallium nitride is (Ioffe Institute 2015a)

$$E_g(T) = E_g(0) - 7.7 \times 10^{-4}T^2/(T + 600)\,\text{eV}, \quad (3.81)$$

where

$$E_g(0) = 3.47\,\text{eV for wurtzite structure,}$$
$$E_g(0) = 3.28\,\text{eV for zinc blende structure.}$$

The Varshini equation for GaN is

$$E_g(T) = E_g(0) - 9.39 \times 10^{-4}T^2/(T + 772)\,\text{eV} \quad \text{where } E_g(0) = 3.427\,\text{eV}. \quad (3.82)$$

Given below are the equations for effective densities of states in conduction and valence bands for wurtzite and zinc blende GaN structures to be used for calculating the intrinsic carrier concentration

$$N_C \approx 4.3 \times 10^{14}T^{1.5}\,(\text{cm}^{-3})\text{ for wurtzite,} \quad (3.83)$$

$$N_C \approx 2.3 \times 10^{14}T^{1.5}\,(\text{cm}^{-3})\text{ for zinc blende,} \quad (3.84)$$

$$N_V \approx 8.9 \times 10^{15}T^{1.5}\,(\text{cm}^{-3})\text{ for wurtzite,} \quad (3.85)$$

$$N_V \approx 8.0 \times 10^{15}T^{1.5}\,(\text{cm}^{-3})\text{ for zinc blende.} \quad (3.86)$$

3.13.4 Diamond

At 300 K, the bandgap variation with temperature is given by (Ioffe Institute 2015c)

$$\text{d}E_g/\text{d}T = -(5.4 \pm 0.5) \times 10^{-5}\,\text{eV K}^{-1}. \quad (3.87)$$

3.14 Discussion and conclusions

When considering operation of semiconductor devices at low temperatures, if we look at the temperature dependence of free carrier concentrations and mobilities, it is necessary to distinguish between two distinct temperatures. These are the liquid-nitrogen temperature (77 K) and liquid-helium temperature (4.2 K). Investigations have shown that carrier concentrations and mobilities do not experience any drastic changes up to the liquid-nitrogen temperature. But below this temperature, the scenario is dissimilar. As the temperature is decreased further down from 77 K, different low-temperature phenomena come into action, notably carrier freeze-out and impurity scattering. Again, when considering the effects of these phenomena on carrier concentrations and mobilities, one must take note of the purity of the semiconductor material, whether undoped or intrinsic, low-doped or heavily doped with impurities. Consideration must also be given to the extent of compensation of the semiconductor, i.e. whether it is totally uncompensated, partially compensated or fully compensated. Knowledge of these aspects is vital because in a pure material, carrier concentration and mobility decrease at low temperatures, the former due to carrier freeze-out and the latter by ionized impurity scattering, even though the amount of impurities present is low. In a heavily doped degenerate material, carrier concentration and mobility are temperature-independent. Uncompensated and compensated semiconductors differ regarding the effect of temperature on mobility. In an uncompensated semiconductor, mobility decreases as the temperature is decreased. But in a compensated semiconductor, there is scarcely any influence of temperature on mobility.

Review exercises

3.1. Why does the energy bandgap of a semiconductor material decrease when the temperature falls, and conversely? Write the equations of Varshini for variation of the bandgap with temperature. Explain the meanings of the symbols used in these equations. What are the values of the fitting coefficients α and β in the model for 4H-SiC?

3.2. Write down the equation for the intrinsic carrier concentration of a semiconductor and indicate the physical parameters in this equation which depend on temperature. What are the effective masses of electrons and holes in gallium arsenide to be used in effective densities of states in conduction and valence bands? Using these values, obtain the equation for temperature dependence of intrinsic carrier concentration in gallium arsenide.

3.3. Consider silicon. Calculate the intrinsic carrier concentration in it at 4.2 K and 600 K. Treat the effective masses of charge carriers and bandgap energy as temperature-dependent.

3.4. Write the equation given by Quay *et al* for the variation of saturation velocity of charge carriers with temperature and mention the values of the two parameters in the model for silicon.

3.5. Write the equation for conductivity of a semiconductor. What parameters in the equation for conductivity change with temperature?

3.6. Write the equations for temperature-dependent free electron concentration $n(T)$ and temperature-dependent free hole concentration $p(T)$ in terms of ionized donor concentration $N_D^+(T)$, ionized acceptor concentration $N_A^-(T)$ and intrinsic carrier concentration $n_i(T)$. Express the ionized donor and acceptor concentrations N_D^+, N_A^- in terms of the total donor and acceptor concentrations N_D, N_A.

3.7. Write the equation for the Fermi–Dirac distribution function $f(E)$. What does this function say regarding the occupancy of an electronic state at energy E by an electron at temperature T? Provide the interpretation for this function at $T = 0$ K for $E = E_F$, $E < E_F$ and $E > E_F$ where E_F is the Fermi energy.

3.8. Applying the Fermi–Dirac distribution function to a non-degenerate semiconductor, show that at a temperature close to absolute zero, there exists an insignificant density of free carriers. Hence, explain the occurrence of carrier freeze-out in a semiconductor at such a low temperature. Does the same phenomenon take place in a degenerate semiconductor? If not, what is the difference between a degenerate and a non-degenerate semiconductor with regard to carrier freeze-out? Give the explanation for the difference.

3.9. Why does carrier freeze-out take place in a non-degenerate semiconductor whereas in a degenerate semiconductor the free carrier density is not a function of temperature?

3.10. What is the temperature regime $100\ \mathrm{K} \leqslant T \leqslant 500\ \mathrm{K}$ called in the context of ionization of impurity atoms in a semiconductor? For this regime of operation, write the relation between the concentration of ionized donor impurity atoms and total concentration of donor impurity atoms.

3.11. What is meant by the intrinsic regime of ionization of impurity atoms in a semiconductor? How does it differ from the extrinsic regime?

3.12. How do lattice and impurity scattering phenomena affect carrier mobility in a semiconductor? How do the effects on mobility produced by these scattering mechanisms depend on temperature? How are the mobilities influenced by the two mechanisms combined together to obtain a resultant mobility of carriers in a semiconductor?

3.13. How does acoustic phonon scattering differ from optical phonon scattering in temperature dependence? Does impurity scattering vary with impurity concentration?

3.14. Why does the density of charged impurities in an uncompensated semiconductor decrease with temperature whereas in a compensated semiconductor, the same is practically unaffected by temperature? What is the implication of this behavioral difference between uncompensated and compensated materials concerning the mobility variation with temperature?

3.15. How is ionized impurity scattering able to decrease the carrier mobility in a non-degenerate semiconductor at low temperatures? Why are high carrier mobilities observed in degenerate semiconductors even at very low temperatures?

3.16. Write down the Arora–Hauser–Roulston equations for electron/hole mobility in a semiconductor and explain the symbols used. What is the temperature range of their application?

3.17. How is lattice scattering mobility represented in: (a) Klaassen equations and (b) the MINIMOS 4 model?

3.18. How does carrier mobility in a MOSFET channel change as the temperature is lowered? Is this variation of mobility with temperature beneficial or disadvantageous for device operation?

3.19. Why do carrier concentration and mobility decease with temperature in a pure semiconductor, whereas in a degenerate semiconductor they are not affected by temperature?

3.20. Comment on how the variation of carrier concentration and mobility with temperature differ in the following temperature ranges: (a) room temperature to liquid-nitrogen temperature and (b) liquid-nitrogen temperature to liquid-helium temperature.
3.21. Define recombination lifetime. What are the three recombination mechanisms in semiconductors? Which of these mechanisms determines the carrier lifetime in silicon at low carrier concentrations? Which one plays a decisive role in controlling lifetime at high carrier densities? Which mechanism is inconsequential for silicon?
3.22. How is the decrease in the recombination lifetime at low temperatures explained in terms of the giant-trap model? What is the typical order of magnitude of the capture cross-sections in this model? To which type of impurity levels, shallow or deep, does this model apply?
3.23. Why is the low-level lifetime not of interest at low temperatures? Write the power law giving the temperature dependence of the high-level lifetime. What are the limitations of using a simple power exponent of 1.5 in the formula?

References

Ali-Omar M and Reggiani L 1987 Drift and diffusion of charge carriers in silicon and their empirical relation to the electric field *Solid-State Electron.* **30** 693–7

Arora N D, Hauser J R and Roulston D J 1982 Electron and hole mobilities in silicon as a function of concentration and temperature *IEEE Trans. Electron Devices* **29** 292–5

Barber H D 1967 Effective mass and intrinsic concentration in silicon *Solid-State Electron.* **10** 1039–51

Bludau W, Onton A and Heinke W 1974 Temperature dependence of the band gap of silicon *J. Appl. Phys.* **45** 1846–8

Caiafa A, Wang X, Hudgins J L, Santi E and Palmer P R 2003 Cryogenic study and modeling of IGBTs *34th Annual IEEE Power Electronics Specialists Conference* (*Acapulco, Mexico, June 2003*) vol 4, paper 59_02, pp 1897–903

Cole D C and Johnson J B 1989 Accounting for incomplete ionization in modeling silicon-based semiconductor devices *Proc. Workshop Low Temperature Semiconductor Electronics* (*7–8 August*) (Piscataway, NJ: IEEE) pp 73–7

Hairapetian A, Gitlin D and Viswanathan C R 1989 Low-temperature mobility measurements on CMOS devices *IEEE Trans. Electron Devices* **36** 1448–55

Hudgins J L, Godbold C V, Portnoy W M and Mueller O M 1994 Temperature effects on GTO characteristics *IEEE–IAS Annual Meeting Rec.* (*October*) pp 1182–6

Ichimura M, Tajiri H, Ito T and Arai E 1998 Temperature dependence of carrier recombination lifetime in Si wafers *J. Electrochem. Soc.* **145** 3265–71

Ioffe Institute 2015a GaAs—gallium arsenide: band structure and carrier concentration *Ioffe Institute* www.ioffe.ru/SVA/NSM/Semicond/GaAs/bandstr.html#Temperature

Ioffe Institute 2015b Physical properties of silicon (Si) Ioffe Institute www.ioffe.rssi.ru/SVA/NSM/Semicond/Si/

Ioffe Institute 2015c C-diamond: band structure and carrier concentration Ioffe Institute www.ioffe.ru/SVA/NSM/Semicond/Diamond/bandstr.html#Temperature

Ioffe Institute 2015d SiC—silicon carbide: band structure and carrier concentration Ioffe Institute www.ioffe.ru/SVA/NSM/Semicond/SiC/bandstr.html#DependenceTemperature

Jacoboni C, Canali C, Ottaviani G and Alberigi Quaranta A 1977 A review of some charge transport properties of silicon *Solid-State Electron.* **20** 77–89

Jonscher A K 1964 Semiconductors at cryogenic temperatures *Proc. IEEE* **52** 1092–104

Klaassen D B M 1992a A unified mobility model for device simulation—I. Model equations and concentration dependence *Solid State Electron.* **35** 953–9

Klaassen D B M 1992b A unified mobility model for device simulation—II. Temperature dependence of carrier mobility and lifetime *Solid State Electron.* **35** 961–7

Lax M 1959 Giant traps *J. Phys. Chem. Solids.* **8** 66–73

Madelung O, Rössler U and Schulz M (ed) 2002 Gallium arsenide (GaAs), intrinsic carrier concentration, electrical and thermal conductivity *Group IV Elements, IV–IV and III–V Compounds. Part b—Electronic, Transport, Optical and Other Properties* (*Landolt–Börnstein—Group III Condensed Matter* vol 41A1b) (Berlin: Springer)

Palmer P R, Santi E, Hudgins J L, Kang X, Joyce J C and Eng P Y 2003 Circuit simulator models for the diode and IGBT with full temperature dependent features *IEEE Trans. Power Electron* **18** 1220–9

Park I S, Haller E E, Grossman E N and Watson D M 1988 Germanium: gallium photoconductors for far infrared heterodyne detection *Appl. Opt.* **27** 4143–50

Park J-M 2004 Novel power devices for smart power applications *Dissertation* (Sydney: Macquarie University) section 3.1.4

Pieper R J and Michael S 2005 An exact analysis for freeze-out and exhaustion in single impurity semiconductors *Proc. 2005 American Society of Engineering Education Annual Conf. and Exposition* (Washington, DC: American Society for Engineering Education)

Quay R, Moglestue C, Palankovski V and Selberherr S 2000 A temperature dependent model for the saturation velocity in semiconductor materials *Mater. Sci. Semicond. Process.* **3** 149–55

Selberherr S, Hänsch W, Seavey M and Slotboom J 1990 The evolution of the MINIMOS mobility model *Solid-State Electron.* **33** 1425–36

Silvaco 2000 Simulation standard: simulating impurity freeze-out during low temperature operation *Silvaco* www.silvaco.com/tech_lib_TCAD/simulationstandard/2000/nov/a1/a1.html

Singh J 1993 *Physics of Semiconductors and their Heterostructures* (New York: McGraw-Hill), 851 pages

Stefanakis D and Zekentes K 2014 TCAD models of the temperature and doping dependence of the bandgap and low field carrier mobility in 4H-SiC *Microelectron. Eng.* **116** 65–71

Sze S M 1981 *Physics of Semiconductor Devices* 2nd edn (New York: Wiley), 868 pages

Van Zeghbroeck B J 1997 *Effective Mass in Semiconductors* http://ecee.colorado.edu/~bart/book/effmass.htm

Van Zeghbroeck B J 2011 *Principles of Semiconductor Devices* (Boulder, CO: University of Colorado)

Varshini P 1967 Temperature dependence of the energy gap in semiconductors *Physica* **34** 149–54

Wilkinson V and Adams A 1993 The effect of temperature and pressure on InGaAs band structure *Properties of Lattice-matched and Strained Indium Gallium Arsenide* (*EMIS Data Reviews* Series vol 8) ed Bhattacharya (Berlin: Wiley) pp 70–5

Chapter 4

Temperature dependence of the electrical characteristics of silicon bipolar devices and circuits

Silicon is the mainstay of electronic discrete devices and ICs, supported by gallium arsenide and wide bandgap materials such as silicon carbide, gallium nitride and others. This chapter recapitulates the essentials of silicon technology, and then makes an enquiry into the variation of vital electrical parameters of common devices of the bipolar family, such as p–n junctions, Schottky diodes and bipolar junction transistors (BJTs). Parameters of interest include the forward voltage drop, reverse breakdown voltage and leakage current of diodes and transistors. Special attention is paid to the current gain of a bipolar transistor. The switching parameters of devices are also delved into. Investigational studies carried out by researchers on bipolar analog and digital circuits are described.

4.1 Properties of silicon

Silicon is a mainstream electronics material. In compliance with the requirements of many military systems, the uppermost borderline signifying thermal cut-off for routinely used silicon devices is 125 °C. However, this does not necessarily imply that silicon devices cease to function at this temperature. Instead, they become more susceptible to failure mechanisms when this temperature is reached. Failure of a device is a complex process involving edge terminations, biasing voltages and leakage current levels, in addition to temperature effects. The thermal boundary of bipolar and CMOS circuits is easily extendable to 150 °C, and further up to 200 °C to 250 °C by reducing supply voltages. The adoption of dielectric isolation (section 4.3.13) reduces leakage currents significantly, extending the operating range up to 300 °C. The properties of silicon relevant to semiconductor device operation are listed in table 4.1. The crystal structure of silicon is depicted in figure 4.1.

doi:10.1088/978-0-7503-1155-7ch4

Table 4.1. Properties of silicon (Si).

Property	Value
Atomic number	14
Atomic mass	28
Classification	Metalloid
Crystal structure	Two interpenetrating face-centered cubic lattices
Color	Gray
Density at 300 K (g cm^{-3})	2.329
Number of atoms cm^{-3}	5×10^{22}
Lattice constant (Å)	5.43
Melting point (°C)	1410
Dielectric constant	11.7
Thermal conductivity (W cm K^{-1})	1.5
Energy bandgap E_g (eV) at 300 K	1.12
Electrical breakdown field (V cm^{-1})	3×10^5
Intrinsic carrier concentration (cm^{-3})	1×10^{10}
Intrinsic resistivity (Ω cm)	3.2×10^5
Electron mobility (cm^2 V s^{-1})	1400
Hole mobility (cm V s^{-1})	450
Electron diffusion coefficient (cm^2 s^{-1})	36
Hole diffusion coefficient (cm^2 s^{-1})	12
Electron saturated velocity (cm s^{-1})	2.3×10^7
Hole saturated velocity (cm s^{-1})	1.65×10^7
Minority-carrier lifetime (s)	10^{-6}

4.2 Intrinsic temperature of silicon

To calculate the intrinsic temperature of silicon, the approximate equation for intrinsic carrier concentration (3.7) is written down, assuming temperature-independent effective masses for the carriers and semiconductor bandgap:

$$\begin{aligned} n_i &= 4.83 \times 10^{15} \times \left(\frac{m_n^* m_p^*}{m_0^2}\right)^{0.75} T^{1.5} \exp\left(-\frac{E_g}{2k_B T}\right) \text{cm}^{-3} \\ &= K m^* T^{1.5} \exp\left(-\frac{E_g}{2k_B T}\right), \end{aligned} \tag{4.1}$$

where K is a constant $= 4.83 \times 10^{15}$; m^* is expressed in terms of electron effective mass m_n^*, hole effective mass m_p^* and electron rest mass m_0 as

$$m^* = \frac{m_n^* m_p^*}{m_0^2} \tag{4.2}$$

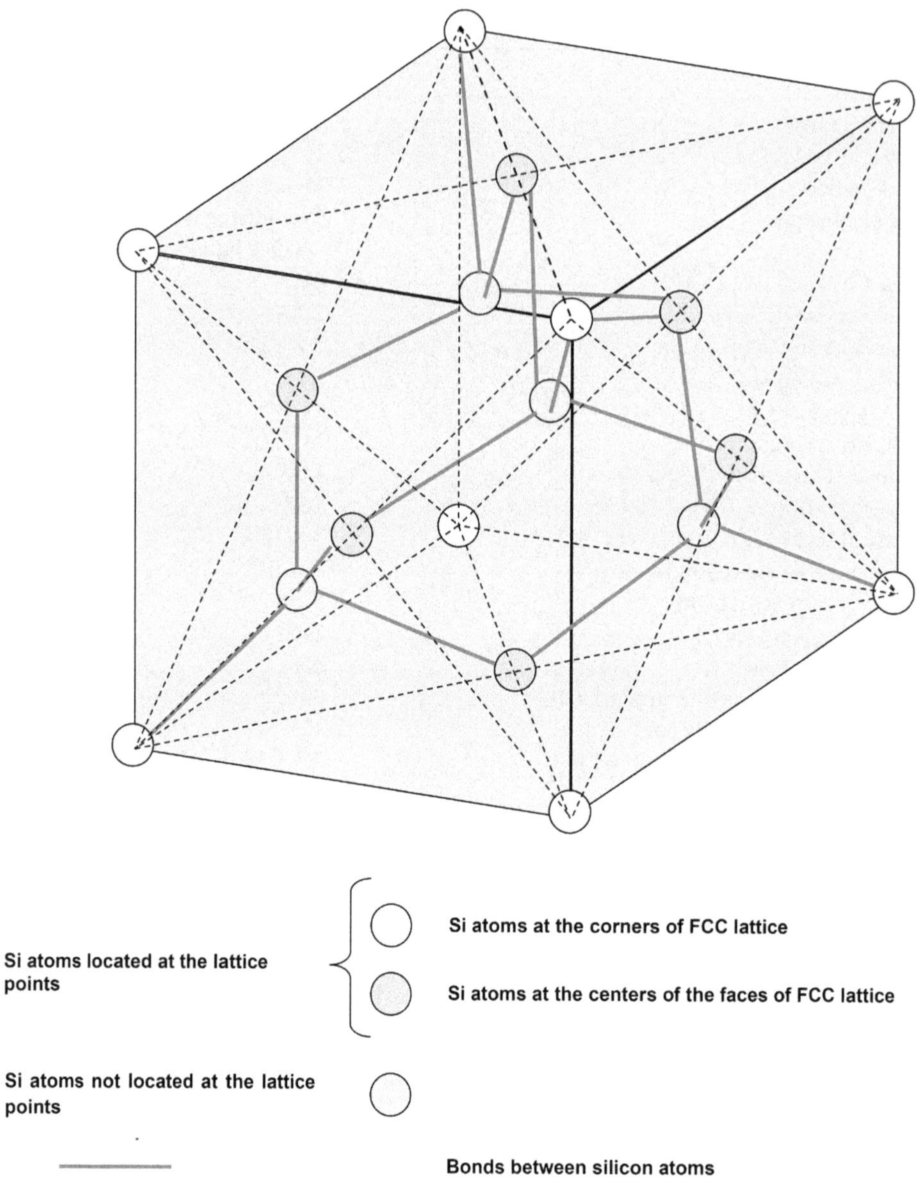

Figure 4.1. Structure of silicon crystal.

where T is temperature, E_g is the bandgap and k_B is the Boltzmann constant. Taking the natural logarithm of both sides of equation (4.1), we obtain

$$\ln(n_i) = \ln(Km^*T^{1.5}) - \frac{E_g}{2k_BT} \tag{4.3}$$

or

$$\ln\{n_i/(Km^*T^{1.5})\} = -\frac{E_g}{2k_BT} \tag{4.4}$$

or

$$T\ln\{n_i/(Km^*T^{1.5})\} = -\frac{E_g}{2k_B} = -\frac{E_g}{2 \times 8.617 \times 10^{-5}} = -5.8025 \times 10^3 E_g. \tag{4.5}$$

At the intrinsic temperature, the intrinsic carrier concentration n_i is taken as the dopant concentration of the starting silicon wafer used for device fabrication ($1 \times 10^{15}\ \mathrm{cm}^{-3}$)

$$\begin{aligned} T\ln\{1 \times 10^{15}/(4.83 \times 10^{15} m^* T^{1.5})\} &= -5.8025 \times 10^3 E_g \\ \therefore\ T\ln\{0.207/(m^*T^{1.5})\} &= -5.8025 \times 10^3 E_g. \end{aligned} \tag{4.6}$$

For Si,

$$\begin{aligned} &m_n^* = 1.08 m_0,\ m_p^* = 0.81 m_0,\ E_g = 1.12\ \text{eV} \\ &\therefore\ m^* = \left(\frac{m_n^* m_p^*}{m_0^2}\right)^{0.75} = \left(\frac{1.08 m_0 \times 0.81 m_0}{m_0^2}\right)^{0.75} = 0.9046. \end{aligned} \tag{4.7}$$

Hence,

$$T\ln\{0.207/(0.9046T^{1.5})\} = -5.8025 \times 10^3 \times 1.12 \tag{4.8}$$

or

$$T\ln(0.229/T^{1.5}) = -6498.8$$

or

$$\begin{aligned} T\ln 0.229 - T\ln T^{1.5} &= -6498.8 \\ \therefore -1.474T - T\ln T^{1.5} &= -6498.8. \end{aligned} \tag{4.9}$$

If we take,

$$\begin{aligned} T &= 588\ \text{K} \\ \text{left-hand side (lhs)} &= -1.474 \times 588 - 588\ln 588^{1.5} \\ &= -866.71 - 5624.27 = -6490.98. \end{aligned} \tag{4.10}$$

If

$$\begin{aligned} &T = 589\ \text{K} \\ &\text{lhs} = -1.474 \times 589 - 589\ln 589^{1.5} = -868.19 - 5635.34 = -6503.53. \end{aligned} \tag{4.11}$$

When

$$\begin{aligned} T &= 588.6 \text{ K} \\ \text{lhs} &= -1.474 \times 588.6 - 588.6 \ln 588.6^{1.5} \\ &= -867.60 - 5630.91 = -6498.51. \end{aligned} \tag{4.12}$$

$$\therefore T \approx 588.63 \text{ K for which lhs} = -1.474 \times 588.63 - 588.635 \ln 588.63^{1.5} \\ = -867.64 - 5631.245 = -6498.885. \tag{4.13}$$

This value represents an approximation to the intrinsic temperature because of the main assumptions underlying the calculations.

4.3 Recapitulating single-crystal silicon wafer technology

4.3.1 Electronic grade polysilicon production

Metallurgical grade (MG) polysilicon containing ~2% impurities is obtained by reducing silica. Silica reduction is performed by heating it with carbon in an electric arc furnace at 1900 °C (Fisher *et al* 2012). MG polysilicon is purified by conversion to trichlorosilane (TCS) with a boiling point 31.8 °C. Conversion to TCS is carried out by reaction with HCl in a fluidized bed reactor at 300 °C:

$$\text{Si (MG)} + 3\text{HCl} \rightarrow \text{SiHCl}_3 \text{ (liquid)} + \text{H}_2. \tag{4.14}$$

The TCS is distilled yielding a highly pure liquid. EG polysilicon is produced by reaction between purified TCS and hydrogen in a CVD reactor at 1000 °C to 1200 °C:

$$\text{SiHCl}_3 \text{ (liquid)} + \text{H}_2 \rightarrow \text{Si (EG)} + 3\text{HCl}. \tag{4.15}$$

An intermediate compound such as SiH_4 can also be decomposed to produce EG polysilicon:

$$\text{SiH}_4 + \text{H}_2 \rightarrow \text{Si (EG)} + 3\text{H}_2. \tag{4.16}$$

4.3.2 Single crystal growth

The Czochralski method

In the Czochralski (CZ) method, polysilicon (EG) obtained as above is melted in a quartz crucible (figure 4.2). A seed crystal is immersed into molten polysilicon (EG). It is slowly pulled upwards. Simultaneously with the pulling, it is also rotated. Through crystal pulling/rotating from the melt, a cylindrical ingot of monocrystalline silicon is formed. Vital parameters that must be stringently monitored during this operation are: the temperature gradient, the pulling rate and speed of rotation of the crystal. The process is carried out in an argon atmosphere maintained inside a quartz chamber.

Float-zone method

A molten region is slowly moved along a silicon bar (figure 4.3). Impurities segregate towards the liquefied side leaving the solidified portion in a purified state.

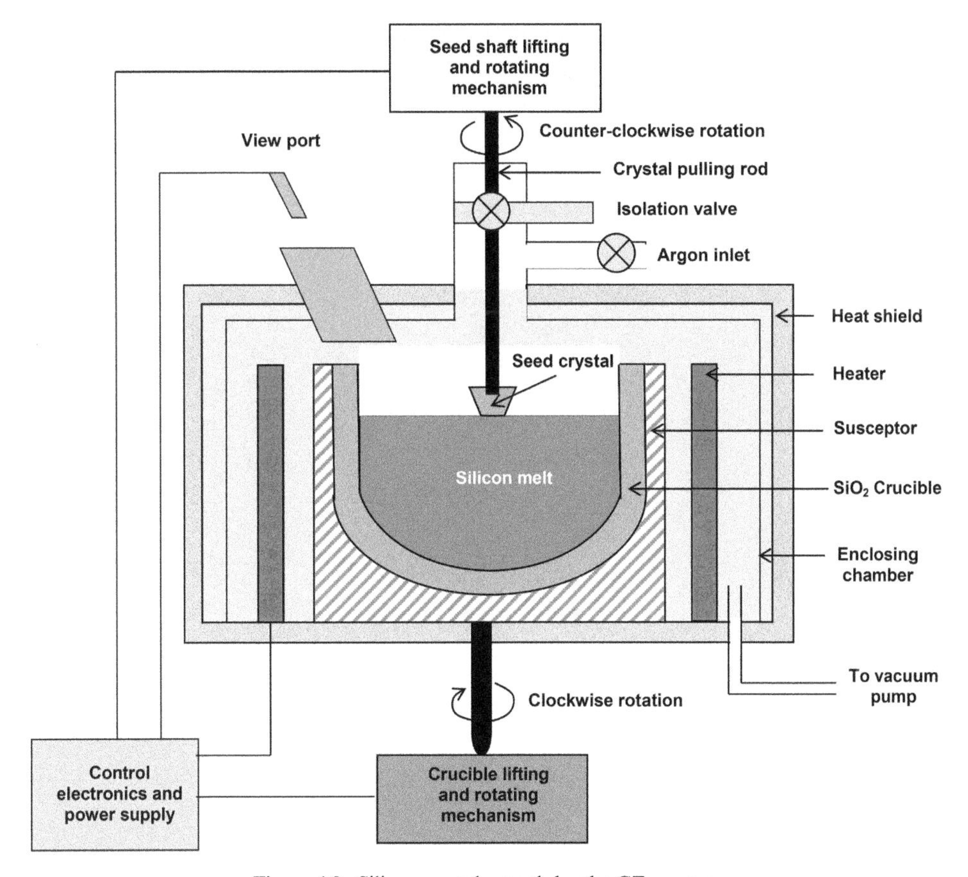

Figure 4.2. Silicon crystal growth by the CZ process.

4.3.3 Photolithography

Several kinds of photolithographic exposure and mask aligning systems are available. Optical lithography has borne the major burden of silicon device and IC manufacturing. This technique is widely used for defining micron and sub-micron geometries up to the wavelength of light that imposes restrictions on minimum feature size. Two types of photosensitive materials called positive and negative photoresists are used. The positive photoresist becomes soft on exposure while the negative photoresist becomes hard in exposed regions. One type of optical mask aligner, namely a projection system, is shown in figure 4.4(a). In the nanometer range, electron-beam (e-beam) lithography (figure 4.4(b)) is employed.

4.3.4 Thermal oxidation of silicon

By exposing silicon wafers to oxygen, either pure oxygen gas or by bubbling oxygen through heated water in a flask and transporting the water vapors to a furnace at a high temperature of ~900 °C to 1200 °C, good quality silicon dioxide films are formed on the wafer surface. When oxygen is used alone, the process is called dry oxidation. If

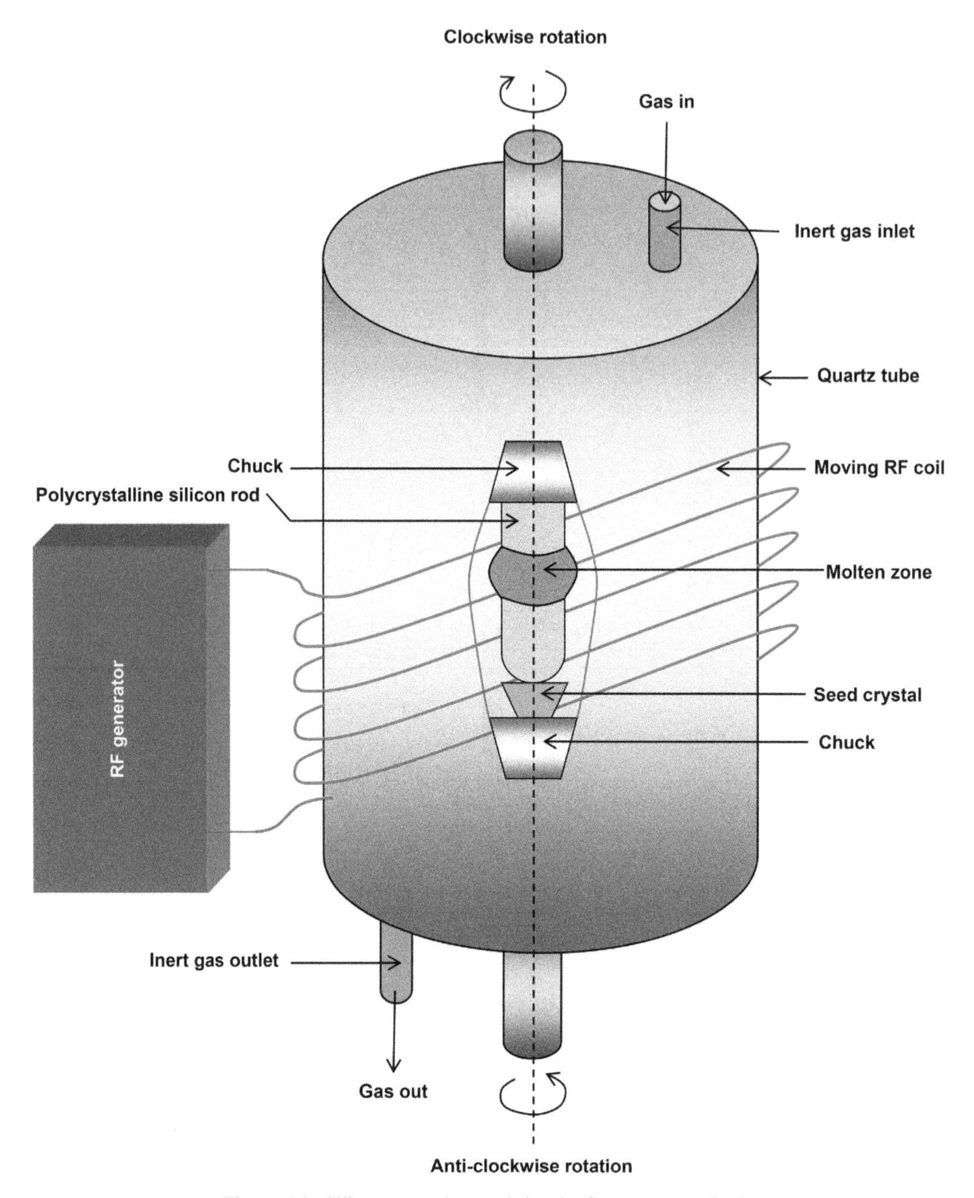

Figure 4.3. Silicon crystal growth by the float zone method.

oxygen is passed through hot water, the process is known as wet oxidation. These oxide films serve as insulators. They are used as diffusion masks to produce windows through which selective diffusion can be performed in defined areas:

$$Si + O_2 = SiO_2, \tag{4.17}$$

$$Si + 2H_2O = SiO_2 + 2H_2\,(g). \tag{4.18}$$

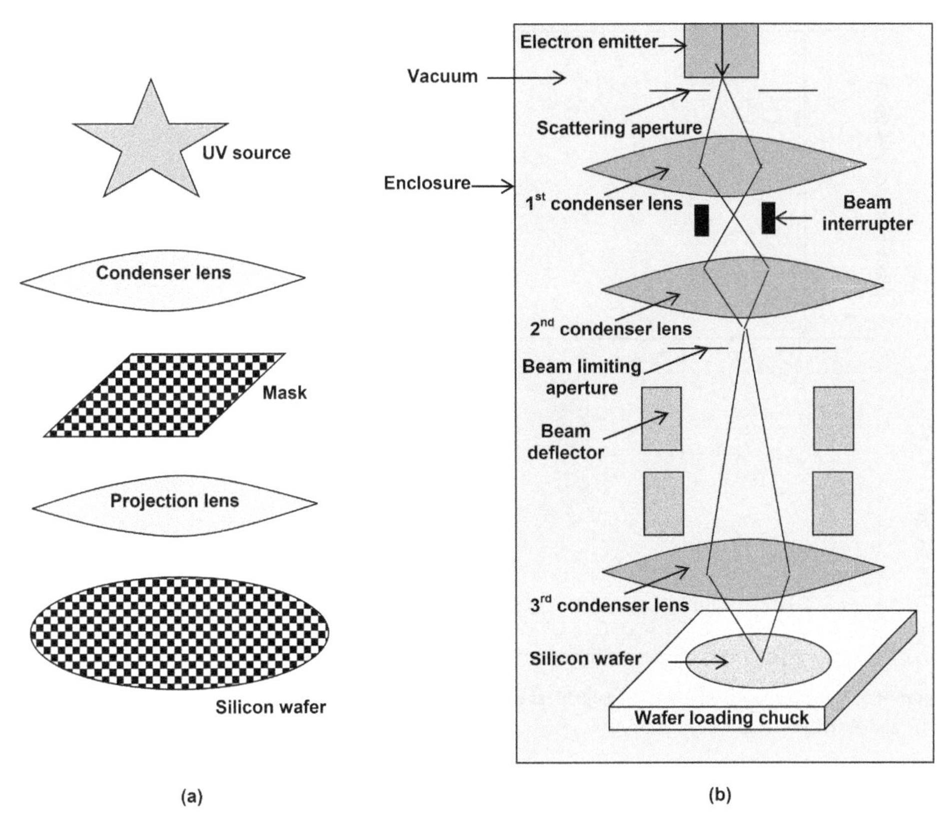

Figure 4.4. Photolithography: (a) optical projection lithography and (b) e-beam lithography.

4.3.5 n-type doping of silicon by thermal diffusion

In figure 4.5, the energy-band diagram of silicon is drawn showing the energy levels of the impurities commonly used as dopants for producing n-type and p-type materials.

Phosphorous oxychloride (figure 4.6) is a liquid dopant source used to form phosphosilicate glass (PSG) on the silicon wafer at 900 °C to 1000 °C:

$$4POCl_3 + 3O_2 \rightarrow 2P_2O_5 + 6Cl_2; 2P_2O_5 + 5Si \rightarrow 4P + 5SiO_2. \quad (4.19)$$

A gaseous source (phosphine or PH_3) may also be used:

$$2PH_3 + 4O_2 \rightarrow P_2O_5 + 3H_2O. \quad (4.20)$$

4.3.6 p-type doping of silicon by thermal diffusion

The boron source is a ceramic wafer of boron nitride. This wafer is oxidized at 1000 °C in an oxygen atmosphere to form B_2O_3 glass. During boron deposition, B_2O_3 evaporates from the boron nitride wafer and condenses on the silicon wafer as a glassy layer (borosilicate glass (BSG)) serving as a source of elemental boron:

$$2B_2O_3 + 3Si \rightarrow 4B + 3SiO_2. \quad (4.21)$$

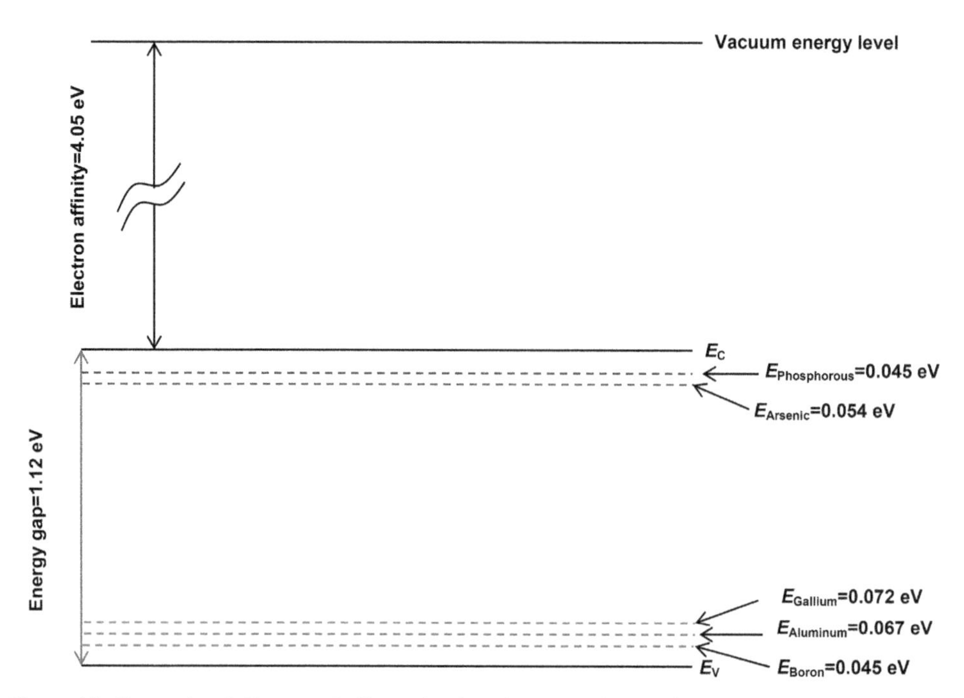

Figure 4.5. Energy band diagram of silicon showing the energy levels of commonly used impurities for n- (P and As) and p-doping (B, Al, Ga).

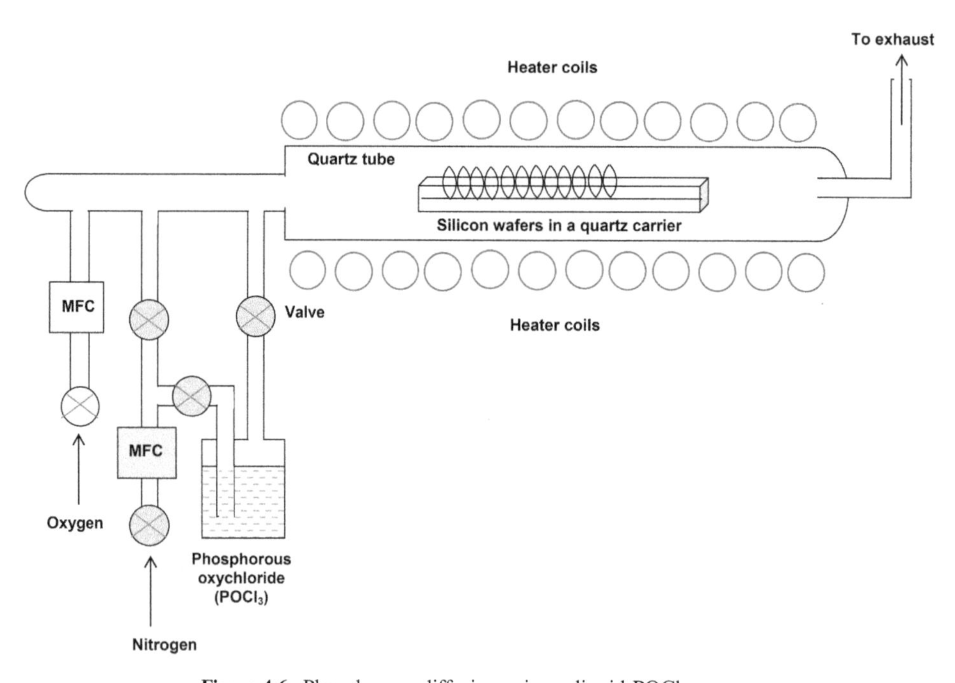

Figure 4.6. Phosphorous diffusion using a liquid $POCl_3$ source.

A gaseous source such as diborane (B_2H_6) is sometimes used to form B_2O_3 glass:

$$B_2H_6 + 3O_2 \rightarrow B_2O_3 + 3H_2O. \tag{4.22}$$

4.3.7 Impurity doping by ion implantation

As an alternative to diffusion, pre-deposition is performed by ion implantation (figure 4.7). It is a room-temperature technique in which a beam of ions obtained from a gaseous source, e.g. boron trifluoride (BF_3) for boron, phosphine (PH_3) for phosphorous, etc, is separated by the charge-to-mass ratio of the ions, focused, accelerated by an electric field and scanned across the silicon wafer. The energy of the ion beam determines the depth, and the dose measured in ions cm^{-2} decides the concentration. A thermal annealing step at 900 °C for 30 min almost always follows the ion implantation. This step is essential for two reasons. First, to anneal the damage caused by ion bombardment and second, to transfer the impurity ions from interstitial to substitutional sites. In order to diffuse the implanted ions deeper into the semiconductor crystal, a high-temperature drive-in step at 1100 °C to 1200 °C may be required. Typical parameters for ion implantation are: accelerator energy = 1–500 keV, dose = 10^{11}–10^{16} cm^{-2}. Ion implanters are classified according to the magnitudes of beam currents used by them. Medium current implanters range in ion beam currents from ~10 μA to 2 mA; high current types can be up to 30 mA.

4.3.8 Low-pressure chemical vapor deposition

For low-pressure CVD (LPCVD) the starting pressure in the quartz tube (figure 4.8) is 0.1 Pa and the working gas (dilution gas + reactive species) is introduced at 10–1000 Pa. It is used for depositing polysilicon, silicon nitride and doped oxides

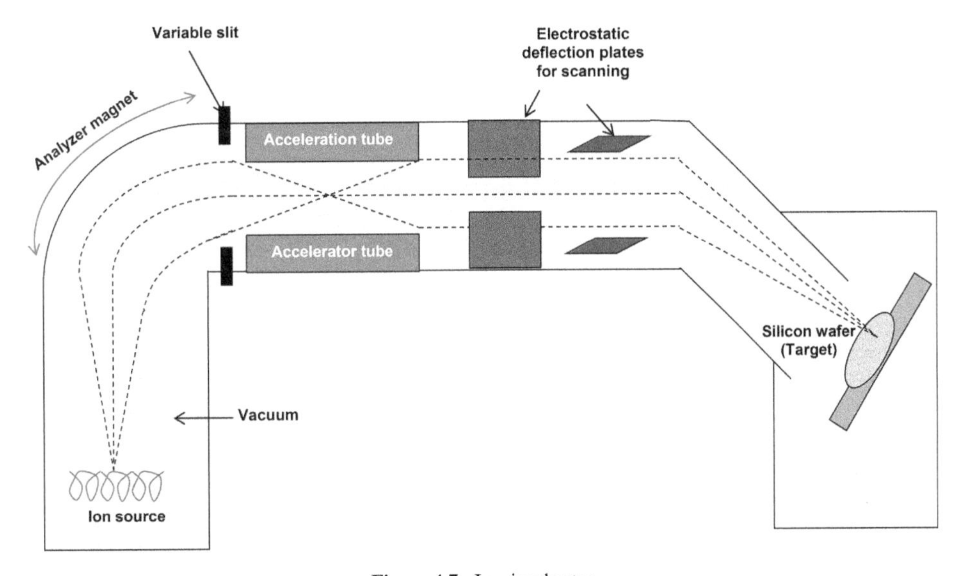

Figure 4.7. Ion implanter.

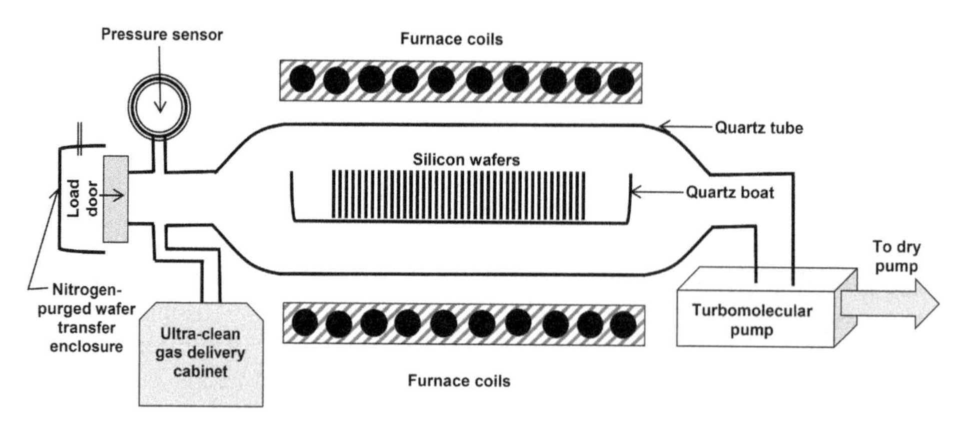

Figure 4.8. LPCVD system.

such as BSG and PSG. Temperatures range from 600 °C to 660 °C for polysilicon to 850 °C to 950 °C for BSG and 950 °C to 1100 °C for PSG.

Polysilicon is obtained by pyrolysis of silane (SiH_4):

$$SiH_4 = Si + 2H_2, \tag{4.23}$$

or from trichlorosilane ($SiHCl_3$):

$$SiHCl_3 + H_2 = Si + 3HCl. \tag{4.24}$$

Silicon dioxide is deposited using silane (SiH_4), dichlorosilane (SiH_2Cl_2) or tetraethylorthosilicate, (TEOS, $Si(OC_2H_5)_4$):

$$SiH_4 + O_2 = SiO_2 + 2H_2 \tag{4.25}$$

$$SiH_2Cl_2 + 2N_2O = SiO_2 + 2N_2 + 2HCl \tag{4.26}$$

$$Si(OC_2H_5)_4 + 12O_2 = SiO_2 + 10H_2O + 8CO_2. \tag{4.27}$$

The reactions for silicon nitride (Si_3N_4) deposition are:

$$3SiH_4 + 4NH_3 = Si_3N_4 + 12H_2 \tag{4.28}$$

$$3SiH_2Cl_2 + 4NH_3 = Si_3N_4 + 6HCl + 6H_2. \tag{4.29}$$

Tungsten can be deposited using tungsten hexafluoride (WF_6) and silane:

$$WF_6 + 2SiH_4 = W + 2SiHF_3 + 3H_2. \tag{4.30}$$

4.3.9 Plasma-enhanced chemical vapor deposition

In plasma-enhanced CVD (PECVD), employing a parallel plate electrode configuration, plasma is excited by a RF signal (figure 4.9). The energy of electrons in cold thermal plasma helps PECVD to operate at relatively lower temperatures (100 °C to 400 °C) than LPCVD. It is widely used for deposition of silicon dioxide (SiO_2), silicon nitride (Si_3N_4) and silicon oxynitride ($Si_xO_yN_z$) films.

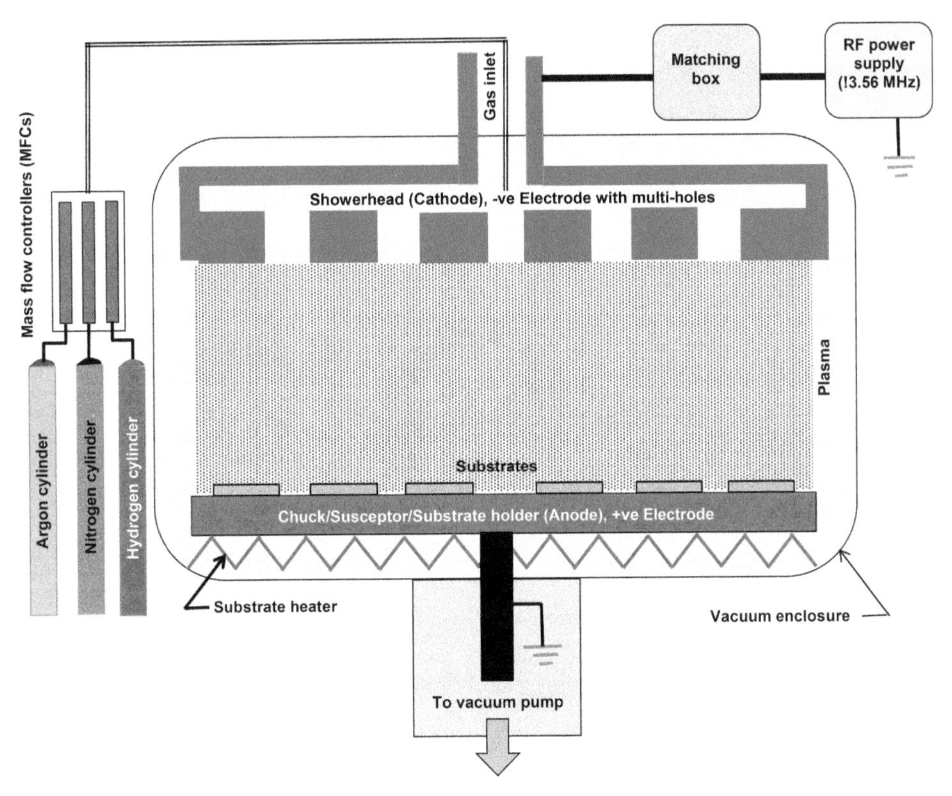

Figure 4.9. PECVD system.

4.3.10 Atomic layer deposition

Atomic layer deposition (ALD) may be referred to as a subclass of CVD or an advanced thin-film coating technique. In this method, ultrathin, conformal, dense, defect-free films (mainly oxides such as Al_2O_3, HfO_2, Ta_2O_5, V_2O_5, ZrO_2, etc) having thicknesses in nanometers, are deposited in a precisely controlled manner. Consecutive atomic layers are formed by the reaction of, generally, two chemical species called precursors with the substrate in a sequential manner. The reaction is broken into two half-reactions and the precursors are always kept separate. The process is also self-limiting.

First the substrate surface is exposed to one precursor. Then any unreacted excess precursor is flushed away by introducing a cleaning gas in the reaction chamber. Precursor removal is followed by exposure of the substrate to the second precursor. Then purging of the reaction chamber is performed again to remove this second precursor. Several such cycles are repeated until the desired film thickness is achieved.

4.3.11 Ohmic (non-rectifying) contacts to Si

Metallization techniques are thermal or e-beam evaporation (figure 4.10) and sputtering (figure 4.11). Examples of materials used include: aluminum on p-type

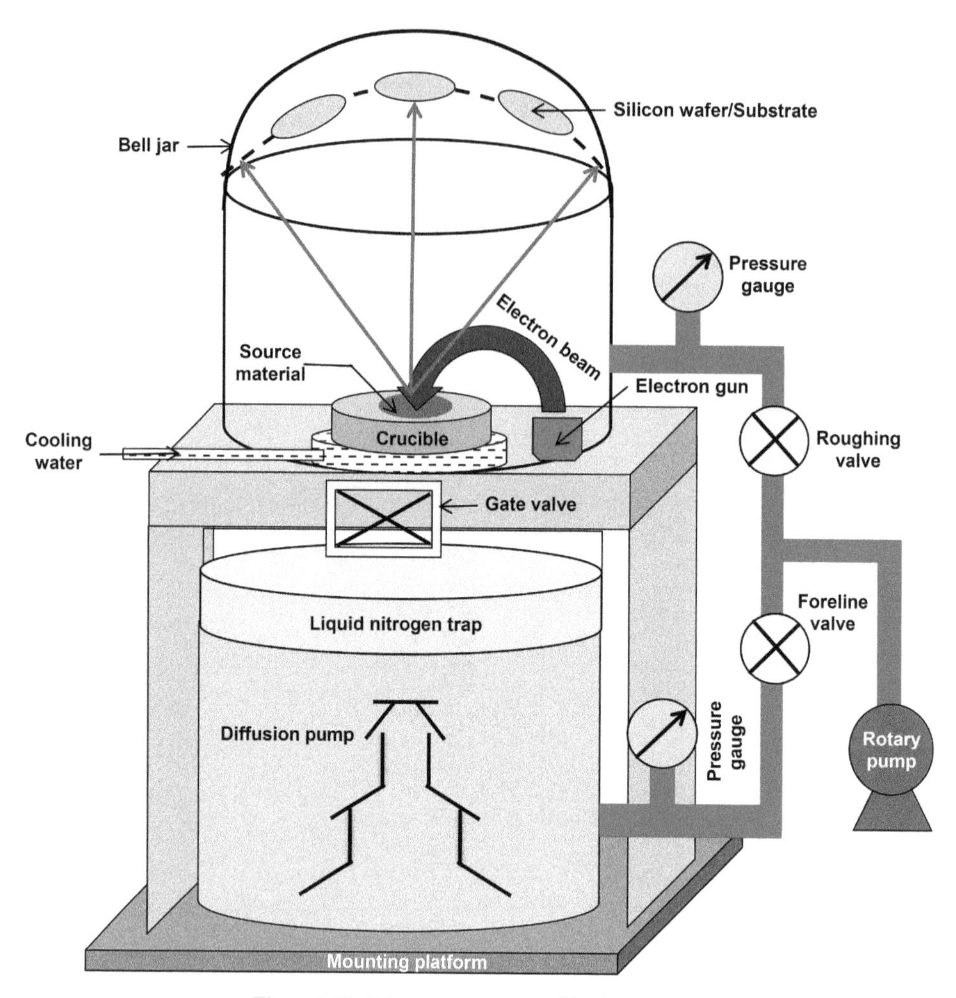

Figure 4.10. E-beam vacuum metallization system.

silicon or heavily doped n-type silicon; silicides ($TiSi_2$, $CoSi_2$, WSi_2, $TaSi_2$, and $MoSi_2$), W, TiN, etc, on n-type silicon (Sinha 1981).

4.3.12 Schottky contacts to Si

An example is aluminum on moderately or low-doped n-type silicon

4.3.13 p–n junction and dielectric isolation in silicon integrated circuits

The wide variety of components fabricated on an IC chip is separated by two main types of silicon technology, namely junction and dielectric isolation. In junction isolation (figure 4.12), n-type islands are formed on a p-type starting substrate. In these islands, the different devices are fabricated. During operation of the circuit, the p-type substrate is maintained at the most negative potential with respect to the remaining chip. In this manner, the reverse-biased p–n diode serves as an electrically

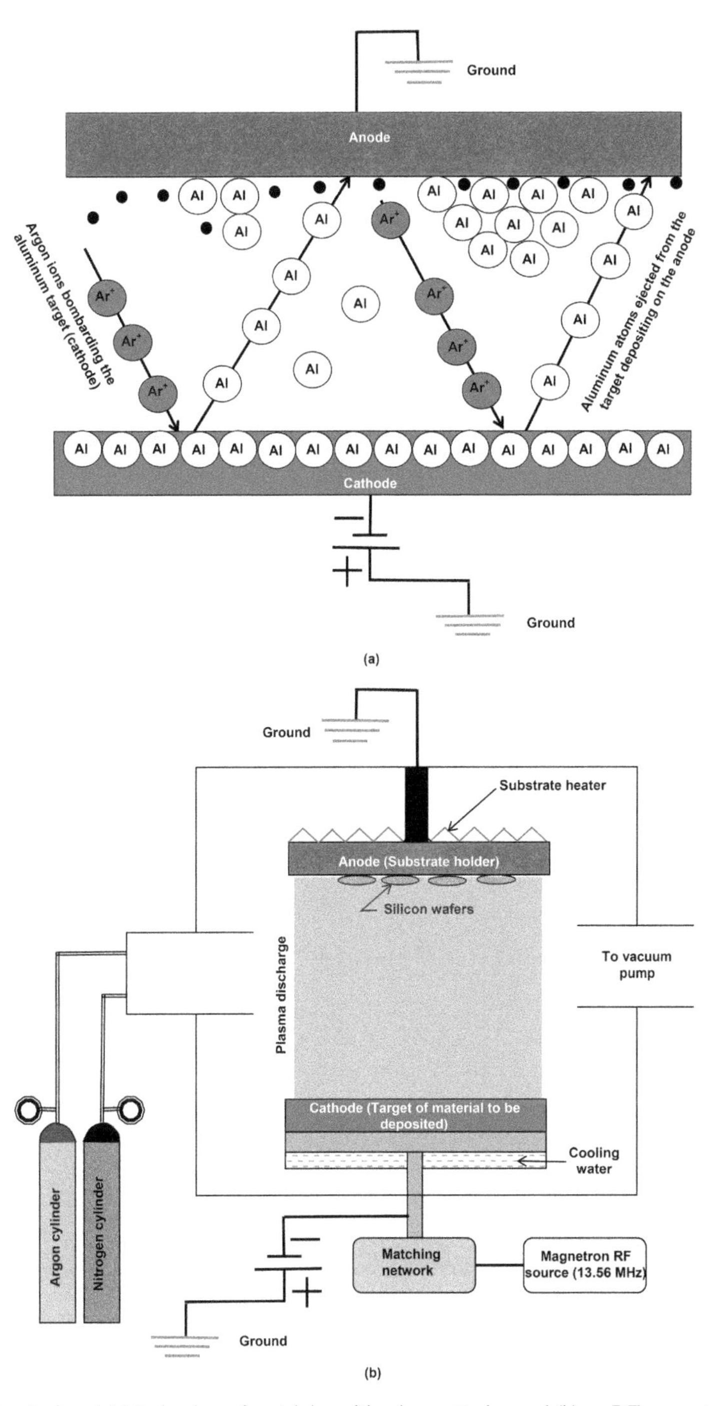

Figure 4.11. Sputtering. (a) Mechanism of metal deposition by sputtering and (b) an RF magnetron sputtering system.

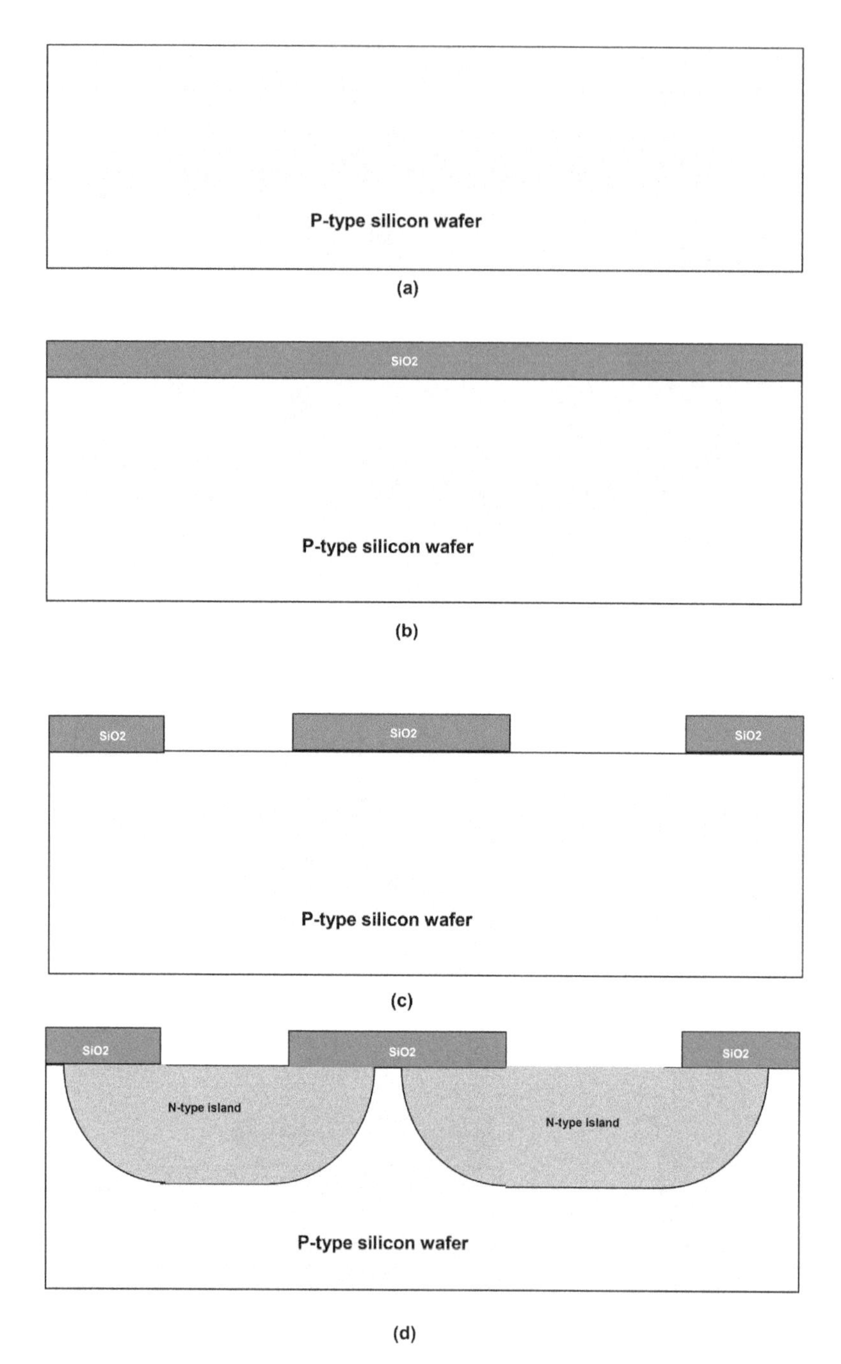

Figure 4.12. p–n junction isolation. (a) Initial p-type silicon wafer, (b) thermal oxidation, (c) photolithography and oxide etching, and (d) n-diffusion.

insulating material. The significant advantage gained with this method is its compatibility with IC processing. Disadvantages include the long diffusion times required to create isolation, the consumption of a significant proportion of chip area by lateral diffusion and the introduction of stray capacitances. At high temperatures, the junction isolation becomes excessively leaky, impairing the performance of the chip. This is further worsened by the EHP generation due to photon irradiation.

Dielectric isolation (figure 4.13) solves both the stray capacitance and leakage current problems. The dielectric constant of silicon dioxide (3.9) is three times smaller than that of silicon (11.7), reducing the capacitive effect. Thermal SiO_2 has a DC resistivity of 10^{14}–10^{16} Ωcm. It need not be kept at a negative potential as required in junction isolation. SOI technology is a semiconductor fabrication technology which does not employ the traditional bulk silicon wafers as the starting material. Instead, the starting material is a composite material comprising a topmost thin layer of single-crystal silicon of several nanometers to micron thicknesses called the active silicon layer, formed on a thin thermally grown SiO_2 insulating layer, ranging in thickness from sub-microns up to 1 μm, known as the buried oxide (BOX) layer, and supported on a bottom silicon layer several hundred microns thick, termed the handle wafer. The transistors and other devices are formed in the active silicon layer. The underlying BOX layer provides the isolation.

4.4 Examining temperature effects on bipolar devices

The requirement of operation over a broad range of temperatures imposed by many applications makes it essential to take into account the variation of the performance parameters of semiconductor devices with temperature throughout device design, in order to guarantee that the device will work satisfactorily during thermal excursions. Interestingly, many device parameters change advantageously with decreasing temperature, while others vary favorably with increasing temperature. Therefore, the influence of temperature change in a given direction cannot always be said to be either beneficial or adverse for the application at hand, and each parameter must be examined separately.

4.4.1 The Shockley equation for the current–voltage characteristics of a p–n junction diode

The simplest semiconductor device is the p–n junction diode comprising the interface or boundary between a p-type and an n-type semiconductor (figure 4.14). It has the property of allowing current flow in one direction only and blocking it in the opposite direction.

The current *I* flowing through a parallel plane p–n junction diode as a function of voltage *v* applied across its terminals is given by the theoretically derived Shockley equation (Leach 2004):

$$i = I_S\left\{\exp\left(\frac{qv}{\eta k_B T}\right) - 1\right\}, \tag{4.31}$$

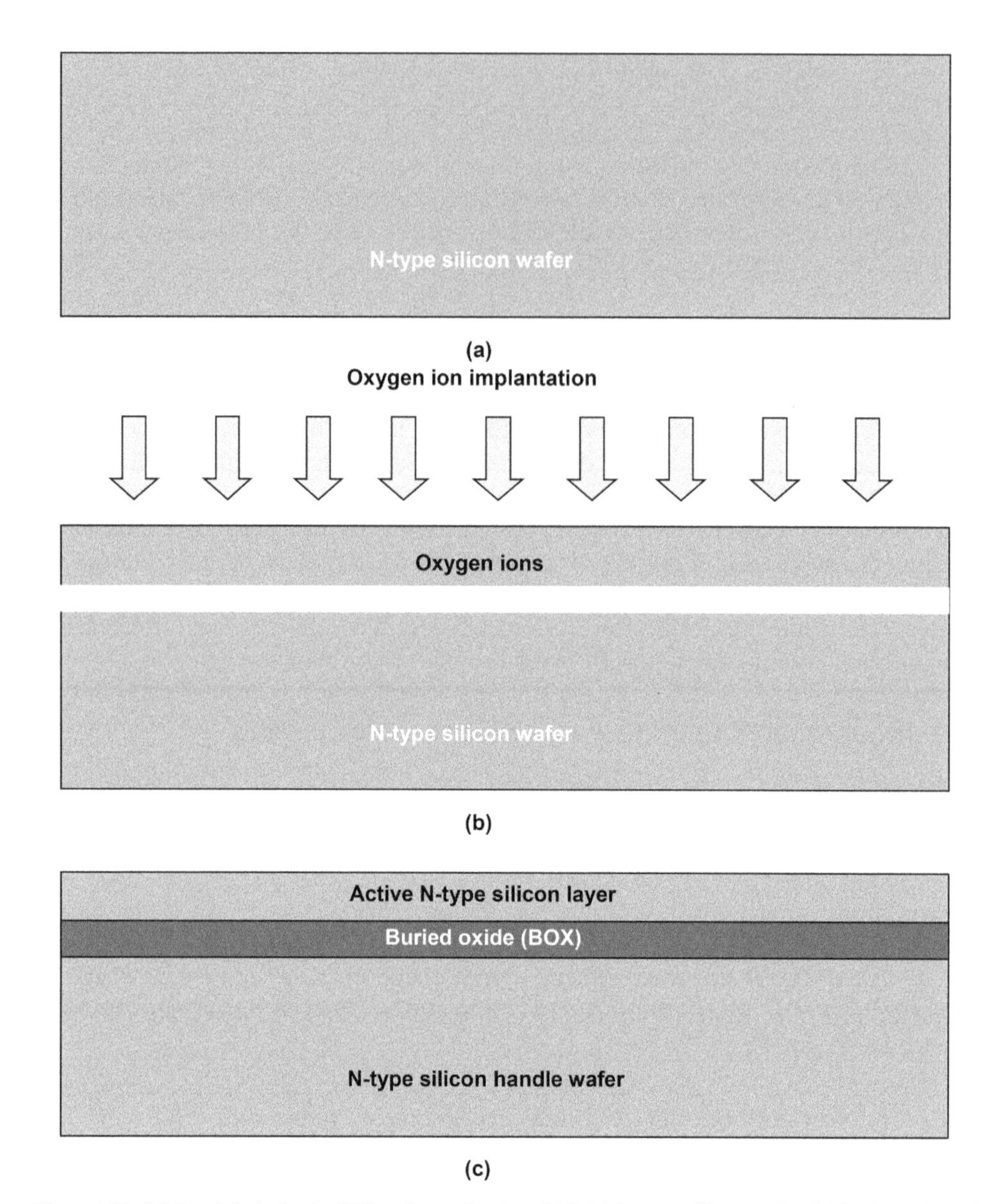

Figure 4.13. Dielectric isolation by SOI wafer production. (a) Initial n-type silicon wafer, (b) implantation of oxygen ions at 180 keV energy and 3×10^{17} to 1.8×10^{18} ions cm^{-2}, and (c) thermal annealing at 1 320 °C. The process is called separation by implantation of oxygen (SIMOX).

where I_S is the reverse saturation current in the dark ($= 1 \times 10^{-12}$ A), η is the emission coefficient or ideality factor ($1 \leqslant \eta \leqslant 2$), q is the electronic charge ($= 1.6 \times 10^{-19}$ C), k_B is the Boltzmann constant ($= 8.62 \times 10^{-5}$ eV K^{-1}) and T is the absolute temperature.

The emission coefficient η accounts for the decrease in current flowing through the diode by recombination of electrons and holes in the depletion region. When this

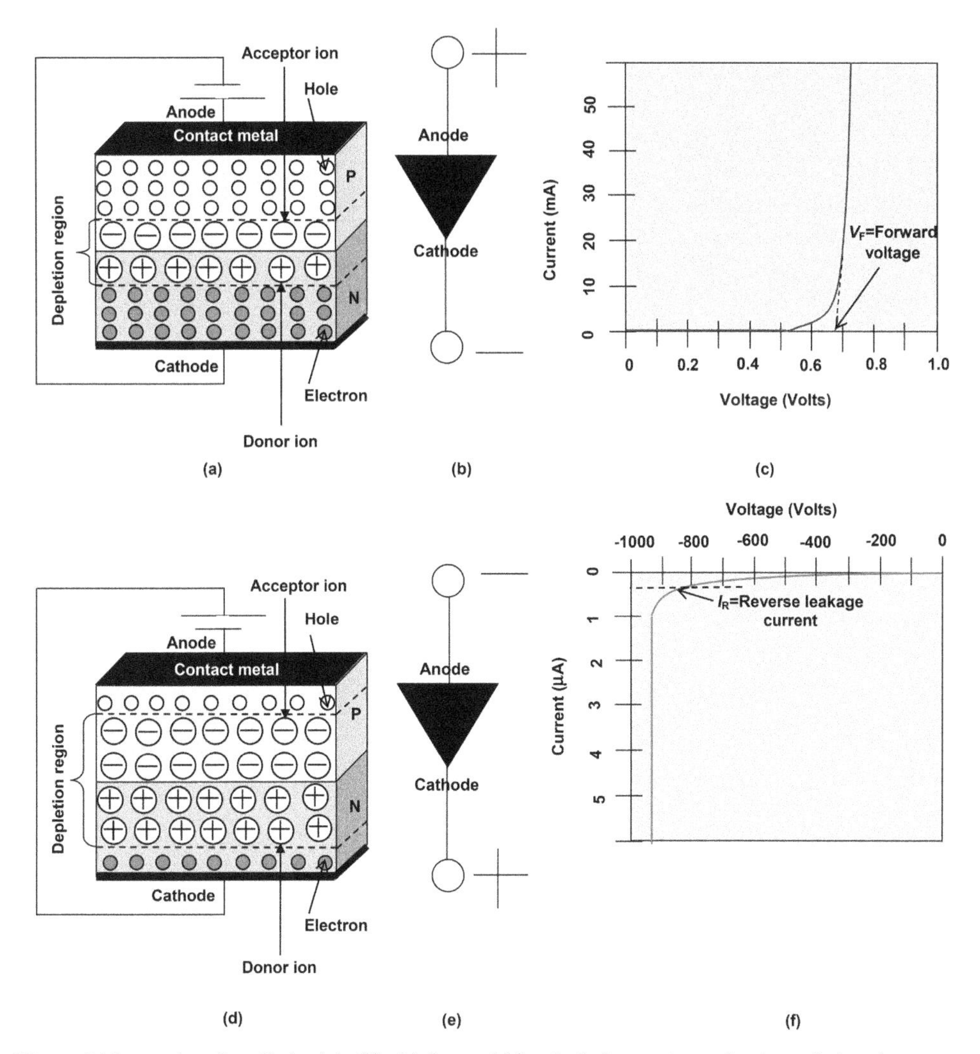

Figure 4.14. p–n junction diode. (a), (b), (c) forward-biased diode structure, circuit symbol and current–voltage characteristics; (d), (e), (f) the same for a reverse-biased diode.

recombination is small, $\eta = 1$. But if appreciable recombination takes place, $\eta = 2$. Therefore, the value of η depends upon two factors: (i) the semiconductor material and (ii) the doping concentrations of the p- and n-regions of the diode. Generally, $\eta = 1$ for indirect bandgap semiconductors, e.g. Si, Ge; $\eta = -2$ for direct bandgap semiconductors, e.g. GaAs, InP. Sometimes for Si diodes which form a sub-component of an IC $\eta = 1$, and for discrete silicon diodes $\eta = 2$.

The sub-expression k_BT/q has the dimensions of voltage. It is called thermal voltage V_{Thermal}. At room temperature

$$V_{\text{Thermal}} = (k_BT)/q = 25.9\ \text{mV}. \tag{4.32}$$

In terms of the thermal voltage V_{Thermal}, the diode equation (4.31) can be written as

$$i = I_S\left\{\exp\left(\frac{v}{\eta V_{\text{Thermal}}}\right) - 1\right\}. \tag{4.33}$$

The saturation current varies with temperature according to the relation (Leach 2004)

$$I_S = KT^{3/\eta}\exp\left(-\frac{E_G}{\eta V_{\text{Thermal}}}\right), \tag{4.34}$$

where K is a constant which is directly proportional to the cross-sectional area of the junction and E_G is the energy bandgap of the semiconductor in which the diode is fabricated.

4.4.2 Forward voltage drop across a p–n junction diode

Figure 4.15 shows the trend of variation of the forward current–voltage characteristics of a p–n junction diode with temperature changes.

Equation (4.33) is solved for the diode voltage v as follows:

$$\frac{i}{I_S} = \exp\left(\frac{v}{\eta V_{\text{Thermal}}}\right) - 1$$

or

$$\frac{i}{I_S} + 1 = \exp\left(\frac{v}{\eta V_{\text{Thermal}}}\right)$$

or

$$\ln\left(\frac{i}{I_S} + 1\right) = \frac{v}{\eta V_{\text{Thermal}}} \tag{4.35}$$

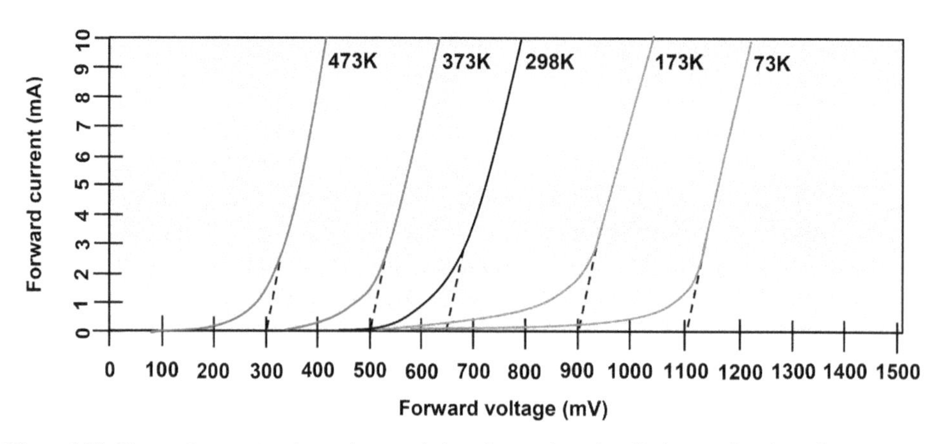

Figure 4.15. Forward current–voltage characteristics of a p–n junction diode as a function of temperature.

$$\therefore v = \eta V_{\mathrm{Thermal}} \ln\left(\frac{i}{I_{\mathrm{S}}} + 1\right) \approx \eta V_{\mathrm{Thermal}} \ln\left(\frac{i}{I_{\mathrm{S}}}\right). \tag{4.36}$$

In this equation, V_{Thermal} and I_{S} are temperature-dependent. Keeping i constant and applying the chain and quotient rules for derivatives, we have

$$\begin{aligned}
\frac{\mathrm{d}v}{\mathrm{d}T} &= \eta \frac{\mathrm{d}V_{\mathrm{Thermal}}}{\mathrm{d}T} \ln\left(\frac{i}{I_{\mathrm{S}}}\right) + \eta V_{\mathrm{Thermal}} \frac{\mathrm{d}}{\mathrm{d}T}\left\{\ln\left(\frac{i}{I_{\mathrm{S}}}\right)\right\} \\
&= \eta \frac{\mathrm{d}V_{\mathrm{Thermal}}}{\mathrm{d}T} \ln\left(\frac{i}{I_{\mathrm{S}}}\right) + \eta V_{\mathrm{Thermal}} \left\{1 \Big/ \left(\frac{i}{I_{\mathrm{S}}}\right)\right\}\left(\frac{-i}{I_{\mathrm{S}}^2}\right)\left(\frac{\mathrm{d}I_{\mathrm{S}}}{\mathrm{d}T}\right) \\
&= \eta \frac{\mathrm{d}V_{\mathrm{Thermal}}}{\mathrm{d}T} \ln\left(\frac{i}{I_{\mathrm{S}}}\right) - \eta V_{\mathrm{Thermal}} \left(\frac{1}{I_{\mathrm{S}}}\right)\left(\frac{\mathrm{d}I_{\mathrm{S}}}{\mathrm{d}T}\right).
\end{aligned} \tag{4.37}$$

For determination of $\mathrm{d}v/\mathrm{d}T$, it is necessary to find $\mathrm{d}V_{\mathrm{Thermal}}/\mathrm{d}T$ and $\mathrm{d}I_{\mathrm{S}}/\mathrm{d}T$,

$$\frac{\mathrm{d}V_{\mathrm{Thermal}}}{\mathrm{d}T} = \frac{\mathrm{d}}{\mathrm{d}T}\left(\frac{k_{\mathrm{B}}T}{q}\right) = \left(\frac{k_{\mathrm{B}}}{q}\right) \tag{4.38}$$

$$\begin{aligned}
\frac{\mathrm{d}I_{\mathrm{S}}}{\mathrm{d}T} &= \frac{\mathrm{d}}{\mathrm{d}T}\left\{KT^{3/\eta} \exp\left(-\frac{E_{\mathrm{G}}}{\eta V_{\mathrm{Thermal}}}\right)\right\} = \frac{\mathrm{d}}{\mathrm{d}T}\left\{KT^{3/\eta} \exp\left(-\frac{qE_{\mathrm{G}}}{\eta k_{\mathrm{B}}T}\right)\right\} \\
&= \frac{3}{\eta} KT^{(3/\eta)-1} \exp\left(-\frac{qE_{\mathrm{G}}}{\eta k_{\mathrm{B}}T}\right) + KT^{3/\eta}\left(\frac{qE_{\mathrm{G}}}{\eta k_{\mathrm{B}}T^2}\right) \exp\left(-\frac{qE_{\mathrm{G}}}{\eta k_{\mathrm{B}}T}\right) \\
&= \frac{3}{\eta T} KT^{3/\eta} \exp\left(-\frac{qE_{\mathrm{G}}}{\eta k_{\mathrm{B}}T}\right) + \left(\frac{qE_{\mathrm{G}}}{\eta k_{\mathrm{B}}T}\right)\left(\frac{1}{T}\right) KT^{3/\eta} \exp\left(-\frac{qE_{\mathrm{G}}}{\eta k_{\mathrm{B}}T}\right) \\
&= \frac{3}{\eta T} KT^{3/\eta} \exp\left\{-\frac{E_{\mathrm{G}}}{\eta\left(\frac{k_{\mathrm{B}}T}{q}\right)}\right\} + \left\{\frac{E_{\mathrm{G}}}{\eta\left(\frac{k_{\mathrm{B}}T}{q}\right)}\right\}\left(\frac{1}{T}\right) KT^{3/\eta} \exp\left\{-\frac{E_{\mathrm{G}}}{\eta\left(\frac{k_{\mathrm{B}}T}{q}\right)}\right\} \\
&= \frac{3}{\eta T} KT^{3/\eta} \exp\left(-\frac{E_{\mathrm{G}}}{\eta V_{\mathrm{Thermal}}}\right) + \left(\frac{E_{\mathrm{G}}}{\eta V_{\mathrm{Thermal}}}\right)\left(\frac{1}{T}\right) KT^{3/\eta} \exp\left(-\frac{E_{\mathrm{G}}}{\eta V_{\mathrm{Thermal}}}\right) \\
&= \frac{3}{\eta T} I_S + \left(\frac{E_{\mathrm{G}}}{\eta V_{\mathrm{Thermal}}}\right)\frac{1}{T} I_S = I_S\left(\frac{3}{\eta T} + \frac{E_{\mathrm{G}}}{\eta T V_{\mathrm{Thermal}}}\right).
\end{aligned} \tag{4.39}$$

Substituting for $\mathrm{d}V_{\mathrm{Thermal}}/\mathrm{d}T$ from equation (4.38) and $\mathrm{d}I_{\mathrm{S}}/\mathrm{d}T$ from equation (4.39) into equation (4.37), we obtain

$$\frac{\mathrm{d}v}{\mathrm{d}T} = \eta\left(\frac{k_B}{q}\right)\ln\left(\frac{i}{I_S}\right) - \eta V_{\text{Thermal}}\left(\frac{1}{I_S}\right)I_S\left(\frac{3}{\eta T} + \frac{E_G}{\eta T V_{\text{Thermal}}}\right). \tag{4.40}$$

But from equation (4.36)

$$v = \eta V_{\text{Thermal}}\ln\left(\frac{i}{I_S}\right) \tag{4.41}$$

$$\begin{aligned}\therefore \frac{\mathrm{d}v}{\mathrm{d}T} &= \frac{1}{T}\eta\left(\frac{k_B T}{q}\right)\ln\left(\frac{i}{I_S}\right) - \eta V_{\text{Thermal}}\left(\frac{3}{\eta T} + \frac{E_G}{\eta T V_{\text{Thermal}}}\right) \\ &= \frac{v}{T} - \frac{3V_{\text{Thermal}}}{T} - \frac{E_G}{T} = \frac{v - (3V_{\text{Thermal}} + E_G)}{T}.\end{aligned} \tag{4.42}$$

For a silicon diode, $v = 0.7$ V, $E_G = 1.11$ eV. Since $V_{\text{Thermal}} = 0.0259$ V, $T = 300$ K at room temperature, we have

$$\begin{aligned}\frac{\mathrm{d}v}{\mathrm{d}T} &= \frac{0.7 - (3 \times 0.0259 + 1.11)}{300} = -\frac{0.4877}{300} \\ &= -0.0016\ \text{V K}^{-1} = -1.6\ \text{mV K}^{-1} \approx -2\ \text{mV K}^{-1}.\end{aligned} \tag{4.43}$$

Thus the forward voltage dropped across a silicon p–n junction diode decreases by ~2 mV K^{-1} or 2 mV °C^{-1}.

The forward voltage drop across a diode increases by 2 mV for every 1 °C fall in temperature (Godse and Bakshi 2009). Hence, if the forward voltage of a Si diode is 0.71 V at 25 °C, and the temperature decreases to −55 °C, the forward voltage will increase by an amount $= 2 \times 10^{-3} \times (25 + 55) = 0.16$ V to $0.71 + 0.16$ V $= 0.87$ V.

4.4.3 Forward voltage of a Schottky diode

Figure 4.16 shows the schematic diagram of an SBD and its circuit diagram symbol. The Schottky trend of variation of the forward voltage with temperature is depicted in figure 4.17.

From classical thermionic emission theory, the static current (I_d)–voltage (V_d) characteristic of a Schottky diode is described by the equation (Cory 2009)

$$I_d = I_{D0}\left[\exp\{(qV_d)/(nk_B T)\} - 1\right] \tag{4.44}$$

where I_{D0} is the reverse saturation current given by

$$I_{D0} = K_{SB}T^2 \exp\{(-q\phi_B)/(k_B T)\} \tag{4.45}$$

where $K_{SB} = A^*A$, A^* is the effective Richardson constant, A is the effective area of the diode and ϕ_B is the height of the Schottky barrier. The constant n is the ideality

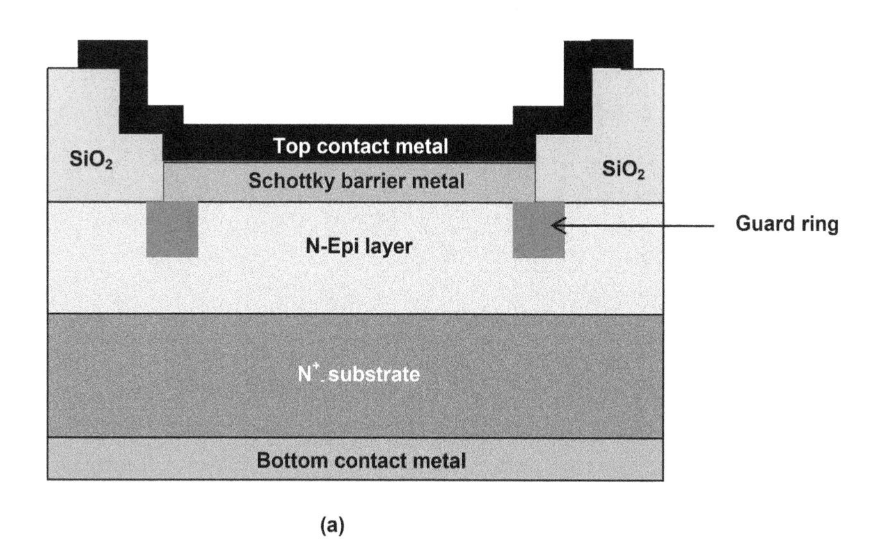

Figure 4.16. Schottky diode. (a) Cross-section and (b) circuit symbol.

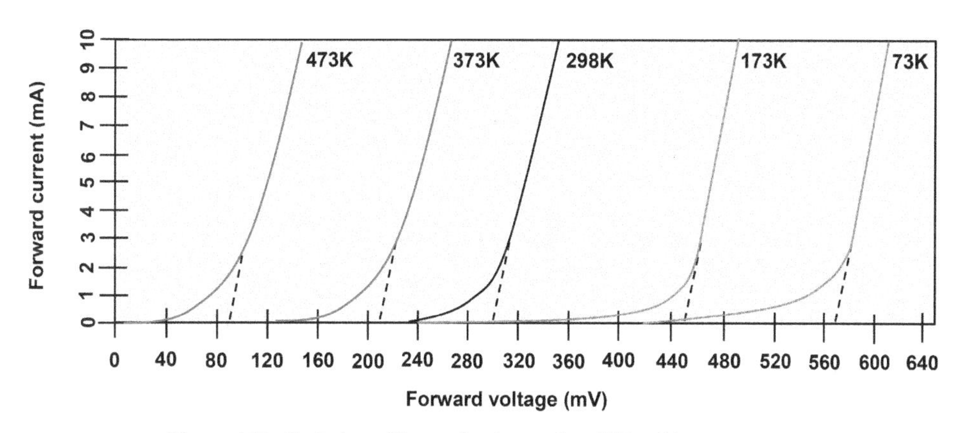

Figure 4.17. Variation of forward voltage of an SBD with temperature.

factor portraying the departure of the practical Schottky diode from theory; $n = 1$ for ideal diodes.

Taking $n = 1$, equation (4.44) is re-arranged as

$$I_d/I_{D0} = \exp\{(qV_d)/(k_BT)\} - 1. \tag{4.46}$$

Neglecting unity on the right-hand side, equation (4.46) reduces to

$$I_d/I_{D0} = \exp\{(qV_d)/(k_BT)\}. \tag{4.47}$$

Taking the natural logarithm of both sides

$$\ln(I_d/I_{D0}) = (qV_d)/(k_BT)$$

or

$$V_d = \{k_BT\ln(I_d/I_{D0})\}/q. \tag{4.48}$$

Differentiating both sides with respect to T, and keeping I_d constant,

$$\begin{aligned}\frac{dV_d}{dT} &= \left(\frac{k_B}{q}\right)\ln(I_d/I_{D0}) + \left(\frac{k_BT}{q}\right)\frac{1}{(I_d/I_{D0})} \times \frac{-I_d(dI_{D0}/dT)}{(I_{D0})^2}\\ &= \frac{1}{T}\left(\frac{k_BT}{q}\right)\ln(I_d/I_{D0}) - \left(\frac{k_BT}{q}\right)\frac{dI_{D0}/dT}{I_{D0}}\\ &= \frac{V_d}{T} - \left(\frac{k_BT}{q}\right)\frac{dI_{D0}/dT}{I_{D0}}.\end{aligned} \tag{4.49}$$

with the help of equation (4.48).

From equation (4.45)

$$\begin{aligned}\frac{dI_{D0}}{dT} &= \frac{d\left[K_{SB}T^2\exp\{(-q\phi_B)/(k_BT)\}\right]}{dT}\\ &= K_{SB} \times 2T\exp\{(-q\phi_B)/(k_BT)\}\\ &\quad + K_{SB}T^2\exp\{(-q\phi_B)/(k_BT)\} \times \left(\frac{-q}{k_B}\right) \times \frac{-\phi_B}{T^2}\\ &= \left(\frac{2}{T}\right)\left[K_{SB} \times T^2\exp\{(-q\phi_B)/(k_BT)\}\right]\\ &\quad + \left(\frac{1}{T}\right)K_{SB}T^2\exp\{(-q\phi_B)/(k_BT)\} \times \left(\frac{q}{k_B}\right) \times \frac{\phi_B}{T}\\ &= \left(\frac{2}{T}\right)I_{D0} + \left(\frac{1}{T}\right)I_{D0} \times \left(\frac{q\phi_B}{k_BT}\right)\end{aligned} \tag{4.50}$$

by using equation (4.45) again. Hence,

$$\frac{dI_{D0}}{dT} = \left(\frac{I_{D0}}{T}\right)\left\{2 + \left(\frac{q\phi_B}{k_BT}\right)\right\}. \tag{4.51}$$

Substituting for $\mathrm{d}I_{\mathrm{D0}}/\mathrm{d}T$ from equation (4.51) into equation (4.49), we obtain

$$\frac{\mathrm{d}V_{\mathrm{d}}}{\mathrm{d}T}=\frac{V_{\mathrm{d}}}{T}-\left(\frac{k_{\mathrm{B}}T}{q}\right)\frac{\left(\frac{I_{\mathrm{D0}}}{T}\right)\left\{2+\left(\frac{q\phi_{\mathrm{B}}}{k_{\mathrm{B}}T}\right)\right\}}{I_{\mathrm{D0}}}=\frac{V_{\mathrm{d}}}{T}-\left(\frac{k_{\mathrm{B}}T}{q}\right)\left(\frac{1}{T}\right)\left\{2+\left(\frac{q\phi_{\mathrm{B}}}{k_{\mathrm{B}}T}\right)\right\}. \tag{4.52}$$

For a Schottky diode fabricated using aluminum Schottky contact on silicon, $\phi_{\mathrm{B}} = 0.7$ eV at $T = 300$ K. If $V_{\mathrm{d}} = 0.4$ V,

$$\begin{aligned}\frac{\mathrm{d}V_{\mathrm{d}}}{\mathrm{d}T}&=\frac{0.4}{300}-(8.62\times10^{-5}\times300)\left(\frac{1}{300}\right)\left\{2+\left(\frac{0.7}{8.62\times10^{-5}\times300}\right)\right\}\\&=1.33\times10^{-3}-(8.62\times10^{-5})(2+27.0688)\\&=1.33\times10^{-3}-2.50573\times10^{-3}\\&=-1.17573\times10^{-3}\ \mathrm{V\ K^{-1}}\\&=-1.17573\ \mathrm{mV\ K^{-1}}\approx-1.2\ \mathrm{mV\ K^{-1}}.\end{aligned} \tag{4.53}$$

Thus the TC of the forward voltage drop for a Schottky diode ~ -1.2 mV K^{-1} is smaller than that for a p–n junction diode ~ -2 mV K^{-1}.

4.4.4 Reverse leakage current of a p–n junction diode

When a diode is reverse biased, a small current flows across the junction. This current is due the diffusion of minority carrier holes on the n-side and minority-carrier electrons on the p-side, across the junction. Hence, its magnitude depends on the diffusion coefficients of the minority carriers. Additionally, the carriers, which are generated in the depletion region, also contribute to this current flow. Within a certain limit, this current remains constant with increase of reverse voltage; hence it is called the reverse saturation current. But it is highly sensitive to temperature variations (figure 4.18). Its temperature sensitivity can be found as follows.

Since

$$\frac{\mathrm{d}\{\ln(I_{\mathrm{S}})\}}{\mathrm{d}T}=\left(\frac{1}{I_{\mathrm{S}}}\right)\frac{\mathrm{d}I_{\mathrm{S}}}{\mathrm{d}T} \tag{4.54}$$

putting the value of $\mathrm{d}I_{\mathrm{S}}/\mathrm{d}T$ from equation (4.39)

$$\begin{aligned}\frac{\mathrm{d}\{\ln(I_{\mathrm{S}})\}}{\mathrm{d}T}&=\left(\frac{1}{I_{\mathrm{S}}}\right)\frac{\mathrm{d}I_{\mathrm{S}}}{\mathrm{d}T}=\left(\frac{1}{I_{\mathrm{S}}}\right)I_{\mathrm{S}}\left(\frac{3}{\eta T}+\frac{E_{\mathrm{G}}}{\eta TV_{\mathrm{Thermal}}}\right)\\&=\frac{3}{\eta T}+\frac{E_{\mathrm{G}}}{\eta TV_{\mathrm{Thermal}}}\end{aligned}$$

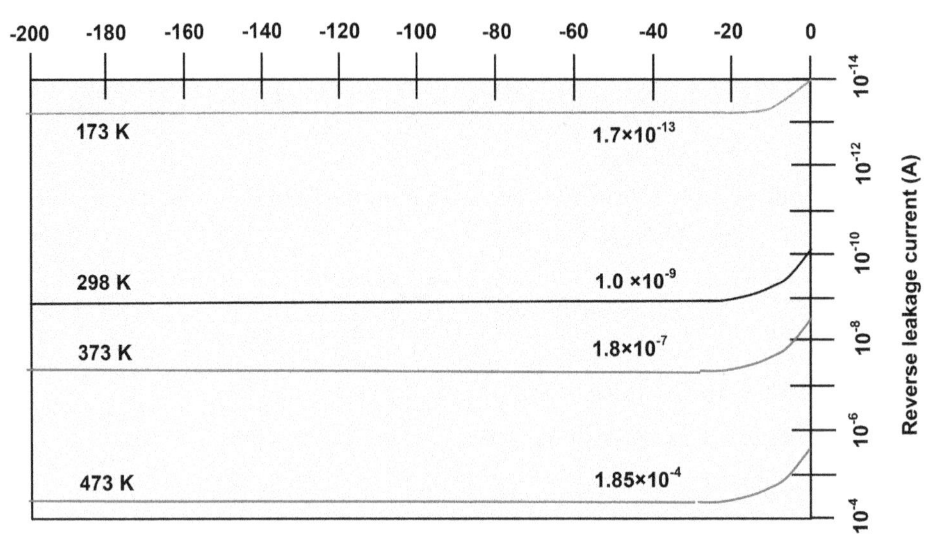

Figure 4.18. Reverse current–voltage characteristics of a p–n junction diode at different temperatures.

or

$$\frac{\Delta\{\ln(I_S)\}}{\Delta T} = \frac{3}{\eta T} + \frac{E_G}{\eta T V_{\text{Thermal}}} \tag{4.55}$$

$$\therefore \Delta\{\ln(I_S)\} = \left(\frac{3}{\eta T} + \frac{E_G}{\eta T V_{\text{Thermal}}}\right)\Delta T. \tag{4.56}$$

If $I_S = I_{S1}$ at temperature T and $I_S = I_{S2}$ at temperature $T + \Delta T$,

$$\ln\left(\frac{I_{S2}}{I_{S1}}\right) = \left(\frac{3}{\eta T} + \frac{E_G}{\eta T V_{\text{Thermal}}}\right)\Delta T$$

or

$$\frac{I_{S2}}{I_{S1}} = \exp\left\{\left(\frac{3}{\eta T} + \frac{E_G}{\eta T V_{\text{Thermal}}}\right)\Delta T\right\}. \tag{4.57}$$

For silicon diodes at room temperature $T = 300$ K, $E_G = 1.11$ eV, $V_{\text{Thermal}} = 0.0259$ V. Suppose $\Delta T = 10\,°\text{C}$, taking $\eta = 1$,

$$\begin{aligned}\frac{I_{S2}}{I_{S1}} &= \exp\left\{\left(\frac{3}{1 \times 300} + \frac{1.11}{1 \times 300 \times 0.0259}\right) \times 10\right\} \\ &= \exp\{(0.01 + 0.1429) \times 10\} \\ &= \exp(1.529) = 4.61.\end{aligned} \tag{4.58}$$

Taking $\eta = 2$,

$$\begin{aligned}\frac{I_{S2}}{I_{S1}} &= \exp\left\{\left(\frac{3}{2 \times 300} + \frac{1.11}{2 \times 300 \times 0.0259}\right) \times 10\right\} \\ &= \exp\{(0.005 + 0.07143) \times 10\} \\ &= \exp(0.7643) = 2.15.\end{aligned} \tag{4.59}$$

Thus, for every 10 °C rise of temperature, the leakage current of a silicon diode quadruples if $\eta = 1$ and doubles if $\eta = 2$. For $\eta = 2$, the reverse leakage current of a Si diode is halved for every 10 °C fall in temperature. Hence, if the reverse current is 1 nA at 25 °C, and the temperature goes down to −55 °C, the reverse current will decrease to 1/2 nA at 15 °C, 1/4 nA at 5 °C, 1/8 nA at −5 °C, 1/16 nA at −15 °C, 1/32 nA at −25 °C, 1/64 nA at −35 °C, 1/128 nA at −45 °C and finally 1/256 nA = 0.0039 nA = 3.9 pA at −55 °C. Note that the temperature decreases by 80 °C = 8 × 10 °C. Hence, the reverse current = $(1/2^8) \times 1$ nA = (1/256) × 1 nA = 3.9 pA.

The doubling of reverse leakage current for each 10 °C increase in temperature is mathematically expressed as

$$I_S(T) = I_S(T_0)2^{(T-T_0)/10°C}, \tag{4.60}$$

where $I_S(T)$ and $I_S(T_0)$ are the leakage currents at temperatures T, T_0, respectively.

4.4.5 Avalanche breakdown voltage of a p–n junction diode

The breakdown voltage of a p–n junction increases with an increase of temperature, as shown in figure 4.19.

At very high electric field called the critical field ($\approx 3 \times 10^5$ V cm^{-1} for Si), electrons (minority carriers) gain energy very fast. The rate of energy acquisition by electrons is higher than the rate at which they can lose energy through the emission of optical phonons in a reverse biased p–n junction. As a result, the high-energy electrons collide with bound electrons in the valence band. By collision, they excite the bonded electrons into the conduction band, creating an EPH. The phenomenon is termed impact ionization (Mukherjee 2011).

Avalanche breakdown is explained in terms of an ionization integral based on an important parameter, the ionization coefficient α. It is the average number of ionizing collisions accomplished by the carrier in traversing unit distance in the direction of electric field. The ionization integral becomes unity at the critical electric field

$$\int_0^W \alpha_p \exp\left\{\int_0^x (\alpha_n - \alpha_p)\mathrm{d}x\right\}\mathrm{d}x = 1, \tag{4.61}$$

where W is the width of the depletion region, and α_n, α_p are the ionization coefficients of electrons and holes respectively. To explore the variation of breakdown voltage with temperature, attention is directed particularly towards the ionization coefficient.

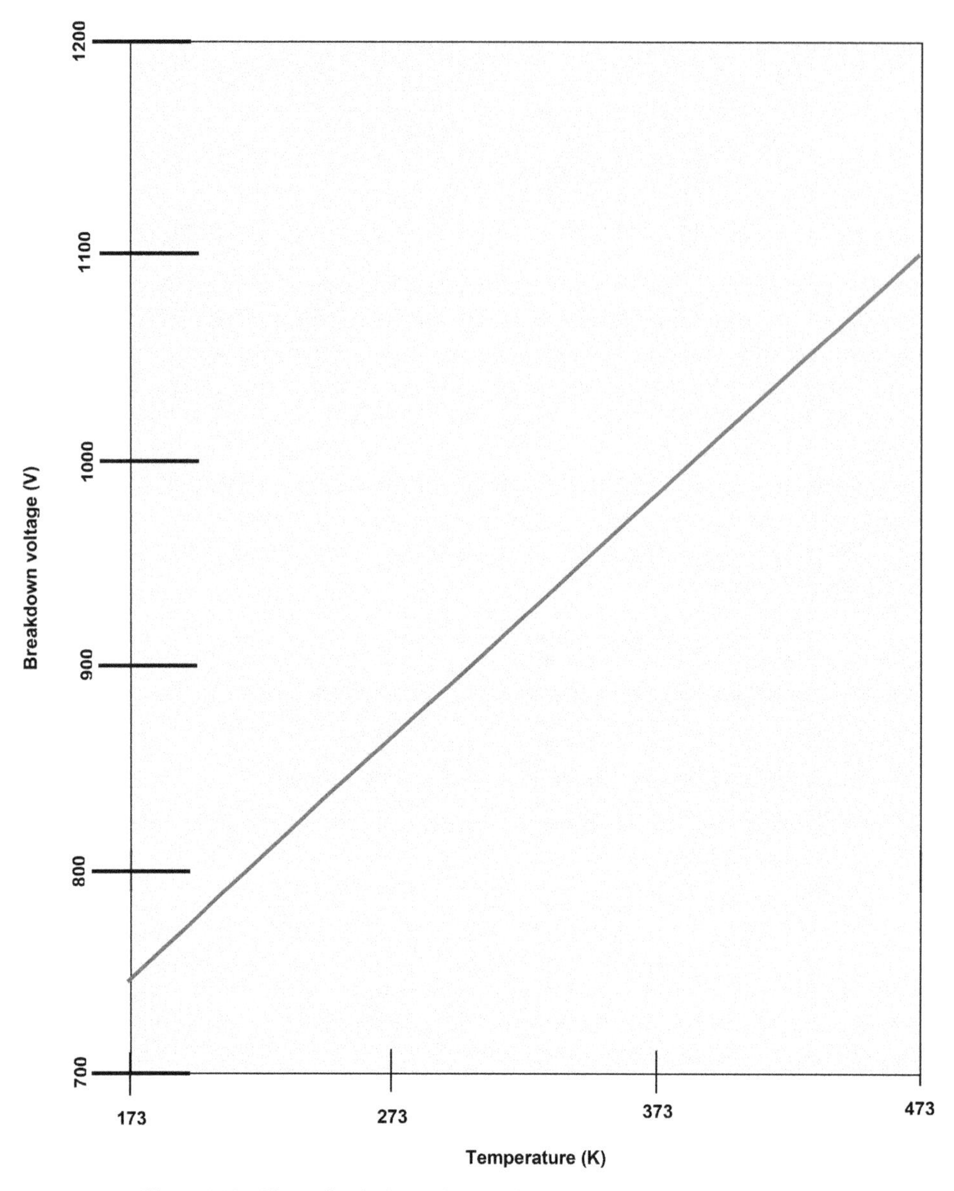

Figure 4.19. Change in the breakdown voltage of a diode with temperature.

Following the analysis by Chang *et al* (1971), the ionization rate α for an electron or a hole can be correlated with three material parameters:

(i) The ionization threshold energy W_I, i.e. the minimum energy required to cause an ionizing collision. It is approximately $1.5E_g$ for a parabolic band structure having the same effective masses for the carrier, if both energy and momentum conservation are taken into account (Maes *et al* 1990):

$$W_I = 1.5E_g. \tag{4.62}$$

(ii) The average energy loss due to optical phonon scattering (E_p):

$$\langle E_p \rangle = E_p \tanh\left(\frac{E_p}{2k_B T}\right), \tag{4.63}$$

where E_p is the optical phonon energy (= 0.063 eV for Si).

(iii) The average carrier mean free path (λ) due to the scattering given by the Cromwell–Sze equation (Crowell and Sze 1966):

$$\lambda = \lambda_0 \tanh\left(\frac{E_p}{2k_B T}\right), \tag{4.64}$$

where λ_0 is the high-energy low-temperature asymptotic phonon mean free path (= 76 Å for Si).

For Si, the average energy losses by scattering at 4.2, 77.2 and 300 K are calculated below:
at 4.2 K,

$$\langle E_p \rangle = 0.063 \tanh\left(\frac{0.063}{2 \times 8.617 \times 10^{-5} \times 4.2}\right) = 0.0630 \text{ eV}, \tag{4.65}$$

at 77.2 K,

$$\langle E_p \rangle = 0.063 \tanh\left(\frac{0.063}{2 \times 8.617 \times 10^{-5} \times 77.2}\right) = 0.06299 \text{ eV}, \tag{4.66}$$

at 300 K,

$$\langle E_p \rangle = 0.063 \tanh\left(\frac{0.063}{2 \times 8.617 \times 10^{-5} \times 300}\right) = 0.05287 \text{ eV}. \tag{4.67}$$

The average energy loss at 300 K is smaller than the average energy loss at 77.2 K, which is smaller than the average energy loss at 4.2 K. The change of average energy loss occurs mainly between 300 K and 77.2 K, and becomes less from 77.2 K to 4.2 K.

The corresponding mean free paths at these three temperatures are:
at 4.2 K,

$$\lambda = 76 \tanh\left(\frac{0.063}{2 \times 8.617 \times 10^{-5} \times 4.2}\right) = 76 \text{ Å}. \tag{4.68}$$

at 77.2 K,

$$\lambda = 76 \tanh\left(\frac{0.063}{2 \times 8.617 \times 10^{-5} \times 77.2}\right) = 75.9883 \text{ Å}, \tag{4.69}$$

at 300 K,

$$\lambda = 76 \tanh\left(\frac{0.063}{2 \times 8.617 \times 10^{-5} \times 300}\right) = 63.78053\ \text{Å}. \tag{4.70}$$

The mean free path at 4.2 K is larger than the mean free path at 77.2 K which is larger than the mean free path at 300 K. As before, the change occurs mainly between 300 K and 77.2 K, but is much smaller from 77.2 K to 4.2 K. Table 4.2 compiles the results of the calculations above.

A carrier gains energy from the electric field and loses part of its energy by optical phonon scattering. At a low temperature, the following events take place:

(i) E_g and hence W_I increase as T decreases: as $T\downarrow$, $W_I\uparrow$. Hence, it is more difficult to cause ionization.

(ii) The mean free path λ of the carrier is longer. As $T\downarrow$, $\lambda\uparrow$. Therefore, a carrier can gain more energy from the field. Hence, ionization becomes easier.

(iii) Average energy loss by scattering is higher. As $T\downarrow$, $\langle E_p\rangle\uparrow$. So, a carrier becomes less effective for ionization.

(iv) The scattering incidents are less frequent. The number of scattering incidents per unit time is smaller. As a result, energy losses through scattering become negligible. The loss of energy by few scattering events, although larger for each event at lower temperature than at higher temperature, cannot prevent the carriers from acquiring high energy values. This considerably offsets the effect of (iii).

The predominance of factors (ii) and (iv) over (i) and (iii) leads to a large increase in energy by acceleration between two consecutive scattering incidents, whereby a greater number of carriers succeed in reaching the ionization threshold energy W_I necessary to dislodge electrons and promote them from the valence band edge to the conduction band. Hence, the breakdown voltage decreases at low temperatures.

Thus the reason that the breakdown voltage decreases with a reduction in temperature is due to the increased mean free path of charge carriers from the reduced frequency of phonon scattering. As a result, higher energy can be gained in the electric field of the depletion region before the charge carriers collide with the lattice structure. The higher energy is the reason for impact ionization and device breakdown at lower temperatures.

Table 4.2. Average energy loss and mean free path at different temperatures.

Temperature (K)	Average energy loss (eV)	Mean free path (Å)
4.2	0.630	76.0
77.2	0.06299	75.99
300	0.053	63.78

4.4.6 Analytical model of temperature coefficient of avalanche breakdown voltage

A first-order model for impact ionization was formulated by Waldron *et al* (2005) starting from the Chynoweth formula (Chynoweth 1958) according to which the ionization coefficients of electrons and holes α_n, α_p are expressed as (Reisch 2003)

$$\alpha_n = a_n \exp(-b_n/E) \tag{4.71}$$

$$\alpha_p = a_p \exp(-b_p/E) \tag{4.72}$$

where a_n, b_n, a_p, b_p are fitting parameters dependent on the material and E is the intensity of the lateral electric field. In the presence of large electric fields perpendicular to the direction of motion of electrons, as in the channel of a MOSFET, the Chynoweth equations are modified as

$$\alpha_n = a_n \exp\{-(b_n|\mathbf{J}_n|)/(|\mathbf{E} \cdot \mathbf{J}_n|)\} \tag{4.73}$$

$$\alpha_p = a_p \exp\{-(b_p|\mathbf{J}_p|)/(|\mathbf{E} \cdot \mathbf{J}_p|)\} \tag{4.74}$$

where $\mathbf{J}_n$, $\mathbf{J}_p$ are the electron and hole current densities. In the present analysis, the same equation

$$\alpha = a \exp(-b/E) \tag{4.75}$$

will be used for both electrons and holes.

The calculation of ionization coefficients with local electric field dependence can be understood in terms of Shockley's lucky electron model of impact ionization. Shockley's model considers two principal mechanisms of energy loss by electrons, namely by optical phonon scattering and by generation of an EHP as soon as the energy of the electron exceeds the ionization threshold energy W_I. Shockley interpreted the parameter b in Chynoweth equations as:

$$b = \frac{W_I(T)}{q\lambda(T)} \tag{4.76}$$

for a mean-free path λ of an electron of charge q. The mean-free path is given by the Cromwell–Sze equation (4.64).

From modeling of threshold ionization energy by Ershov and Ryzhii (1995), based on the assumption of its proportionality to the bandgap energy of silicon, we can write

$$W_I(T) = C_1 + C_2T + C_3T^2, \tag{4.77}$$

where

$C_1 = 1.1785$ eV,

$C_2 = -9.025 \times 10^{-5}$ eV K^{-1},

$C_3 = -3.05 \times 10^{-7}$ eV K^{-2} for $T > 170$ K.

The temperature dependence of the ionization coefficient is defined by the TC as

$$\mathrm{TC} = (1/\alpha)(\mathrm{d}\alpha/\mathrm{d}T). \tag{4.78}$$

Substituting for α from equation (4.75) into equation (4.78), we obtain

$$\begin{aligned}
\mathrm{TC} &= \frac{1}{a\exp(-b/E)} \times \frac{\mathrm{d}}{\mathrm{d}T}\{a\exp(-b/E)\} \\
&= \frac{1}{a\exp(-b/E)} \times a\exp(-b/E) \times \frac{\mathrm{d}}{\mathrm{d}T}(-b/E) \\
&= \frac{\mathrm{d}}{\mathrm{d}T}(-b/E) = -\frac{1}{E}\left(\frac{\mathrm{d}b}{\mathrm{d}T}\right) = -\frac{1}{E}\frac{\mathrm{d}}{\mathrm{d}T}\left\{\frac{W_{\mathrm{I}}(T)}{q\lambda(T)}\right\} \\
&= -\frac{1}{qE}\frac{\mathrm{d}}{\mathrm{d}T}\left\{\frac{W_{\mathrm{I}}(T)}{\lambda(T)}\right\} \\
&= -\frac{1}{qE}\frac{(\mathrm{d}W_{\mathrm{I}}/\mathrm{d}T)\lambda - W_{\mathrm{I}}(\mathrm{d}\lambda/\mathrm{d}T)}{\lambda^2} \\
&= -\left(\frac{1}{qE}\right)\left\{\frac{(\mathrm{d}W_{\mathrm{I}}/\mathrm{d}T)\lambda}{\lambda^2}\right\} + \left(\frac{1}{qE}\right)\left\{\frac{W_{\mathrm{I}}(\mathrm{d}\lambda/\mathrm{d}T)}{\lambda^2}\right\} \\
&= \left(\frac{W_{\mathrm{I}}}{qE\lambda}\right)\left\{\frac{1}{\lambda}\left(\frac{\mathrm{d}\lambda}{\mathrm{d}T}\right) - \frac{1}{W_{\mathrm{I}}}\left(\frac{\mathrm{d}W_{\mathrm{I}}}{\mathrm{d}T}\right)\right\}.
\end{aligned} \tag{4.79}$$

The first term inside the curly brackets in equation (4.79) is the TC of λ. From equation (4.64)

$$\begin{aligned}
\frac{1}{\lambda}\left(\frac{\mathrm{d}\lambda}{\mathrm{d}T}\right) &= \left(\frac{1}{\lambda}\right)\frac{\mathrm{d}}{\mathrm{d}T}\left\{\lambda_0 \tanh\left(\frac{E_{\mathrm{p}}}{2k_{\mathrm{B}}T}\right)\right\} \\
&= \left(\frac{\lambda_0}{\lambda}\right)\frac{E_{\mathrm{p}}}{2k_{\mathrm{B}}}\left\{1 - \tanh^2\left(\frac{E_{\mathrm{p}}}{2k_{\mathrm{B}}T}\right)\right\} \times \frac{0-1}{T^2} \\
&= -\left(\frac{\lambda_0}{\lambda}\right)\frac{E_{\mathrm{p}}}{2k_{\mathrm{B}}T^2}\,\mathrm{sech}^2\left(\frac{E_{\mathrm{p}}}{2k_{\mathrm{B}}T}\right)
\end{aligned} \tag{4.80}$$

since

$$\tanh^2 x + \mathrm{sech}^2 x = 1. \tag{4.81}$$

The second term inside curly brackets in equation (4.79) is the TC of W_{I}. From equation (4.77)

$$\begin{aligned}
\frac{1}{W_{\mathrm{I}}}\left(\frac{\mathrm{d}W_{\mathrm{I}}}{\mathrm{d}T}\right) &= \left(\frac{1}{W_{\mathrm{I}}}\right)\frac{\mathrm{d}}{\mathrm{d}T}(C_1 + C_2T + C_3T^2) \\
&= \left(\frac{1}{W_{\mathrm{I}}}\right)(0 + C_2 \times 1 + C_3 \times 2T^{2-1}) = \left(\frac{1}{W_{\mathrm{I}}}\right)(C_2 + 2C_3T).
\end{aligned} \tag{4.82}$$

Substituting the values of the TC of λ and W_I from equations (4.80) and (4.82) into equation (4.79), we have

$$\begin{aligned}\mathrm{TC} &= -\left(\frac{W_I}{qE\lambda}\right)\left\{\left(\frac{\lambda_0}{\lambda}\right)\left(\frac{E_p}{2k_BT^2}\right)\mathrm{sech}^2\left(\frac{E_p}{2k_BT}\right) + \left(\frac{1}{W_I}\right)(C_2 + 2C_3T)\right\} \\ &= -\left(\frac{1}{E}\right)\left[\left(\frac{W_I\lambda_0E_p}{2q\lambda^2k_BT^2}\right)\mathrm{sech}^2\left(\frac{E_p}{2k_BT}\right) + \left(\frac{1}{q\lambda}\right)(C_2 + 2C_3T)\right]\end{aligned} \tag{4.83}$$

TC of λ is negative. Because both C_2 and C_3 are negative, the TC of W_I is also negative. Hence, the sign of TC of breakdown voltage is determined by the comparative magnitudes of the two terms in the above equation. Whether TC is positive or negative will be clear by performing an example calculation.

For Si at room temperature $T = 300$ K

$$\begin{aligned}W_I(T) &= 1.1785 + (-9.025 \times 10^{-5}) \times 300 + (-3.05 \times 10^{-7}) \times (300)^2 \\ &= 1.1785 - 2.7075 \times 10^{-2} - 2.745 \times 10^{-2} = 1.123975\ \mathrm{eV}\end{aligned} \tag{4.84}$$

$$\lambda_0 = 7.6\mathrm{nm},\ E_p = 0.053\mathrm{eV},\ q = 1.6 \times 10^{-19}\mathrm{C},$$
$$\lambda = 6.378\mathrm{nm},\ k_B = 1.381 \times 10^{-23}J\mathrm{K}^{-1}.$$

Putting the above values in equation (4.83)

$$\begin{aligned}\mathrm{TC} &= -\left(\frac{1}{E}\right)\left[\left\{\frac{1.123975 \times 1.6 \times 10^{-19} \times 7.6 \times 10^{-9} \times 0.053 \times 1.6 \times 10^{-19}}{2 \times 1.6 \times 10^{-19} \times (6.378 \times 10^{-9})^2 \times (1.381 \times 10^{-23})(300)^2}\right\}\right. \\ &\quad \mathrm{sech}^2\left\{\frac{0.053 \times 1.6 \times 10^{-19}}{2 \times (1.381 \times 10^{-23}) \times 300}\right\} \\ &\quad +\left(\frac{1}{1.6 \times 10^{-19} \times 6.378 \times 10^{-9}}\right)\left\{-9.025 \times 10^{-5} \times 1.6 \times 10^{-19}\right. \\ &\quad \left.\left.+(2 \times -3.05 \times 10^{-7} \times 1.6 \times 10^{-19} \times 300)\right\}\right] \\ &= -\left(\frac{1}{E}\right)\left[\left\{\frac{1.159 \times 10^{-47}}{1.6179 \times 10^{-53}}\right\}\mathrm{sech}^2\left\{\frac{8.48 \times 10^{-21}}{8.286 \times 10^{-21}}\right\}\right. \\ &\quad \left.+\left(\frac{1}{1.02048 \times 10^{-27}}\right)\left\{-1.444 \times 10^{-23} - 2.928 \times 10^{-23}\right\}\right] \\ &= -\left(\frac{1}{E}\right)\left[(7.163607145 \times 10^5)\mathrm{sech}^2(1.023412985759)\right. \\ &\quad \left.+ (9.799310128 \times 10^{26})(-4.372 \times 10^{-23})\right] \\ &= -\left(\frac{1}{E}\right)\left[(7.163607145 \times 10^5)(0.6365287)^2 - (42842.58388)\right] \\ &= -\left(\frac{1}{E}\right)[290247.000977 - 42842.58388] \\ &= -\left(\frac{1}{E}\right)\left(\frac{247404.4171}{E}\right)\mathrm{V\ m^{-1}\ K^{-1}} \approx -\left(\frac{1}{E}\right)2474\ \mathrm{V\ cm^{-1}\ K^{-1}}.\end{aligned} \tag{4.85}$$

The calculation shows that (i) silicon has a negative TC of breakdown voltage and (ii) the small value of the TC reveals that the temperature dependence is a weak one.

4.4.7 Zener breakdown voltage of a diode

The reverse breakdown voltage of an avalanche diode decreases with decrease in temperature while that of a Zener diode shows the opposite behavior (table 4.3).

4.4.8 Storage time (t_s) of a p^+–n junction diode

The storage time is a principal figure of merit for characterizing the transient behavior of a diode. The storage time t_s of a p^+–n diode is expressed in terms of the forward current I_F, reverse current I_R and hole lifetime τ_p as (Dokić and Blanuša 2015)

$$t_s = \tau_p \ln\left(1 + \frac{I_F}{I_R}\right). \quad (4.86)$$

Table 4.3. Temperature dependence of avalanche and Zener breakdown voltages.

Feature	Avalanche breakdown	Zener breakdown
Type of junction	Lightly doped	Heavily doped
Width of depletion region	Large	Small
Electric field across depletion region	Low	High
Range	Breakdown voltage $>6\ E_g/q = 6 \times 1.11 = 6.66$ V	Breakdown voltage $< 4E_g/q = 4 \times 1.11 = 4.44$ V
TC of breakdown voltage	Positive and increases with an increase in breakdown voltage, e.g. from 3 to 6 mV K^{-1} for an 8 V diode to 13–18 mV K^{-1} for an 18 V diode.	Negative and independent of the breakdown voltage rating; typically −3 mV K^{-1}.
Thermal mechanism	As the temperature rises, the vibrational displacements of lattice atoms about their equilibrium positions increase. Hence, the carriers suffer more collisions from lattice atoms, and cannot be easily accelerated to high velocities to cause ionizations of lattice atoms. Greater voltage is required to trigger an avalanche.	With increasing temperature, the energies of valence electrons in silicon atoms increase. They can easily tear off from the covalent bonds. Less voltage is required to liberate these electrons from the bonds to participate in conduction. Therefore, the breakdown voltage decreases as the temperature increases because the bandgap decreases.

The storage time is small if the hole lifetime is low; the forward current I_F is low because a smaller amount of charge is to be removed; and the reverse current I_R is high because the stored charge can be swept away at a faster rate. Hole lifetimes $\tau_p(T)$, $\tau_p(T_0)$ at two different temperatures T, T_0 are inter-related as (Dokić and Blanuša 2015)

$$\tau_p(T) = \tau_p(T_0)(T/T_0)^r \tag{4.87}$$

where r is a constant = 3.5 for Si and 2.2 for Ge diodes at low levels of carrier injection. For a temperature increase from −55 °C to +175 °C, the carrier lifetime

$$\tau_p(175 + 273\ \text{K}) = \tau_p(-55 + 273)\{(175 + 273)/(-55 + 273)\}^{3.5}$$

or

$$\tau_p(448\ \text{K}) = \tau_p(218)(448/218)^{3.5} = 12.44\ \tau_p(218). \tag{4.88}$$

Since, the carrier lifetime in silicon increases by a factor larger than 12, the storage time is considerably lengthened. This stretching of storage time is manifested as an increase in switching loss.

4.4.9 Current gain of a bipolar junction transistor

Figure 4.20 presents the schematic diagram and circuit symbols of a BJT.

The DC current gain–collector curves of a BJT show a decline in current gain with decreasing temperature, as shown in figure 4.21.

Buhanan performed a comprehensive theoretical analysis of the current gain equation using the piece-by-piece methodology (Buhanan 1969). For examination of the temperature dependence of current gain, the common-base current gain α serves as a convenient platform. Since the current gain β of an n–p–n BJT in the common-emitter configuration,

$$\beta = \text{collector current } (I_C)/\text{base current } (I_B), \tag{4.89}$$

is related to its current gain α in the common-base connection

$$\alpha = \text{collector current } (I_C)/\text{emitter current } (I_E) \tag{4.90}$$

by the familiar equation

$$\beta = \alpha/(1 - \alpha), \tag{4.91}$$

the temperature dependence for β becomes evident from that of α. The common-base current gain α is a product of three factors:

(i) The emitter injection efficiency γ given by Kauffman and Bergh (1968)

$$\gamma = I_{nB}/\left(I_{nB} + I_{pE}\right), \tag{4.92}$$

where I_{nB} is the electron current from the n^+-emitter to the p-base and I_{pE} is the hole current from the p-base to the n^+-emitter, so that the injection efficiency is the ratio of electron current flowing from emitter to base to the total current (electron current + hole current) moving across the emitter–base junction. The notation used for currents designates the transistor

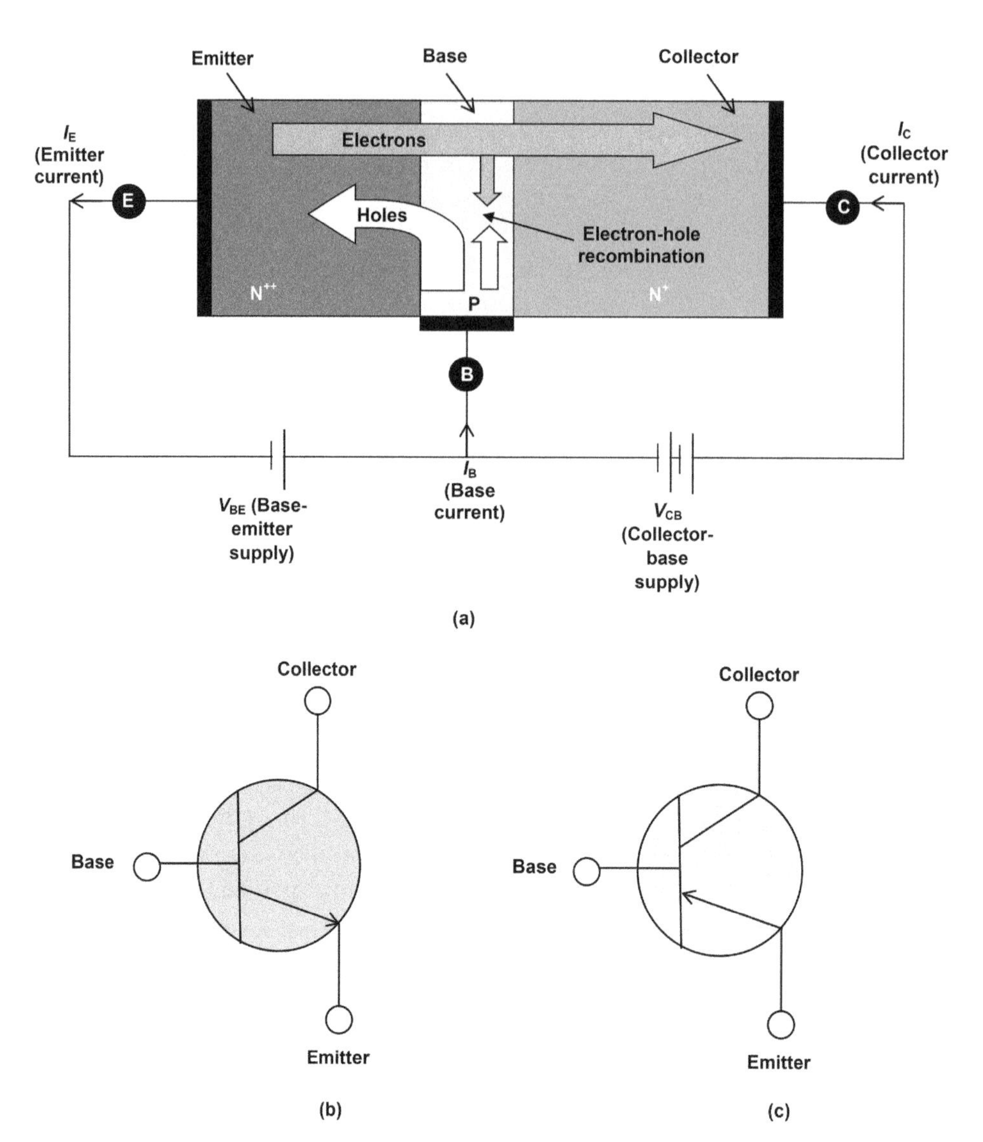

Figure 4.20. BJT. (a) Schematic cross-section of an n–p–n transistor and (b) circuit diagram symbols of n–p–n and p–n–p transistors, respectively.

terminal (emitter, base or collector) into which the electrons (n)/holes (p) are injected.

(ii) The base transport factor α_T written as

$$\alpha_T = I_{nC}/I_{nB} \tag{4.93}$$

where I_{nC} is the electron current in the collector.

(iii) The collector multiplication ratio M expressed as

$$M = I_C/I_{nC} = \left(I_{nC} + I_{pC}\right)/I_{nC}, \tag{4.94}$$

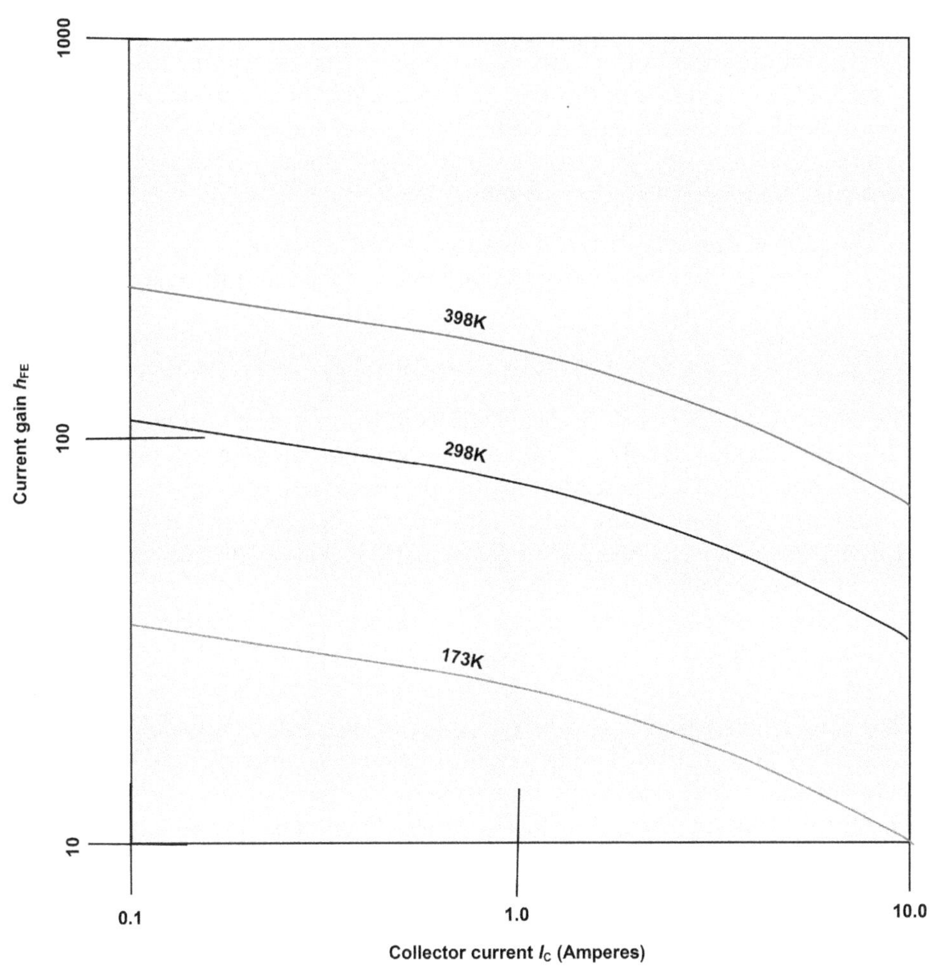

Figure 4.21. Shifts in the plots of DC current gain versus collector current of a Si bipolar transistor with temperature.

where I_{pC} is the hole current in the collector. Thus

$$\alpha = \gamma \alpha_T M. \tag{4.95}$$

At voltages $\ll BV_{CEO}$, $M = 1$; hence

$$\alpha = \gamma \alpha_T = \left\{1 + (D_{pE}/D_{nB})(p_{nE}/n_{pB})(W/L_{pE})\right\}^{-1} \times \{\mathrm{sech}(W/L_{nB})\}, \tag{4.96}$$

where the first reciprocal term in curly brackets preceding the multiplication sign (×) represents γ while the second term after this sign gives α_T. The symbols have the following meanings: D_{pE} = the diffusion constant of minority-carrier holes in the n$^+$-emitter; D_{nB} = the diffusion constant of minority-carrier electrons in the p-base; p_{nE} = the number of holes in the n$^+$-emitter; n_{pB} = the number of electrons in the p-base; W = the base width; L_{pE} = the diffusion length of holes in the emitter and L_{nB} = the diffusion length of electrons in the base. Let us examine the behavior of each term and every factor in this equation individually to understand the effect of temperature on the current gain. Combining together the effects of temperature on each term separately will help us to visualize the influence of temperature on the current gain when considered as a whole.

(i) The ratio of minority-carrier diffusion constants (D_{pE}/D_{nB})
(a) D_{pE} is related to the mobility μ_{pE} of holes in the emitter through the Einstein equation

$$D_{pE} = \mu_{pE}\{(k_B T)/q\}. \tag{4.97}$$

The emitter of a transistor is a heavily doped region with electron concentration $>10^{19}\ cm^{-3}$. In this heavily doped region, impurity scattering will dominate over lattice scattering. In lattice scattering, mobility decreases as temperature increases, but impurity scattering shows the reverse trend. Therefore, the predominance of impurity scattering will make the hole mobility increase with temperature. Thus the diffusion constant increases with temperature.
(b) D_{nB} is connected with the mobility of electrons μ_{nB} in the base via the equation

$$D_{nB} = \mu_{nB}\{(k_B T)/q\}. \tag{4.98}$$

The base of a transistor is a moderately doped region. So, lattice scattering will foreshadow impurity scattering. Lattice scattering increases as temperature increases, and thereby decreases the mobility. So, the mobility μ_{nB} will decrease with temperature increase. As the D_{nB} equation contains the factor T, which is itself increasing, the decrease of μ_{nB} will be compensated by the increase of T. So, D_{nB} will remain constant.
In view of (a) and (b) above, the ratio (D_{pE}/D_{nB}) will increase with an increase in temperature.

(ii) The ratio of minority carrier concentrations (p_{nE}/n_{pB})
A high impurity concentration in the emitter produces deformations in the lattice and other defects. These defects lower the silicon bandgap in the emitter by an amount $\Delta\xi_g$ from the initial value ξ_g to the final value $\xi_g - \Delta\xi_g$. Without bandgap narrowing, the minority-carrier concentrations are

$$p_{nE} = n_i^2/N_D, \tag{4.99}$$

where N_D is the donor concentration in the emitter, and

$$n_{pB} = n_i^2/N_A, \tag{4.100}$$

where N_A is the acceptor concentration in the base. Also, the intrinsic carrier concentration n_i is

$$n_i^2 = \text{constant} \times T^3 \exp\{-\xi_g/(2k_B T)\}. \tag{4.101}$$

After bandgap reduction, the intrinsic carrier concentration expression changes to

$$n_i^2 = \text{constant} \times T^3 \exp\left\{-\left(\xi_g - \Delta\xi_g\right)/(2k_B T)\right\}. \tag{4.102}$$

Hence, after bandgap narrowing

$$p_{nE} = \left[\text{constant } A \times T^3 \exp\left\{-\left(\xi_g - \Delta\xi_g\right)/(2k_B T)\right\}\right]/N_D. \tag{4.103}$$

But unlike the emitter, there is no bandgap narrowing in the moderately doped base region. Hence, the equation for n_{pB} contains the initial bandgap ξ_g:

$$n_{pB} = \left[\text{constant } B \times T^3 \exp\left\{-\xi_g/(2k_B T)\right\}\right]/N_A. \tag{4.104}$$

So, the minority carrier concentration ratio is

$$\begin{aligned} p_{nE}/n_{pB} &= \frac{\left[\text{constant } A \times T^3 \exp\left\{-\left(\xi_g - \Delta\xi_g\right)/(2k_B T)\right\}\right]/N_D}{\left[\text{constant } B \times T^3 \exp\left\{-\xi_g/(2k_B T)\right\}\right]/N_A} \\ &= \frac{\text{constant } A}{\text{constant } B} \times \exp\left\{\left(-\xi_g + \Delta\xi_g + \xi_g\right)/(2k_B T)\right\} \\ &\times \frac{N_A}{N_D} = \frac{AN_A}{BN_D} \times \exp\left\{\Delta\xi_g/(2k_B T)\right\} \\ &= K \exp\left\{\Delta\xi_g/(2k_B T)\right\}, \end{aligned} \tag{4.105}$$

where $K = AN_A/(BN_D)$. This equation shows that as the temperature decreases, the minority carrier concentration ratio increases exponentially, thereby causing a substantial reduction in current gain. As temperature increases, the current gain is enhanced. The exponential nature of variation lends this factor a prominent role in determining the temperature-induced variation of current gain.

(iii) Ratio of base width to diffusion length of minority carriers in the emitter (W/L_{pE})
The base width W is determined by the base profile. The base profile depends on the ionization of diffused impurity atoms in the base region. The acceptor level introduced by boron is very proximate to the edge of the valence band, around 0.045 eV above the valence band edge. Up to 100 K or −173 °C, all the impurity atoms may be considered to be ionized so that the free carrier concentration equals the concentration of diffused impurity atoms. So, at temperatures above 100 K, it may be safely assumed that all the impurity atoms are ionized and the base width is constant. For this temperature range, base width W has no role to play in the temperature dependence of current gain. The diffusion length L_{pE} is given by

$$L_{pE} = \sqrt{D_{pE}\tau_{pE}}, \tag{4.106}$$

where τ_{pE} is the minority-carrier lifetime of holes in the emitter. As already discussed, the diffusion constant D_{pE} varies directly with temperature, decreasing with a fall in temperature. Further, τ_{pE} also decreases with temperature. Therefore, the overall effect is that the factor (W/L_{pE}) increases as temperature decreases. Hence, the emitter injection efficiency and thereby the current gain are degraded.

(iv) Hyperbolic secant of the ratio of base width to diffusion length of minority carriers in the base {sech(W/L_{nB})}

The diffusion length L_{nB} is

$$L_{nB} = \sqrt{D_{nB}\tau_{nB}}. \tag{4.107}$$

As already pointed out above, the base is a lightly doped region in a transistor structure. Hence, lattice scattering is preponderant and impurity scattering is less pronounced. Lattice scattering decreases as temperature falls. Therefore, the mobility μ_{nB} increases with decrease of temperature. But the temperature T itself decreases. Accordingly, the diffusion constant D_{nB} given by equation (4.98) remains roughly constant. Since, τ_{nB} decreases when temperature is decreased, the parameter L_{nB} also declines. Hence, (W/L_{nB}) increases. Since we know that hyperbolic secant of a function decreases as the value of the function increases, the current gain will decrease.

The results of the above analysis (i)–(iv) are summarized in table 4.4.

Thus, on the whole, decreasing the temperature has a deprecating influence on the current gain of a bipolar transistor. In practice, the current gain falls by 0.3–0.6% K^{-1}. It was also inferred by Buhanan (1969) that as transistor manufacturing was confined within a few orders of magnitude of doping concentrations, all the effects, whether advantageous or disadvantageous, acquire a secondary role in comparison to that of p_{nE}/n_{pB}. Therefore, the current gain degradation produced by heavy doping-induced bandgap narrowing (BGN) in the emitter by way of modifying the intrinsic carrier concentration is worthy of further attention.

4.4.10 Approximate analysis

Admittedly, the reduction in current gain with falling temperature is a result of the smaller bandgap in the emitter than in the base. The base current is typically limited by hole injection into the emitter, and its temperature dependence is largely determined by the effective bandgap of the emitter, which is reduced because of a

Table 4.4. Effects of decreasing temperature on a transistor's parameters and its current gain.

Parameter	Effect of decreasing temperature	Effect on current gain
D_{pE}/D_{nB}	Decreases	Enhancement
p_{nE}/n_{pB}	Increases exponentially	Strong degradation
W/L_{pE}	Increases	Degradation
sech(W/L_{nB})	Decreases	Degradation

combination of heavy doping effects. The collector current, on the other hand, consists of electrons traversing the base and its temperature dependence is therefore determined by the bandgap in the base. The difference in bandgap between emitter and base causes the ideal gain to decrease exponentially with temperature by 2 to 3 orders of magnitude between room (~300 K) and liquid-nitrogen (=80 K) temperatures.

Neglecting the space charge recombination current (non-ideal base current) and assuming unity for the base transport factor, we can write the following equations:

$$I_{\mathrm{nB}} = I_{\mathrm{C}} \tag{4.108}$$

$$I_{\mathrm{pE}} = I_{\mathrm{B}} \tag{4.109}$$

$$h_{\mathrm{fE}} = \frac{I_{\mathrm{C}}}{I_{\mathrm{B}}} = \frac{I_{\mathrm{nB}}}{I_{\mathrm{pE}}}. \tag{4.110}$$

The temperature dependence of both I_{nB} and I_{pE} is dominated by the temperature dependence of the intrinsic carrier concentration n_{i} in the emitter and base regions, respectively.

But

$$n_{\mathrm{i}}^2 \propto \exp\left(-\frac{\xi_{\mathrm{g}}}{k_{\mathrm{B}}T}\right), \tag{4.111}$$

where ξ_{g} is the bandgap, k_{B} is Boltzmann's constant ($= 8.617 \times 10^{-5}$ eV K^{-1}), and T is absolute temperature.

Therefore, we can write

$$I_{\mathrm{C}}(T) = I_{\mathrm{nB}}(T) \propto \exp\left(-\frac{\xi_{\mathrm{gB}}}{k_{\mathrm{B}}T}\right), \tag{4.112}$$

$$I_{\mathrm{B}}(T) = I_{\mathrm{pE}}(T) \propto \exp\left(-\frac{\xi_{\mathrm{gE}}}{k_{\mathrm{B}}T}\right), \tag{4.113}$$

where ξ_{gB}, ξ_{gE} denote the bandgaps of the base and emitter materials, respectively. Hence, the common-emitter current gain

$$\begin{aligned} h_{\mathrm{fE}} &= \frac{I_{\mathrm{C}}(T)}{I_{\mathrm{B}}(T)} = \frac{I_{\mathrm{nB}}(T)}{I_{\mathrm{pE}}(T)} \\ &= \frac{\exp\left(-\frac{\xi_{\mathrm{gB}}}{k_{\mathrm{B}}T}\right)}{\exp\left(-\frac{\xi_{\mathrm{gE}}}{k_{\mathrm{B}}T}\right)} = \exp\left\{-\left(\frac{\xi_{\mathrm{gB}} - \xi_{\mathrm{gE}}}{k_{\mathrm{B}}T}\right)\right\} = \exp\left(-\frac{\Delta\xi_{\mathrm{gBE}}}{k_{\mathrm{B}}T}\right). \end{aligned} \tag{4.114}$$

If $\Delta\xi_{gBE} = 0$, then $h_{FE} = 1 = h_{fE0}$ (assume). Then for any value of $\Delta\xi_{gBE}$

$$h_{fE} = h_{fE0} \exp\left(-\frac{\Delta\xi_{gBE}}{k_B T}\right). \tag{4.115}$$

For a typical value of $\Delta\xi_{gBE} = 0.05$ eV, the current gain at 300 K is $(h_{fE})_{300\text{ K}} = 0.1445$, while that at 77.2 K is $(h_{fE})_{77.2\text{ K}} = 5.4421 \times 10^{-4}$; decreasing by a factor of 265.5225. This shows that for a finite value of $\Delta\xi_{gBE}$, as the temperature decreases h_{fE} also decreases.

For a higher value of $\Delta\xi_{gBE} = 0.1$ eV, the current gain at 300 K is $(h_{fE})_{300\text{ K}} = 0.02089$ while that at 77.2 K is $(h_{fE})_{77.2\text{ K}} = 2.9617 \times 10^{-7}$; decreasing by a factor of 9.7579×10^4.

Table 4.5 presents the values of current gains at specified temperatures and also the decreasing factors, i.e. the ratios of current gains at two selected temperatures. It is observed that the greater ΔE_{gBE} is, the greater is the reduction in h_{FE} with a decrease in temperature.

From the above considerations, homojunction bipolar transistors are generally not considered for operation at liquid-nitrogen temperatures, because of insufficient DC current gain (Stork *et al* 1987).

4.4.11 Saturation voltage of a bipolar junction transistor

Figure 4.22 illustrates the translation of the curve drawn between the collector current and the collector–emitter voltage of a bipolar transistor with rising temperature.

The collector–emitter voltage v_{CES} during operation of a bipolar transistor in the saturation mode is given by the difference of base–emitter voltage v_{BE} and base–collector voltage v_{BC}

$$v_{CES} = v_{BE} - v_{BC}. \tag{4.116}$$

Differentiating both sides of equation (4.116) with respect to temperature, we find that

$$\frac{dv_{CES}}{dT} = \frac{dv_{BE}}{dT} - \frac{dv_{BC}}{dT}. \tag{4.117}$$

Table 4.5. Current gains and related decreasing factors.

Current gain (h_{fE})/decreasing factors	$\Delta\xi_{gBE} = 0.05$ eV	$\Delta\xi_{gBE} = 0.1$ eV
$(h_{fE})_{300\text{ K}}$	0.1445	0.02089
$(h_{fE})_{77.2\text{ K}}$	5.4421×10^{-4}	2.9617×10^{-7}
$(h_{fE})_{4.2\text{ K}}$	1.007×10^{-60}	1.0015×10^{-120}
$(h_{fE})_{300\text{ K}}/(h_{fE})_{77.2\text{ K}}$	265.5225	9.7579×10^4
$(h_{fE})_{300\text{ K}}/(h_{fE})_{4.2\text{ K}}$	1.435×10^{59}	2.0859×10^{118}

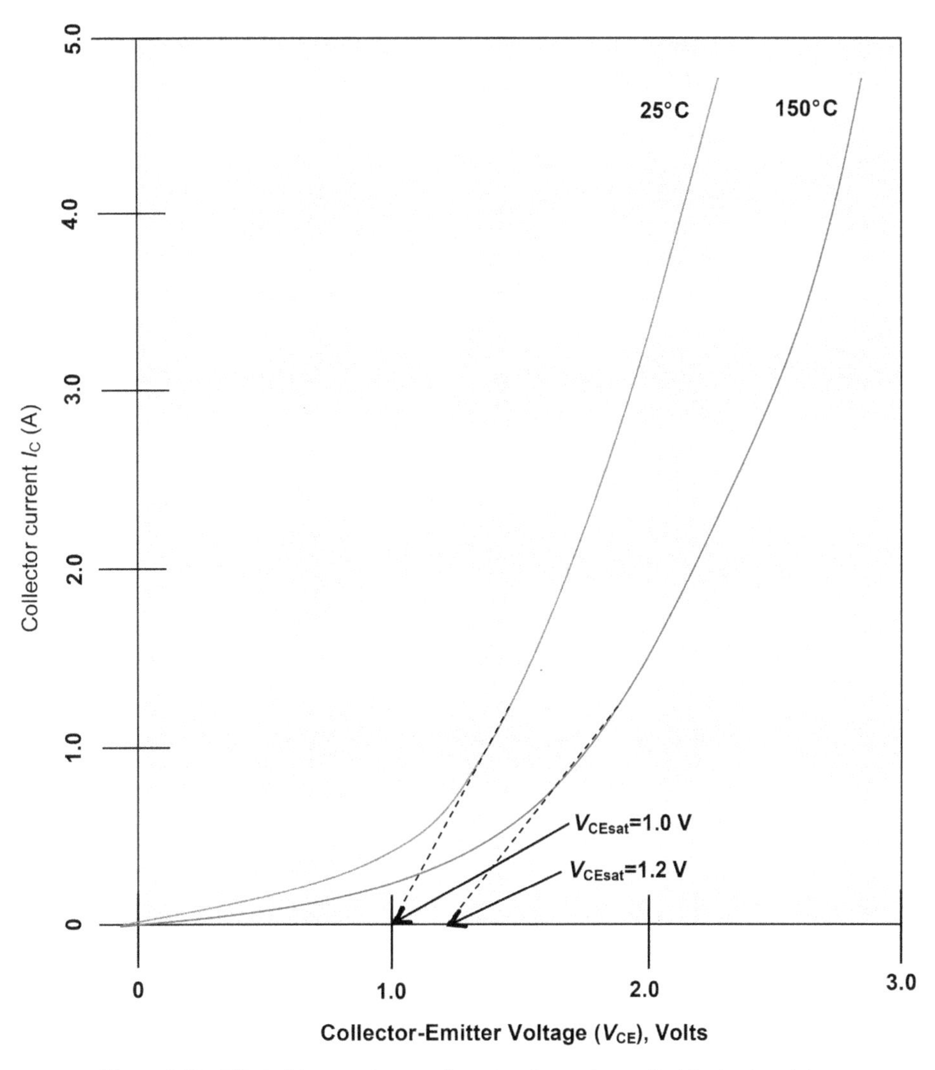

Figure 4.22. Effect of temperature on the saturation voltage of a bipolar transistor.

This equation shows that the TC of the collector–emitter voltage is equal to the TC of the base–emitter voltage minus the TC of the base–collector voltage. Applying equation (4.42) for the TC of the forward voltage of a p–n junction diode successively to each of the two diodes formed by the base–emitter and base–collector junctions, we obtain

$$\begin{aligned} \frac{\mathrm{d}v_{\mathrm{CES}}}{\mathrm{d}T} &= \frac{\mathrm{d}v_{\mathrm{BE}}}{\mathrm{d}T} - \frac{\mathrm{d}v_{\mathrm{BC}}}{\mathrm{d}T} \\ &= \frac{v_{\mathrm{BE}} - (3V_{\mathrm{Thermal}} + E_{\mathrm{G}})}{T} - \frac{v_{\mathrm{BC}} - (3V_{\mathrm{Thermal}} + E_{\mathrm{G}})}{T} = \frac{v_{\mathrm{BE}}}{T} - \frac{v_{\mathrm{BC}}}{T}. \end{aligned} \tag{4.118}$$

The built-in potential φ of a p–n junction diode with acceptor doping concentration N_A, donor doping concentration N_D is

$$\varphi_{bi} = V_{Thermal}\ln\left(\frac{N_A N_D}{n_i^2}\right), \tag{4.119}$$

where $V_{Thermal}$ is the thermal voltage = 25.9 mV at room temperature and n_i is the intrinsic carrier concentration. Using this equation to express the built-in potentials v_{BE} and v_{BC} of the base–emitter and base–collector diodes, we obtain for an n–p–n transistor

$$\begin{aligned}\frac{dv_{CES}}{dT} &= \frac{V_{Thermal}}{T}\ln\frac{(N_D)_{Emitter}(N_A)_{Base}}{n_i^2} - \frac{V_{Thermal}}{T}\ln\frac{(N_A)_{Base}(N_D)_{Collector}}{n_i^2} \\ &= \frac{V_{Thermal}}{T}\ln\left[\left\{\frac{(N_D)_{Emitter}(N_A)_{Base}}{n_i^2}\right\}\Big/\left\{\frac{(N_A)_{Base}(N_D)_{Collector}}{n_i^2}\right\}\right] \\ &= \frac{V_{Thermal}}{T}\ln\left\{\frac{(N_D)_{Emitter}}{(N_D)_{Collector}}\right\}.\end{aligned} \tag{4.120}$$

At $T = 300$ K, for a transistor with $(N_D)_{Emitter} = 1 \times 10^{20}$ cm^{-3} and $(N_D)_{Collector} = 1 \times 10^{15}$ cm^{-3},

$$\frac{dv_{CES}}{dT} = \frac{25.9\ \text{mV}}{300}\ln\left(\frac{1 \times 10^{20}}{1 \times 10^{15}}\right) = 0.086333\ln(1 \times 10^5) = 0.994\ \text{mV K}^{-1}. \tag{4.121}$$

For a high-voltage power transistor with $(N_D)_{Emitter} = 1 \times 10^{20}$ cm^{-3} and $(N_D)_{Collector} = 1 \times 10^{14}$ cm^{-3}

$$\frac{dv_{CES}}{dT} = \frac{25.9\ \text{mV}}{300}\ln\left(\frac{1 \times 10^{20}}{1 \times 10^{14}}\right) = 0.086333\ln(1 \times 10^6) = 1.1927\ \text{mV K}^{-1}. \tag{4.122}$$

Comparing the numerical values of TCs, it is found that the TC of the saturation voltage of a bipolar transistor is lower than the TC of the forward voltage of a diode. Looking at the signs of the TC, it is evident that the TC of the saturation voltage of a bipolar transistor is positive in sign, in contrast to that of the forward voltage of a diode, which was negative in sign. Practically, V_{CEsat} varies at a rate of +0.2–0.4% °C^{-1}.

4.4.12 Reverse base and emitter currents of a bipolar junction transistor (I_{CBO} and I_{CEO})

I_{CBO} and I_{CEO} denote collector currents with the collector junction reverse-biased, but in I_{CBO}, the emitter is open-circuited whereas in I_{CEO}, the base is open-circuited. They are inter-related as

$$I_{CEO} = I_{CBO}/(1 - \alpha) = (\beta + 1)I_{CBO} \tag{4.123}$$

where α, β are the current gains of the transistor in the common-base and common-emitter configurations, respectively. Both I_{CBO} and I_{CEO} are temperature-dependent. I_{CBO} doubles for every 10 °C increment of temperature. For I_{CBO}, the following equation holds

$$I_{\mathrm{CBO}}(T) = I_{\mathrm{CBO}}(T_0)2^{(T-T_0)/10^{\circ}\mathrm{C}}. \tag{4.124}$$

Regarding I_{CEO}, equation (4.123) can be rewritten as

$$I_{\mathrm{CEO}} = (\beta + 1)I_{\mathrm{CBO}} \approx \beta I_{\mathrm{CBO}}. \tag{4.125}$$

A growing exponential function is used to represent the variation of current gain β with temperature. However, within a limited range of temperatures (250–330 K), the nature of the graph is almost linear,

$$\beta(T) = \beta(T_0) + \varsigma\{1 + \beta(T_0)\}(T - T_0), \tag{4.126}$$

where ς is the TC:

$$\varsigma = \frac{1}{\beta(T_0)}\frac{\mathrm{d}\beta}{\mathrm{d}T} = 0.1 - 1.0 \text{ for Si transistors.} \tag{4.127}$$

Hence,

$$I_{\mathrm{CEO}}(T) = \left[\beta(T_0) + \varsigma\{1 + \beta(T_0)\}(T - T_0)\right]\left\{I_{\mathrm{CBO}}(T_0)2^{(T-T_0)/10\,^{\circ}\mathrm{C}}\right\}. \tag{4.128}$$

As an example, for a transistor of $\beta(300\text{ K}) = 50$ and $I_{\mathrm{CBO}}(300\text{ K}) = 1$ nA, taking $\zeta = 0.5$, I_{CEO} values at $T = 300$ K and $T = 320$ K are:

$$\begin{aligned} I_{\mathrm{CEO}}(300) &= [50 + 0.5\{1 + 50\}(300 - 300)]\{1 \times 2^{(300-300)/10\,^{\circ}\mathrm{C}}\} \\ &= 50 \times 1 = 50 \text{ nA} \end{aligned} \tag{4.129}$$

$$\begin{aligned} I_{\mathrm{CEO}}(320) &= [50 + 0.5\{1 + 50\}(320 - 300)]\{1 \times 2^{(320-300)/10\,^{\circ}\mathrm{C}}\} \\ &= 560 \times 2^2 = 2240 \text{ nA}. \end{aligned} \tag{4.130}$$

But

$$I_{\mathrm{CBO}}(320) = 1 \times 2^{(320-300)/10\,^{\circ}\mathrm{C}} = 1 \times 2^2 = 4 \text{ nA}. \tag{4.131}$$

Thus I_{CBO} increases only four times but I_{CEO} is multiplied by 2240/50 = 44.8 times on raising the temperature from 300 K to 320 K.

4.4.13 Dynamic response of a bipolar transistor

The switching behavior of a bipolar transistor is determined by the charge stored in the base region. After reversal of the base current, the collector current continues flowing as long as sufficient charge persists in the base region. This charge decays by carrier recombination. Only when the excess charge in the base has been removed, is the base–emitter junction discharged so that the transistor is turned off. Thus the

Table 4.6. Enhancing and degrading effects of temperature on device parameters.

Sl. No.	Device	Parameter	Effect of decreasing temperature	Remarks	Effect of increasing temperature	Remarks
1	Diode	Forward voltage drop	Increases	Detrimental	Decreases	Beneficial
2		Reverse leakage current	Decreases	Beneficial	Increases	Detrimental
3		Breakdown voltage	Decreases	Detrimental	Increases	Beneficial
		Storage time	Decreases	Beneficial	Increases	Detrimental
4	Bipolar transistor	Current gain	Decreases	Detrimental	Increases	Beneficial
		Saturation voltage	Decreases	Beneficial	Increases	Detrimental
		Reverse breakdown voltages of E–B and C–B junctions	Decrease	Detrimental	Increase	Beneficial
		Leakage currents	Decrease	Beneficial	Increase	Detrimental
		Turn-off time	Decreases	Beneficial	Increases	Detrimental

minority-carrier lifetime in the base plays a vital role in transistor switching, as in a p–n junction diode. An increase in temperature is associated with a longer turn-off time and hence slower turn-off, which degrades transistor performance (table 4.6).

4.5 Bipolar analog circuits in the 25 °C to 300 °C range

The process and design considerations of a quad operational amplifier (OP-AMP) working in the 25 °C to 300 °C range were described by Beasom and Patterson (1982). The fabrication process employed dielectric isolation in place of conventional junction isolation. A vertical p–n–p transistor structure complemented the high-performance lateral n–p–n device for operational stability over a broad temperature range. Vulnerability to interconnection pitting was avoided by using deeper (3.5 μm n-emitter and 4.5 μm p-emitter) structures.

Before embarking on the actual experiment, the researchers performed prior characterization of circuit components from 25 °C to 300 °C to ascertain the key issues and to decide what suitable corrective measures could be applied. Obviously, the parameter playing the pivotal role in a high-temperature circuit design is the leakage current. To this end, the characteristics of n–p–n and p–n–p transistors were measured at 25 °C, 200 °C and 300 °C. The characteristics at 300 °C were found to be offset by $I_B = 2$ μA for p–n–p and $I_B = 1.5$ μA for n–p–n transistors. The offset occurred because the collector–base leakage current I_{CBO} flows in the opposite direction to the base current I_B. As temperature increases from 200 °C to 300 °C,

I_{CBO} increases to the extent that it exceeds the intrinsic base current I_B. At the same time, the incremental current gain h_{FE} is found to increase. Another noteworthy effect is the reduction in base–emitter voltage V_{BE} with increasing temperature. The V_{BE} change followed the rate -2 mV °C^{-1} up to 300 °C, at which V_{BE} was typically 50–100 mV. The diffused resistor values increased linearly with positive TC of resistance (TCR). For both p–n–p and n–p–n transistors, the unity-gain frequency f_T decreased to one-half at 300 °C from its value at 25 °C.

Particular emphasis was placed on the performance of aluminum metallization during heating from room temperature to 300 °C. The following facts were noted. (i) No failures were observed in Al test structures up to 500 h at 325 °C and 3.3×10^4 A cm^{-2}. (ii) No serious changes took place in contact resistances in p^+- and n^+-doped Si test structures up to 500 h at 325 °C. (iii) No deterioration was noticed in Al-metallized transistors when they were kept at 325 °C for 500 h after applying $V_{CB} = 30$ V at $I_C = 1$ mA.

Learning from the above observations, circuit design adopted the following rules: (i) diode-connected transistors were considered unviable due to the low voltage magnitudes of forward-biased junctions; (ii) because of the reversal of the base current, the base voltage nodes for chains of current sources were to be provided with capabilities of current sourcing and sinking; and (iii) the values of diffused resistors at 300 °C were chosen as double their values at 25 °C to offset changes in forward junction voltages with temperature.

A combination of linear dependences of a Zener diode, a diffused resistor and several base–emitter voltages was exploited to obtain a stable current from the bias network over a wide temperature range. The circuit, consisting of a Zener diode in parallel with four temperature compensation diodes and a resistor, supplied an output current varying by $< \pm 0.2$ fraction in the -55 °C to $+300$ °C temperature range.

The currents produced by the bias network are coupled to the positive and negative supply rails through transistors, which are repeated in each of the four amplifiers of QUAD OP-AMP. The emitter resistors of these transistors provide negative feedback to supply the source and sink currents of each amplifier to counter the temperature gradients caused by the inequality of the power dissipations of the output stages. Biasing resistors of correct values are connected to transistors which are likely to saturate at high temperatures when the collector resistance increases and V_{BE} decreases. Provision is made for a sourcing/sinking current to/from the positive/negative bias lines upon reversal of direction of the base current. Proper scaling of the geometries of leakage compensation diodes is performed to match the sum of areas of sourcing/sinking devices.

In the input and gain stages, p–n–p transistors are favored as the input pair because their lower I_{CBO} results in smaller input bias and offset currents. Input currents at high temperatures are brought down by as much as 5–10 times by I_{CBO} compensation transistors. The collectors of the input pair of p–n–p transistors convey the signal to the grounded base of the n–p–n transistor pair, which translates the signal to the current mirror and to the high impedance point where voltage gain is produced. Protection against reversal of base current is provided by two p–n–p transistors. A Zener diode reduces the effect of output conductance on the offset voltage. This reduction has a major impact at high temperatures when the collector–base leakage current increases.

In the output stage, the voltage generated at the high impedance point is transmitted to the output through pairs of emitter followers. I_{CBO} is compensated through two transistors. This compensation reduces the leakage current which is reflected to the input as offset voltage.

The main performance parameters of the IC OP-AMP were presented by Beasom and Patterson (1982) for the temperature range −55 to +300 °C. The power supply current variation was <10% except at 300 °C. The influence of the leakage current at 300 °C was manifested in the form of an increase in the offset voltage and the input offset current. The decrease in the open-loop gain at 300 °C was caused by degradation of the output conductance. The gain bandwidth product decreased with increasing temperature but still remained higher at 300 °C than that of several OP-AMPs at 25 °C. This is ascribed to the use of vertical p–n–p transistors with $f_T = 100$ MHz in the circuit. A very low value of input noise (= 4.2 nV $Hz^{-1/2}$) was achieved.

4.6 Bipolar digital circuits in the 25 °C to 340 °C range

Prince *et al* (1980) carried out extensive studies of the DC and dynamic characteristics of bipolar digital ICs in the range of temperatures from 25 °C to 340 °C. Transistor–transistor logic (TTL) is a type of saturating logic. In this logic, gold doping is used to decrease the storage time through reduction of the minority carrier lifetime. As an alternative, a Schottky-barrier clamping diode is connected between the base and collector terminals. Saturation does not take place because a Schottky diode has a lower forward voltage than that of the collector–base junction. Excess base current is routed through the Schottky diode. As the Schottky diode is a majority carrier device, it has negligible storage of minority carriers. Using Schottky diode-clamped transistors, storage times of 1–2 ns are obtained. These values are lower than 5–10 ns for gold-doped devices. Further, the gain of non-gold doped transistors is higher than that of gold-doped ones.

Prince *et al* (1980) noted a difference between gold-doped TTL and Schottky-clamped TTL. The former remained functional up to 250 °C, whereas the latter worked up to 325 °C. The ICs tested consisted of two-input and four-input NAND gates. Apart from the lower temperature limits of gold-doped devices and the scaling behaviors of some parameters, both gold-doped and Schottky-clamped TTL devices performed equally well at high temperatures. The failure was observed with respect to noise margins. In particular, the output high-voltage V_{OH} decreased as the temperature was increased. By analyzing the voltage transfer characteristics of the ICs as a function of temperature, it was found that the primary reason for diminution in V_{OH} was the leakage current across the collector–base junction of the phase splitter transistor. This current flows through the phase splitter collector resistor. Failure mechanisms at high temperature were revealed by inspecting the behaviors of the power supply current I_{CCH} for output high conditions and output high voltage versus output high sourcing current for the IC, and correlating these data with the behavior of input low current I_{IL} at high temperatures. It was noticed that a large portion of the high-temperature increase in I_{CCH} could be explained by the increase in I_{IL} at zero input voltage. This increase in current is produced in the phase splitter transistor. It appears as an additional component of I_{IL}. This current flows through the collector resistor of the phase splitter transistor. The voltage drop resulting from this leakage current flow decreases V_{OH} because at medium output and supply currents, V_{OH} tracks the voltage drop across the collector resistor. Thus the leakage current and resistor values combine to yield an upper bound on functionality for Schottky-clamped TTL devices in the neighborhood of 325 °C.

The ICs showed a fan-out capability >1 at temperatures close to maximum operating temperatures. The current sourcing capability decreased owing to the increase in circuit resistance at high temperatures. The current sinking capability increased through augmentation of the current sink transistor gain at these temperatures.

These experiments led to the view that bipolar TTL logic gates can function properly at temperatures near 325 °C.

4.7 Discussion and conclusions

Data sheets of semiconductor components such as diode and bipolar transistors provide the specifications of the devices at a particular temperature. These values of the components listed at a given temperature are liable to vary with changes in temperature. Since the components rarely operate at the extreme temperature mentioned in the data sheets, it is important to know how their values vary at the actual working temperature in an application. In several cases, it is difficult to obtain the correct values as the procedure may require computer simulations. However, in many cases, the trends or directions of changes can be guessed from simple analytical models. The intent of this chapter was to give an overview of the available models in the literature which could help the reader at making a first-cut speculation.

The changes in the electrical parameters of discrete devices impact the performance of circuits fabricated using them as components. Examples showing the thermal effects of component parameter changes on circuit behavior were presented above. Often, the temperature-dependent behavior can be pre-considered at the design stage and the necessary correction applied to ensure proper functioning at high temperatures.

Review exercises

4.1. What is the upper temperature bound on silicon devices? Does this bound imply that a silicon device stops functioning at this temperature? How far is this bound stretchable by suitable techniques?

4.2. Is silicon a metal or a non-metal or does it belong to some other class of materials? What is the resistivity of silicon in its intrinsic state? What is its intrinsic carrier concentration? What is, typically, the ratio of electron mobility/hole mobility for silicon?

4.3. Calculate the intrinsic temperature of silicon for a doping concentration of 1×10^{15} cm^{-3}. Assume that the effective masses of electrons and holes as well as the bandgap of silicon do not vary with temperature.

4.4. How is EG polysilicon obtained from MG polysilicon? Name two techniques used for growth of single-crystal silicon from EG polysilicon. Discuss their main features.

4.5. How is a silicon wafer doped as n- or p-type using the thermal diffusion process? Describe with relevant chemical equations.

4.6. How does ion implantation differ from thermal diffusion of impurities? Name the dopant gases commonly used. What are the typical ranges of ion energies and doses used in ion implantation for fabrication of semiconductor devices? Why is the ion implantation process followed by a thermal annealing step?

4.7. What are the main techniques used for depositing contact metals on semiconductor devices? What material is commonly used to form an ohmic contact on heavily doped n-type silicon? What type of contact will be formed if n-type silicon is lightly doped?

4.8. How are the different components in an IC separated by the p–n junction isolation technique? What are the advantages and disadvantages of this technique?

4.9. How are components of an IC separated by the dielectric isolation technique? What type of wafer is used for IC fabrication if this technique is adopted? What are the special features of these wafers? How do these wafers provide dielectric isolation?

4.10. Write down the Shockley equation for the current–voltage characteristics of a p–n diode, and explain the meanings of the symbols used. What is the significance of the emission coefficient η? What are the typical values of η? What are the factors on which its value depends?

4.11. Why is the expression $(k_B T/q)$ referred to as thermal voltage? What is the value of thermal voltage at room temperature?

4.12. Does the reverse saturation current I_S of a diode vary with temperature T? If so, how? Write the relevant equation. Show that

$$\frac{dI_S}{dT} = I_S\left(\frac{3}{\eta T} + \frac{E_G}{\eta T V_{Thermal}}\right),$$

where η is the emission coefficient, E_G is the bandgap and $V_{Thermal}$ is the thermal voltage.

4.13. If E_G denotes the bandgap and $V_{Thermal}$ the thermal voltage, show that the TC dv/dT of forward voltage drop across a p–n diode is given by

$$\frac{dv}{dT} = \frac{v - (3V_{Thermal} + E_G)}{T}.$$

4.14. Prove that the TC $\mathrm{d}V_\mathrm{d}/\mathrm{d}T$ of a Schottky diode is expressed as

$$\frac{\mathrm{d}V_\mathrm{d}}{\mathrm{d}T} = \frac{V_\mathrm{d}}{T} - \left(\frac{k_\mathrm{B}T}{q}\right)\left(\frac{1}{T}\right)\left\{2 + \left(\frac{q\phi_\mathrm{B}}{k_\mathrm{B}T}\right)\right\},$$

where ϕ_B is the Schottky barrier height.

4.15. If I_{S1}, I_{S2} are the reverse leakage currents of a p–n diode at temperature T, $T + \Delta T$, η is the emission coefficient, E_G is the bandgap and V_{Thermal} is thermal voltage, show that

$$\frac{I_{\mathrm{S2}}}{I_{\mathrm{S1}}} = \exp\left\{\left(\frac{3}{\eta T} + \frac{E_\mathrm{G}}{\eta T V_{\mathrm{Thermal}}}\right)\Delta T\right\}.$$

Taking $\eta = 2$, prove that for silicon diodes the leakage current doubles for every 10 °C rise in temperature.

4.16. What is the critical electric field of silicon? What happens at this field? Explain the phenomenon of impact ionization.

4.17. What is meant by the ionization coefficients of electrons and holes? Write the equation for the ionization integral. What is its value at the critical electric field?

4.18. What is the ionization threshold energy? How is it related to the energy bandgap? Write the Cromwell–Sze equation for the average carrier mean free path.

4.19. Explain why the breakdown voltage of a p–n junction decreases with temperature on the basis of the influence of temperature on ionization threshold energy, energy loss by scattering, the mean free path of carriers and the frequency of scattering events.

4.20. Write the Chynoweth equations for ionization coefficients α_n, α_p of electrons and holes. How did Shockley interpret the fitting parameter b in these equations?

4.21. Derive the following equation for TC of ionization coefficient

$$\mathrm{TC} = \left(\frac{W_\mathrm{I}}{qE\lambda}\right)\left\{\frac{1}{\lambda}\left(\frac{\mathrm{d}\lambda}{\mathrm{d}T}\right) - \frac{1}{W_\mathrm{I}}\left(\frac{\mathrm{d}W_\mathrm{I}}{\mathrm{d}T}\right)\right\},$$

where W_I is the ionization threshold energy, E is the intensity of the lateral electric field, λ is the mean free path of an electron, q is the electronic charge and T is the temperature on the Kelvin scale. From the TCs of λ and W_I express the above equation in the form

$$\mathrm{TC} = -\left(\frac{1}{E}\right)\left[\left(\frac{W_\mathrm{I}\lambda_0 E_\mathrm{p}}{2q\lambda^2 k_\mathrm{B}T^2}\right)\mathrm{sech}^2\left(\frac{E_\mathrm{p}}{2k_\mathrm{B}T}\right) + \left(\frac{1}{q\lambda}\right)(C_2 + 2C_3T)\right],$$

where λ_0 is the high-energy low-temperature asymptotic phonon mean free path, E_p is the average energy loss due to optical phonon scattering, k_B is Boltzmann's constant; and C_2, C_3 are constants in the Ershov–Ryzhii model.

4.22. Distinguish between the avalanche and Zener breakdown mechanisms of a p–n junction diode with regard to the TCs. Elaborate the reasons for the opposite behaviors of the two mechanisms.
4.23. Why does the storage time of a p^+–n diode increase with an increase of temperature? What is the effect on the switching loss?
4.24. How is the current gain of a bipolar transistor in the common-emitter configuration (β) related to that in the common-base connection (α)?
4.25. Define the following terms for a bipolar transistor: (a) the emitter injection efficiency γ, (b) the base transport factor α_T, and (c) the collector multiplication ratio M. How is the common-base current gain α of a bipolar transistor related to γ, α_T and M?
4.26. Explain with reasons the effect of temperature on the term $\{1 + (D_{pE}/D_{nB})\}$ in the equation for common-base current gain α. Here, D_{pE} denotes the diffusion constant of holes in the n^+-emitter and D_{nB} represents the diffusion constant of electrons in the p-base.
4.27. Show that for a bipolar transistor, the ratio of the number of holes in the n^+-emitter (p_{nE}) to the number of electrons in the p-base (n_{pB}) is given by $p_{nE}/n_{pB} = K \exp\{\Delta\xi_g/(2k_B T)\}$ where K is a constant, $\Delta\xi_g$ is the bandgap reduction of the emitter caused by heavy doping-induced defects, k_B is Boltzmann's constant and T is temperature. Using this equation, explain the significant reduction in current gain with a decrease of temperature due to the effect of temperature on the ratio of minority carrier concentrations.
4.28. Discuss the effect of temperature on the ratio of the base width to the diffusion length of minority carriers in the emitter (W/L_{pE}). Hence explain the decrease in current gain with a decrease in temperature.
4.29. If h_{fE} is the common-emitter current gain of a BJT at a temperature T, h_{fE0} is the common-emitter current gain of a BJT at a temperature $T = 0$ K, and $\Delta\xi_{gBE} = \xi_{gB} - \xi_{gE}$ = the difference in bandgaps between the emitter and base, show that

$$h_{fE} = h_{fE0} \exp\left(-\frac{\Delta\xi_{gBE}}{k_B T}\right).$$

Hence argue that the reduction in bandgap of emitter with respect to that of base ($\Delta\xi_{gBE}$) is the main reason responsible for the pronounced deterioration in current gain of a bipolar transistor with falling temperature.
4.30. Show that the TC of the saturation voltage of a bipolar transistor is given by

$$\frac{dv_{CES}}{dT} = \frac{V_{Thermal}}{T} \ln\left\{\frac{(N_D)_{Emitter}}{(N_D)_{Collector}}\right\},$$

where $V_{Thermal}$ is the thermal voltage, T is the temperature, and $(N_D)_{Emitter}$ and $(N_D)_{Collector}$ are the donor doping concentrations of emitter and collector, respectively.
4.31. What do I_{CBO} and I_{CEO} of a bipolar transistor stand for? Which is more sensitive to changes of ambient temperature?

4.32. Describe, after Beasom and Patterson, the process and design considerations of a quad operational amplifier (OP-AMP) working in the 25 °–300 °C range.
4.33. What experiments led to the inference that bipolar TTL logic gates can function properly at temperatures near 325 °C?

References

Beasom J D and Patterson R B 1982 Process characteristics and design methods for a 300 °C quad operational amplifier *IEEE Trans. Ind. Electron.* **29** 112–7

Buhanan D 1969 Investigation of current-gain temperature dependence in silicon transistors *IEEE Trans. Electron. Devices* **16** 117–24

Chang C Y, Chiu S S and Hsu L P 1971 Temperature dependence of breakdown voltage in silicon abrupt p–n junctions *IEEE Trans. Electron Devices* **18** 391–3

Chynoweth A G 1958 Ionization rates for electron and holes in silicon *Phys. Rev.* **109** 1537

Cory R 2009 Schottky diodes *Skyworks Solutions* February, pp 1–5 www.skyworksinc.com/downloads/press_room/published_articles/MPD_022009.pdf

Crowell C R and Sze S M 1966 Temperature dependence of avalanche multiplication in semiconductors *Appl. Phys. Lett.* **92** 42–244

Dokić B L and Blanuša B 2015 Diodes and transistors *Power Electronics Converters and Regulators* (Cham: Springer) ch 4 pp 43–141

Ershov M and Ryzhii V 1995 Temperature dependence of the electron impact ionization coefficient in silicon *Semicond. Sci. Technol.* **10** 138–42

Fisher G, Seacrist M R and Standley R W 2012 Silicon crystal growth and wafer technologies *Proc. IEEE* **100** 1454–74

Godse A P and Bakshi U A 2009 *Basic Electronics*, Technical Publications, Pune, vol 1 pp 3–30

Kauffman W L and Bergh A A 1968 The temperature dependence of ideal gain in double diffused silicon transistors *IEEE Trans. Electron Devices* **15** 732–5

Leach W M 2004 Chapter 2: the junction diode *Lecture Notes* http://users.ece.gatech.edu/mleach/ece3040/notes/chap02.pdf

Maes W, De Meyer K and Van Overstraeten R 1990 Impact ionization in silicon: a review and update *Solid-State Electron.* **33** 705–18

Mukherjee M 2011 SiC devices on different polytypes: prospects and challenges *Silicon Carbide—Materials, Processing and Applications in Electronic Devices* ed M Mukherjee (Rijeka: InTech), pp 337–68

Prince J L, Draper B L, Rapp E A, Kronberg J N and Fitch L T 1980 Performance of digital integrated circuit technologies at very high temperatures *IEEE Trans. Compon. Hybrids Manuf. Technol.* **3** 571–9

Reisch M 2003 *High-Frequency Bipolar Transistors: Physics, Modeling, Applications* (Berlin: Springer) pp. 149–50

Sinha A K 1981 Refractory metal silicides for VLSI applications *J. Vac. Sci. Technol.* **19** 778

Stork J M C, Harame D L, Meyerson B S and Nguyen T N 1987 High performance operation of silicon bipolar transistors at liquid nitrogen temperature *IEEE–IEDM Tech. Dig.* pp. 405–8

Waldron N S, Pitera A J, Lee M L, Fitzgerald E A and del Alamo J A 2005 Positive temperature coefficient of impact ionization in strained-Si *IEEE Trans. Electron Devices* **52** 1627–33

Chapter 5

Temperature dependence of electrical characteristics of silicon MOS devices and circuits

Depending on the gate voltage, a MOSFET can be operated as a negative, zero or positive TC device. This kind of MOSFET behavior owes its origin to the presence of opposing factors in its conduction mechanism, mainly the increase in carrier concentration with temperature being counterbalanced by the decrease in carrier mobility. At large gate bias values, when the drain current increases to excessively high magnitudes, it is the decrease in carrier mobility which helps in thwarting thermal runaway, and thus serves as a built-in protective mechanism. In this chapter, the effects of temperature on the critical electrical parameters of a MOSFET, e.g. threshold voltage, on-resistance, transconductance and breakdown voltages are examined. The effects of temperature on the dynamical response of the MOSFET are also explored. As the temperature increases or descends below room temperature, some parameters are adversely affected while changes in others are encouraging. Therefore, the variation of MOSFET parameters with temperature impacts the performance of MOS analog and digital circuits in both directions. The decrease in the latching susceptibility of CMOS circuits at low temperatures is a significant advantage derived by MOS circuit operation in a cool environment.

5.1 Introduction

We begin by considering how temperature impacts the operation of a MOSFET in comparison to a bipolar transistor. In a bipolar transistor of common-emitter current gain β, the collector current I_C is expressed in terms of base current I_B and collector-base leakage current I_{CBO} with the emitter open as

$$I_C = \beta I_B + (\beta + 1)I_{CBO}. \tag{5.1}$$

The leakage current I_{CBO} is strongly temperature-dependent, doubling for every 10 °C increase of temperature. As the temperature increases, I_{CBO} increases and

doi:10.1088/978-0-7503-1155-7ch5

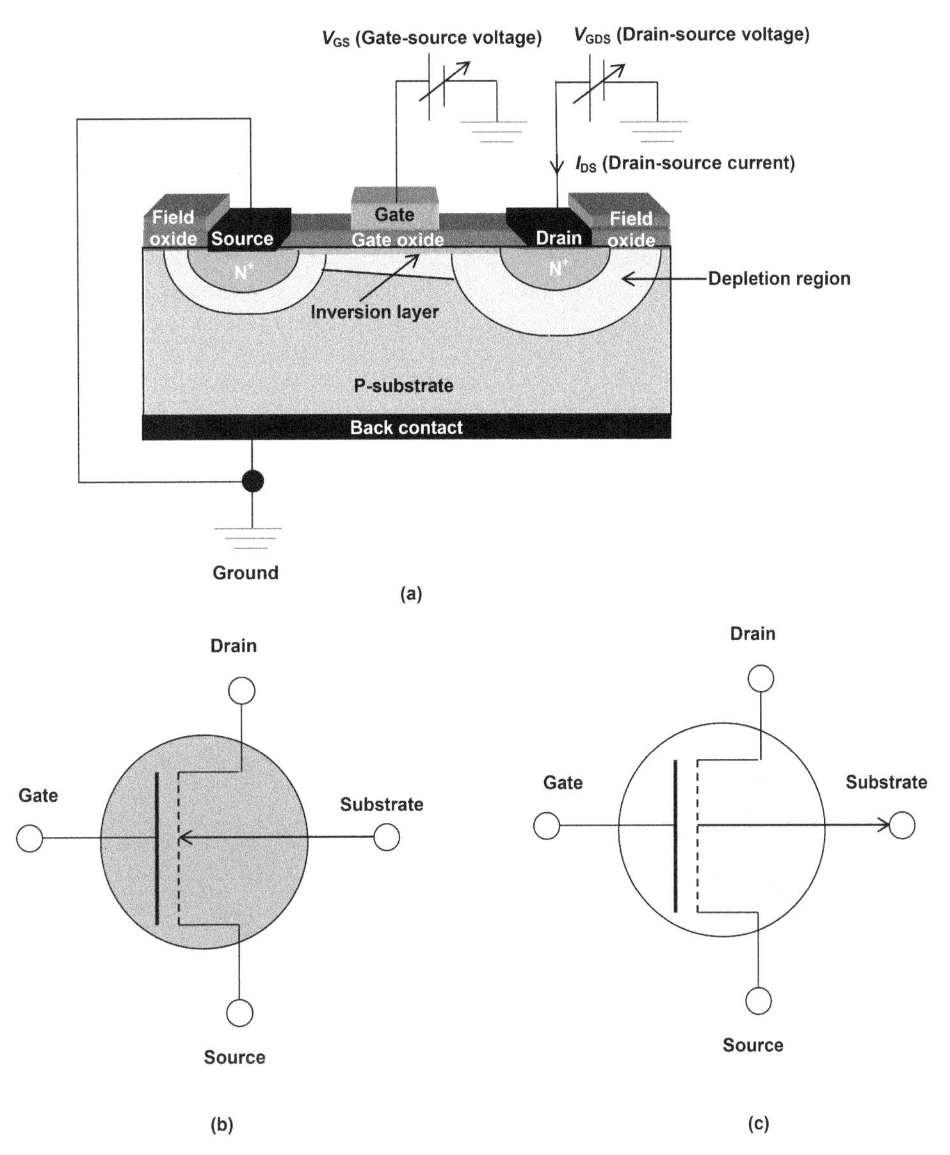

Figure 5.1. MOSFET. (a) Cross-sectional diagram of an n-channel enhancement-mode MOSFET, (b) a circuit diagram symbol of an n-channel device and (c) the symbol for a p-channel device.

collector current I_C climbs appreciably by a factor of $(\beta + 1)I_{CBO}$. Consequent upon this steep rise in collector current I_C, the transistor is heated, increasing I_C still further through the heating effect. This regenerative or reinforcing cycle progresses cumulatively until I_C increases dangerously and the transistor burns out. This self-destruction of the transistor is known as thermal runaway.

Temperature affects the performance of a MOSFET quite differently from a bipolar transistor because the MOSFET is inherently thermally stable and not prone to the uncontrolled positive feedback mechanism, i.e. thermal runaway, which takes place as the temperature or drain current increases. Figure 5.1 shows the cross-section and circuit symbols of a MOS transistor.

5.2 Threshold voltage of an n-channel enhancement mode MOSFET

The threshold voltages of n-channel and p-channel MOSFETs change in opposite directions with an increase of temperature, as shown in figure 5.2.

For an n-channel enhancement mode MOS transistor, the formula for threshold voltage is (Wang *et al* 1971)

$$V_{\mathrm{Th}} = V_{\mathrm{FB}} + 2\phi_{\mathrm{F}} + \frac{\sqrt{2\varepsilon_0\varepsilon_{\mathrm{s}}qN_{\mathrm{A}}\left(2\phi_{\mathrm{F}} + V_{\mathrm{SB}}\right)}}{C_{\mathrm{ox}}}, \tag{5.2}$$

where V_{FB} = flat-band voltage, ϕ_{F} = bulk potential, ε_0 = free-space permittivity, ε_{s} = dielectric constant of silicon, q = electronic charge, N_{A} = acceptor doping concentration, V_{SB} = substrate bias and C_{ox} = oxide capacitance per unit area. Also,

$$V_{\mathrm{FB}} = \phi_{\mathrm{ms}} - \frac{Q_{\mathrm{f}}}{C_{\mathrm{ox}}} - \frac{1}{\varepsilon_0\varepsilon_{\mathrm{ox}}}\int_0^{t_{\mathrm{ox}}} \rho_{\mathrm{ox}}(x)x\mathrm{d}x, \tag{5.3}$$

where ϕ_{ms} is the metal–semiconductor work function, Q_{f} is the fixed charge in the oxide, $\varepsilon_{\mathrm{ox}}$ is the relative permittivity of silicon dioxide, $\rho_{\mathrm{ox}}(x)$ is the charge density in an oxide of thickness t_{ox} varying with distance x in the oxide. For an n-type polysilicon gate (STMicroelectronics 2006)

$$\phi_{\mathrm{ms}} = -\left(\frac{k_{\mathrm{B}}T}{q}\right)\ln\left(\frac{N_{\mathrm{D(g)}}N_{\mathrm{A}}}{n_{\mathrm{i}}^2}\right), \tag{5.4}$$

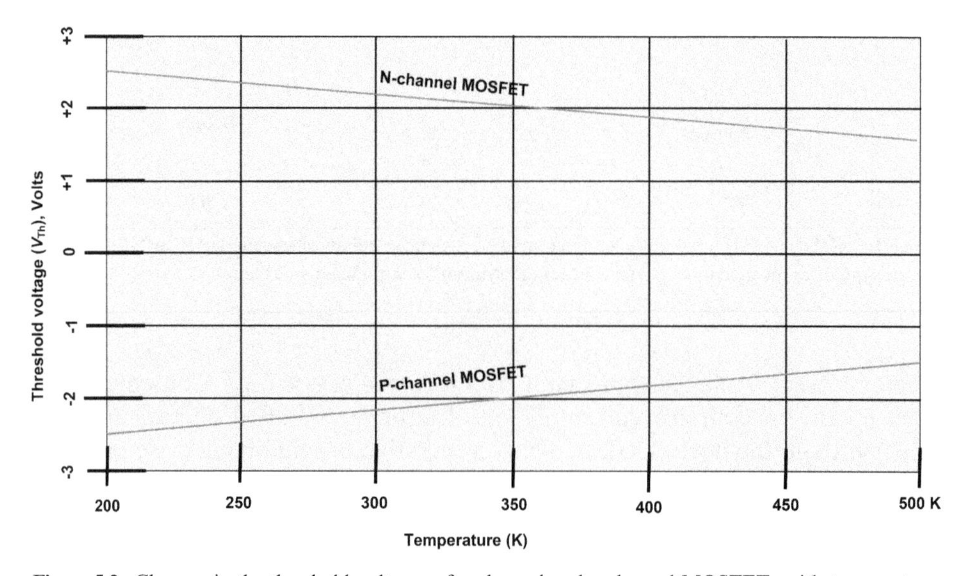

Figure 5.2. Changes in the threshold voltages of n-channel and p-channel MOSFETs with temperature.

where k_B is the Boltzmann constant, T is the temperature (K), q is electronic charge, $N_{D(g)}$ is the doping concentration of the polysilicon gate, N_A is the acceptor concentration of the semiconductor and n_i is the intrinsic carrier concentration. For the p-substrate,

$$\phi_F = \left(\frac{k_BT}{q}\right)\ln\left(\frac{N_A}{n_i}\right). \tag{5.5}$$

From equations (5.2) and (5.3), the threshold voltage can be written in approximate form as

$$V_{Th} \approx \phi_{ms} + 2\phi_F + \frac{\sqrt{2\varepsilon_0\varepsilon_s qN_A(2\phi_F + 0)}}{C_{ox}} = \phi_{ms} + 2\phi_F + \frac{2\sqrt{\varepsilon_0\varepsilon_s qN_A\phi_F}}{C_{ox}}, \tag{5.6}$$

where it has been assumed that oxide charges are negligible and V_{SB} has been taken as zero. Differentiating both sides of equation (5.6) with respect to temperature,

$$\begin{aligned}
\frac{\partial V_{Th}}{\partial T} &= \frac{\partial\phi_{ms}}{\partial T} + 2\left(\frac{\partial\phi_F}{\partial T}\right) + 2 \times \frac{1}{2}\frac{\left(\varepsilon_0\varepsilon_s qN_A\phi_F\right)^{1/2-1}}{C_{ox}} \times \left(\varepsilon_0\varepsilon_s qN_A\right) \times \frac{\partial\phi_F}{\partial T} \\
&= \frac{\partial\phi_{ms}}{\partial T} + 2\left(\frac{\partial\phi_F}{\partial T}\right) + \frac{\left(\varepsilon_0\varepsilon_s qN_A\phi_F\right)^{-1/2}}{C_{ox}} \times \left(\varepsilon_0\varepsilon_s qN_A\right) \times \frac{\partial\phi_F}{\partial T} \\
&= \frac{\partial\phi_{ms}}{\partial T} + 2\left(\frac{\partial\phi_F}{\partial T}\right) + \left(\frac{1}{C_{ox}}\right)\sqrt{\frac{\varepsilon_0\varepsilon_s qN_A}{\phi_F}}\left(\frac{\partial\phi_F}{\partial T}\right) \\
&= \frac{\partial\phi_{ms}}{\partial T} + \left(\frac{\partial\phi_F}{\partial T}\right)\left\{2 + \left(\frac{1}{C_{ox}}\right)\sqrt{\frac{\varepsilon_0\varepsilon_s qN_A}{\phi_F}}\right\}.
\end{aligned} \tag{5.7}$$

Thus the TC of threshold voltage V_{Th} depends on those of the metal–semiconductor work function ϕ_{ms} and bulk potential ϕ_F. Now from equation (5.4),

$$\begin{aligned}
\frac{\partial\phi_{ms}}{\partial T} &= -\frac{\partial}{\partial T}\left[\left(\frac{k_BT}{q}\right)\ln\left(\frac{N_{D(g)}N_A}{n_i^2}\right)\right] \\
&= -\left(\frac{k_B}{q}\right)\ln\left(\frac{N_{D(g)}N_A}{n_i^2}\right) - \left[\left(\frac{k_BT}{q}\right)\frac{1}{\left(\frac{N_{D(g)}N_A}{n_i^2}\right)}\left(N_{D(g)}N_A\right) \times -2n_i^{-2-1} \times \frac{\partial n_i}{\partial T}\right] \\
&= -\{(k_B)/q\}\ln\left(\frac{N_{D(g)}N_A}{n_i^2}\right) - \left[\left(\frac{k_BT}{q}\right) \times n_i^2 \times -2n_i^{-3} \times \frac{\partial n_i}{\partial T}\right] \\
&= -\{(k_B)/q\}\ln\left(\frac{N_{D(g)}N_A}{n_i^2}\right) + 2\left(\frac{k_BT}{qn_i}\right)\left(\frac{\partial n_i}{\partial T}\right) \\
&= \frac{\phi_{ms}}{T} + \left(\frac{2k_BT}{qn_i}\right)\left(\frac{\partial n_i}{\partial T}\right),
\end{aligned} \tag{5.8}$$

where equation (5.4) has been applied again. The intrinsic carrier concentration n_i is

$$\begin{aligned} n_i &= \sqrt{N_C N_V}\, T^{3/2} \exp\left(-\frac{E_g}{2k_B T}\right) \\ &= \sqrt{2.81 \times 10^{19} \times 1.83 \times 10^{19}}\, T^{3/2} \exp\left(-\frac{E_g}{2k_B T}\right), \end{aligned} \tag{5.9}$$

where N_C is the effective densities of states in conduction band (= 2.81×10^{19} cm^{-3}), N_V is the effective densities of states in the valence band (= 1.83×10^{19} cm^{-3}). On computation of the square root term in equation (5.9), we have

$$n_i = 2.26766 \times 10^{19} T^{3/2} \exp\left(-\frac{E_g}{2k_B T}\right) = K T^{3/2} \exp\left(-\frac{E_g}{2k_B T}\right), \tag{5.10}$$

where $K = 2.26766 \times 10^{19}$ cm^{-3}. From equation (5.10),

$$\begin{aligned} \frac{\partial n_i}{\partial T} &= \frac{\partial}{\partial T}\left\{K T^{3/2} \exp\left(-\frac{E_g}{2k_B T}\right)\right\} \\ &= K \times \frac{3}{2} \times T^{3/2-1} \exp\left(-\frac{E_g}{2k_B T}\right) \\ &\quad + K T^{3/2} \exp\left(-\frac{E_g}{2k_B T}\right) \times \left(-\frac{E_g}{2k_B}\right) \times -T^{-1-1} \\ &= K \times \frac{3}{2} \times T^{1/2} \exp\left(-\frac{E_g}{2k_B T}\right) + n_i \times \frac{E_g}{2k_B} \times T^{-2} \\ &= \frac{3}{2T} K T^{3/2} \exp\left(-\frac{E_g}{2k_B T}\right) + \frac{n_i E_g}{2k_B T^2} \\ &= \frac{3n_i}{2T} + \frac{n_i E_g}{2k_B T^2} = \left(\frac{n_i}{2T}\right)\left(3 + \frac{E_g}{k_B T}\right), \end{aligned} \tag{5.11}$$

where equation (5.10) for n_i has been re-applied. Substituting for $\partial n_i/\partial T$ from equation (5.11) into equation (5.8),

$$\frac{\partial \phi_{ms}}{\partial T} = \frac{\phi_{ms}}{T} + \left(\frac{2k_B T}{q n_i}\right)\left(\frac{n_i}{2T}\right)\left(3 + \frac{E_g}{k_B T}\right) = \frac{\phi_{ms}}{T} + \left(\frac{k_B}{q}\right)\left(3 + \frac{E_g}{k_B T}\right). \tag{5.12}$$

Looking at $\partial\phi_F/\partial T$, we find from equation (5.5)

$$\frac{\partial\phi_F}{\partial T}=\frac{\partial}{\partial T}\left\{\left(\frac{k_BT}{q}\right)\ln\left(\frac{N_A}{n_i}\right)\right\}=\left(\frac{k_B}{q}\right)\ln\left(\frac{N_A}{n_i}\right)+\left(\frac{k_BT}{q}\right)$$
$$\times\frac{1}{\left(\frac{N_A}{n_i}\right)}\times N_A\times -n_i^{-1-1}\times\frac{\partial n_i}{\partial T}$$
$$=\left(\frac{k_B}{q}\right)\ln\left(\frac{N_A}{n_i}\right)-\left(\frac{k_BT}{qn_i}\right)\times\frac{\partial n_i}{\partial T}$$
$$=\left(\frac{k_B}{q}\right)\ln\left(\frac{N_A}{n_i}\right)-\left(\frac{k_BT}{qn_i}\right)\times\left(\frac{n_i}{2T}\right)\left(3+\frac{E_g}{k_BT}\right) \tag{5.13}$$
$$=\left(\frac{k_B}{q}\right)\ln\left(\frac{N_A}{n_i}\right)-\left(\frac{k_B}{q}\right)\times\left(\frac{1}{2}\right)\left(3+\frac{E_g}{k_BT}\right)$$
$$=\frac{1}{T}\left\{\left(\frac{k_BT}{q}\right)\ln\left(\frac{N_A}{n_i}\right)\right\}-\frac{3}{2}\left(\frac{k_B}{q}\right)-\frac{E_g}{2qT}$$
$$=\frac{\phi_F}{T}-\frac{3}{2}\left(\frac{k_B}{q}\right)-\frac{E_g}{2qT},$$

where equations (5.11) and (5.5) for $\partial n_i/\partial T$ and ϕ_F have been used.

The temperature dependence of threshold voltage is obtained by substituting the expressions for $\partial\phi_{ms}/\partial T$ and $\partial\phi_F/\partial T$ from equations (5.12) and (5.13), respectively, into equation (5.7) for $\partial V_{Th}/\partial T$

$$\frac{\partial V_{Th}}{\partial T}=\frac{\phi_{ms}}{T}+\left(\frac{k_B}{q}\right)\left(3+\frac{E_g}{k_BT}\right)+\left\{\frac{\phi_F}{T}-\frac{3}{2}\left(\frac{k_B}{q}\right)-\frac{E_g}{2qT}\right\}$$
$$\left\{2+\left(\frac{t_{ox}}{\varepsilon_0\varepsilon_{ox}}\right)\sqrt{\frac{\varepsilon_0\varepsilon_s qN_A}{\phi_F}}\right\}, \tag{5.14}$$

where we have put

$$C_{ox}=\varepsilon_0\varepsilon_{ox}/t_{ox} \tag{5.15}$$

where ε_{ox} is the dielectric constant of silicon dioxide and t_{ox} is oxide thickness.

Taking a typical example of a MOSFET,

$N_{D(g)}=1\times10^{20}\ \mathrm{cm}^{-3}$, $N_A=2\times10^{17}\ \mathrm{cm}^{-3}$, $T=300$ K,

$E_g=1.12$ eV, $\varepsilon_0=8.854\times10^{-14}\ \mathrm{F\ cm}^{-1}$, $\varepsilon_s=11.9$

$\varepsilon_{ox}=3.9$, $t_{ox}=50\ \mathrm{nm}=50\times10^{-9}\ \mathrm{m}=50\times10^{-9}\times100\ \mathrm{cm}=5\times10^{-6}\ \mathrm{cm}$

with values

$$k_{\rm B} = 8.62 \times 10^{-5}\,{\rm eV\,K^{-1}}, \quad n_{\rm i} = 1.45 \times 10^{10}\,{\rm cm^{-3}},$$

$$\phi_{\rm ms} = -\left(8.62 \times 10^{-5} \times 300\right)\ln\left\{\frac{1 \times 10^{20} \times 2 \times 10^{17}}{(1.45 \times 10^{10})^2}\right\} \tag{5.16}$$

$$= -0.02586\ln\left(9.512485 \times 10^{16}\right) = -1.011\ {\rm V}$$

$$\phi_{\rm F} = \left(8.62 \times 10^{-5} \times 300\right)\ln\left(\frac{2 \times 10^{17}}{1.45 \times 10^{10}}\right) = 0.02586\ln\left(1.37931 \times 10^{7}\right) \tag{5.17}$$

$$= 0.42513\ {\rm V}$$

$$\frac{\partial V_{\rm Th}}{\partial T} = \left[-\frac{1.011}{300} + \left(8.62 \times 10^{-5}\right)\left(3 + \frac{1.12}{8.62 \times 10^{-5} \times 300}\right)\right.$$

$$+\left\{\frac{0.42513}{300} - \frac{3}{2}\left(8.62 \times 10^{-5}\right) - \frac{1.12}{2 \times 300}\right\}$$

$$\times\left\{2 + \left(\frac{5 \times 10^{-6}}{8.854 \times 10^{-14} \times 3.9}\right)\right.$$

$$\left.\left.\times\sqrt{\frac{8.854 \times 10^{-14} \times 11.9 \times 1.6 \times 10^{-19} \times 2 \times 10^{17}}{0.42513}}\right\}\right]{\rm V\,K^{-1}} \tag{5.18}$$

$$= \left\{-0.00337 + \left(8.62 \times 10^{-5}\right)(3 + 43.31)\right.$$

$$+\left(1.4171 \times 10^{-3} - 1.293 \times 10^{-4} - 1.867 \times 10^{-3}\right)$$

$$\left.\times\left(2 + 1.44799 \times 10^{7} \times 2.816 \times 10^{-7}\right)\right\}\ {\rm V\,K^{-1}}$$

$$= \left\{-0.00337 + \left(8.62 \times 10^{-5}\right)(46.31) + \left(-5.792 \times 10^{-4}\right) \times 6.07754\right\}\ {\rm V\,K^{-1}}$$

$$= \left(-0.00337 + 3.9919 \times 10^{-3} - 3.5201 \times 10^{-3}\right){\rm V\,K^{-1}}$$

$$= -2.8982 \times 10^{-3}\ {\rm V\,K^{-1}} = -2.9\ {\rm mV\,K^{-1}}.$$

Figure 5.3 shows how the plot of the subthreshold current curve of a MOSFET with respect to gate voltage is affected by temperature.

5.3 On-resistance ($R_{\rm DS(ON)}$) of a double-diffused vertical MOSFET

The on-resistance of a MOSFET increases when its temperature rises. This variation is shown in figure 5.4.

A vertical MOSFET structure is shown in figure 5.5. Looking at the diagram, we find that the on-resistance $R_{\rm DS(ON)}$ of this vertical MOSFET is the combination of many resistive elements. As these resistors are connected in series, the on-resistance

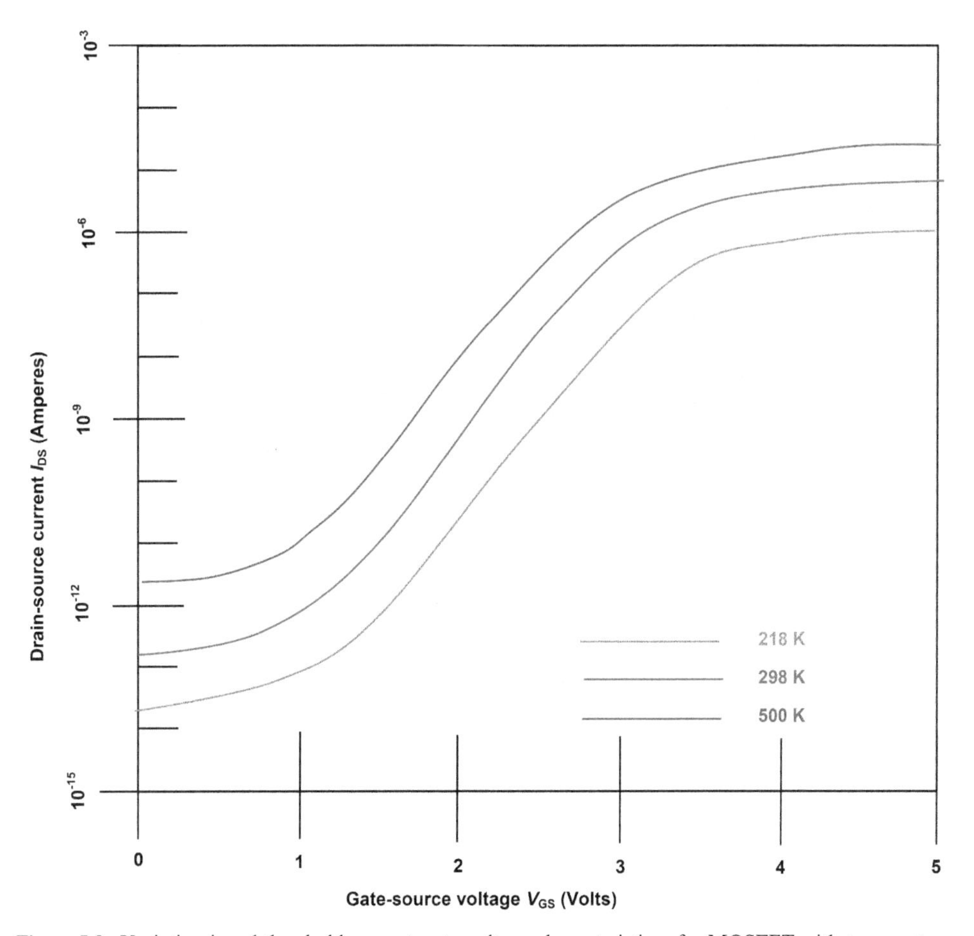

Figure 5.3. Variation in subthreshold current–gate voltage characteristics of a MOSFET with temperature.

is obtained by summation of values of the individual resistors. It can be described by the simple equation

$$\begin{aligned} R_{DS(ON)} = {} & R_{N^+SOURCE} + R_{CHANNEL} + R_{ACCUMULATION} + R_{JFET} + R_{DRIFT} \\ & + R_{SUBSTRATE} + R_{METALLIZATION+WIRE+LEADFRAME} \end{aligned} \tag{5.19}$$

where $R_{N^+SOURCE}$ is the resistance of the heavily doped source region (usually negligible); $R_{CHANNEL}$ is the resistance of the channel region dependent on the ratio of channel width to length, gate oxide thickness and gate drive voltage; $R_{ACCUMULATION}$ is the resistance of the accumulation layer formed in the n^--epi layer and joining the channel with the JFET region (n^--epi region between the p-bodies); R_{JFET} is the resistance of the JFET region with the p-body acting as the gate of the JFET; R_{DRIFT} is the resistance of the drift region; $R_{SUBSTRATE}$ is the resistance of the substrate and $R_{METALLIZATION+WIRE+LEADFRAME}$ is the contact resistance of source/drain metal layer, the resistance of the connecting wires and the resistance of the lead frames, the metal layers inside the package.

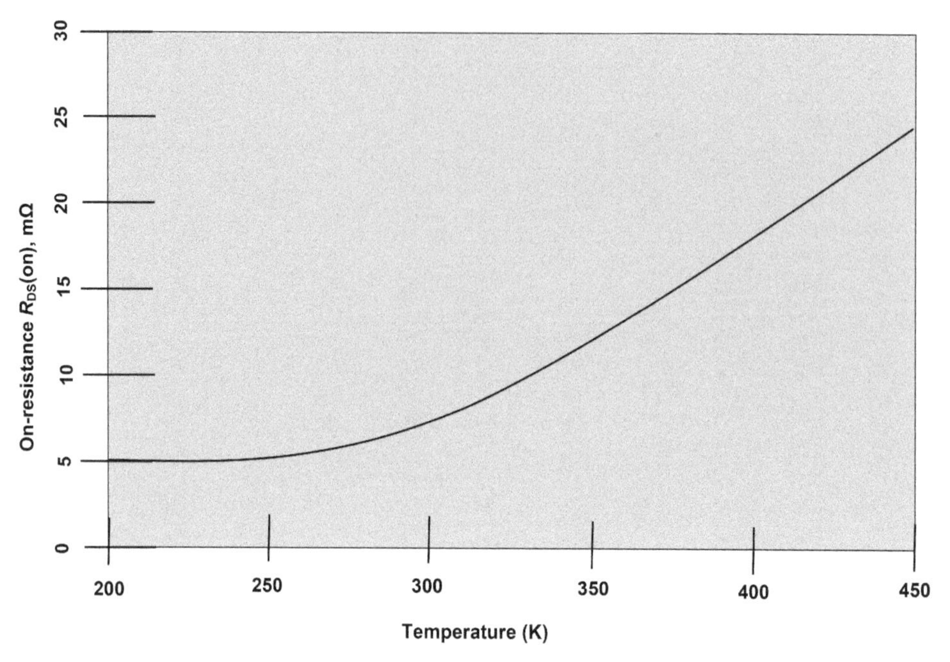

Figure 5.4. Increase in the on-resistance of a MOSFET with temperature.

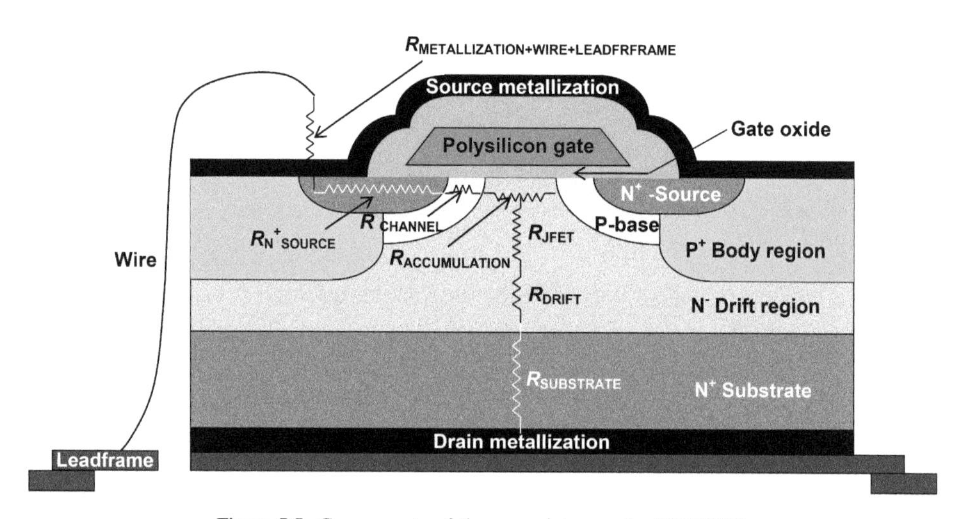

Figure 5.5. Components of the on-resistance of a MOSFET.

The components $R_{CHANNEL}$ and R_{DRIFT} dominate in contribution over other components, and so the equation for on-resistance reduces to the form

$$R_{DS(ON)} = R_{CHANNEL} + R_{DRIFT}. \tag{5.20}$$

To find $R_{CHANNEL}$, let us recall the equation for the static characteristic of a conventional long-channel MOSFET of channel width W, channel length L, electron mobility μ_n in the channel and oxide capacitance C_{ox} per unit area,

$$I_{DS} = (1/2)\mu_n C_{ox}(W/L)\left\{2(V_{GS} - V_{Th})V_{DS} - V_{DS}^2\right\} \text{ for } V_{DS} < (V_{GS} - V_{Th}). \quad (5.21)$$

The on-resistance is a parameter characterizing the on-state operation of a MOSFET switch. In this state, the voltage V_{DS} is $\ll (V_{GS} - V_{Th})$. Hence, V_{DS}^2 term can be neglected, obtaining

$$I_{DS} = (1/2)\mu_n C_{ox}(W/L)\{2(V_{GS} - V_{Th})V_{DS}\} \text{ for } V_{DS} \ll (V_{GS} - V_{Th}). \quad (5.22)$$

The channel resistance is

$$\begin{aligned} R_{CHANNEL} &= \frac{dV_{DS}}{dI_{DS}} = \frac{1}{\frac{dI_{DS}}{dV_{DS}}} = \frac{1}{\frac{d}{dV_{DS}}[(1/2)\mu_n C_{ox}(W/L)\{2(V_{GS} - V_{Th})V_{DS}\}]} \\ &= \frac{2}{\mu_n C_{ox}(W/L)\{2(V_{GS} - V_{Th})\}} \\ &= \frac{1}{\mu_n C_{ox}(W/L)(V_{GS} - V_{Th})}. \end{aligned} \quad (5.23)$$

In this equation, the electron mobility and threshold voltage are temperature-dependent parameters. The mobility $\mu_n(T)$ at temperature T is related to mobility $\mu_n(T_0)$ at temperature T_0 by the equation

$$\mu_n(T) = \mu_n(T_0)(T/T_0)^{-n} \quad (5.24)$$

where $1.5 < n < 2.5$.

The threshold voltage has a TC $\kappa = -2$–4 mV K^{-1}, and the variation of threshold voltage with temperature is written as

$$V_{Th}(T) = V_{Th}(T_0) - \kappa(T - T_0) \quad (5.25)$$

$$\begin{aligned} \therefore R_{CHANNEL}(T) &= \frac{1}{\mu_n(T_0)(T/T_0)^{-n}C_{ox}(W/L)\{V_{GS} - V_{Th}(T_0) + \kappa(T - T_0)\}} \\ &= \frac{(T/T_0)^n}{\mu_n(T_0)C_{ox}(W/L)\{V_{GS} - V_{Th}(T_0)\}\left\{1 + \frac{\kappa(T - T_0)}{V_{GS} - V_{Th}(T_0)}\right\}} \\ &= \frac{1}{\mu_n(T_0)C_{ox}(W/L)\{V_{GS} - V_{Th}(T_0)\}} \times \frac{(T/T_0)^n}{1 + \frac{\kappa(T - T_0)}{V_{GS} - V_{Th}(T_0)}} \\ &= R_{CHANNEL}(T_0) \times \frac{(T/T_0)^n}{1 + \frac{\kappa(T - T_0)}{V_{GS} - V_{Th}(T_0)}}. \end{aligned} \quad (5.26)$$

The resistance

$$R_{\mathrm{DRIFT}} = \text{Resistivity } (\rho) \times \frac{\text{Length } (d)}{\text{Cross sectional area } (S)}$$
$$= \frac{1}{\text{Electronic charge } (q) \times \text{Electron mobility } (\mu_{\mathrm{n}}) \times \text{Donor concentration } (N_{\mathrm{D}})} \times \frac{d}{S} \quad (5.27)$$
$$= \frac{d}{q\mu_{\mathrm{n}} N_{\mathrm{D}} S},$$

where d is the thickness of the drift region, N_{D} is the donor concentration in this region and S is the surface area of the MOSFET. Incorporating the temperature dependence of mobility, we have

$$\begin{aligned} R_{\mathrm{DRIFT}}(T) &= \frac{d}{q\mu_{\mathrm{n}}(T_0)(T/T_0)^{-n} N_{\mathrm{D}} S} \\ &= \frac{d}{q\mu_{\mathrm{n}}(T_0)S} \times (T/T_0)^n = R_{\mathrm{DRIFT}}(T_0) \times (T/T_0)^n. \end{aligned} \quad (5.28)$$

Thus

$$R_{\mathrm{DS(ON)}}(T) = R_{\mathrm{CHANNEL}}(T_0) \times \frac{(T/T_0)^n}{1 + \dfrac{\kappa(T - T_0)}{V_{\mathrm{GS}} - V_{\mathrm{Th}}(T_0)}} + R_{\mathrm{DRIFT}}(T_0) \times (T/T_0)^n. \quad (5.29)$$

Thus on raising the temperature, the on-resistance of the MOSFET is augmented because the channel as well as drift region components of on-resistance increase. Since

$$\frac{\kappa(T - T_0)}{V_{\mathrm{GS}} - V_{\mathrm{Th}}(T_0)} \ll 1 \quad (5.30)$$

$$\begin{aligned} R_{\mathrm{DS(ON)}}(T) &= R_{\mathrm{CHANNEL}}(T_0) \times (T/T_0)^n + R_{\mathrm{DRIFT}}(T_0) \times (T/T_0)^n \\ &= (R_{\mathrm{CHANNEL}} + R_{\mathrm{DRIFT}})(T/T_0)^n = R_{\mathrm{DS(ON)}}(T_0)(T/T_0)^n. \end{aligned} \quad (5.31)$$

The $R_{\mathrm{DS(ON)}}(T)$ of an n- or p-channel power MOSFET at a given temperature T can be calculated from its value $R_{\mathrm{DS(ON)}}(T_0)$ at room temperature $T_0 = 27$ °C by using the approximate equation (Fairchild Semiconductor Corporation 2000)

$$R_{\mathrm{DS(ON)}}(T) = R_{\mathrm{DS(ON)}}(300)(T/300)^{2.3}. \quad (5.32)$$

In table 5.1, the $R_{\mathrm{DS(ON)}}$ values of a power MOSFET at different operating temperatures T are estimated from the formula (5.32) in terms of its $R_{\mathrm{DS(ON)}}$ value at 300 K.

An increase in the on-resistance of a MOSFET with temperature is a boon when paralleling several MOSFETs. A MOSFET passing more current is heated up. Heating of the MOSFET elevates its on-resistance. The increased on-resistance of this particular MOSFET automatically lowers the current being carried by it, and prevents it from being damaged by overdriving. Thus a MOSFET has a positive TC

Table 5.1. Calculated factors by which the on-resistance of a MOSFET changes with respect to its room-temperature value (T = 27 °C).

Temperature T (K)	4.2	77.2	150	300	600
$R_{DS(ON)}(T)/R_{DS(ON)}$ (T = 300 K)	5.4462×10^{-5}	4.406978×10^{-2}	0.2030631	1	4.92457765

of on-resistance. The positive coefficient limits the current conducted by individual MOSFETs connected in a parallel arrangement and ensures proper sharing of current amongst the MOSFETs.

5.4 Transconductance (g_m) of a MOSFET

In many circuit applications of MOSFETs, the input signal is the gate–source voltage V_{GS} and the output signal is the drain–source current I_{DS}. For an applied drain–source voltage V_{DS}, the capability of a MOSFET to amplify the input signal is measured by its transconductance g_m defined as the ratio

$$g_m = \left(\frac{\partial I_{DS}}{\partial V_{GS}}\right)_{V_{DS}}. \tag{5.33}$$

If the value of the applied drain–source voltage provides MOSFET operation in the saturation region, the transconductance is said to be saturated transconductance. The transconductance–drain current curves of a MOSFET vary with temperature as shown in figure 5.6.

From equations (5.22) and (5.33), keeping V_{DS} constant,

$$\begin{aligned} g_m &= \frac{\partial}{\partial V_{GS}}\left[(1/2)\mu_n C_{ox}(W/L)\{2(V_{GS} - V_{Th})V_{DS}\}\right] \\ &= (1/2)\mu_n C_{ox}(W/L)\{2(1-0)V_{DS}\} = \mu_n C_{ox}(W/L)V_{DS}. \end{aligned} \tag{5.34}$$

Making mobility temperature-dependent

$$\begin{aligned} g_m(T) &= \mu_n(T_0)(T/T_0)^{-n}C_{ox}(W/L)V_{DS} = \mu_n(T_0)C_{ox}(W/L)V_{DS} \times (T/T_0)^{-n} \\ &= g_m(T_0) \times (T/T_0)^{-n}. \end{aligned} \tag{5.35}$$

When the temperature increases, the mobility of carriers decreases. Hence, the transconductance of a MOSFET falls. Like the on-resistance of a MOSFET, a crude relation representing this deterioration of transconductance $g_m(T_0)$ at temperature T_0 to the new value $g_m(T)$ at temperature T is (Fairchild Semiconductor Corporation 2000)

$$g_m(T) = g_m(300)(T/300)^{-2.3}. \tag{5.36}$$

This equation predicts the changes in transconductance by different factors relative to the value of transconductance at 300 K. A rough estimate of the trends can be seen from table 5.2.

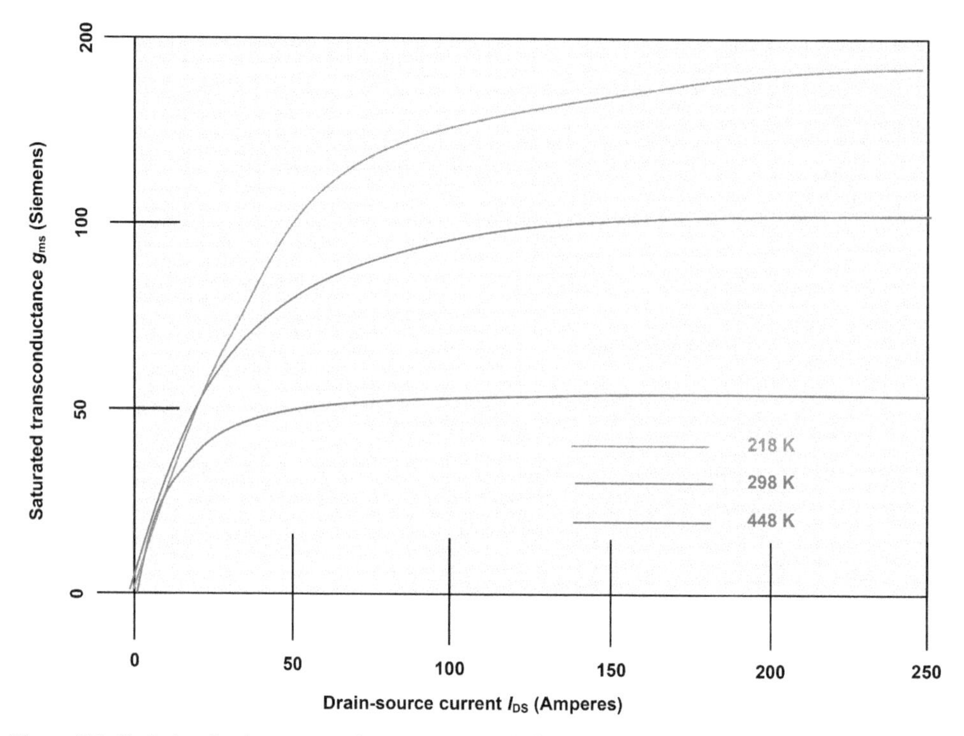

Figure 5.6. Variation in the transconductance versus drain current characteristics of a MOSFET with temperature.

Table 5.2. Effect of temperature on the transconductance of a MOSFET.

Temperature T (K)	4.2	77.2	150	300	600
$g_m(T)/g_m$ (300)	1.8136×10^4	22.69	4.92458	1	0.203063

5.5 BV_{DSS} and I_{DSS} of a MOSFET

BV_{DSS} stands for the drain–source breakdown voltage of a MOSFET with the gate shorted to the source. It is the maximum drain–source voltage which the device can withstand without the body–drain diode reaching the avalanche breakdown condition. I_{DSS} is the corresponding drain–source leakage current (Fairchild Semiconductor Corporation 2000). The astute reader may analyze the breakdown voltage and leakage current variations in MOSFET from the understanding of the corresponding parameters for p–n junction diodes and bipolar transistors.

5.6 Zero temperature coefficient biasing point of MOSFET

One possibility to operate a MOSFET-based circuit at a high temperature utilizes the property that the drain–source current exhibits zero or very small changes with

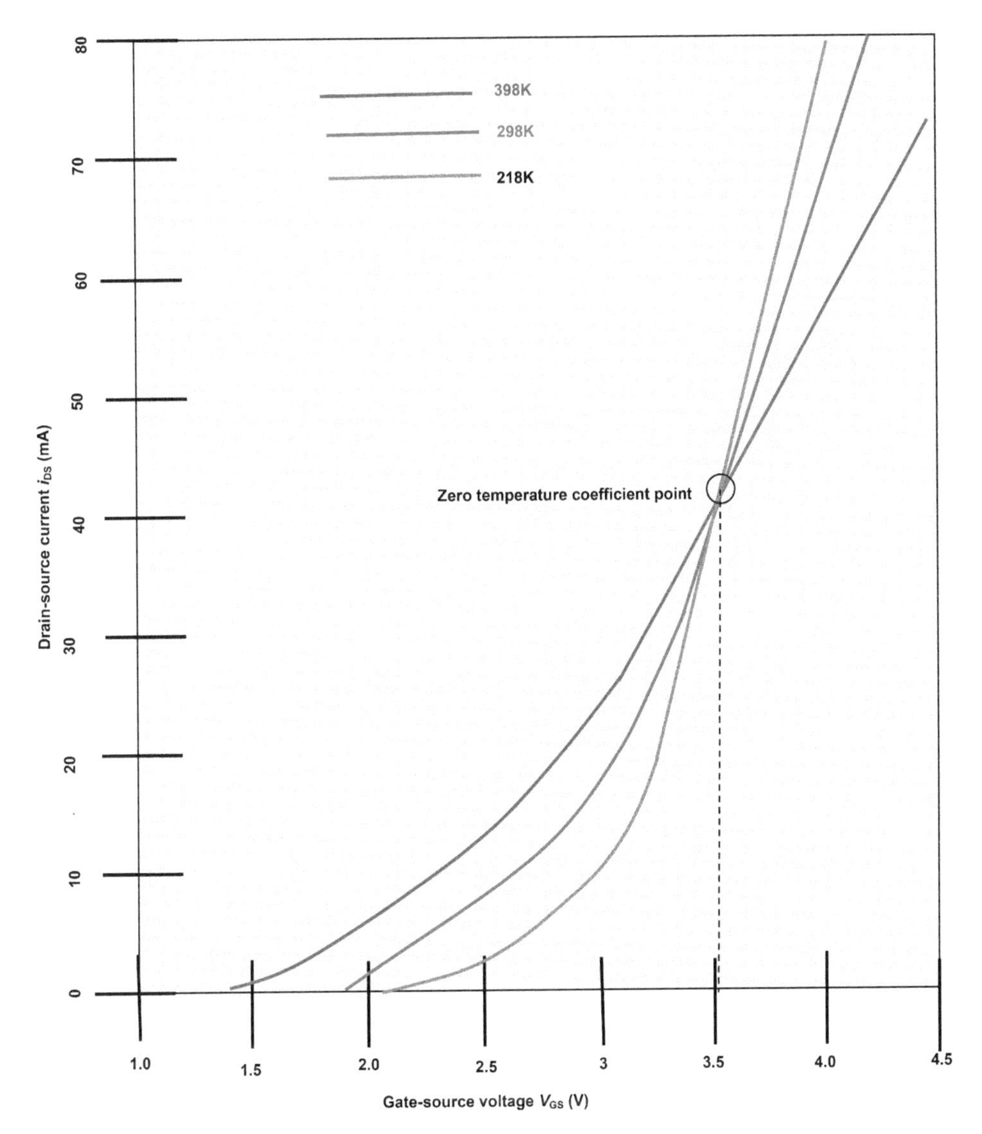

Figure 5.7. Transfer characteristics of a MOSFET at three temperatures.

temperature if the device is operated at a particular gate-source voltage, V_{GSO}. This value of gate–source voltage is called the zero TC biasing point (ZTC); see figure 5.7. Thus the ZTC of a MOSFET is the gate–source voltage at which its drain–source current is almost invariant with respect to temperature.

To find the ZTC of a MOSFET, we perform a simplified analysis in which the equation for the drain–source current I_{DS} is differentiated with respect to temperature, keeping $V_{GS} = \text{constant}$ and $V_{DS} = \text{constant}$. We rewrite the MOSFET current equation (5.22) in saturation mode operation by substituting the temperature-dependent expressions for mobility and threshold voltage in equations (5.24) and (5.25):

$$\begin{aligned} I_{DS}(T) &= (1/2)\mu_n(T_0)(T/T_0)^{-n}C_{ox}(W/L)\left[2\{V_{GS} - V_{Th}(T_0) + \kappa(T - T_0)\}V_{DS}\right] \\ &= \mu_n(T_0)(T/T_0)^{-n}C_{ox}(W/L)\{V_{GS} - V_{Th}(T_0)\}V_{DS} \\ &\quad + \mu_n(T_0)(T/T_0)^{-n}C_{ox}(W/L)\kappa T V_{DS} \\ &\quad - \mu_n(T_0)(T/T_0)^{-n}C_{ox}(W/L)\kappa T_0 V_{DS} \end{aligned} \tag{5.37}$$

$$\begin{aligned} \therefore \frac{dI_{DS}}{dT} &= \mu_n(T_0)\left(\frac{1}{T_0}\right)^{-n}(-nT^{-n-1})C_{ox}(W/L)\{V_{GS} - V_{Th}(T_0)\}V_{DS} \\ &\quad + \mu_n(T_0)\left(\frac{1}{T_0}\right)^{-n}(-nT^{-n-1})C_{ox}(W/L)\kappa T V_{DS} + \mu_n(T_0)(T/T_0)^{-n}C_{ox}(W/L)\kappa V_{DS} \\ &\quad - \mu_n(T_0)\left(\frac{1}{T_0}\right)^{-n}(-nT^{-n-1})C_{ox}(W/L)\kappa T_0 V_{DS} \\ &= -n\mu_n(T_0)C_{ox}(W/L)\left(\frac{T}{T_0}\right)^{-n}(T^{-1})\{V_{GS} - V_{Th}(T_0)\}V_{DS} \\ &\quad - n\mu_n(T_0)C_{ox}(W/L)\left(\frac{T}{T_0}\right)^{-n}(T^{-1})\kappa T V_{DS} + \kappa\mu_n(T_0)(T/T_0)^{-n}C_{ox}(W/L)V_{DS} \\ &\quad + n\mu_n(T_0)C_{ox}(W/L)\left(\frac{T}{T_0}\right)^{-n}(T^{-1})\kappa T_0 V_{DS} \\ &= -\frac{n}{T}\mu_n(T_0)C_{ox}(W/L)\left(\frac{T}{T_0}\right)^{-n}\{V_{GS} - V_{Th}(T_0)\}V_{DS} \\ &\quad - n\mu_n(T_0)C_{ox}(W/L)\left(\frac{T}{T_0}\right)^{-n}\kappa V_{DS} + \kappa\mu_n(T_0)(T/T_0)^{-n}C_{ox}(W/L)V_{DS} \\ &\quad + \left(\frac{nT_0}{T}\right)\mu_n(T_0)C_{ox}(W/L)\left(\frac{T}{T_0}\right)^{-n}\kappa V_{DS} \\ &= \mu_n(T_0)C_{ox}(W/L)\left(\frac{T}{T_0}\right)^{-n}V_{DS}\left[\left(-\frac{n}{T}\right)\{V_{GS} - V_{Th}(T_0)\} - n\kappa + \kappa + \left(\frac{nT_0}{T}\right)\kappa\right] \\ &= \mu_n(T_0)C_{ox}(W/L)\left(\frac{T}{T_0}\right)^{-n}V_{DS}\left[\left(-\frac{n}{T}\right)\{V_{GS} - V_{Th}(T_0)\} + \left\{-n + 1 + \left(\frac{nT_0}{T}\right)\right\}\kappa\right] \\ &= \mu_n(T_0)C_{ox}(W/L)\left(\frac{T}{T_0}\right)^{-n}V_{DS}\left[\left\{1 - n + \left(\frac{nT_0}{T}\right)\right\}\kappa - \frac{n}{T}\{V_{GS} - V_{Th}(T_0)\}\right]. \end{aligned} \tag{5.38}$$

For the ZTC

$$\frac{dI_{DS}}{dT} = 0 \tag{5.39}$$

Hence,

$$\left\{1 - n + \left(\frac{nT_0}{T}\right)\right\}\kappa - \frac{n}{T}\{V_{GS} - V_{Th}(T_0)\} = 0 \tag{5.40}$$

or

$$\begin{aligned}
\frac{n}{T}\{V_{GS} - V_{Th}(T_0)\} &= \left\{1 - n + \left(\frac{nT_0}{T}\right)\right\}\kappa \\
V_{GS} - V_{Th}(T_0) &= \frac{T}{n}\left\{1 - n + \left(\frac{nT_0}{T}\right)\right\}\kappa = \frac{T\kappa}{n} - T\kappa + T_0\kappa = \frac{T\kappa}{n} - \kappa(T - T_0) \\
\text{Or, } V_{GS} &= V_{Th}(T_0) - \kappa(T - T_0) + \frac{T\kappa}{n} \\
V_{GS} &= V_{Th}(T) + \frac{T\kappa}{n},
\end{aligned} \tag{5.41}$$

using equation (5.25). Since this V_{GS} value corresponds to the ZTC, it will be designated by $V_{GS}(\text{ZTC})$. Hence,

$$V_{GS}(\text{ZTC}) = V_{Th}(T) + \frac{T\kappa}{n}. \tag{5.42}$$

Taking $n = 2$, $\kappa = -4$ mV K^{-1}, for a $V_{Th} = 1$ V MOSFET at room temperature $T =$ 300 K,

$$V_{GS}(\text{ZTC}) = 1 + \frac{300 \times (4 \times 10^{-3})}{2} = 1.60 \text{ V}. \tag{5.43}$$

5.7 Dynamic response of a MOSFET

MOSFETs exhibit faster switching speeds than bipolar devices. The reason is that there is no delay in switching caused by storage of minority carriers, as experienced in p–n junction diodes and bipolar transistors. Primarily the intrinsic capacitance C_g and intrinsic resistance R_g of a MOSFET determine its switching characteristics (Sattar and Tsukanov 2007). The main components of intrinsic capacitance C_g are:

(i) input capacitance C_{iss} consisting of capacitance C_{gs} between the gate and source, and capacitance C_{gd} between the gate and drain (figure 5.8),

$$C_{iss} = C_{gs} + C_{gd}; \tag{5.44}$$

(ii) output capacitance comprising drain–source capacitance C_{ds} and gate–drain capacitance C_{gd},

$$C_{oss} = C_{ds} + C_{gd}; \tag{5.45}$$

(iii) reverse-transfer capacitance

$$C_{rss} = C_{gd}. \tag{5.46}$$

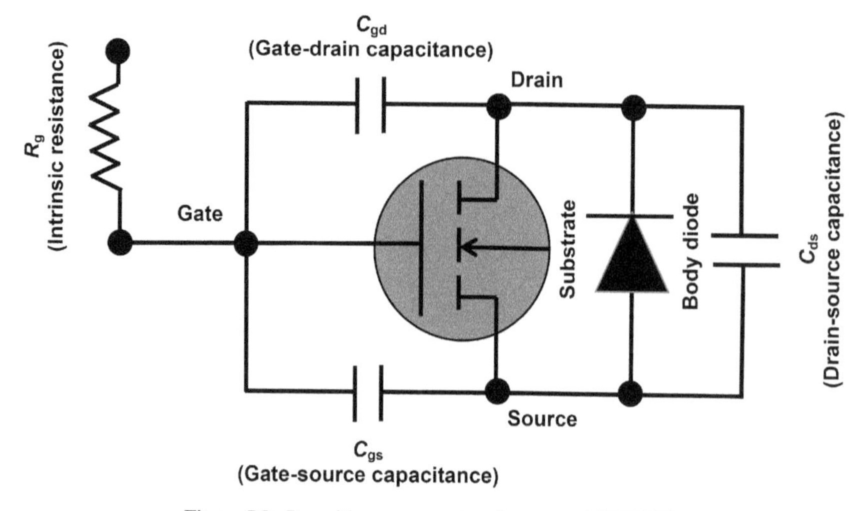

Figure 5.8. Parasitic components of a power MOSFET.

> Intrinsic resistance R_g is a component of the total gate resistance, which includes externally connected resistances to the gate including the driver resistance. Due to its small magnitude, it exerts an imperceptible effect on the switching time.

The resistance–capacitance network formed between R_g and the input capacitance C_{iss} of the MOSFET has a time constant τ given by

$$\tau = R_g C_{iss}. \tag{5.47}$$

The charging and discharging times of the capacitance C_{iss}, i.e. the supply or removal of charges on these capacitances, through the resistance R_g decide how fast the voltage is established on the gate insulator to cause a current flow between the source and drain terminals. As the input capacitance is not affected by temperature variations, the switching speed of the MOSFET remains practically unaltered when the ambient temperature changes. Any subtle changes in C_{iss} or R_g may give rise to slight changes in switching parameters, but these are generally negligible. Naturally, the switching loss of the MOSFET is also invariant with temperature.

5.8 MOS analog circuits in the 25 °C to 300 °C range

Shoucair (1986) showed that analog CMOS circuits can be designed for operation up to 250 °C using commercial CMOS technologies, and considering design parameters such as transconductance, output conductance and gain. Below we explain the detailed analysis of the thermal behavior of a MOSFET, after Shoucair. In this analysis, the symbol for the TC of threshold voltage is p_0 in place of the κ used previously. To a good approximation, the variation of threshold voltage with temperature is described by a first-order polynomial of the form

$$V_{Th}(T) = p_0 T + q_0 \tag{5.48}$$

where

$$p_0 = \mathrm{d}V_{\mathrm{Th}}/\mathrm{d}T = -2.4\ \mathrm{mV\ K^{-1}},\ q_0 = +1.72\ \mathrm{V\ for\ an\ n\text{-}channel\ MOSFET} \quad (5.49)$$

$$p_0 = \mathrm{d}V_{\mathrm{Th}}/\mathrm{d}T = +2.4\ \mathrm{mV\ K^{-1}},\ q_0 = -1.72\mathrm{V\ for\ a\ p\text{-}channel\ MOSFET}. \quad (5.50)$$

Compared to the room-temperature (25 °C) value, V_{Th} changes by ~0.55 V at 250 °C for either polarity of MOSFET. The change in threshold voltage affects the analog circuits through a shift in biasing voltages with the exception of the ZTC biasing points described below. For comparison, it may be recalled that in digital circuits, there is a decrease in the noise margins. The average carrier mobility follows the $T^{-1.5}$ relationship:

$$\mu(T) = \mu(T_0)\left(\frac{T}{T_0}\right)^{-1.5}. \quad (5.51)$$

At a ZTC point, the drain current remains constant in spite of temperature changes. Hence, the normalized derivative of the drain–source current I_{DS} with respect to temperature T can be equated to zero. For the linear region, this derivative is

$$\begin{aligned}\frac{\mathrm{d}I_{\mathrm{DS}}}{\mathrm{d}T} &= \frac{\mathrm{d}}{\mathrm{d}T}\left[\mu(T)C_{\mathrm{ox}}\left(\frac{W}{L}\right)\{V_{\mathrm{GS(ZTC)}} - V_{\mathrm{Th}}(T)\}\,V_{\mathrm{DS}}\right]\\ &= \frac{\mathrm{d}\mu(T)}{\mathrm{d}t}C_{\mathrm{ox}}\left(\frac{W}{L}\right)\{V_{\mathrm{GS(ZTC)}} - V_{\mathrm{Th}}(T)\}\,V_{\mathrm{DS}}\\ &\quad + \mu C_{\mathrm{ox}}\left(\frac{W}{L}\right)\left(0 - \frac{\mathrm{d}V_{\mathrm{Th}}(T)}{\mathrm{d}T}\right)V_{\mathrm{DS}}\\ &= \frac{\mathrm{d}\mu}{\mathrm{d}t}C_{\mathrm{ox}}\left(\frac{W}{L}\right)\{V_{\mathrm{GS(ZTC)}} - V_{\mathrm{Th}}(T)\}\,V_{\mathrm{DS}} - \mu C_{\mathrm{ox}}\left(\frac{W}{L}\right)\left\{\frac{\mathrm{d}V_{\mathrm{Th}}(T)}{\mathrm{d}T}\right\}V_{\mathrm{DS}}.\end{aligned} \quad (5.52)$$

Normalizing this derivative, we have

$$\frac{1}{I_{\mathrm{DS}}}\left(\frac{\mathrm{d}I_{\mathrm{DS}}}{\mathrm{d}T}\right) = \frac{1}{\mu}\left(\frac{\mathrm{d}\mu}{\mathrm{d}t}\right) - \frac{1}{V_{\mathrm{GS(ZTC)}} - V_{\mathrm{Th}}(T)}\left\{\frac{\mathrm{d}V_{\mathrm{Th}}(T)}{\mathrm{d}T}\right\}. \quad (5.53)$$

Here,

$$\frac{1}{\mu}\left(\frac{\mathrm{d}\mu}{\mathrm{d}t}\right) = \frac{1}{\mu(T_0)\left(\frac{T}{T_0}\right)^{-1.5}} \times \mu(T_0) \times -1.5\left(\frac{T}{T_0}\right)^{-1.5-1} \times \frac{T_0 - 0}{T_0^2} = -\frac{1.5}{T_0}. \quad (5.54)$$

Putting

$$\frac{\mathrm{d}V_{\mathrm{Th}}}{\mathrm{d}T} = p_0 \quad (5.55)$$

$$\frac{1}{I_{\mathrm{DS}}}\left(\frac{\mathrm{d}I_{\mathrm{DS}}}{\mathrm{d}T}\right) = -\frac{1.5}{T_0} - \frac{p_0}{V_{\mathrm{GS(ZTC)}} - V_{\mathrm{Th}}(T)} \tag{5.56}$$

$$\therefore -\frac{1.5}{T_0} - \frac{p_0}{V_{\mathrm{GS(ZTC)}} - V_{\mathrm{Th}}(T)} = 0, \tag{5.57}$$

or

$$\frac{p_0}{V_{\mathrm{GS(ZTC)}} - V_{\mathrm{Th}}} = -\frac{1.5}{T_0} \tag{5.58}$$

$$\therefore V_{\mathrm{GS(ZTC)}} - V_{\mathrm{Th}} = -\frac{T_0}{1.5}p_0 \tag{5.59}$$

$$\therefore I_{\mathrm{DS}} = \mu(T)C_{\mathrm{ox}}\left(\frac{W}{L}\right)\left\{-\frac{T_0}{1.5}p_0\right\}V_{\mathrm{DS}}. \tag{5.60}$$

In the saturation region, the temperature derivative of drain current is

$$\frac{\mathrm{d}I_{\mathrm{DS}}}{\mathrm{d}T} = \frac{\mathrm{d}}{\mathrm{d}T}\left[(1/2)\mu(T)C_{\mathrm{ox}}\left(\frac{W}{L}\right)\left\{V_{\mathrm{GS(ZTC)}} - V_{\mathrm{Th}}(T)\right\}^m\right] \tag{5.61}$$

where $m = 2$,

$$\begin{aligned}\frac{\mathrm{d}I_{\mathrm{DS}}}{\mathrm{d}T} &= \frac{\mathrm{d}\mu(T)}{\mathrm{d}t}(1/2)C_{\mathrm{ox}}\left(\frac{W}{L}\right)\left\{V_{\mathrm{GS(ZTC)}} - V_{\mathrm{Th}}(T)\right\}^m \\ &\quad + (1/2)\mu(T)C_{\mathrm{ox}}\left(\frac{W}{L}\right)m\left\{V_{\mathrm{GS(ZTC)}} - V_{\mathrm{Th}}(T)\right\}^{m-1} \times -\frac{\mathrm{d}V_{\mathrm{Th}}(T)}{\mathrm{d}T}\end{aligned} \tag{5.62}$$

$$\begin{aligned}\therefore \frac{1}{I_{\mathrm{DS}}}\frac{\mathrm{d}I_{\mathrm{DS}}}{\mathrm{d}T} &= \frac{1}{\mu(T)}\left\{\frac{\mathrm{d}\mu(T)}{\mathrm{d}t}\right\} - \frac{m}{V_{\mathrm{GS(ZTC)}} - V_{\mathrm{Th}}(T)}\frac{\mathrm{d}V_{\mathrm{Th}}(T)}{\mathrm{d}T} \\ &= -\frac{1.5}{T_0} - \frac{m}{V_{\mathrm{GS(ZTC)}} - V_{\mathrm{Th}}(T)} \times p_0\end{aligned} \tag{5.63}$$

using equation (5.54). Setting

$$\frac{1}{I_{\mathrm{DS}}}\frac{\mathrm{d}I_{\mathrm{DS}}}{\mathrm{d}T} = 0 \tag{5.64}$$

$$-\frac{1.5}{T_0} - \frac{m}{V_{\mathrm{GS(ZTC)}} - V_{\mathrm{Th}}(T)} \times p_0 = 0, \tag{5.65}$$

or

$$\frac{mp_0}{V_{GS(ZTC)} - V_{Th}(T)} = -\frac{1.5}{T_0}$$
$$V_{GS(ZTC)} - V_{Th}(T) = -\frac{mT_0}{1.5}p_0 \tag{5.66}$$

$$\therefore I_{DS} = (1/2)\mu(T)C_{ox}\left(\frac{W}{L}\right)\left\{-\frac{mT_0}{1.5}p_0\right\}^m. \tag{5.67}$$

For a MOSFET with known parameters (p_0, q_0) over the temperature range (T_1, T_2), there exist two separate ZTC gate–source biasing voltages V_{GS}(ZTC), one located in the linear region and other in the saturation region, at which the drain current of the MOSFET exhibits the least sensitivity to temperature. These biasing voltages obey the following analytical equation derived by least squares minimization over (T_1, T_2) (Shoucair 1986)

$$V_{GS}(\text{ZTC}) = -(1/6)p_0(T_1 + T_2)(2m - 3) + q_0, \tag{5.68}$$

where m is the index in the approximate equation for drain–source current I_{DS}

$$I_{DS} \sim (V_{GS} - V_{Th})^m. \tag{5.69}$$

The index $m = 1$ in the linear region and $m = 2$ in the saturation region of MOSFET operation under ideal square law conditions. The equation (5.68) for V_{GS}(ZTC) has two forms corresponding to the two values of $m = 1, 2$:

$$\begin{aligned} V_{GS}(\text{ZTC})_{|\text{Linear}(m=1)} &\approx -(1/6)p_0(T_1 + T_2)(2 \times 1 - 3) + q_0 \\ &= +(1/6)p_0(T_1 + T_2) + q_0 \end{aligned} \tag{5.70}$$

$$\begin{aligned} V_{GS}(\text{ZTC})_{|\text{Saturation}(m=2)} &\approx -(1/6)p_0(T_1 + T_2)(2 \times 2 - 3) + q_0 \\ &= -(1/6)p_0(T_1 + T_2) + q_0. \end{aligned} \tag{5.71}$$

As an example, for an n-channel MOSFET operated in the saturation region in the temperature range 273–523 K,

$$\begin{aligned} V_{GS}(\text{ZTC})_{|\text{Saturation}(m=2)} &= -(1/6) \times -2.4 \times 10^{-3} \times (273 + 523) + 1.72 \\ &= -0.4 \times 10^{-3} \times 796 + 1.72 \\ &= -0.3184 + 1.72 = 1.4016\ \text{V}. \end{aligned} \tag{5.72}$$

Drain currents for V_{GS} (ZTC) values are obtained from the drain current equations in the linear and saturation regions. Measurements of transfer characteristics of MOSFETs showed that there was a pronounced upswing in drain current above 250 °C. This originated from the marked increase in drain–body leakage current to the extent that it became comparable in magnitude with the forward MOSFET current. Leakage currents of source or drain junctions comprise contributions from generation and diffusion components, depending upon temperatures. For CMOS processes, a transition temperature $T_{trans} \sim 130\,°C$ to $150\,°C$ was found. Below the transition temperature, generation current dominated:

$$I_{gen}(T) = (qAw/\tau)n_i(T) \text{ for } T = 25\,°C \text{ to } 150\,°C, \quad (5.73)$$

where A is the effective area of the p–n junction, w is the depletion region width and τ is the carrier lifetime.

Above the transition temperature, the diffusion component was predominant. For an n-channel MOSFET

$$I_{diff}(T) = (qAD_n/L_n)\{n_i^2(T)/N_A\} \text{ for } T = 150\,°C \text{ to } 300\,°C, \quad (5.74)$$

where D_n, L_n are the diffusion coefficient and diffusion length of electrons, and N_A is the acceptor concentration. An identical equation applies to the p-channel device. The leakage currents and hence leakage conductances increase by five orders of magnitude in the temperature range 25 °C to 300 °C. Consequently, gain is degraded. In analog circuits, the biasing points are shifted, whereas in digital circuits the noise margins decrease due to leakage current upswing. One possible reason for the higher leakage currents shown by n-channel MOSFETs is the additional processing required during their fabrication, namely, p-well formation, which introduces more defects, thereby decreasing carrier lifetime. Guard rings surrounding the source and drain diffusions help in preventing the minority carriers from reaching the crucial regions, and consequently avert damaging CMOS latchups.

Transconductance in the saturation region halves in the range 25 °C to 300 °C. It is given by

$$g_m(T) = 2I_{DS}(T)/\{V_{GS} - V_{Th}\}. \quad (5.75)$$

The body-effect conductance g_{mb} is expressed in terms of transconductance g_m as

$$g_{mb}(T) \approx \{\partial V_{Th}(T)/\partial V_{sub}\}g_m(T), \quad (5.76)$$

where V_{sub} is the substrate voltage. Like transconductance, the body-effect transconductance is also halved in the above temperature range.

Also halved is the channel output conductance (Shoucair and Early 1984)

$$g_d(T) = I_D(T)/V_a, \quad (5.77)$$

where V_a is the Early voltage. The overall impact is that the gain–bandwidth product is halved.

Shoucair (1986) laid out the guidelines for designing a two-stage topology of a CMOS OP-AMP. For stabilizing the circuit operation against temperature variations, the two gain stages are biased at their ZTC drain current values in the saturation region by applying a voltage $V_{GS}(\text{ZTC})|_{sat}$ to the current source biasing each stage, either by employing a voltage dividing string of MOSFETs or by using passive devices like polysilicon resistors. In the differential input stage of the CMOS OP-AMP, a diode is added to compensate for leakage currents. The low-frequency small-signal gain depends on the output conductances besides the transconductance and small-signal leakage-induced conductances, which become appreciably high at temperatures above 200 °C. In the output stage, provision is made to allow that the two drain–body junctions leak equal amounts of current. Leakage areas are matched if the leakage current densities are equal.

The MOSFET capacitance consists of overlap and junction capacitances. The overlap capacitances arise from overlap of metallization over the gate–source and gate–drain regions whereas junction capacitances originate from the source–body and drain–body junctions. The overlap capacitances have a small TC ~25 ppm °C^{-1} causing a smaller than 5% capacitance increase in the temperature range 25 °C to 300 °C while junction capacitances have much larger TCs ~100–1500 ppm °C^{-1} producing a 5%–50% increase in this range of temperatures. The former can be described as relatively weak temperature effect and the latter as a weak temperature effect. The influence of the increase in capacitance on analog circuits is observed as a slowing down of the circuit speed.

Considering that the source is shorted to the body, $V_{\text{Body-Source}} = 0$. Then the capacitance C_{DS} is the drain–body junction capacitance. If this junction is assumed to be an abrupt one,

$$C_{\text{DS}} = \sqrt{\frac{q\varepsilon_0\varepsilon_{\text{Si}}(N_{\text{A}} + N_{\text{D}})}{2(V_{\text{bi}} + V_{\text{R}})}}, \tag{5.78}$$

where ε_{Si} is the dielectric constant of Si, N_{A} is the acceptor concentration, N_{D} is the donor concentration, V_{bi} is the built-in potential and V_{R} is the applied reverse bias. The potential V_{bi} is given by

$$V_{\text{bi}} = \left(\frac{k_{\text{B}}T}{q}\right)\ln\left\{\frac{N_{\text{A}}N_{\text{D}}}{n_{\text{i}}^2(T)}\right\}. \tag{5.79}$$

Differentiating both sides of the equation (5.78) for C_{DS}, we obtain (Shoucair *et al* 1984)

$$\begin{aligned}
\frac{\mathrm{d}C_{\text{DS}}(T)}{\mathrm{d}T} &= \left(\frac{1}{2}\right)\left\{\frac{q\varepsilon_0\varepsilon_{\text{Si}}(N_{\text{A}} + N_{\text{D}})}{2(V_{\text{bi}} + V_{\text{R}})}\right\}^{1/2-1} \\
&\quad \times \frac{0 - 2(\mathrm{d}V_{\text{bi}}/\mathrm{d}T)\{q\varepsilon_0\varepsilon_{\text{Si}}(N_{\text{A}} + N_{\text{D}})\}}{\{2(V_{\text{bi}} + V_{\text{R}})\}^2} \\
&= \left(\frac{1}{2}\right)\left\{\frac{2(V_{\text{bi}} + V_{\text{R}})}{q\varepsilon_0\varepsilon_{\text{Si}}(N_{\text{A}} + N_{\text{D}})}\right\}^{1/2} \times \frac{-2(\mathrm{d}V_{\text{bi}}/\mathrm{d}T)\{q\varepsilon_0\varepsilon_{\text{Si}}(N_{\text{A}} + N_{\text{D}})\}}{\{2(V_{\text{bi}} + V_{\text{R}})\}^2} \\
&= -(\mathrm{d}V_{\text{bi}}/\mathrm{d}T)\frac{\{q\varepsilon_0\varepsilon_{\text{Si}}(N_{\text{A}} + N_{\text{D}})\}^{1/2}}{\{2(V_{\text{bi}} + V_{\text{R}})\}^{1.5}}
\end{aligned} \tag{5.80}$$

$$\therefore \left(\frac{1}{C_{\text{DS}}}\right)\frac{\mathrm{d}C_{\text{DS}}(T)}{\mathrm{d}T} = -\frac{1}{2(V_{\text{bi}} + V_{\text{R}})}(\mathrm{d}V_{\text{bi}}/\mathrm{d}T), \tag{5.81}$$

but

$$\frac{\mathrm{d}V_{\mathrm{bi}}}{\mathrm{d}T} = \left(\frac{k_{\mathrm{B}}}{q}\right)\ln\left\{\frac{N_{\mathrm{A}}N_{\mathrm{D}}}{n_{\mathrm{i}}^2(T)}\right\} + \left(\frac{k_{\mathrm{B}}T}{q}\right)\left\{\frac{n_{\mathrm{i}}^2(T)}{N_{\mathrm{A}}N_{\mathrm{D}}}\right\} \times \frac{0 - 2n_{\mathrm{i}}(T)(\mathrm{d}n_i/\mathrm{d}T)N_{\mathrm{A}}N_{\mathrm{D}}}{\left\{n_{\mathrm{i}}^2(T)\right\}^2}$$
$$= \left(\frac{k_{\mathrm{B}}}{q}\right)\ln\left\{\frac{N_{\mathrm{A}}N_{\mathrm{D}}}{n_{\mathrm{i}}^2(T)}\right\} - \left(\frac{k_{\mathrm{B}}T}{q}\right)\frac{2(\mathrm{d}n_i/\mathrm{d}T)}{n_{\mathrm{i}}(T)}. \tag{5.82}$$

Since,

$$n_{\mathrm{i}}(T) = 4.68 \times 10^{15}T^{3/2}\exp\left(-\frac{E_{\mathrm{g}}}{2k_{\mathrm{B}}T}\right) \tag{5.83}$$

$$\therefore \frac{\mathrm{d}n_{\mathrm{i}}(T)}{\mathrm{d}T} = 4.68 \times 10^{15}\left(\frac{3}{2}\right)T^{3/2-1}\exp\left(-\frac{E_{\mathrm{g}}}{2k_{\mathrm{B}}T}\right) + 4.68$$
$$\times 10^{15}T^{3/2}\left\{\exp\left(-\frac{E_{\mathrm{g}}}{2k_{\mathrm{B}}T}\right)\right\} \times \frac{+2k_{\mathrm{B}}E_{\mathrm{g}}}{(2k_{\mathrm{B}}T)^2}$$
$$= 4.68 \times 10^{15}\left(\frac{3}{2}\right)T^{1/2}\exp\left(-\frac{E_{\mathrm{g}}}{2k_{\mathrm{B}}T}\right) + 4.68 \tag{5.84}$$
$$\times 10^{15}\left\{\exp\left(-\frac{E_{\mathrm{g}}}{2k_{\mathrm{B}}T}\right)\right\} \times \frac{E_{\mathrm{g}}}{2k_{\mathrm{B}}T^{1/2}}$$
$$= 4.68 \times 10^{15}\exp\left(-\frac{E_{\mathrm{g}}}{2k_{\mathrm{B}}T}\right)\left\{\left(\frac{3}{2}\right)T^{1/2} + \frac{E_{\mathrm{g}}}{2k_{\mathrm{B}}T^{1/2}}\right\}$$

$$\frac{2(\mathrm{d}n_i/\mathrm{d}T)}{n_{\mathrm{i}}(T)} = \left[2 \times 4.68 \times 10^{15}\exp\left(-\frac{E_{\mathrm{g}}}{2k_{\mathrm{B}}T}\right)\left\{\left(\frac{3}{2}\right)T^{\frac{1}{2}} + \frac{E_{\mathrm{g}}}{2k_{\mathrm{B}}T^{\frac{1}{2}}}\right\}\right]\Big/$$
$$\left\{4.68 \times 10^{15}T^{\frac{3}{2}}\exp\left(-\frac{E_{\mathrm{g}}}{2k_{\mathrm{B}}T}\right)\right\}$$
$$= 2T^{-3/2}\left\{\left(\frac{3}{2}\right)T^{1/2} + \frac{E_{\mathrm{g}}}{2k_{\mathrm{B}}T^{1/2}}\right\} = T^{-3/2}\left(3T^{1/2} + \frac{E_{\mathrm{g}}}{k_{\mathrm{B}}T^{1/2}}\right)$$
$$= \left(\frac{3}{T} + \frac{E_{\mathrm{g}}}{k_{\mathrm{B}}T^2}\right) = \frac{1}{T}\left(3 + \frac{E_{\mathrm{g}}}{k_{\mathrm{B}}T}\right). \tag{5.85}$$

Combining equations (5.81), (5.82) and (5.85) we obtain

$$\left(\frac{1}{C_{DS}}\right)\frac{dC_{DS}(T)}{dT} = -\frac{1}{2(V_{bi}+V_R)}\left[\left(\frac{k_B}{q}\right)\ln\left\{\frac{N_A N_D}{n_i^2(T)}\right\} - \left(\frac{k_B T}{q}\right)\frac{2(dn_i/dT)}{n_i(T)}\right]$$
$$= -\frac{1}{2(V_{bi}+V_R)}\left(\frac{k_B}{q}\right)\left[\ln\left\{\frac{N_A N_D}{n_i^2(T)}\right\} - T\times\frac{1}{T}\left(3+\frac{E_g}{k_B T}\right)\right] \quad (5.86)$$
$$= -\frac{1}{2(V_{bi}+V_R)}\left(\frac{k_B}{q}\right)\left[\ln\left\{\frac{N_A N_D}{n_i^2(T)}\right\} - \left(3+\frac{E_g}{k_B T}\right)\right].$$

At $T = 300$ K, using $V_{bi} = 0.7$ V, $V_R = 0$ V, $k_B/q = 8.62\times10^{-5}$ eV K^{-1}, $N_A = 1\times10^{16}$ cm^{-3}, $N_D = 2\times10^{15}$ cm^{-3}, $n_i = 1.45\times10^{10}$ cm^{-3}, $E_g = 1.12$ eV, we have

$$\left(\frac{1}{C_{DS}}\right)\frac{dC_{DS}(T)}{dT} = -\frac{1}{2(0.7+0)}(8.62\times10^{-5})$$
$$\times\left[\ln\left\{\frac{1\times10^{16}\times2\times10^{15}}{(1.45\times10^{10})^2}\right\} - \left\{3+\frac{1.12}{8.62\times10^{-5}\times300}\right\}\right] \quad (5.87)$$
$$= -6.157\times10^{-5}[\ln(9.51\times10^{10})-3-43.31]$$
$$= -6.157\times10^{-5}[25.28-46.31] = 1.295\times10^{-4}/°C.$$

At $V_R = 5$ V,

$$\left(\frac{1}{C_{DS}}\right)\frac{dC_{DS}(T)}{dT} = -\frac{1}{2(0.7+5)}(8.62\times10^{-5})[25.28-46.31] \quad (5.88)$$
$$= 1.59\times10^{-4}/°C.$$

At $V_R = 0$ V, when the temperature rises by 300 °C, the junction capacitance increases by 39% whereas at $V_R = 5$ V, it does so by 4.8%.

5.9 Digital CMOS circuits in −196 °C to 270 °C range

Investigations by Prince and co-workers showed that satisfactory static and dynamic performance characteristics are obtainable from CMOS ICs, both unbuffered and buffered types, such as NAND gates, up to 270 °C (Prince *et al* 1980). Above this temperature, the high magnitude of the p-well substrate leakage current was the main hurdle. The cause of failure was the large value of the output low voltage V_{OL} at high temperatures arising from the inability of n-channel transistors to sink the leakage currents.

Enormous benefits are derived by operating MOSFET digital ICs below room temperature. Some of the advantages are: (i) an increase in speed, which is made possible by enhanced mobility and conductance; (ii) slow-down of degradation mechanisms, such as chemical reactions and electromigration, which are promoted by thermal energy; (iii) noise reduction at low temperatures; and (iv) tighter packing density due to easy heat withdrawal (Gaensslen *et al* 1977).

Particular interest in CMOS circuits is focused towards the increase in opposition to latchup phenomena as the temperature decreases (Estreich and Dutton 1982). Latchup is suppressed by using special structures such as guard rings, deep-trench isolations or epitaxial layers.

Latchup triggering starts by one of the following conditions: overvoltage stress, capacitive coupling or transitory irradiation. Latchup sustenance requires that a minimum current level called the holding current be maintained by the supply. On bringing the temperature down from 300 K to 77 K, it was found that the sustaining current increased by a factor of 2–4 (Dooley and Jaeger 1984). However, temperatures lower than 77 K may be necessary if complete disappearance of latchup is desired. Many n- and p-type epitaxial twin tub CMOS structures examined in the temperature range 77 K to 400 K were deemed as latchup-free between 100 K and 200 K (Sangiorgi *et al* 1986).

The vulnerability of a CMOS circuit to latchup is decided by two important parameters: (i) the current gains of the parasitic bipolar transistors and (ii) related distributed base–emitter shunting resistances. Both these parameters decrease as the temperature falls (Shoucair 1988).

5.10 Discussion and conclusions

An increase in temperature affects the MOS device behavior favorably in some respects and unfavorably in others. The circuit designer must keep these variational trends in mind, along with their physical origins. In many situations, these trends can be turned to one's advantage and gainfully utilized in establishing proper circuit operation. A proper empathy of thermal limitations of MOSFET paves the way towards successful circuit design.

Review exercises

5.1. Write down the equation for the threshold voltage of a MOSFET and explain the meanings of the symbols used. Applying this equation show that the TC of threshold voltage V_{Th} of a MOSFET depends on the TC of the metal–semiconductor work function ϕ_{ms} and that of the bulk potential ϕ_{F} of the semiconductor.

5.2. Show that the intrinsic carrier concentration n_{i} in a semiconductor of bandgap E_{g} varies with temperature T according to the equation

$$\frac{\partial n_{\mathrm{i}}}{\partial T} = \left(\frac{n_{\mathrm{i}}}{2T}\right)\left(3 + \frac{E_{\mathrm{g}}}{k_{\mathrm{B}}T}\right),$$

where k_{B} is Boltzmann's constant.

5.3. Prove that the variation of the metal–semiconductor work function ϕ_{ms} with temperature T can be described by the equation

$$\frac{\partial \phi_{ms}}{\partial T} = \frac{\phi_{ms}}{T} + \left(\frac{k_B}{q}\right)\left(3 + \frac{E_g}{k_B T}\right),$$

where E_g is the bandgap of the semiconductor and k_B is Boltzmann's constant.

5.4. Show that the bulk potential ϕ_F of a semiconductor varies with temperature T according to the equation

$$\frac{\partial \phi_F}{\partial T} = \frac{\phi_F}{T} - \frac{3}{2}\left(\frac{k_B}{q}\right) - \frac{E_g}{2qT},$$

where q is the electronic charge and E_g is the bandgap of the semiconductor.

5.5. Prove that the temperature dependence of the threshold voltage of an n-channel enhancement-mode MOSFET can be expressed by the equation

$$\frac{\partial V_{Th}}{\partial T} = \frac{\phi_{ms}}{T} + \left(\frac{k_B}{q}\right)\left(3 + \frac{E_g}{k_B T}\right) + \left\{2 + \left(\frac{t_{ox}}{\varepsilon_0 \varepsilon_{ox}}\right)\sqrt{\frac{\varepsilon_0 \varepsilon_s q N_A}{\phi_F}}\right\}.$$

The symbols are explained as follows: T denotes the temperature in absolute scale, ϕ_{ms} is the metal–semiconductor work function, k_B is the Boltzmann's constant, q is the electronic charge, E_g is the bandgap of the semiconductor, t_{ox} is the thickness of the gate oxide, ϕ_F is the bulk potential of the semiconductor and N_A is the doping concentration of p-substrate. The three epsilons ε_0, ε_{ox}, ε_s denote the permittivity of free space, the relative permittivity of silicon dioxide (SiO_2) and the relative permittivity of silicon (Si), respectively. Taking a representative example of a MOSFET with typical structural parameters, show that $\partial V_{Th}/\partial T = -2.9\ \text{mV K}^{-1}$.

5.6. What are the seven components of the on-resistance of a MOSFET? Which two of these seven components are predominant?

5.7. Show that the channel resistance $R_{CHANNEL}(T)$ of a MOSFET at a temperature T is given by

$$R_{CHANNEL}(T) = R_{CHANNEL}(T_0) \times \frac{(T/T_0)^n}{1 + \dfrac{\kappa(T - T_0)}{V_{GS} - V_{Th}(T_0)}}$$

where κ is the TC of threshold voltage V_{Th}.

5.8. Taking into account the temperature dependence of the resistances of the channel and drift regions of a MOSFET, show that the temperature dependence of the on-resistance $R_{DS(ON)}$ of a MOSFET can be expressed as

$$R_{DS(ON)}(T) = R_{DS(ON)}(T_0)(T/T_0)^n.$$

5.9. Prove that the transconductance g_m of a MOSFET decreases with temperature T according to the relation

$$g_m(T) = g_m(T_0) \times (T/T_0)^{-n}.$$

5.10. Comment on the variation of BV_{DSS} and I_{DSS} of a MOSFET with temperature.

5.11. What is meant by the ZTC of a MOSFET? Using the equation for drain–source current I_{DS} of a MOSFET in the saturation region, and applying the ZTC condition $dI_{DS}/dT = 0$, derive the following equation for the voltage bias V_{GS}(ZTC)

$$V_{GS}(\mathrm{ZTC}) = V_{Th}(T) + \frac{T\kappa}{n},$$

where κ is the TC of threshold voltage V_{Th} and n is the index in the equation for the temperature dependence of mobility.

5.12. Why is a MOSFET faster in speed than a bipolar transistor? Explain why a MOSFET experiences much smaller excursions in switching parameters with temperature.

5.13. Following Shoucair's analysis, show that there are two ZTC points for a MOSFET. One ZTC point lies in the linear region and the other ZTC point is in the saturation region of current–voltage characteristics. Write the analytical equation derived by him for ZTC gate–source biasing voltages V_{GS}(ZTC) applied to a MOSFET with known parameters (p_0, q_0) over the temperature range (T_1, T_2).

5.14. Prove, following Shoucair, that in the linear and saturation regions of MOSFET operation, the equations for drain-source current I_{DS} are respectively

$$I_{DS} = \mu(T)C_{ox}\left(\frac{W}{L}\right)\left\{-\frac{T_0}{1.5}p_0\right\}V_{DS}$$

$$I_{DS} = (1/2)\mu(T)C_{ox}\left(\frac{W}{L}\right)\left\{-\frac{mT_0}{1.5}p_0\right\}^m,$$

where p_0 denotes the TC of the threshold voltage and index $m = 2$. The remaining symbols have their usual connotations.

5.15. Explain the term 'transition temperature' in the context of changeover of leakage current of a MOSFET from the generation to diffusion component. Write the equations for the generation and diffusion components of leakage current. What is the typical transition temperature for a CMOS device? Why does an n-channel MOSFET show a higher leakage current than a p-channel MOSFET?

5.16. What are the guidelines laid out by Shoucair for designing a two-stage topology of CMOS OP-AMP?

5.17. What are the overlap and junction capacitances in a MOSFET? Which of the two capacitances is more temperature sensitive?

5.18. Show that the drain–source capacitance C_{DS} of a MOSFET changes with temperature T according to the following equation

$$\left(\frac{1}{C_{\mathrm{DS}}}\right)\frac{\mathrm{d}C_{\mathrm{DS}}(T)}{\mathrm{d}T} = -\frac{1}{2(V_{\mathrm{bi}} + V_{\mathrm{R}})}\left(\frac{k_{\mathrm{B}}}{q}\right)\left[\ln\left\{\frac{N_{\mathrm{A}}N_{\mathrm{D}}}{n_{\mathrm{i}}^{2}(T)}\right\} - \left(3 + \frac{E_{\mathrm{g}}}{k_{\mathrm{B}}T}\right)\right],$$

where V_{bi} is the built-in potential, V_{R} is the applied reverse voltage, k_{B} is Boltzmann's constant, q is the electronic charge, n_{i} is the intrinsic carrier concentration, E_{g} is the energy gap, and N_{A}, N_{D} are the acceptor and donor concentrations, respectively.

5.19. Mention three advantages of operating MOSFET digital ICs below room temperature. Discuss the effect of temperature on latchup in a CMOS circuit.

References

Dooley J G and Jaeger R C 1984 Temperature dependence of latchup in CMOS circuits *IEEE Electron. Device Lett.* **5** 41–3

Estreich D B and Dutton R W 1982 Modeling latch-up in CMOS integrated circuits *IEEE Trans. Comput.-Aided Des. Integr. Circuits Syst.* **4** 157–62

Fairchild Semiconductor Corporation 2000 MOSFET basics *Fairchild Semiconductor Corporation* www.fairchildsemi.com/application-notes/AN/AN-9010.pdf

Gaensslen F H, Rideout V L, Walker E J and Walker J J 1977 Very small MOSFETs for low-temperature operation *IEEE Trans. Electron Devices.* **24** 218–29

Prince J L, Draper B L, Rapp E A, Kronberg J N and Fitch L T 1980 Performance of digital integrated circuits at very high temperatures *IEEE Trans. Components Hybrids Manuf. Technol.* **3** 571–9

Sangiorgi E, Johnston R L, Pinto M R, Bechtold P F and Fichtner W 1986 Temperature dependence of latch-up phenomena in scaled CMOS structures *IEEE Electron Device Lett.* **7** 28–31

Sattar A and Tsukanov V 2007 MOSFETs withstand stress of linear-mode operation *Power Electronics Technology* April, pp 34–39

Shoucair F S and Early J M 1984 High-temperature diffusion leakage current-dependent MOSFET small signal conductance *IEEE Trans. Electron Devices* **31** 1866–72

Shoucair F S, Hwang W and Jain P 1984 Electrical characteristics of large scale integration (LSI) MOSFETs at very high temperatures: Parts I and II *Microelectron. Rel.* **24** 465–85 and 487–510

Shoucair F S 1986 Design considerations in high temperature analog CMOS integrated circuits *IEEE Trans. Components Hybrids Manuf. Technol.* **9** 242–51

Shoucair F S 1988 High-temperature latchup characteristics in VLSI CMOS circuits *IEEE Trans. Electron Devices* **35** 2424–6

STMicroelectronics 2006 How to achieve the threshold voltage thermal coefficient of the MOSFET acting on design parameter *STMicroelectronics Application Note* www.bdtic.com/DownLoad/ST/Application_Note/AN2386.pdf

Wang R, Dunkley J, DeMassa T A and Jelsma L F 1971 Threshold voltage variations with temperature in MOS transistors *IEEE Trans. Electron Devices* 386–8

Chapter 6

The influence of temperature on the performance of silicon–germanium heterojunction bipolar transistors

The constraints faced in designing homojunction BJTs of the desired current gain and switching frequency necessitate trade-offs and compromise solutions. In this chapter, the design flexibility offered by the heterojunction structure is explained, and the HBT fabrication process is described. A comparative analysis between silicon BJTs and Si/SiGe HBTs is presented. The superior cryogenic performance of HBTs compared to BJTs in terms of current gain and frequency response characteristics is elaborated.

6.1 Introduction

A conventional BJT, which is made of the same indistinguishable silicon material throughout, is often referred to as a homojunction BJT because of its homogeneous composition. Two important parameters of a BJT are its current gain and switching speed. In order to maximize these parameters, the available device structural parameters to be varied are the emitter and base doping concentrations and the base width. To increase the current gain, the ratio of these doping concentrations must be large. Only then will there be a large concentration gradient from the emitter to base, so that a high emitter injection efficiency will be achieved. For a large ratio between the emitter and base doping concentrations, the base doping density must be low. This means that the depletion region extends to a larger depth on the base side at a low voltage when the collector–base junction is reverse biased. An obvious drawback is that this junction becomes susceptible to punch-through breakdown at a low voltage. To increase the punch-through breakdown voltage, a larger base width is necessary. But in a transistor with a large base width, the carriers injected from the emitter into the broader base take a longer time to cross the base and enter the collector, i.e. the transit time of carriers through the base region is

doi:10.1088/978-0-7503-1155-7ch6

lengthened. As a result, the transistor becomes slower in speed. Thus we must either sacrifice the punch-through breakdown voltage capability or the switching speed if we insist on keeping the current gain high.

Considering an alternative route, to increase the speed of the device the base must be thin. If the base is thin, its doping concentration must be high to prevent spreading of the depletion region across the base to the emitter edge causing punch-through breakdown. But a high doping concentration of the base lowers the emitter injection efficiency, and hence the current gain of the transistor. In effect, the transistor design involves a trade-off between different parameters. In view of these constraints, a compromise must be made between the electrical parameters to achieve the best solution suitable for a given application.

As the name suggests, a HBT contains a heterojunction. This heterojunction is a junction between two different materials. One material comprising this heterojunction is silicon and the other material is silicon–germanium, an alloy of silicon and germanium with the chemical formula $Si_{1-x}Ge_x$ where x is the mole fraction of germanium in the alloy with a value from 0 to 1 (Ioffe Institute 2015). Figure 6.1 illustrates the crystalline structure of silicon–germanium.

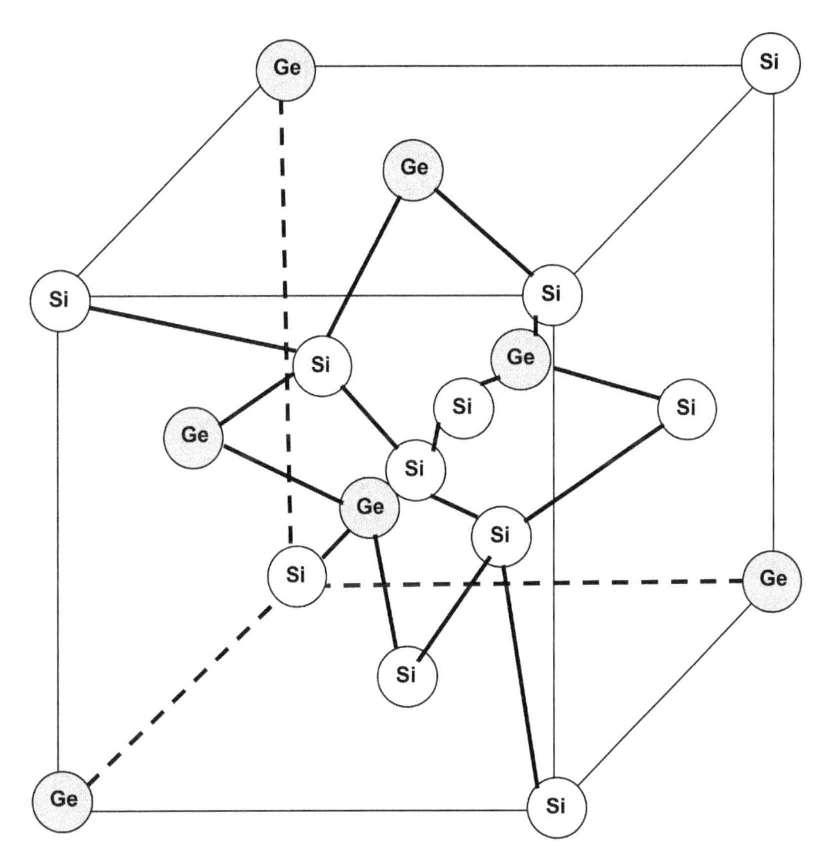

Figure 6.1. The structure of a silicon–germanium crystal.

The silicon–germanium alloy is an indirect bandgap semiconductor. Its bandgap is given by the equations (Virginia Semiconductor 2002)

$$E_{g(\mathrm{indirect})} = 1.155 - 0.43x + 0.206x^2 \text{ eV at } T = 300 \text{ K, for } x < 0.85 \quad (6.1)$$

$$E_{g(\mathrm{indirect})} = 2.010 - 1.27x \text{ eV at } T = 300 \text{ K, for } x > 0.85. \quad (6.2)$$

The bandgap of SiGe is smaller than that of Si. The amount by which the SiGe bandgap is smaller than the silicon bandgap depends on the value of x. In an HBT, two regions, the emitter and collector, are made of silicon. The third region, which is the HBT base, is made of the lower bandgap material SiGe. The fact that the material used for base (SiGe) has a smaller bandgap than the emitter material (Si) has a favorable outcome. If an n–p–n transistor is considered, the injection of holes from the p-base into the n-emitter is prevented. In this way, the emitter injection efficiency is enhanced. The HBT structure with near-lattice matched or pseudomorphic epitaxial layers is well suited for n–p–n HBT fabrication because the band offset takes place exclusively in the valence band. No barrier is introduced opposing the injection of electrons from the emitter into the base (Shiraki and Usami 2011). The difference in operation between a Si BJT and a SiGe HBT can be understood by reference to figure 6.2.

In consequence of the above efforts, one does not need to rely on the ratio of doping concentrations of the emitter and base for achieving the required emitter injection efficiency for a high current gain value. Instead, the difference of bandgaps between the emitter and base regions takes care of this efficiency. This means that the limitation imposed on the base carrier concentration from the viewpoint of current gain is removed. A high carrier concentration can be used for this region. Obviously, the base region can be made thin without losing the capability for punch-through breakdown. Hence, the high current gain and punch-through breakdown capabilities can be provided in a device along with a high switching speed. By using a graded profile for germanium incorporation in the base, with the germanium concentration decreasing from the emitter edge towards the collector edge, an accelerating electric field is created across the base layer. This electric field helps the carriers injected from the emitter into the base to be swept fast into the collector by a field-assisted transport mechanism. Thus the speed of the transistor is further increased. In this way, the HBT structure provides an all-in-one solution for transistor designers for complying with the specifications of high current gain, large breakdown voltage and ultrafast speed.

6.2 HBT fabrication

Figure 6.3 shows the schematic cross-section of SiGe HBT. The high-quality p-type SiGe base layer is grown epitaxially. The epitaxy is carried out either using reduced pressure CVD (RPCVD) or ultra-high vacuum CVD (UHVCVD). There are two modalities of epitaxial growth: selective and non-selective epitaxy. Selective epitaxy enables the fabrication of self-aligned structures with small overlaps, minimizing parasitic effects. Non-selective epitaxy is a simpler process and provides better

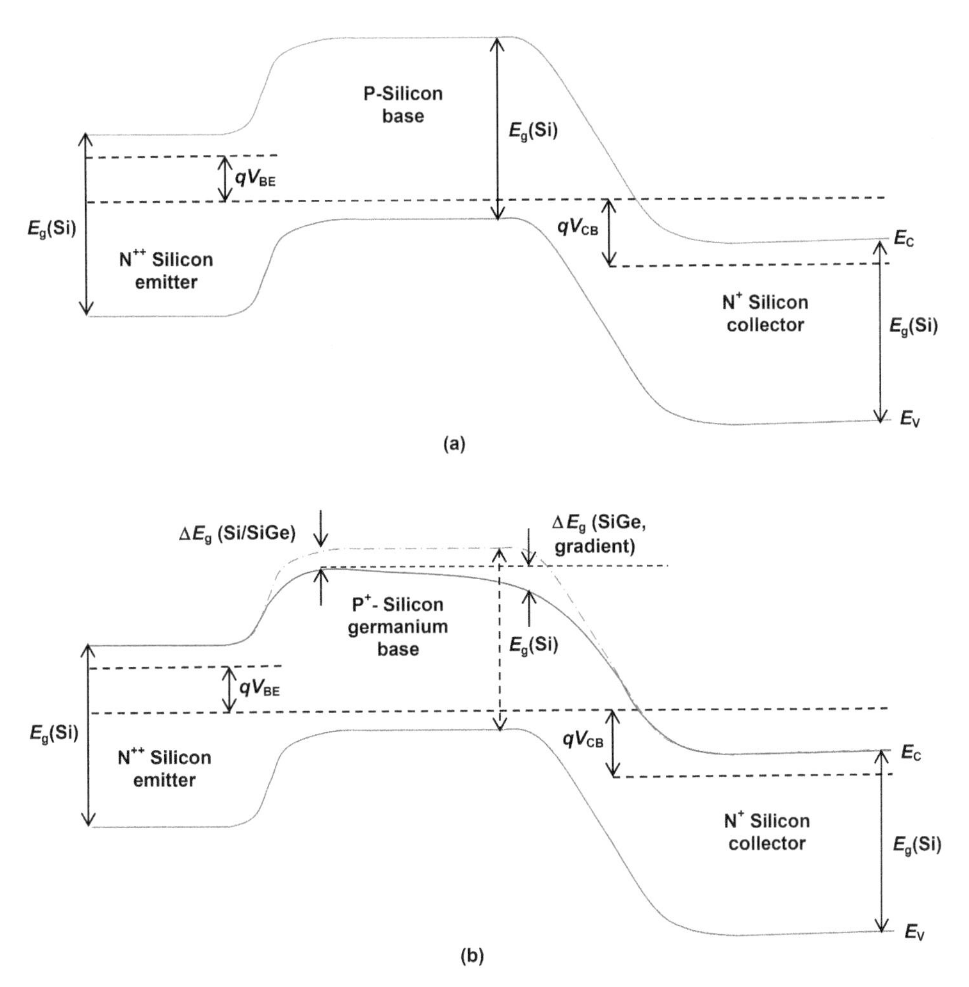

Figure 6.2. Energy band diagrams. (a) Silicon bipolar transistor and (b) silicon–germanium HBT compared to a silicon bipolar transistor. ΔE_g(Si, SiGe) is the bandgap difference between Si and SiGe at the emitter edge of the base to be denoted in terms of position x in the base as $\Delta E_{gB,Ge}(x = 0)$ and ΔE_g(SiGe, gradient) is the bandgap difference in the SiGe base from the emitter edge to collector edge to be represented by $\Delta E_{gB,Ge}(x = W_B)$ where W_B is base width. Then $\Delta E_{gB,Ge}(x = W_B) - \Delta E_{gB,Ge}(x = 0) = \Delta E_{g,Ge}$(grade).

control over the thickness uniformity of the SiGe layer. Boron doping is performed *in situ* during epitaxial layer growth. Incorporating a small concentration of carbon $\leqslant 1\%$ in the SiGe layer is found to suppress boron out-diffusion appreciably during subsequent thermal processing (Hållstedt 2004). It thus prevents the base layer from enlarging and preserves the narrow base width necessary for a small transit time (Mitrovic et al 2005). After the base layer deposition, the polysilicon emitter layer is formed.

6.3 Current gain and forward transit time of $Si/Si_{1-x}Ge_x$ HBT

Consider an n–p–n HBT with uniformly doped emitter and base regions. Let N_{DE} be the doping concentration in the n-emitter layer of width W_E and N_{AB} be the acceptor

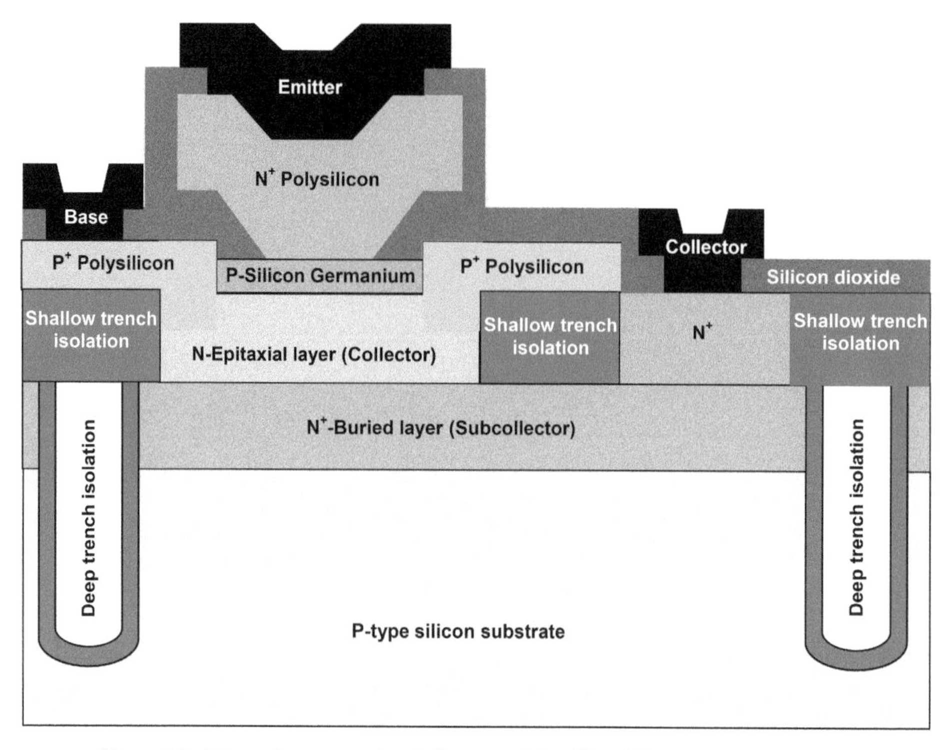

Figure 6.3. Schematic cross-sectional diagram of the silicon/silicon–germanium HBT.

concentration in the p-base layer of width W_B. Let $D_n(T)$ be the diffusion coefficient of electrons at temperature T; electrons are minority carriers in the base. Let D_p be the diffusion coefficient of holes at temperature T; holes are minority carriers in the emitter. Let ΔE_g be the energy bandgap difference between the emitter and base semiconductor materials. Then the common-emitter current gain β is (Basu and Sarkar 2011)

$$\beta = \left(\frac{N_{DE}}{N_{AB}}\right)\left(\frac{W_E}{W_B}\right)\left\{\frac{D_n(T)}{D_p(T)}\right\}\exp\left(\frac{\Delta E_g}{k_B T}\right), \tag{6.3}$$

where, in accordance with Einstein's equation, the electron and hole diffusion coefficients $D_n(T)$, $D_p(T)$ are related to corresponding mobilities $\mu_n(T)$, $\mu_p(T)$ as

$$D_n(T) = \left(\frac{k_B T}{q}\right)\mu_n(T) \tag{6.4}$$

$$D_p(T) = \left(\frac{k_B T}{q}\right)\mu_p(T). \tag{6.5}$$

The forward transit time τ_F through the HBT is equal to (emitter transit time (τ_E) + base transit time (τ_B)), written as

$$\tau_F = \tau_E + \tau_B, \tag{6.6}$$

where

$$\tau_E(T) = W_E^2/\{2\beta D_p(T)\} \tag{6.7}$$

$$\tau_B(T) = W_B^2/\{2D_n(T)\}. \tag{6.8}$$

For Si/Si$_{1-x}$Ge$_x$ HBT, the equation for current gain $(\beta)_{Si/SiGe}$ is

$$(\beta)_{Si/SiGe} = \left(\frac{N_{DE}}{N_{AB}}\right)\left(\frac{W_E}{W_B}\right)\left\{\frac{D_{n,SiGe}(T)}{D_{p,Si}(T)}\right\}\exp\left(\frac{\Delta E_g}{k_B T}\right). \tag{6.9}$$

The electron mobility $\mu_{n,Si}$ in Si is

$$\mu_{n,Si}(T) \propto T^{-2.42}$$

or

$$\mu_{n,Si}(T) = K_3 T^{-2.42} \tag{6.10}$$

$$\therefore K_3 = \mu_{n,Si}(T)/T^{-2.42} = \mu_{n,Si}(T) \times T^{2.42}. \tag{6.11}$$

Since $\mu_{n,Si} = 1350\ \text{cm}^2\ \text{V}^{-1}\ \text{s}^{-1}$ at $T = 300$ K,

$$\therefore K_3 = 1350 \times T^{2.42} = 1350 \times (300)^{2.42} = 9.8772 \times 10^5\ \text{cm}^2\ \text{V}^{-1}\ \text{s}^{-1}\ \text{K}^{2.42} \tag{6.12}$$

$$\therefore \mu_{n,Si}(T) = 9.88 \times 10^5 T^{-2.42}. \tag{6.13}$$

The electron mobility in Ge is

$$\mu_{n,Ge}(T) \propto T^{-1.66}$$

or

$$\mu_{n,Ge}(T) = K_4 T^{-1.66} \tag{6.14}$$

$$\therefore K_4 = \mu_{n,Ge}(T)/T^{-1.66} = \mu_{n,Ge}(T) \times T^{1.66}. \tag{6.15}$$

Since $\mu_{n,Ge} = 3900\ \text{cm}^2\ \text{V}^{-1}\ \text{s}^{-1}$ at $T = 300$ K,

$$\therefore K_4 = 3900 \times T^{1.66} = 3900 \times (300)^{1.66} = 1.294 \times 10^4\ \text{cm}^2\ \text{V}^{-1}\ \text{s}^{-1}\ \text{K}^{1.66} \tag{6.16}$$

$$\therefore \mu_{n,Ge}(T) = 1.29 \times 10^4 T^{-1.66}. \tag{6.17}$$

Now,

$$\begin{aligned}\mu_{n,SiGe}(T) = (1 - x)\mu_{n,Si}(T) + x\mu_{n,Ge}(T) = 9.88 \times 10^5(1 - x)T^{-2.42}\\ + 1.29 \times 10^4 xT^{-1.66}\end{aligned} \tag{6.18}$$

$$\therefore D_{n,SiGe}(T) = \left(\frac{k_B T}{q}\right)\{9.88 \times 10^5(1 - x)T^{-2.42} + 1.29 \times 10^4 xT^{-1.66}\}. \tag{6.19}$$

The hole mobility in Si is

$$\mu_{p,Si}(T) \propto T^{-2.2}$$

or

$$\mu_{p,Si}(T) = K_5 T^{-2.2}. \tag{6.20}$$

Since $\mu_{p,Si} = 500\ \text{cm}^2\ \text{V}^{-1}\ \text{s}^{-1}$ at $T = 300$ K,

$$\begin{aligned}\therefore K_5 = \mu_{p,Si}(T)/T^{-2.2} = \mu_{p,Si}(T) \times T^{2.2} = 500 \times (300)^{2.2}\\ = 1.408 \times 10^8\ \text{cm}^2\ \text{V}^{-1}\ \text{s}^{-1}\ \text{K}^{2.2}\end{aligned} \tag{6.21}$$

$$\therefore \mu_{p,Si}(T) = 1.408 \times 10^8 T^{-2.2} \tag{6.22}$$

$$\therefore D_{p,Si}(T) = \left(\frac{k_B T}{q}\right)1.408 \times 10^8 T^{-2.2}. \tag{6.23}$$

The difference between energy bandgaps of the Si emitter and Si_xGe_{1-x} base is (Basu and Sarkar 2011, Shur 1995)

$$\left(\Delta E_g\right)_{Si/SiGe} = 0.43x - 0.0206x^2. \tag{6.24}$$

Further, the emitter and base transit times are written as

$$\begin{aligned}\{\tau_E(T)\}_{Si/SiGe} &= W_E^2\Big/\left[2\beta D_{p,Si}(T)\right] = W_E^2\Big/\left[2(\beta)_{Si/SiGe} D_{p,Si}(T)\right]\\ &= W_E^2\Bigg/\left[2\left(\frac{N_{DE}}{N_{AB}}\right)\left(\frac{W_E}{W_B}\right)\left\{\frac{D_{n,SiGe}(T)}{D_{p,Si}(T)}\right\}\exp\left(\frac{\Delta E_g}{k_B T}\right)D_{p,Si}(T)\right]\\ &= W_E^2\Bigg/\left[2\left(\frac{N_{DE}}{N_{AB}}\right)\left(\frac{W_E}{W_B}\right)D_{n,SiGe}(T)\exp\left(\frac{\Delta E_g}{k_B T}\right)\right]\end{aligned} \tag{6.25}$$

$$\{\tau_{\rm B}(T)\}_{\rm Si/SiGe} = W_{\rm B}^2/\{2D_{\rm n,SiGe}(T)\}. \tag{6.26}$$

The model allows the prediction of HBT device performance with respect to current gain and forward transit time as a function of different emitter/base compositions and over a wide range of operating temperatures.

6.4 Comparison between Si BJT and Si/SiGe HBT

Figures 6.4 and 6.5 show the shifts in current-voltage characteristics and current gain-collector current curves of a SiGe HBT with temperature. In a silicon BJT, the temperature dependence of current gain is mainly controlled by the doping-induced bandgap narrowing of the emitter and base regions. To achieve high emitter injection efficiency, the doping level of the emitter region is kept much higher than that of the base region. Hence, the bandgap narrowing in the emitter region is much more pronounced than in the base region. If $\Delta E_{\rm gE}$ is the bandgap narrowing of

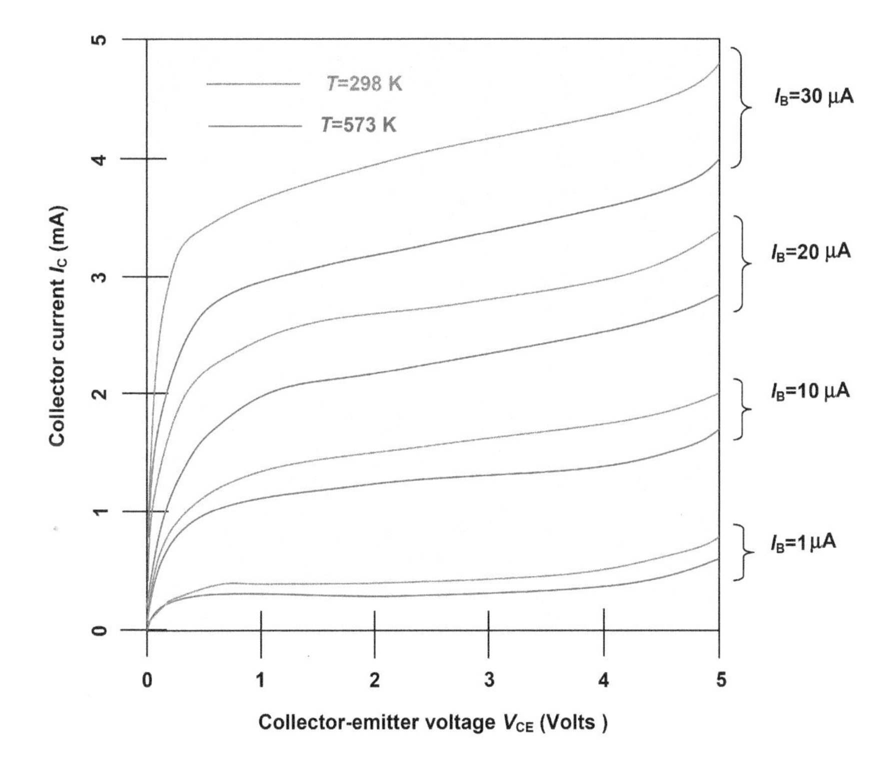

Figure 6.4. Change in current–voltage characteristics of a Si/SiGe HBT with temperature.

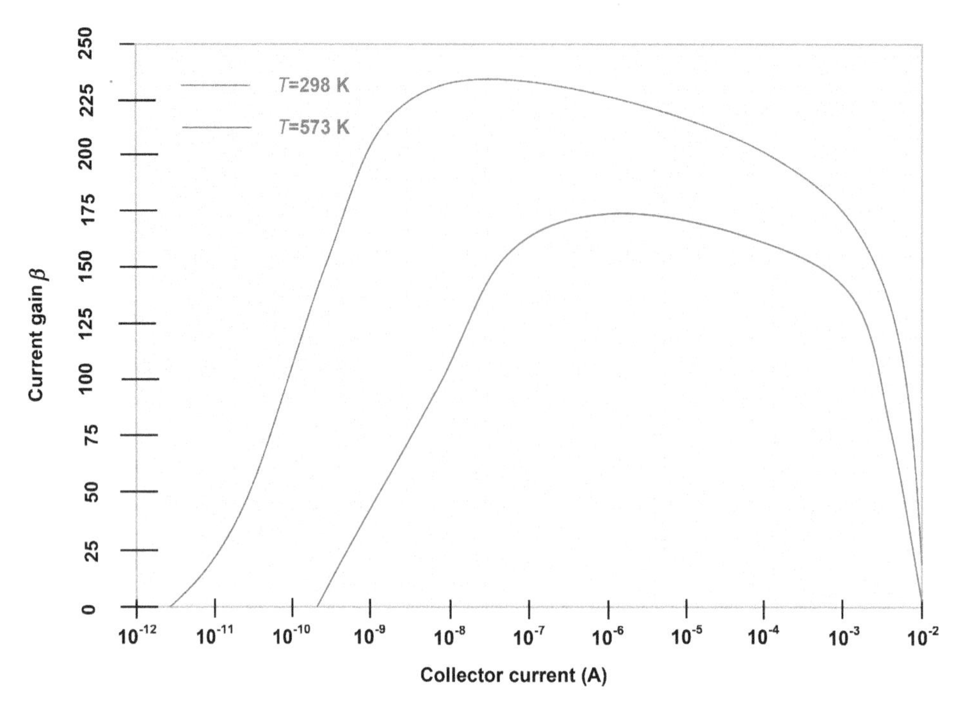

Figure 6.5. Shift in current gain–collector current curves of a silicon/silicon germanium HBT with temperature.

the emitter and ΔE_{gB} is the bandgap narrowing of the base, the difference between the bandgap narrowing of emitter and base

$$= \Delta E_{g\mathrm{E}} - \Delta E_{g\mathrm{B}} \Delta E_{g\mathrm{EB}}. \tag{6.27}$$

The current gain of silicon BJT is

$$\beta_{\mathrm{Si}}(T) \propto \exp\left\{-\left(\Delta E_{\mathrm{gE}} - \Delta E_{\mathrm{gB}}\right)/(k_{\mathrm{B}}T)\right\} \propto \exp\left\{-\left(\Delta E_{\mathrm{gEB}}\right)/(k_{\mathrm{B}}T)\right\}. \tag{6.28}$$

Consequently, the current gain of a silicon BJT exponentially decreases with temperature, resulting in a substantial reduction of gain when a temperature $T = 77$ K is reached.

In a Si/SiGe HBT, the inclusion of germanium in the silicon of the base layer to form the SiGe base layer increases the bandgap narrowing ΔE_{gB} of base. Thus the bandgap narrowing effects in the emitter ΔE_{gE} and base ΔE_{gB} become comparable. The difference $\Delta E_{\mathrm{gE}} - \Delta E_{\mathrm{gB}} = \Delta E_{\mathrm{gEB}}$ becomes smaller. Therefore, the degradation of current gain $\beta_{\mathrm{Si/SiGe}}(T)$ of Si/SiGe HBT with a decrease in temperature becomes slower compared to $\beta_{\mathrm{Si}}(T)$ of Si BJT. Stated in other words, the bandgap reduction occurring due to the presence of germanium in the SiGe base layer of Si/SiGe HBT serves to compensate the bandgap narrowing effect of heavy doping in the emitter on the bandgap of the emitter, whereby the current gain of the Si/SiGe HBT at a given temperature is higher than that of Si BJT.

For a deeper introspection into the operation of HBT, let us recall the Moll–Ross relationship, generalized by Kroemer. This relationship, valid for low injection conditions, is an important relationship because it shows that the collector current density is not affected by the nature of doping in the base but by the total quantity of charges present in the base. It relates the collector current density J_C of a bipolar transistor having a uniformly or non-uniformly doped base with the applied base-emitter voltage V_{BE} at a given temperature T as (Cressler and Niu 2003)

$$J_C = \frac{q\left\{\exp\left(\dfrac{qV_{BE}}{k_B T}\right) - 1\right\}}{\displaystyle\int_{x=0}^{x=W_B} \frac{p_B(x)dx}{D_{nB}(x)n_{iB}^2(x)}} \tag{6.29}$$

where $x = 0$ is the boundary of the base on the emitter side, $x = W_B$ is the base boundary on the collector side, $p_B(x)$ is the position-dependent hole concentration in the base varying with position x, $D_{nB}(x)$ is the position-dependent diffusion coefficient of electrons in the base and $n_{iB}(x)$ is the position-dependent intrinsic carrier concentration in the base. The position-dependence of $D_{nB}(x)$ originates from the variation of electron mobility $\mu_{nB}(x)$ in the base due to the germanium profile through the equation

$$D_{nB}(x) = \left(\frac{k_B T}{q}\right)\mu_{nB}(x). \tag{6.30}$$

The position-dependence of $n_{iB}(x)$ arises from the bandgap changes caused by germanium. The square of intrinsic carrier concentration in SiGe is

$$n_{iB}^2(x) = N_{C,SiGe}(x)N_{V,SiGe}(x)\exp\left\{-\frac{E_{gB}(x)}{k_B T}\right\}, \tag{6.31}$$

where $N_{C,SiGe}(x)$ is the position-dependent effective density of states in the conduction band of SiGe, $N_{V,SiGe}(x)$ is the position-dependent effective density of states in the valence band of SiGe and $E_{gB}(x)$ is the position-dependent energy bandgap of the SiGe base layer. In an HBT fabricated with a linearly graded base layer, suppose E_{gB0} = silicon bandgap at zero doping = 1.12 eV, $\Delta E_{gB,A}$ is the bandgap narrowing of the base layer due to the acceptor impurity doping effect, $\Delta E_{gB,Ge}(x = 0)$ is the bandgap offset of the base layer at $x = 0$ and $\Delta E_{gB,Ge}(x = W_B)$ is the bandgap offset of the base layer at $x = W_B$. Using these symbols, we can write

$$\begin{aligned} E_{gB}(x) = E_{gB0} - \Delta E_{gB,A} + \{\Delta E_{gB,Ge}(x = 0) - \Delta E_{gB,Ge}(x = W_B)\} \times (x/W_B) \\ - \Delta E_{gB,Ge}(x = 0) \end{aligned} \tag{6.32}$$

$$\begin{aligned}
\therefore n_{\mathrm{ib}}^2(x) &= N_{\mathrm{C,SiGe}}(x) N_{\mathrm{V,SiGe}}(x) \\
&\times \exp\left[-\frac{E_{\mathrm{gB0}} - \Delta E_{\mathrm{gB,A}} + \{\Delta E_{\mathrm{gB,Ge}}(x=0) - \Delta E_{\mathrm{gB,Ge}}(x=W_{\mathrm{B}})\} \times (x/W_{\mathrm{B}}) - \Delta E_{\mathrm{gB,Ge}}(x=0)}{k_{\mathrm{B}}T}\right] \\
&= N_{\mathrm{C,SiGe}}(x) N_{\mathrm{V,SiGe}}(x) \\
&\times \exp\left[\frac{-E_{\mathrm{gB0}} + \Delta E_{\mathrm{gB,A}} - \{\Delta E_{\mathrm{gB,Ge}}(x=0) - \Delta E_{\mathrm{gB,Ge}}(x=W_{\mathrm{B}})\} \times (x/W_{\mathrm{B}}) + \Delta E_{\mathrm{gB,Ge}}(x=0)}{k_{\mathrm{B}}T}\right] \\
&= N_{\mathrm{C,SiGe}}(x) N_{\mathrm{V,SiGe}}(x) \exp\left(-\frac{E_{\mathrm{gB0}}}{k_{\mathrm{B}}T}\right) \exp\left(\frac{\Delta E_{\mathrm{gB,A}}}{k_{\mathrm{B}}T}\right) \\
&\times \exp\left[\left\{\frac{\Delta E_{\mathrm{gB,Ge}}(x=W_{\mathrm{B}}) - \Delta E_{\mathrm{gB,Ge}}(x=0)}{k_{\mathrm{B}}T}\right\} \times (x/W_{\mathrm{B}})\right] \exp\left\{\frac{\Delta E_{\mathrm{gB,Ge}}(x=0)}{k_{\mathrm{B}}T}\right\} \\
&= \frac{N_{\mathrm{C,SiGe}}(x) N_{\mathrm{V,SiGe}}(x)}{N_{\mathrm{C,Si}}(x) N_{\mathrm{V,Si}}(x)} N_{\mathrm{C,Si}}(x) N_{\mathrm{V,Si}}(x) \exp\left(-\frac{E_{\mathrm{gB0}}}{k_{\mathrm{B}}T}\right) \exp\left(\frac{\Delta E_{\mathrm{gB,A}}}{k_{\mathrm{B}}T}\right) \\
&\times \exp\left[\left\{\frac{\Delta E_{\mathrm{gB,Ge}}(x=W_{\mathrm{B}}) - \Delta E_{\mathrm{gB,Ge}}(x=0)}{k_{\mathrm{B}}T}\right\} \times (x/W_{\mathrm{B}})\right] \exp\left\{\frac{\Delta E_{\mathrm{gB,Ge}}(x=0)}{k_{\mathrm{B}}T}\right\}
\end{aligned} \tag{6.33}$$

Let us define

$$\frac{N_{\mathrm{C,SiGe}}(x) N_{\mathrm{V,SiGe}}(x)}{N_{\mathrm{C,Si}}(x) N_{\mathrm{V,Si}}(x)} = N \tag{6.34}$$

$$\Delta E_{\mathrm{gB,Ge}}(x = W_{\mathrm{B}}) - \Delta E_{\mathrm{gB,Ge}}(x = 0) = \Delta E_{\mathrm{g,Ge}}(\mathrm{grade}). \tag{6.35}$$

Obviously,

$$N_{\mathrm{C,Si}}(x) N_{\mathrm{V,Si}}(x) \exp\left(-\frac{E_{\mathrm{gB0}}}{k_{\mathrm{B}}T}\right) = n_{\mathrm{i,Si}}^2. \tag{6.36}$$

Hence,

$$\begin{aligned}
\therefore n_{\mathrm{ib}}^2(x) = N n_{\mathrm{i,Si}}^2 \exp\left(\frac{\Delta E_{\mathrm{gB,A}}}{k_{\mathrm{B}}T}\right) \exp\left[\left\{\frac{\Delta E_{\mathrm{g,Ge}}(\mathrm{grade})}{k_{\mathrm{B}}T}\right\}(x/W_{\mathrm{B}})\right] \\
\times \exp\left\{\frac{\Delta E_{\mathrm{gB,Ge}}(x=0)}{k_{\mathrm{B}}T}\right\}.
\end{aligned} \tag{6.37}$$

Substituting the expression for $n_{\mathrm{ib}}^2(x)$ from (6.37) into (6.29) for J_{C}, we obtain

$$\begin{aligned}
J_{\mathrm{C,SiGe}} &= \frac{q\left\{\exp\left(\frac{qV_{\mathrm{BE}}}{k_{\mathrm{B}}T}\right) - 1\right\}}{\int_{x=0}^{x=W_{\mathrm{B}}} \frac{p_{\mathrm{B}}(x)\mathrm{d}x}{D_{\mathrm{nB}}(x) N n_{\mathrm{i,Si}}^2 \exp\left(\frac{\Delta E_{\mathrm{gB,A}}}{k_{\mathrm{B}}T}\right) \exp\left[\left\{\frac{\Delta E_{\mathrm{g,Ge}}(\mathrm{grade})}{k_{\mathrm{B}}T}\right\}(x/W_{\mathrm{B}})\right] \exp\left\{\frac{\Delta E_{\mathrm{gB,Ge}}(x=0)}{k_{\mathrm{B}}T}\right\}}} \\
&= \frac{q\left\{\exp\left(\frac{qV_{\mathrm{BE}}}{k_{\mathrm{B}}T}\right) - 1\right\}}{\left(\frac{1}{\tilde{D}_{\mathrm{nB}}}\right)\left(\frac{1}{\tilde{N}}\right) N_{\mathrm{A,B}} \left\{\frac{1}{n_{\mathrm{i,Si}}^2 \exp\left(\frac{\Delta E_{\mathrm{gB,A}}}{k_{\mathrm{B}}T}\right)}\right\} \left\{\frac{1}{\exp\left\{\frac{\Delta E_{\mathrm{gB,Ge}}(x=0)}{k_{\mathrm{B}}T}\right\}}\right\} \int_{x=0}^{x=W_{\mathrm{B}}} \frac{\mathrm{d}x}{\exp\left[\left\{\frac{\Delta E_{\mathrm{g,Ge}}(\mathrm{grade})}{k_{\mathrm{B}}T}\right\}(x/W_{\mathrm{B}})\right]}},
\end{aligned} \tag{6.38}$$

where $\tilde{D}_{nB}$ is defined as the position-averaged diffusion coefficient across the base profile. $\tilde{N}$ is defined as the position-averaged ratio of the effective densities of states in SiGe and Si, across the base profile. $N_{A,B}$ is the acceptor dopant concentration in the base:

$$\begin{aligned}\therefore J_{C,SiGe} &= q\frac{\tilde{D}_{nB}\tilde{N}}{N_{A,B}} \frac{n_{i,Si}^2 \exp\left(\frac{\Delta E_{gB,A}}{k_B T}\right)\exp\left\{\frac{\Delta E_{gB,Ge}(x=0)}{k_B T}\right\}\left\{\exp\left(\frac{qV_{BE}}{k_B T}\right)-1\right\}}{\int_{x=0}^{x=W_B}\exp\left[\{-\Delta E_{g,Ge}(\text{grade})\}\{x/(W_B k_B T)\}\right]dx} \\ &= q\frac{\tilde{D}_{nB}\tilde{N}}{N_{A,B}} \frac{n_{i,Si}^2 \exp\left(\frac{\Delta E_{gB,A}}{k_B T}\right)\exp\left\{\frac{\Delta E_{gB,Ge}(x=0)}{k_B T}\right\}\left\{\exp\left(\frac{qV_{BE}}{k_B T}\right)-1\right\}}{\left[-\frac{1}{\{\Delta E_{g,Ge}(\text{grade})/(W_B k_B T)\}}\exp[\{-\Delta E_{g,Ge}(\text{grade})\}\{x/(W_B k_B T)\}]\right]_{x=0}^{x=W_B}}\end{aligned} \tag{6.39}$$

because

$$\int \exp(-ax)dx = -(1/a)\exp(-ax).$$

Therefore,

$$\begin{aligned} J_{C,SiGe} &= q\frac{\tilde{D}_{nB}\tilde{N}}{N_{A,B}} \\ &\times \frac{n_{i,Si}^2 \exp\left(\frac{\Delta E_{gB,A}}{k_B T}\right)\exp\left\{\frac{\Delta E_{gB,Ge}(x=0)}{k_B T}\right\}\left\{\exp\left(\frac{qV_{BE}}{k_B T}\right)-1\right\}}{\left[-\frac{1}{\{\Delta E_{g,Ge}(\text{grade})/(W_B k_B T)\}}\right]\left[\exp\left[\{-\Delta E_{g,Ge}(\text{grade})\}\{W_B/(W_B k_B T)\}\right]-\exp\left[\{-\Delta E_{g,Ge}(\text{grade})\}\{0/(W_B k_B T)\}\right]\right]} \\ &= q\frac{\tilde{D}_{nB}\tilde{N}}{N_{A,B}} \frac{n_{i,Si}^2 \exp\left(\frac{\Delta E_{gB,A}}{k_B T}\right)\exp\left\{\frac{\Delta E_{gB,Ge}(x=0)}{k_B T}\right\}\left\{\exp\left(\frac{qV_{BE}}{k_B T}\right)-1\right\}}{\{-(W_B k_B T)/\Delta E_{g,Ge}(\text{grade})\}\left[\exp\left[\{-\Delta E_{g,Ge}(\text{grade})\}\{1/(k_B T)\}\right]-\exp(0)\right]} \\ &= q\frac{\tilde{D}_{nB}\tilde{N}}{N_{A,B}W_B} \frac{n_{i,Si}^2 \exp\left(\frac{\Delta E_{gB,A}}{k_B T}\right)\{-\Delta E_{g,Ge}(\text{grade})/(k_B T)\}\exp\left\{\frac{\Delta E_{gB,Ge}(x=0)}{k_B T}\right\}\left\{\exp\left(\frac{qV_{BE}}{k_B T}\right)-1\right\}}{\left[\exp\{-\Delta E_{g,Ge}(\text{grade})/(k_B T)\}-1\right]} \\ &= q\frac{\tilde{D}_{nB}\tilde{N}}{N_{A,B}W_B} n_{i,Si}^2 \exp\left(\frac{\Delta E_{gB,A}}{k_B T}\right)\left\{\exp\left(\frac{qV_{BE}}{k_B T}\right)-1\right\} \times \frac{\{\Delta E_{g,Ge}(\text{grade})/(k_B T)\}\exp\left\{\frac{\Delta E_{gB,Ge}(x=0)}{k_B T}\right\}}{\left[1-\exp\{-\Delta E_{g,Ge}(\text{grade})/(k_B T)\}\right]}. \end{aligned} \tag{6.40}$$

In this equation, looking at the second factor after the multiplication sign (×), it is evident that the collector current density $J_{C,SiGe}$ is directly proportional to the degree of bandgap grading $\Delta E_{g,Ge}(\text{grade})$ and exponentially dependent on the band offset $\Delta E_{gB,Ge}(x = 0)$ caused by Ge at the edge of the base towards the emitter side. These two dependences empower the HBT designer with the capability to easily achieve the desired gain at any temperature.

If we focus attention on a Si HBT device which is fabricated using similar process parameters as the SiGe HBT, and assume that the germanium profile on the emitter side of the neutral base region does not stretch far enough into the emitter to significantly alter the base current density, it is logical to infer that the base current density J_B of the Si HBT will be identical to that for SiGe HBT. Since, $\beta = J_C/J_B$, one can write for similarly constructed SiGe and Si HBTs (Cressler 1998)

$$\frac{\beta_{\mathrm{SiGe}}}{\beta_{\mathrm{Si}}}=\frac{J_{\mathrm{C,SiGe}}}{J_{\mathrm{C,Si}}}=\left\{\frac{\tilde{N}_{\mathrm{C,SiGe}}(x)\tilde{N}_{\mathrm{V,SiGe}}(x)}{\tilde{N}_{\mathrm{C,Si}}(x)\tilde{N}_{\mathrm{V,Si}}(x)}\right\}\left(\frac{\tilde{D}_{\mathrm{nB,SiGe}}}{\tilde{D}_{\mathrm{nB,Si}}}\right)\times\frac{\{\Delta E_{\mathrm{g,Ge}}(\mathrm{grade})/(k_{\mathrm{B}}T)\}\exp\left\{\frac{\Delta E_{\mathrm{gB,Ge}}(x=0)}{k_{\mathrm{B}}T}\right\}}{\left[1-\exp\{-\Delta E_{\mathrm{g,Ge}}(\mathrm{grade})/(k_{\mathrm{B}}T)\}\right]},\tag{6.41}$$

where the superscript ~ over a variable means that position averaging has been performed over its value. The ratio $\beta_{\mathrm{SiGe}}/\beta_{\mathrm{Si}}$ represents the enhancement in current gain by use of SiGe over a silicon transistor.

Assuming constant doping in the base region, the forward base transit time of a SiGe HBT is given by Krömer's formula (Harame et al 1995)

$$\begin{aligned}
\tau_{\mathrm{b,SiGe}}&=\int_{x=0}^{x=W_{\mathrm{B}}}\frac{n_{\mathrm{iB}}^2(x)}{N_{\mathrm{B}}(x)}\left[\int_{y=x}^{y=W_{\mathrm{B}}}\left\{\frac{N_{\mathrm{B}}(y)\mathrm{d}y}{D_{\mathrm{nb}}(y)n_{\mathrm{iB}}^2(y)}\right\}\right]\mathrm{d}x\\
&=\int_{x=0}^{x=W_{\mathrm{B}}}\frac{n_{\mathrm{iB}}^2(x)}{N_{\mathrm{B}}(x)}\left[\frac{N_{\mathrm{B}}(y)}{D_{\mathrm{nb}}(y)}\int_{y=x}^{y=W_{\mathrm{B}}}\left\{\frac{\mathrm{d}y}{n_{\mathrm{iB}}^2(y)}\right\}\right]\mathrm{d}x\\
&=\int_{x=0}^{x=W_{\mathrm{B}}}\frac{n_{\mathrm{iB}}^2(x)}{N_{\mathrm{B}}(x)}\left[\frac{N_{\mathrm{B}}(y)}{D_{\mathrm{nb}}(y)}\int_{y=x}^{y=W_{\mathrm{B}}}\left\{\frac{\mathrm{d}y}{Nn_{\mathrm{i,Si}}^2\mathrm{e}^{\frac{\Delta E_{\mathrm{gB,A}}}{k_{\mathrm{B}}T}}\mathrm{e}^{\frac{\Delta E_{\mathrm{g,Ge}}(\mathrm{grade})y}{W_{\mathrm{B}}k_{\mathrm{B}}T}}\mathrm{e}^{\frac{\Delta E_{\mathrm{gB,Ge}}(x=0)}{k_{\mathrm{B}}T}}}\right\}\right]\mathrm{d}x\\
&=\int_{x=0}^{x=W_{\mathrm{B}}}\frac{n_{\mathrm{iB}}^2(x)}{N_{\mathrm{B}}(x)}\left[\frac{N_{\mathrm{B}}(y)}{D_{\mathrm{nb}}(y)}\int_{y=x}^{y=W_{\mathrm{B}}}\left\{\frac{1}{Nn_{\mathrm{i,Si}}^2}\mathrm{e}^{-\frac{\Delta E_{\mathrm{gB,A}}}{k_{\mathrm{B}}T}}\mathrm{e}^{-\frac{\Delta E_{\mathrm{g,Ge}}(\mathrm{grade})y}{W_{\mathrm{B}}k_{\mathrm{B}}T}}\mathrm{e}^{-\frac{\Delta E_{\mathrm{gB,Ge}}(x=0)}{k_{\mathrm{B}}T}}\mathrm{d}y\right\}\right]\mathrm{d}x\\
&=\int_{x=0}^{x=W_{\mathrm{B}}}\frac{n_{\mathrm{iB}}^2(x)}{N_{\mathrm{B}}(x)}\left[\frac{N_{\mathrm{B}}(y)}{D_{\mathrm{nb}}(y)}\times\frac{\mathrm{e}^{-\frac{\Delta E_{\mathrm{gB,A}}}{k_{\mathrm{B}}T}}\mathrm{e}^{-\frac{\Delta E_{\mathrm{gB,Ge}}(x=0)}{k_{\mathrm{B}}T}}}{Nn_{\mathrm{i,Si}}^2}\int_{y=x}^{y=W_{\mathrm{B}}}\left\{\mathrm{e}^{-\frac{\Delta E_{\mathrm{g,Ge}}(\mathrm{grade})y}{W_{\mathrm{B}}k_{\mathrm{B}}T}}\mathrm{d}y\right\}\right]\mathrm{d}x\\
&=\int_{x=0}^{x=W_{\mathrm{B}}}\frac{n_{\mathrm{iB}}^2(x)N_{\mathrm{B}}(y)\mathrm{e}^{-\frac{\Delta E_{\mathrm{gB,A}}}{k_{\mathrm{B}}T}}\mathrm{e}^{-\frac{\Delta E_{\mathrm{gB,Ge}}(x=0)}{k_{\mathrm{B}}T}}}{N_{\mathrm{B}}(x)D_{\mathrm{nb}}(y)Nn_{\mathrm{i,Si}}^2}\left[-\frac{1}{\frac{\Delta E_{\mathrm{g,Ge}}(\mathrm{grade})}{W_{\mathrm{B}}k_{\mathrm{B}}T}}\mathrm{e}^{-\frac{\Delta E_{\mathrm{g,Ge}}(\mathrm{grade})y}{W_{\mathrm{B}}k_{\mathrm{B}}T}}\right]_{y=x}^{y=W_{\mathrm{B}}}\mathrm{d}x\\
&=\int_{x=0}^{x=W_{\mathrm{B}}}\frac{n_{\mathrm{iB}}^2(x)N_{\mathrm{B}}(y)\mathrm{e}^{-\frac{\Delta E_{\mathrm{gB,A}}}{k_{\mathrm{B}}T}}\mathrm{e}^{-\frac{\Delta E_{\mathrm{gB,Ge}}(x=0)}{k_{\mathrm{B}}T}}}{N_{\mathrm{B}}(x)D_{\mathrm{nb}}(y)Nn_{\mathrm{i,Si}}^2}\left[-\frac{W_{\mathrm{B}}k_{\mathrm{B}}T}{\Delta E_{\mathrm{g,Ge}}(\mathrm{grade})}\mathrm{e}^{-\frac{\Delta E_{\mathrm{g,Ge}}(\mathrm{grade})y}{W_{\mathrm{B}}k_{\mathrm{B}}T}}\right]_{y=x}^{y=W_{\mathrm{B}}}\mathrm{d}x\\
&=\int_{x=0}^{x=W_{\mathrm{B}}}\frac{n_{\mathrm{iB}}^2(x)N_{\mathrm{B}}(y)\mathrm{e}^{-\frac{\Delta E_{\mathrm{gB,A}}}{k_{\mathrm{B}}T}}\mathrm{e}^{-\frac{\Delta E_{\mathrm{gB,Ge}}(x=0)}{k_{\mathrm{B}}T}}}{N_{\mathrm{B}}(x)D_{\mathrm{nb}}(y)Nn_{\mathrm{i,Si}}^2}\left[-\frac{W_{\mathrm{B}}k_{\mathrm{B}}T}{\Delta E_{\mathrm{g,Ge}}(\mathrm{grade})}\mathrm{e}^{-\frac{\Delta E_{\mathrm{g,Ge}}(\mathrm{grade})W_{\mathrm{B}}}{W_{\mathrm{B}}k_{\mathrm{B}}T}}\right.\\
&\qquad\left.+\frac{W_{\mathrm{B}}k_{\mathrm{B}}T}{\Delta E_{\mathrm{g,Ge}}(\mathrm{grade})}\mathrm{e}^{-\frac{\Delta E_{\mathrm{g,Ge}}(\mathrm{grade})x}{W_{\mathrm{B}}k_{\mathrm{B}}T}}\right]\mathrm{d}x\\
&=\int_{x=0}^{x=W_{\mathrm{B}}}\frac{n_{\mathrm{iB}}^2(x)N_{\mathrm{B}}(y)W_{\mathrm{B}}k_{\mathrm{B}}T\mathrm{e}^{-\frac{\Delta E_{\mathrm{gB,A}}}{k_{\mathrm{B}}T}}\mathrm{e}^{-\frac{\Delta E_{\mathrm{gB,Ge}}(x=0)}{k_{\mathrm{B}}T}}}{N_{\mathrm{B}}(x)D_{\mathrm{nb}}(y)Nn_{\mathrm{i,Si}}^2\Delta E_{\mathrm{g,Ge}}(\mathrm{grade})}\left[\mathrm{e}^{-\frac{\Delta E_{\mathrm{g,Ge}}(\mathrm{grade})x}{W_{\mathrm{B}}k_{\mathrm{B}}T}}-\mathrm{e}^{-\frac{\Delta E_{\mathrm{g,Ge}}(\mathrm{grade})W_{\mathrm{B}}}{W_{\mathrm{B}}k_{\mathrm{B}}T}}\right]\mathrm{d}x.
\end{aligned}\tag{6.42}$$

Substituting for $n_{\text{iB}}^2(x)$ from (6.37) into (6.42), we obtain

$$\tau_{\text{b,SiGe}} = \int_{x=0}^{x=W_\text{B}} \frac{Nn_{\text{i,Si}}^2 \text{e}^{\frac{\Delta E_{\text{gB,A}}}{k_\text{B}T}} \text{e}^{\left\{\frac{\Delta E_{\text{g,Ge}}(\text{grade})}{k_\text{B}T}\right\}(x/W_\text{B})} \text{e}^{\frac{\Delta E_{\text{gB,Ge}}(x=0)}{k_\text{B}T}}}{N_\text{B}(x)D_{\text{nb}}(y)Nn_{\text{i,Si}}^2\, \Delta E_{\text{g,Ge}}(\text{grade})} \tag{6.43}$$

$$N_\text{B}(y)W_\text{B}k_\text{B}T\text{e}^{-\frac{\Delta E_{\text{gB,A}}}{k_\text{B}T}}\text{e}^{-\frac{\Delta E_{\text{gB,Ge}}(x=0)}{k_\text{B}T}} \times \left[\text{e}^{-\frac{\Delta E_{\text{g,Ge}}(\text{grade})x}{W_\text{B}k_\text{B}T}} - \text{e}^{-\frac{\Delta E_{\text{g,Ge}}(\text{grade})W_\text{B}}{W_\text{B}k_\text{B}T}}\right]\text{d}x$$

$$\begin{aligned}
&= \int_{x=0}^{x=W_\text{B}} \frac{Nn_{\text{i,Si}}^2 \text{e}^{\frac{\Delta E_{\text{gB,A}}}{k_\text{B}T}} \text{e}^{-\frac{\Delta E_{\text{gB,A}}}{k_\text{B}T}} \text{e}^{\frac{\Delta E_{\text{gB,Ge}}(x=0)}{k_\text{B}T}} \text{e}^{-\frac{\Delta E_{\text{gB,Ge}}(x=0)}{k_\text{B}T}}}{N_\text{B}(x)D_{\text{nb}}(y)Nn_{\text{i,Si}}^2\Delta E_{\text{g,Ge}}(\text{grade})} N_\text{B}(y)W_\text{B}k_\text{B}T \\
&\quad \text{e}^{\left\{\frac{\Delta E_{\text{g,Ge}}(\text{grade})}{k_\text{B}T}\right\}(x/W_\text{B})} \\
&\quad \times \left[\text{e}^{-\frac{\Delta E_{\text{g,Ge}}(\text{grade})x}{W_\text{B}k_\text{B}T}} - \text{e}^{-\frac{\Delta E_{\text{g,Ge}}(\text{grade})W_\text{B}}{W_\text{B}k_\text{B}T}}\right]\text{d}x \\
&= \int_{x=0}^{x=W_\text{B}} \frac{N_\text{B}(y)W_\text{B}k_\text{B}T}{N_\text{B}(x)D_{\text{nb}}(y)\Delta E_{\text{g,Ge}}(\text{grade})} \text{e}^{\left\{\frac{\Delta E_{\text{g,Ge}}(\text{grade})}{k_\text{B}T}\right\}(x/W_\text{B})} \\
&\quad \times \left[\text{e}^{-\frac{\Delta E_{\text{g,Ge}}(\text{grade})x}{W_\text{B}k_\text{B}T}} - \text{e}^{-\frac{\Delta E_{\text{g,Ge}}(\text{grade})W_\text{B}}{W_\text{B}k_\text{B}T}}\right]\text{d}x \\
&= \int_{x=0}^{x=W_\text{B}} \frac{N_\text{B}(y)W_\text{B}k_\text{B}T}{N_\text{B}(x)D_{\text{nb}}(y)\Delta E_{\text{g,Ge}}(\text{grade})} \left[1 - \text{e}^{\left\{\frac{\Delta E_{\text{g,Ge}}(\text{grade})}{k_\text{B}T}\right\}(x/W_\text{B})} \text{e}^{-\frac{\Delta E_{\text{g,Ge}}(\text{grade})}{k_\text{B}T}}\right]\text{d}x \\
&= \frac{W_\text{B}}{D_{\text{nb}}(y)} \frac{k_\text{B}T}{\Delta E_{\text{g,Ge}}(\text{grade})} \int_{x=0}^{x=W_\text{B}} \left[1 - \text{e}^{\left\{\frac{\Delta E_{\text{g,Ge}}(\text{grade})}{k_\text{B}T}\right\}(x/W_\text{B})} \text{e}^{-\frac{\Delta E_{\text{g,Ge}}(\text{grade})}{k_\text{B}T}}\right]\text{d}x
\end{aligned} \tag{6.44}$$

taking

$$N_\text{B}(y) = N_\text{B}(x).$$

Now,

$$
\begin{aligned}
\tau_{\mathrm{b,SiGe}} &= \frac{W_{\mathrm{B}}}{D_{\mathrm{nb}}(y)}\frac{k_{\mathrm{B}}T}{\Delta E_{\mathrm{g,Ge}}(\mathrm{grade})}\left[\int_{x=0}^{x=W_{\mathrm{B}}}\mathrm{d}x\right.\\
&\quad\left.-\int_{x=0}^{x=W_{\mathrm{B}}}\mathrm{e}^{\left\{\frac{\Delta E_{\mathrm{g,Ge}}(\mathrm{grade})}{k_{\mathrm{B}}T}\right\}(x/W_{\mathrm{B}})}\mathrm{e}^{-\frac{\Delta E_{\mathrm{g,Ge}}(\mathrm{grade})}{k_{\mathrm{B}}T}}\mathrm{d}x\right]\\
&= \frac{W_{\mathrm{B}}}{D_{\mathrm{nb}}(y)}\frac{k_{\mathrm{B}}T}{\Delta E_{\mathrm{g,Ge}}(\mathrm{grade})}[x]_{=0}^{x=W_{\mathrm{B}}} - \frac{W_{\mathrm{B}}}{D_{\mathrm{nb}}(y)}\\
&\quad \frac{k_{\mathrm{B}}T}{\Delta E_{\mathrm{g,Ge}}(\mathrm{grade})}\int_{x=0}^{x=W_{\mathrm{B}}}\mathrm{e}^{\left\{\frac{\Delta E_{\mathrm{g,Ge}}(\mathrm{grade})}{k_{\mathrm{B}}T}\right\}(x/W_{\mathrm{B}})}\mathrm{e}^{-\frac{\Delta E_{\mathrm{g,Ge}}(\mathrm{grade})}{k_{\mathrm{B}}T}}\mathrm{d}x\\
&= \frac{W_{\mathrm{B}}}{D_{\mathrm{nb}}(y)}\frac{k_{\mathrm{B}}T}{\Delta E_{\mathrm{g,Ge}}(\mathrm{grade})}[W_{\mathrm{B}}] - \frac{W_{\mathrm{B}}}{D_{\mathrm{nb}}(y)}\frac{k_{\mathrm{B}}T\mathrm{e}^{-\frac{\Delta E_{\mathrm{g,Ge}}(\mathrm{grade})}{k_{\mathrm{B}}T}}}{\Delta E_{\mathrm{g,Ge}}(\mathrm{grade})}\\
&\quad\times\int_{x=W_{\mathrm{B}}}^{x=0}\mathrm{e}^{\left\{\frac{\Delta E_{\mathrm{g,Ge}}(\mathrm{grade})}{k_{\mathrm{B}}T}\right\}(x/W_{\mathrm{B}})}\mathrm{d}x\\
&= \frac{W_{\mathrm{B}}^2}{D_{\mathrm{nb}}(y)}\frac{k_{\mathrm{B}}T}{\Delta E_{\mathrm{g,Ge}}(\mathrm{grade})} - \frac{W_{\mathrm{B}}}{D_{\mathrm{nb}}(y)}\frac{k_{\mathrm{B}}T\mathrm{e}^{-\frac{\Delta E_{\mathrm{g,Ge}}(\mathrm{grade})}{k_{\mathrm{B}}T}}}{\Delta E_{\mathrm{g,Ge}}(\mathrm{grade})}\frac{1}{\left\{\frac{\Delta E_{\mathrm{g,Ge}}(\mathrm{grade})}{k_{\mathrm{B}}T}\right\}(1/W_{\mathrm{B}})} \qquad (6.45)\\
&\quad\times\left[\mathrm{e}^{\left\{\frac{\Delta E_{\mathrm{g,Ge}}(\mathrm{grade})}{k_{\mathrm{B}}T}\right\}(x/W_{\mathrm{B}})}\right]_{x=0}^{x=W_{\mathrm{B}}}\\
&= \frac{W_{\mathrm{B}}^2}{D_{\mathrm{nb}}(y)}\frac{k_{\mathrm{B}}T}{\Delta E_{\mathrm{g,Ge}}(\mathrm{grade})} - \frac{W_{\mathrm{B}}^2}{D_{\mathrm{nb}}(y)}\frac{k_{\mathrm{B}}T}{\Delta E_{\mathrm{g,Ge}}(\mathrm{grade})}\\
&\quad\times\frac{k_{\mathrm{B}}T}{\Delta E_{\mathrm{g,Ge}}(\mathrm{grade})}\mathrm{e}^{-\frac{\Delta E_{\mathrm{g,Ge}}(\mathrm{grade})}{k_{\mathrm{B}}T}}\left[\mathrm{e}^{\frac{\Delta E_{\mathrm{g,Ge}}(\mathrm{grade})}{k_{\mathrm{B}}T}} - \mathrm{e}^{0}\right]\\
&= \frac{W_{\mathrm{B}}^2}{D_{\mathrm{nb}}(y)}\frac{k_{\mathrm{B}}T}{\Delta E_{\mathrm{g,Ge}}(\mathrm{grade})} - \frac{W_{\mathrm{B}}^2}{D_{\mathrm{nb}}(y)}\frac{k_{\mathrm{B}}T}{\Delta E_{\mathrm{g,Ge}}(\mathrm{grade})}\\
&\quad\times\frac{k_{\mathrm{B}}T}{\Delta E_{\mathrm{g,Ge}}(\mathrm{grade})}\left\{1 - \mathrm{e}^{-\frac{\Delta E_{\mathrm{g,Ge}}(\mathrm{grade})}{k_{\mathrm{B}}T}}\right\}\\
&= \frac{W_{\mathrm{B}}^2}{D_{\mathrm{nb}}(y)}\frac{k_{\mathrm{B}}T}{\Delta E_{\mathrm{g,Ge}}(\mathrm{grade})}\left[1 - \frac{k_{\mathrm{B}}T}{\Delta E_{\mathrm{g,Ge}}(\mathrm{grade})}\left\{1 - \mathrm{e}^{-\frac{\Delta E_{\mathrm{g,Ge}}(\mathrm{grade})}{k_{\mathrm{B}}T}}\right\}\right].
\end{aligned}
$$

But

$$\tau_{\mathrm{b,Si}} = \frac{W_{\mathrm{B}}^2}{2D_{\mathrm{nb}}(y)} \qquad (6.46)$$

$$\therefore \tau_{b,SiGe} = 2 \times \frac{W_B^2}{2D_{nb}(y)} \times \frac{k_B T}{\Delta E_{g,Ge}(grade)} \times \left[1 - \frac{k_B T}{\Delta E_{g,Ge}(grade)}\left\{1 - e^{-\frac{\Delta E_{g,Ge}(grade)}{k_B T}}\right\}\right] \tag{6.47}$$

$$= 2\tau_{b,Si} \times \frac{k_B T}{\Delta E_{g,Ge}(grade)}\left[1 - \frac{k_B T}{\Delta E_{g,Ge}(grade)}\left\{1 - e^{-\frac{\Delta E_{g,Ge}(grade)}{k_B T}}\right\}\right]$$

or

$$\frac{\tau_{b,SiGe}}{\tau_{b,Si}} = \frac{2k_B T}{\Delta E_{g,Ge}(grade)}\left[1 - \frac{k_B T}{\Delta E_{g,Ge}(grade)}\left\{1 - e^{-\frac{\Delta E_{g,Ge}(grade)}{k_B T}}\right\}\right]. \tag{6.48}$$

Taking $\Delta E_{g,Ge}(grade) = 75$ meV $= 75 \times 10^{-3}$ eV $= 0.075$ eV, $k_B = 8.62 \times 10^{-5}$ eV K^{-1}, at $T = 300$ K

$$\left(\frac{\tau_{b,SiGe}}{\tau_{b,Si}}\right)_{300K} = \frac{2 \times 8.62 \times 10^{-5} \times 300}{0.075} \times \left[1 - \frac{8.62 \times 10^{-5} \times 300}{0.075}\left\{1 - e^{-\frac{0.075}{8.62 \times 10^{-5} \times 300}}\right\}\right] \tag{6.49}$$

$$= 0.6896(1 - 0.3448 \times 0.9449895) = 0.4649.$$

At $T = 4.2$ K,

$$\left(\frac{\tau_{b,SiGe}}{\tau_{b,Si}}\right)_{4.2K} = \frac{2k_B \times 4.2}{\Delta E_{g,Ge}(grade)}\left[1 - \frac{k_B \times 4.2}{\Delta E_{g,Ge}(grade)}\left\{1 - e^{-\frac{\Delta E_{g,Ge}(grade)}{k_B \times 4.2}}\right\}\right]$$

$$= \frac{2 \times 8.62 \times 10^{-5} \times 4.2}{0.075}\left[1 - \frac{8.62 \times 10^{-5} \times 4.2}{0.075}\left\{1 - e^{-\frac{0.075}{8.62 \times 10^{-5} \times 4.2}}\right\}\right] \tag{6.50}$$

$$= 0.0096544(1 - 0.0048272 \times 1) = 0.009608$$

$$\left(\frac{\tau_{b,SiGe}}{\tau_{b,Si}}\right)_{300K} / \left(\frac{\tau_{b,SiGe}}{\tau_{b,Si}}\right)_{4.2K} = \frac{0.4649}{0.009608} = 48.38676. \tag{6.51}$$

The above calculations show that at 300 K, $\tau_{b,SiGe} = 0.5 \times \tau_{b,Si}$. At 4.2 K, $\tau_{b,SiGe} = 0.01 \times \tau_{b,Si}$ becoming considerably shorter. The device becomes faster. Hence, the decrease in temperature favors the SiGe HBT, making it much faster than the Si HBT.

6.5 Discussion and conclusions

The outperformance of HBT compared to homojunction BJT at cryogenic temperatures provides a welcome relief to design engineers working on analog, digital and mixed signal circuits for low-temperature operation.

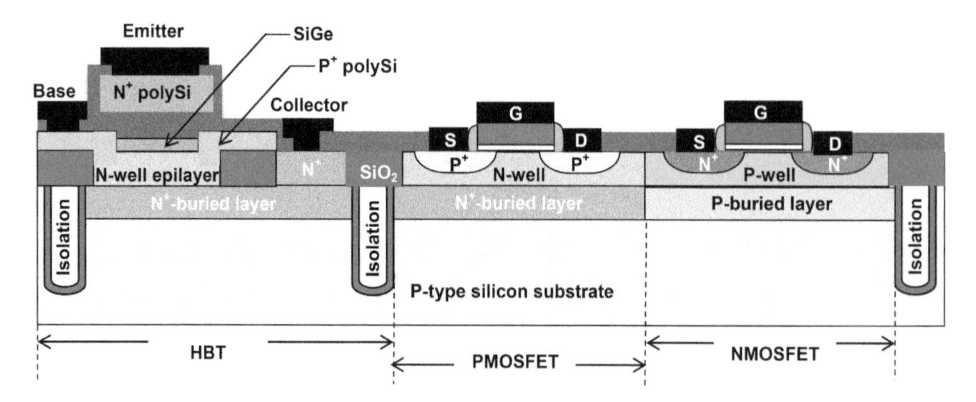

Figure 6.6. BiCMOS integration on a single platform.

The SiGe HBT is compatible and easily integrable with Si CMOS structures, yielding bipolar CMOS (BiCMOS) circuits for wired/wireless RF communication networks and computing applications (figure 6.6). The SiGe BiCMOS is a flexible technological platform for mixed-signal RF and microwave circuits. By merging the high gain and speed provided by bipolar technology with the low power consumption logic gates of CMOS technology, it is possible to build high-performance circuits combining the best features of both technologies (Harame *et al* 2001).

Review exercises

6.1. How does a homojunction BJT differ from a heterojunction BJT? Argue why you cannot design a homojunction BJT having a high current gain together with fast switching capability. What structural parameters of the BJT are to be traded off in this design?

6.2. How does the use of a silicon–germanium heterojunction structure help in resolving the conflict between current gain and switching speed in a homojunction BJT? Explain how the heterojunction BJT achieves a high gain and low base transit time simultaneously.

6.3. How does the lower bandgap of the silicon–germanium base in comparison to the silicon emitter of a heterojunction BJT help in increasing the emitter injection efficiency of the heterostructure?

6.4. How does a graded profile of the silicon–germanium base in an HBT increase its switching speed?

6.5. Which has a smaller bandgap: silicon or germanium? Write the formula for the bandgap of the silicon–germanium alloy as a function of germanium percentage.

6.6. What is meant by BiCMOS technology? What is the principal advantage offered by this technology in comparison to either bipolar technology alone or CMOS technology alone?

6.7. How is the SiGe base layer of a Si/SiGe HBT formed? Discuss the relative advantages and disadvantages of selective epitaxy and non-selective epitaxy.

6.8. Why is it necessary to incorporate carbon in the base layer of a Si/SiGe HBT? How large is the proportion of carbon with respect to SiGe?

6.9. Write down the equation for the current gain $(\beta)_{Si/SiGe}$ of a $Si/Si_{1-x}Ge_x$ HBT and explain the symbols used. Using this equation for $(\beta)_{Si/SiGe}$, write the equation for the forward transit time through the emitter of a $Si/Si_{1-x}Ge_x$ HBT.

6.10. Given that electron mobility $\mu_{n,Si}$ in Si is

$$\mu_{n,Si}(T) \propto T^{-2.42}$$

and electron mobility in Ge is

$$\mu_{n,Ge}(T) \propto T^{-1.66}$$

derive the following equation for diffusion coefficient of electrons in $Si_{1-x}Ge_x$

$$D_{n,SiGe}(T) = \left(\frac{k_B T}{q}\right)\{9.88 \times 10^5(1 - x)T^{-2.42} + 1.29 \times 10^4 x T^{-1.66}\}.$$

if $\mu_{n,Si} = 1350\ \text{cm}^2\ \text{V}^{-1}\ \text{s}^{-1}$ at $T = 300$ K, $\mu_{n,Ge} = 3900\ \text{cm}^2\ \text{V}^{-1}\ \text{s}^{-1}$ at $T =$ 300 K.

6.11. When the temperature of a Si BJT is lowered down from room temperature towards absolute zero, its current gain decreases rapidly with temperature. However, when a Si/SiGe heterojunction BJT is subjected to a similar fall of temperature, its current gain degradation with temperature is much less severe. Explain, giving reasons, this difference in the current gain behaviors of a BJT and an HBT at low temperatures.

6.12. Starting from the Moll–Ross relationship connecting the collector current density J_C of a bipolar transistor with the applied base-emitter voltage V_{BE} at a given temperature T, derive the equation for the ratio of current gains β_{SiGe}/β_{Si} of the silicon/silicon–germanium HBT and silicon BJT in terms of the degree of bandgap grading $\Delta E_{g,Ge}(\text{grade})$ and the band offset $\Delta E_{g,Ge}(x = 0)$ caused by Ge at the edge of the base towards the emitter side. What is the importance of this equation to an HBT designer trying to design an HBT for a specific operating temperature?

6.13. Explain, with the help of the equation for the ratio of current gains β_{SiGe}/β_{Si} of the silicon/silicon–germanium HBT and silicon BJT, why it is possible to design a silicon/silicon–germanium HBT with the required current gain β_{SiGe} at a given temperature T by choosing the appropriate degree of bandgap grading $\Delta E_{g,Ge}(\text{grade})$ and the band offset $\Delta E_{g,Ge}(x = 0)$ caused by Ge at the edge of the base towards the emitter side. Hence, elaborate the greater design flexibility afforded by an HBT over a BJT for low-temperature operation.

6.14. Derive the equation for the ratio of forward base transit times $(\tau_{b,SiGe}/\tau_{b,Si})$ of a silicon–germanium HBT and a Si BJT in terms of the degree of bandgap grading $\Delta E_{g,Ge}(\text{grade})$. Hence, show that a SiGe HBT becomes much faster than a Si BJT when the temperature is decreased from 300 K to 4.2 K.

6.15. Explain, with reference to the current gain and base transit time parameters, the superiority of a SiGe HBT over a Si BJT for operation at low temperatures.

References

Basu S and Sarkar P 2011 Analytical modeling of AlGaAs/GaAs and Si/SiGe HBTs including the effect of temperature *J. Electron Devices* **9** 325–9

Cressler J D 1998 SiGe HBT technology: a new contender for Si-based RF and microwave circuit applications *IEEE Trans. Microw. Theory Tech.* **46** 572–89

Cressler J D and Niu G 2003 *Silicon–Germanium Heterojunction Bipolar Transistors* (Boston, MA: Artech House) pp 98–104

Hållstedt J 2004 Epitaxy and characterization of SiGeC layers grown by reduced pressure chemical vapor deposition *Licentiate Thesis* Stockholm Royal Institute of Technology (KTH), Stockholm, 39 pages

Harame D L, Ahlgren D C, Coolbaugh D D, Dunn J S and Freeman G G *et al* 2001 Current status and future trends of SiGe BiCMOS technology *IEEE Trans. Electron Devices* **48** 2575–94

Harame D L, Comfort J H, Cressler J D, Crabbé E F, Sun J Y-C, Meyerson B S and Tice T 1995 *IEEE Trans. Electron Devices* **42** 455–68

Ioffe Institute 2015 SiGe—silicon germanium: band structure and carrier concentration *Ioffe Institute* www.ioffe.ru/SVA/NSM/Semicond/SiGe/bandstr.html

Mitrovic I Z, Buiu O, Hall S, Bagnall D M and Ashburn P 2005 Review of SiGe HBTs on SOI *Solid-State Electron* **49** 1556–67

Shiraki Y and Usami N 2011 *Silicon–Germanium (Si–Ge) Nanostructures: Production, Properties and Applications in Electronics* (Cambridge: Woodhead) p 424

Shur M 1995 *Physics of Semiconductor Devices* (New Delhi: Prentice Hall India), 704 pages

Virginia Semiconductor 2002 The general properties of Si, Ge, SiGe, SiO_2 and Si_3N_4 *Virginia Semiconductor* www.virginiasemi.com/pdf/generalpropertiesSi62002.pdf

Chapter 7

The temperature-sustaining capability of gallium arsenide electronics

Gallium arsenide ranks second to silicon with regard to two primary aspects; it is a wider bandgap semiconductor suitable for high-temperature operation and with technological maturity. The high electron mobility in GaAs together with its direct bandgap make this material a strong competitor to silicon for the fabrication of high-temperature microwave and optoelectronic devices and circuits. GaAs circuits fabricated with existing state-of-the-art technologies were found to succumb prematurely to elevated temperatures due to reasons for which GaAs itself was not directly responsible. For realization of the full thermal capability of GaAs prescribed by its fundamental physical limits, it is necessary to introduce thermally stable contact metallization, modify many processes, and formulate innovative device structures and designs so that failure from any cause unassociated with GaAs is completely eliminated.

7.1 Introduction

The wider bandgap of GaAs, by 1.424 to 1.12 eV = 0.304 eV, in comparison to silicon pushes the operating temperature range for GaAs electronics from between 200 °C and 300 °C for silicon to 400 °C for GaAs. Although, theoretically, the permissible range for GaAs extends up to approximately 500 °C, practical limitations have restricted the operation to lower temperatures. Table 7.1 compiles the properties of GaAs of interest to device and circuit designers. In figure 7.1, the crystal structure of gallium arsenide is displayed.

Some striking fundamental dissimilarities between gallium arsenide and silicon or germanium need attention. (i) Both germanium and silicon are elemental semiconductors. A cursory glance at the periodic classification of chemical elements shows that group IV is the abode of these elements. Unlike elemental Ge and Si, gallium arsenide is a III–V compound semiconductor, formed by the union of

doi:10.1088/978-0-7503-1155-7ch7

Table 7.1. Properties of gallium arsenide.

Property	Value	Property	Value	Property	Value
Chemical formula	GaAs	Melting point (°C)	1238	Hole mobility ($cm^2\ Vs^{-1}$)	400
Molecular mass	144.65	Dielectric constant	12.9	Electron diffusion coefficient ($cm^2\ s^{-1}$)	200
Classification	III–V compound semiconductor	Thermal conductivity ($W\ cmK^{-1}$)	0.46	Hole diffusion coefficient ($cm^2\ s^{-1}$)	10
Crystal structure	Zinc blende	Energy bandgap E_g (eV) at 300 K	1.424	Electron saturated velocity ($cm\ s^{-1}$)	4.4×10^7
Color	Dark red	Electrical breakdown field ($V\ cm^{-1}$)	4×10^5	Hole saturated velocity ($cm\ s^{-1}$)	1.8×10^7
Density at 300 K ($g\ cm^{-3}$)	5.32	Intrinsic carrier concentration (cm^{-3})	2.1×10^6	Minority-carrier lifetime (s)	10^{-8}
Number of atoms cm^{-3}	4.42×10^{22}	Intrinsic resistivity (Ωcm)	3.3×10^8		
Lattice constant (Å)	5.65	Electron mobility ($cm^2\ Vs^{-1}$)	8500		

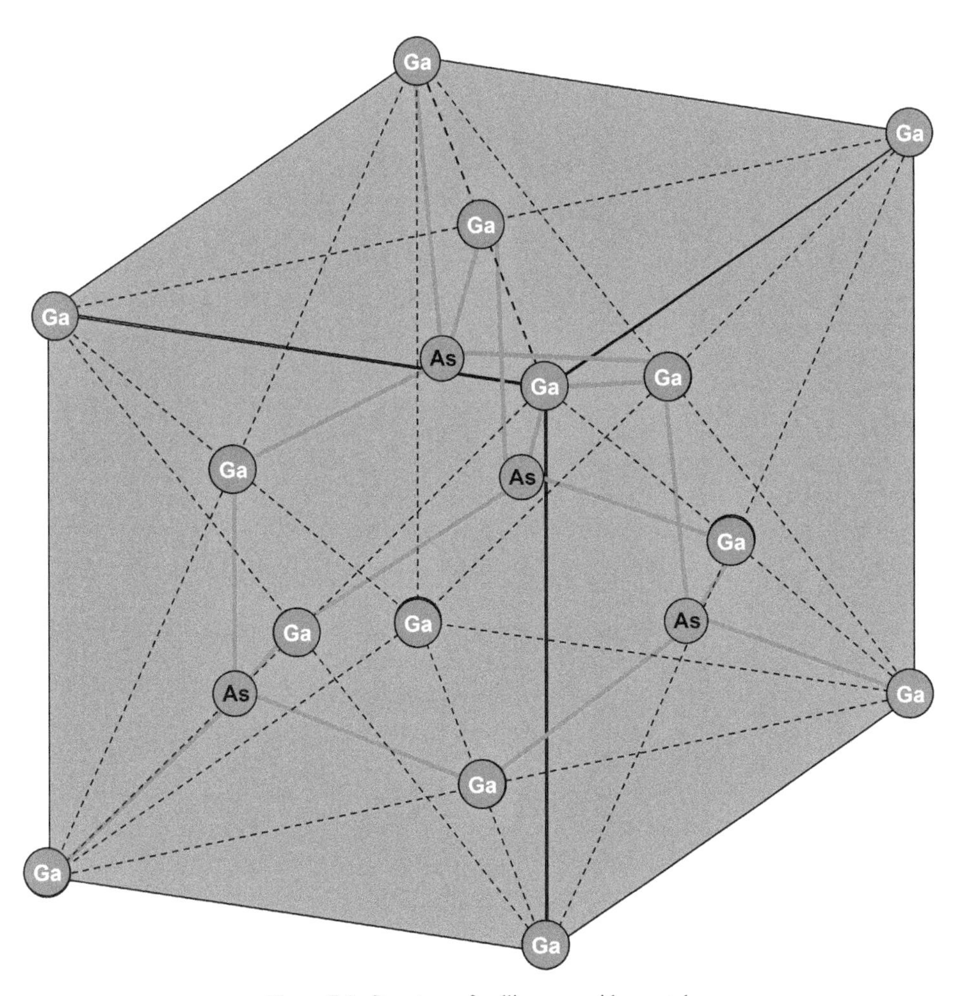

Figure 7.1. Structure of gallium arsenide crystal.

elements which are members of group III and group V. It is so called because it comprises two distinct elements, gallium and arsenic. Gallium resides in group III of the periodic table while arsenic is housed in group V of the periodic table. (ii) Another contrasting feature of gallium arsenide with germanium and silicon pertains to the crystal structure. It is well known that Ge and Si crystallize in the familiar diamond structure. But the crystal configuration of gallium arsenide is not like diamond, it has a zinc blende structure. A gallium arsenide crystal is made up of two face-centered cubic (fcc) sublattices. These fcc sublattices are displaced with respect to each other by a distance equal to half the length of the diagonal of the fcc sublattice. (iii) In contrast to the indirect bandgap property of silicon, gallium arsenide is a direct bandgap semiconductor, enabling its utilization for fabrication of light-emitting diodes and lasers. In addition, the high electron mobility in GaAs makes it possible to use it for microwave and high-frequency devices. (iv) A thin silicon dioxide film of thickness 1 nm spontaneously grows on a silicon surface due to the oxygen present in the atmosphere. This oxide layer, called the native oxide, is beneficial for microelectronics. On a GaAs surface, the available oxides Ga_2O_3, As_2O_3 and As_2O_5 are more problematic than useful.

Other noteworthy differences between silicon and GaAs ICs are as follows. (i) In GaAs circuits, the primary device is a metal–semiconductor FET (MESFET). It plays the same role in GaAs circuits as the MOSFET in silicon technology. The absence of MOSFETs in GaAs technology is due to the non-availability of a tough adherent oxide layer comparable to the thermal SiO_2, which is easily grown in the fabrication of silicon circuits. Consequently, the gating action in MESFETs is provided through an SBD. The reverse leakage current of an SBD in the MESFET is higher by orders of magnitude than that through the gate oxide of the MOSFET. Additionally, MESFET operation is influenced by electron injection from the channel into the substrate at the high electric fields prevalent near the drain. Despite being favored by an oxide dielectric, silicon MOSFETs are limited by leakage current constraints to lower temperatures than GaAs MESFETs, because diffusion-dominated leakage currents become predominant in GaAs at much higher temperatures than in silicon due to the higher bandgap of GaAs. (ii) In GaAs circuits, there is no comparable dielectric isolation process to the SOI technology. For medium and low-temperature operation, the GaAs circuits are fabricated on semi-insulating substrates, which cannot be used at high temperatures due to an increase in their conductivity. Naturally, junction isolation must be resorted to. (iii) Surface passivation is a tricky issue. The problem is overcome by using either a silicon nitride or silicon dioxide layer as the passivant.

Apart from GaAs MESFETs, another important device in GaAs circuits is the GaAs HBT. HBTs are now backing up MESFET technology. Circuits using GaAs MESFETs and HBT circuits together constitute the next milestone to silicon in the march towards elevated temperature operation. GaAs MESFET-based analog and digital circuits for microwave and optoelectronic applications have benefitted the fields of HTE occupying the niches where temperatures a little higher than those allowed by silicon technology are required.

7.2 The intrinsic temperature of GaAs

In analogy to calculations performed for silicon, the calculations for the intrinsic temperature of GaAs are carried out herein. For GaAs, we have

$$\begin{gathered} m_{\mathrm{n}}^{*} = 0.85m_0,\ m_{\mathrm{p}}^{*} = 0.53m_0,\ E_{\mathrm{g}} = 1.424\ \mathrm{eV}. \\ T\ln\{0.207/(m^{*}T^{1.5})\} = -5.8025 \times 10^{3}E_{\mathrm{g}} \end{gathered} \tag{7.1}$$

Equation (7.1) is to be recast for GaAs (refer to equation (4.6)).

Since, for GaAs

$$m^{*} = \left(\frac{m_{\mathrm{n}}^{*}m_{\mathrm{p}}^{*}}{m_0^2}\right)^{0.75} = \left(\frac{0.85m_0 \times 0.53m_0}{m_0^2}\right)^{0.75} = 0.5499 \tag{7.2}$$

we have

$$T \ln\{0.207/(0.5499T^{1.5})\} = -5.8025 \times 10^3 \times 1.424 \tag{7.3}$$

or

$$T \ln (0.3764/T^{1.5}) = -8262.76$$

or

$$T \ln 0.3764 - T \ln T^{1.5} = -8262.76 \tag{7.4}$$

$$\therefore -0.9771T - T \ln T^{1.5} = -8262.76. \tag{7.5}$$

When

$$\begin{aligned} T &= 756.5\text{K} \\ \text{lhs} &= -0.9771 \times 756.5 - 756.5 \ln 756.5^{1.5} = -739.176 - 7521.92 \\ &= -8261.096. \end{aligned} \tag{7.6}$$

If

$$\begin{aligned} T &= 756.6\ \text{K} \\ \text{lhs} &= -0.9771 \times 756.6 - 756.6 \ln 756.6^{1.5} = -739.274 - 7523.0645 \\ &= -8262.3385. \end{aligned} \tag{7.7}$$

If

$$\begin{aligned} T &= 756.7\ \text{K} \\ \text{lhs} &= -0.9771 \times 756.7 - 756.7 \ln 756.7^{1.5} = -739.3716 - 7524.2088 \\ &= -8263.58. \end{aligned} \tag{7.8}$$

$$\therefore T = 756.6\ \text{K}. \tag{7.9}$$

The intrinsic temperature of GaAs (756.6 K) is higher by 756.6 – 588.63 = 167.97 K than that for Si (588.63 K).

7.3 Growth of single-crystal gallium arsenide

For microelectronic device fabrication, single-crystal wafers of GaAs are required. In the liquid encapsulated Czochralski (LEC) method (figure 7.2), elemental Ga and As are placed inside a pyrolytic boron nitride (pBN) crucible. A pellet of boron trioxide (B_2O_3) is also added. The crucible is heated inside a high-pressure crystal puller. When the temperature reaches 460 °C, the boron trioxide melts. The molten boron trioxide (B_2O_3) is a thick liquid with high viscosity. The B_2O_3 layer surrounds the melt from all sides building an enclosure all around it. B_2O_3 lies between the crucible and the melt and thus is separated from the crucible. The top surface of the melt is fully covered by B_2O_3, forming a protective cover. Thus the GaAs melt is completely enclosed by liquid B_2O_3. Hence, this method is called LEC.

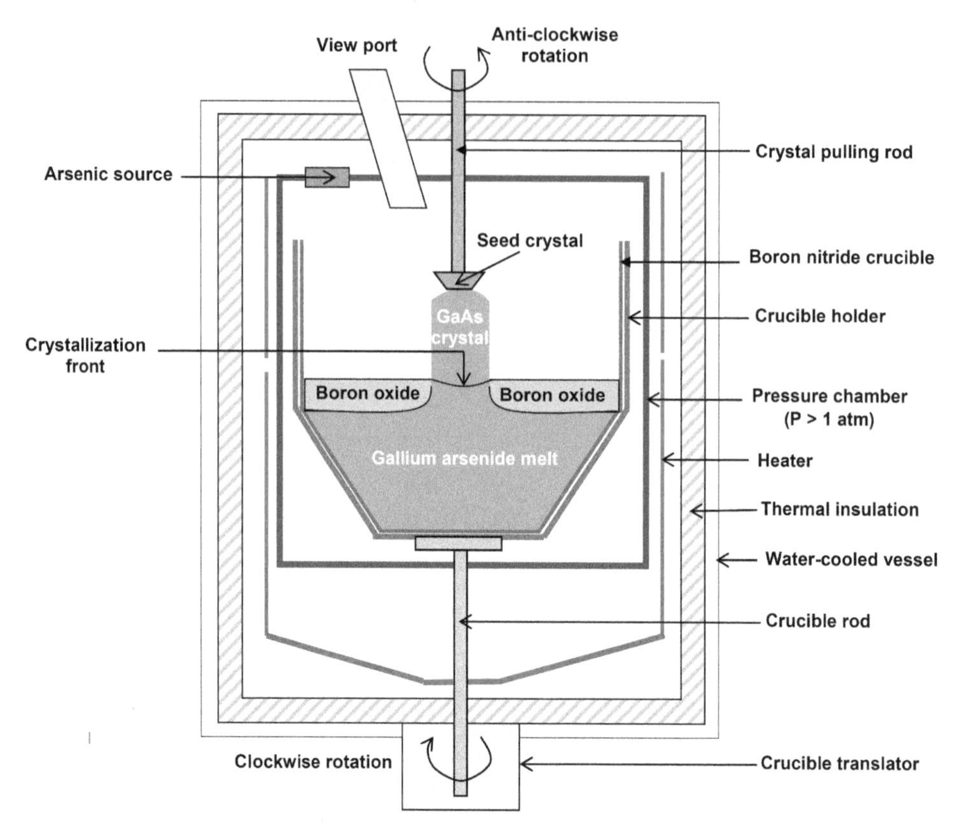

Figure 7.2. The LEC process for the growth of single-crystal gallium arsenide.

Encapsulation of the GaAs melt by liquid B_2O_3 is necessary to prevent sublimation of the volatile As. If As sublimes, the melt will become gallium-rich and its stoichiometry will be disturbed. For crystal growth, a seed crystal is lowered into the melt, passing through the B_2O_3 layer and touching the GaAs surface. The crystal is slowly lifted upwards. During withdrawal, the crystal is also continuously rotated. As a result of this motion, a single crystal of GaAs propagates from the melt.

7.4 Doping of GaAs

Figure 7.3 presents the energy band diagram of GaAs showing the energy levels of impurities used for doping. For n-type doping of GaAs, group IV or group VI elements are used as the impurities (Shur 1987). Common group IV impurities are silicon (Si) and tin (Sn). These elements must occupy Ga sites to act as donors. Since the covalent radius of gallium (1.26 Å) is larger than that of arsenic (1.14 Å), group IV impurities (silicon, 1.11 Å and tin, 1.45 Å) tend to occupy gallium sites. Group VI impurities include sulfur (S), tellurium (Te) and selenium (Se). These elements must occupy As sites to act as acceptors. The doping efficiency is defined as the ratio of doping density to the implanted ion density. It is enhanced if the sample is heated during ion implantation. A lower dose gives a higher doping efficiency owing to a

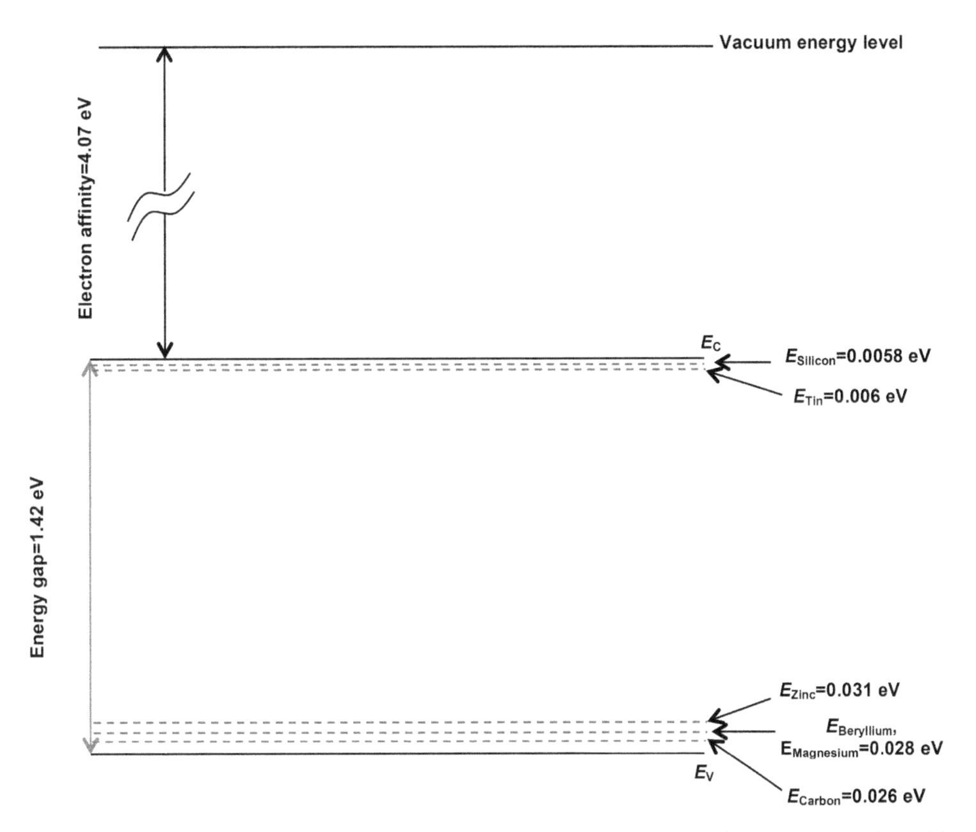

Figure 7.3. Energy band diagram of gallium arsenide showing the energy levels of commonly used impurities for N- (Si, Sn) and P-doping (Zn, Be, Mg, C).

comparatively smaller damage to the crystal lattice. The implantation step is followed by an annealing step between 850 °C and 950 °C. PECVD silicon nitride is used to passivate the GaAs surface, otherwise it loses As when heated at temperatures higher than 600 °C.

For p-type doping, beryllium (Be), zinc (Zn), carbon (C) or magnesium (Mg) ions are implanted into GaAs. The doping efficiency varies with the value of dose. It is 100% at low values of dose up to 10^{14} cm^{-2}, but decreases at higher doses. This decrease is due to the constraint of the solid solubility limit. The annealing temperature also affects the doping efficiency.

Chromium introduces an acceptor energy level near the middle of the bandgap. Cr-doping is performed to produce semi-insulating GaAs substrates (SI-GaAs: Cr) with conductivity in the dark of 3×10^{-9} S cm^{-1} at 300 K, close to the intrinsic conductivity of GaAs.

7.5 Ohmic contacts to GaAs

7.5.1 Au–Ge/Ni/Ti contact to n-type GaAs for room-temperature operation

Au is a metal to which wire can be bonded. The Au–Ge eutectic temperature is 356 °C. Gold reacts with GaAs, producing the vacancies for Ge to move into GaAs.

During alloying, Ga out-diffuses into the contact metal and Ge diffuses into the lattice sites. It is a dopant element and dopes GaAs degeneratively, forming an n^+ semiconductor layer. It also initiates the melting of the contact metal during alloying: Au + GaAs → Au–Ga + As. Nickel is a metal catalyzing the alloying process by driving Ge into GaAs and promoting uniformity: Ni + As → Ni–As. Titanium is a diffusion barrier, preventing excess diffusion of Au. The inter-diffusion of metals and formation of intermetallic compounds depends on the alloying conditions. Rapid thermal annealing (RTA) is best suited for alloying because undesired compounds may be formed if heating and cooling are prolonged. Primarily due to the metal inter-diffusion effects, the thermal stability of this metallization system is poor. However, it yields low contact resistance and is good for room-temperature operation.

7.5.2 High-temperature ohmic contacts to n-type GaAs

Unusual problems associated with these contacts are discussed by Eun and Cooper (1993). A high-temperature metallization scheme has a layered structure (Fricke *et al* 1989), a 200 Å thick Ge film/25 Å thick Au film/100 Å thick Ni film/1000 Å thick diffusion barrier comprising nine alternating layers of e-beam evaporated W and Si with a 50 Å thick Ti film interposed between the fourth Si and fifth W layers/500 Å thick Au film (figure 7.4). The total W–Si layer thickness above the Ti film is 500 Å. The layer thickness below Ti is the same. As before, Ge dopes GaAs n-type, the

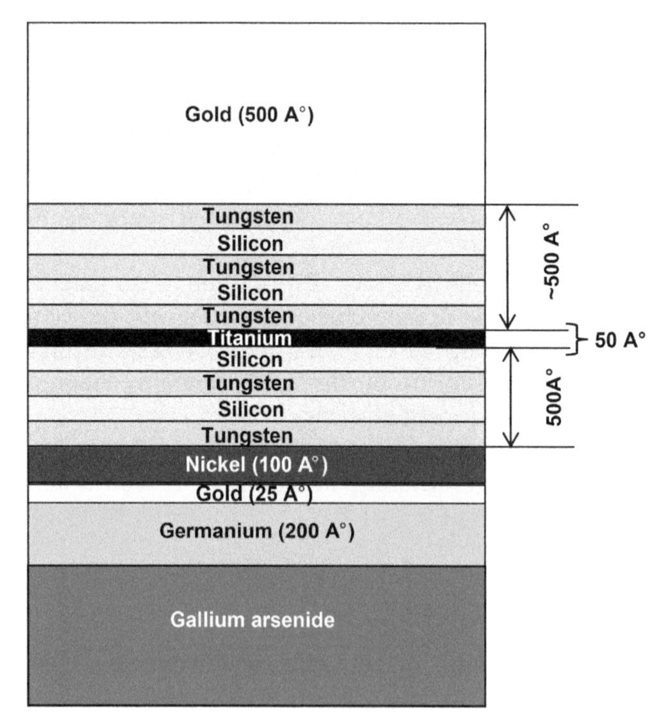

Figure 7.4. Metallization scheme for n-type GaAs.

reaction between Au and Ga leads to the formation of vacancies for the diffusion of Ge into GaAs, and Ni also drives Ge into GaAs during the annealing step. Due to the addition of Au, the annealing temperature is lowered to 590 °C. All Au must be consumed during annealing. If not consumed, any residual Au may cause degradation later because inter-diffusion between Au and GaAs is the main cause of degradation of contacts at high temperatures. After RTA at 640 °C, the contact resistance was $5 \times 10^{-6}\ \Omega\ cm^2$ on GaAs with a donor concentration of $1 \times 10^{17}\ cm^{-3}$. The contact showed no deterioration after storage at 300 °C for over 1000 h. MESFETs with this contact could be operated at 400 °C after the above high-temperature storage. The lifetime at 400 °C was over 100 h.

7.6 Schottky contacts to GaAs

A common multilayer contact scheme is Ti–Pt–Au (Fricke *et al* 1989), where Ti is the adhesion layer, Pt is the diffusion barrier and Au is the high-conductivity layer. The Ti–Pt–Au gate MESFETs did not degrade after storage at 300 °C for over 1000 h. They have a barrier height of 0.78 eV and an ideality factor of 1.1. Sputtering of WSi_2 on the GaAs followed by Ti–Pt–Au metal deposition improved the thermal stability of the contact. However, the barrier height decreases to 0.7 eV and the ideality factor becomes 1.4.

Au–LaB_6 Schottky contacts to GaAs (figure 7.5) yield a larger barrier height, ~0.9 eV, assuring a low leakage current during operation at high temperatures (Würfi *et al* 1990). The reliability of this contact depends on the processing and annealing conditions. The best reliability was achieved on moderately annealed samples (20 h for 20 min). The Schottky contacts were found to be stable after thermal stressing at 400 °C for several hundred hours.

7.7 Commercial GaAs device evaluation in the 25 °C to 400 °C temperature range

As an initial step, the performance of the then state-of-the-art commercial GaAs devices was studied in the temperature range 25 °C to 400 °C (Shoucair and Ojala

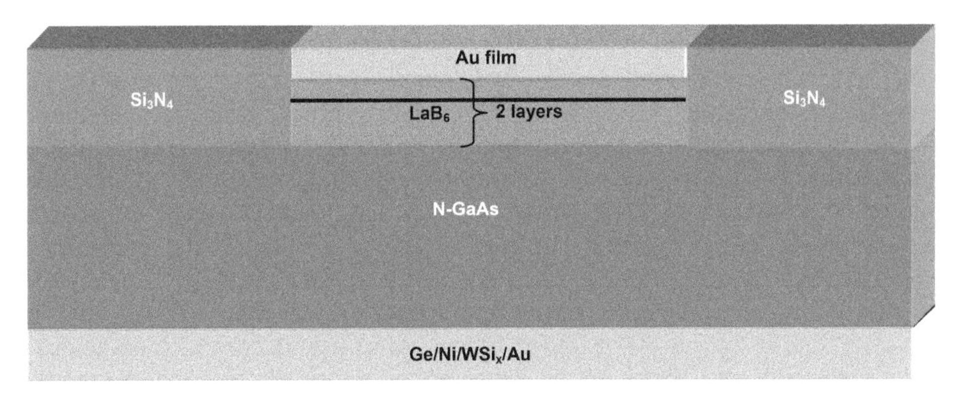

Figure 7.5. A LaB_6 diode.

1992). The tested MESFET devices had Au–Ge/Ni/Au ohmic contacts and tungsten nitride (WN_x) Schottky contacts. Fabricated using LEC semi-insulating substrates, these MESFETs contained a p-type buried layer underneath the n-channels for isolation from the substrate. The electrical behavior of these GaAs MESFETs at elevated temperatures was also compared with that of Si MOSFETs. The study brought to light that, in general, the GaAs devices exhibited identical degradation of electrical parameters to Si MOSFETs. Furthermore, they were riddled with complex difficulties, which necessitated immediate attention.

The temperature dependence of the transconductance parameter β of MESFET devices was dominated by variations in the mobility and saturation velocity of charge carriers. Like Si MOSFETs, the electron mobility followed the familiar power law dependence

$$\mu = \mu_0(T/T_0)^{-n} \tag{7.10}$$

on temperature with the exponent n lying between 1.6 and 2. The saturation velocity

$$v_{sat} \propto T^{-1}. \tag{7.11}$$

Both for enhancement- and depletion-mode MESFETs, the threshold voltage variation with temperature is modeled by a similar equation to Si MOSFETs:

$$V_{Th}(T) = p_0 T + q_0 \text{ with } p_0 = -1.2 \text{ mV K}^{-1} \text{ for } 25 \leqq T \leqq 250\ ^\circ\text{C}, \tag{7.12}$$

where q_0 is the value of V_{Th} at 0 K, as found by extrapolation. In analogy to Si MOSFETs, the existence of ZTC bias points, one in the linear region and another in the saturation region, was confirmed.

While the above characteristics displayed similarity with Si MOSFETs, in opposition to Si MOSFETs, there was a significant leakage current flow from the gate. This was because the MESFET gate relied on the Schottky barrier junction in the reverse-biased mode whereas the Si MOSFET gate depended on the SiO_2 dielectric layer, which allowed imperceptible leakage current flow. To the gate leakage current was added the substrate leakage current arising from the injection of electrons from the channel into the substrate under the influence of the high electric field prevailing near the drain of the MESFET. The substrate leakage current accounted for as much as 50% of the drain current. It was also affected by side-gating and back-gating. All throughout the temperature range 25 °C to 400 °C, the predominant component of leakage current was the generation-recombination current. It showed a direct proportionality relationship with intrinsic carrier concentration. In GaAs, there is transition from the generation–recombination component dominated leakage current behavior at a much higher temperature because of the larger bandgap of this material. In silicon and therefore in Si MOSFETs, this transition occurs at a temperature between 125 °C and 150 °C. The ratio of on- and off-state currents was inferior by 2–3 orders of magnitude for GaAs MESFETs than Si MOSFETs. Transconductance decreased monotonically with temperature and drain output resistance increased as the temperature was raised over the temperature range 25 °C to 250 °C and frequency range 1–10^6 Hz.

The gate input resistance for a depletion-mode MESFET decreased by 4–6 orders of magnitude from 25 °C to 400 °C. The Schottky barrier height fell from 0.65 eV at 25 °C to 0.55 at 400 °C. On the whole, the study showed that performance-wise, GaAs MESFETs were in many respects worse than Si MOSFETs for high-temperature operation. Despite the larger bandgap of GaAs than Si, GaAs devices were succumbing to damage at lower temperatures than equivalent Si devices. This premature failure of GaAs devices fabricated by existing technology showed that considerable research and development efforts were necessary for the improvement of characteristics (Shoucair and Ojala 1992).

7.8 Structural innovations for restricting the leakage current of GaAs MESFET up to 300 °C

Computer simulations of the electrical characteristics of n-channel MESFETs show that an increase in temperature was accompanied by decrease in threshold voltage and transconductance together with an increase in leakage current (Kacprzak and Materka 1983, Wilson and O'Neill 1995). Among these three parameters, the leakage current increase was the issue of the most serious concern as it leads to a loss of transistor action. In GaAs, the crossover from the generation–recombination component of leakage current to the diffusion-related component takes place around 250 °C instead of the 150 °C for Si devices. Two types of leakage current were identified as crucial to device current. (i) Gate leakage: this leakage current is determined by the Schottky barrier height, which is generally ~0.7 eV and decreases as temperature increases, leading to enhanced injection of electrons from the metal gate into the channel. (ii) Drain leakage: in the off-state, the channel is depleted of carriers. With an increase in temperature, the background carrier concentration increases exponentially. Drift current flows through the conducting path between the drain, gate and source. Another path is formed towards the substrate contributing to additional leakage. The substrate leakage part constitutes a substantial chunk of the drain leakage.

Principally due to the above two factors, which increase the leakage current to alarming proportions, originating from the change in device structure from MOSFET to MESFET, the GaAs device lost favor to the Si device. These leakage currents in GaAs MESFETs must be subjugated or stopped from upsetting device operation at high temperatures by modifications in device design. Isolating the gate will prevent the thermionic emission of carriers from the gate into the channel. The conduction path to the substrate must be blocked to reduce substrate leakage. By introducing heterojunction barriers, one between the gate and channel and another between the channel and substrate, both leakages are suppressible. From these considerations, wider bandgap AlGaAs barriers are incorporated both above and below the MESFET channel, sealing the conduction paths from the gate and towards the substrate. Such barriers are called heterojunction barriers because they consist of two semiconductors of different bandgaps. The lower AlGaAs/GaAs barrier is called the backwall barrier. The heterojunction transistor design raises the upper limit of reliable operation of GaAs MESFETs to 300 °C. Figure 7.6 shows the original and improved GaAs MESFET structures.

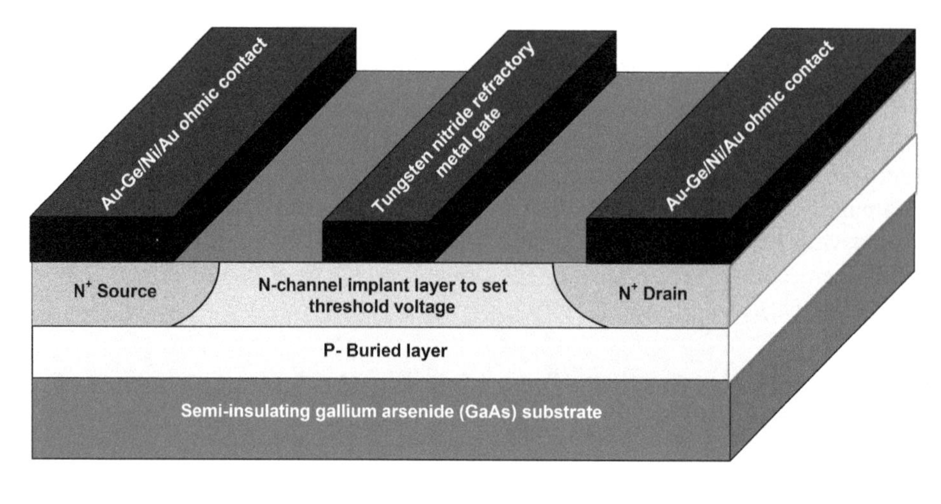

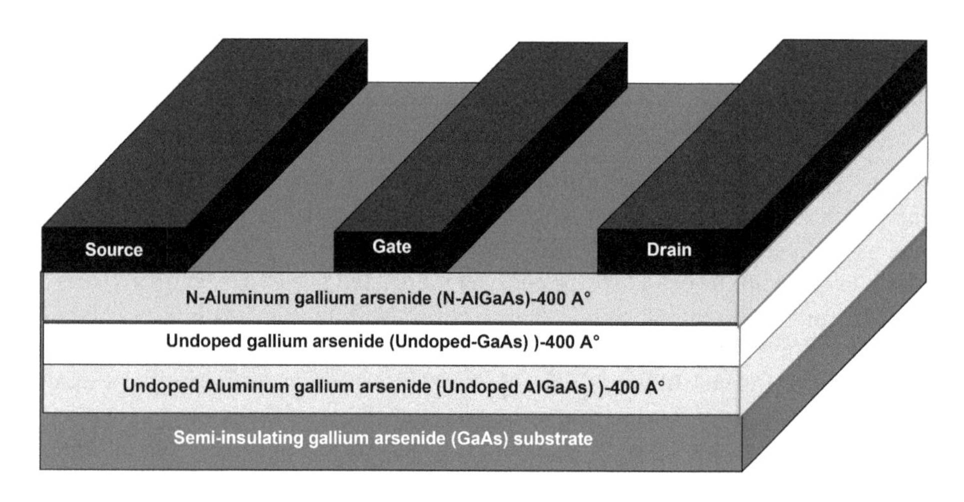

Figure 7.6. GaAs MESFETs. (a) Original structure and (b) improved structure incorporating two AlGaAs layers: the first layer at the gate interface and a second layer serving as a buffer layer between the active region and the substrate.

In a separate experimental study, a stable MESFET technology was developed for operation up to 300 °C by inserting WSi_2 diffusion barriers in the ohmic contacts, supported by optimization of the surface passivation technique using PECVD Si_3N_4 to prevent out-diffusion of Au (Fricke *et al* 1989).

7.9 Won *et al* threshold voltage model for a GaAs MESFET

The drain current of a MESFET (figure 7.7) of channel width W, channel length L and threshold voltage V_{Th}, working under a drain–source voltage V_{DS} and a gate–source voltage V_{GS}, is given by (Won *et al* 1999)

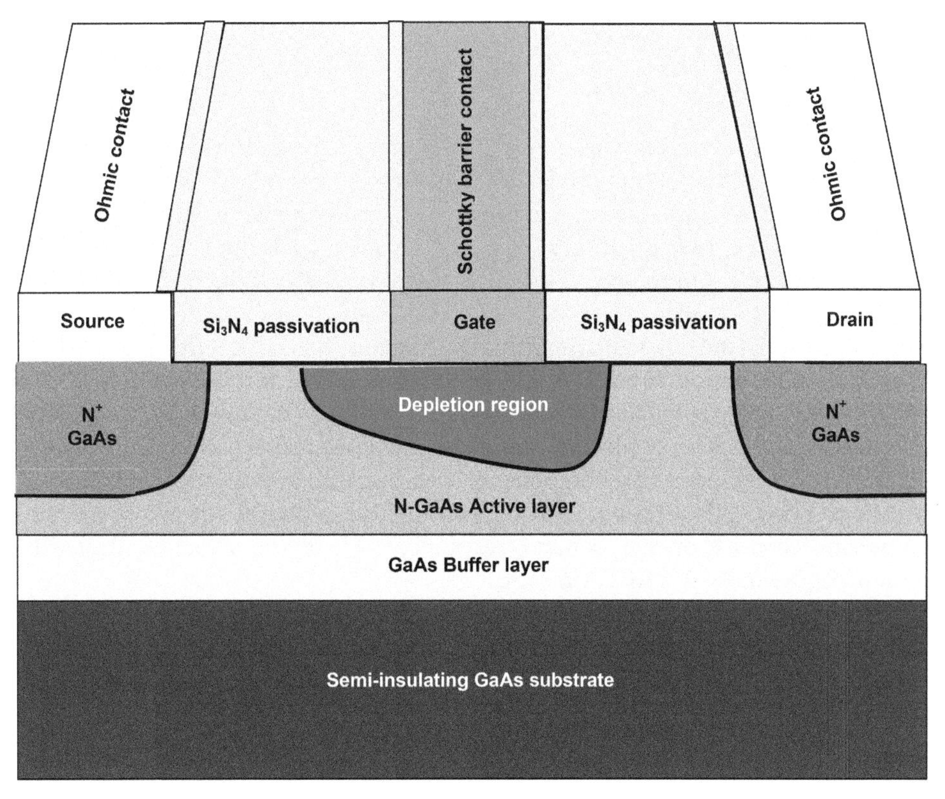

Figure 7.7. A gallium arsenide MESFET.

$$I_{DS} = \beta\{2(V_{GS} - V_{Th})V_{DS} - V_{DS}^2\}(1 + \lambda V_{DS}) \text{ for } V_{DS} \leqslant (V_{GS} - V_{Th}) \tag{7.13}$$

$$I_{DS} = \beta(V_{GS} - V_{Th})^2(1 + \lambda V_{DS}) \text{ for } V_{DS} \geqslant (V_{GS} - V_{Th}) \tag{7.14}$$

where β is the transconductance parameter containing the electron mobility μ_n and λ is the channel length modulation parameter.

The drain current becomes temperature dependent through the decrease in both mobility and threshold voltage with temperature. A linear relation is used for the degradation of mobility with temperature. The decrease in threshold voltage with increase of temperature needs cautious inspection. Considering a depletion-mode MESFET, the threshold voltage is the voltage applied on the gate electrode to produce a depletion layer having a thickness equaling the depth a of the active region. Hence, threshold voltage expression consists of three terms contributed by different voltages present in the device as stated below:

$$\begin{aligned} V_{Th} = {} & \text{built-in potential of the Schottky barrier } (V_{bi}) \\ & -\text{pinch-off voltage } (V_{po}) \\ & -\text{voltage drop due to leakage current } (V_l) \end{aligned} \tag{7.15}$$

$$= \left\{ \Phi_{bn} - \left(\frac{k_B T}{q}\right) \ln\left(\frac{N_C}{N_D}\right) \right\} - \left(\frac{q N_D a^2}{2\varepsilon_0 \varepsilon_s}\right) - \left\{ R_g WLA^* T^2 \exp\left(\frac{q\Phi_{bn}}{k_B T}\right) \right\}. \quad (7.16)$$

The symbols in this equation need explanation. Φ_{bn} is the Schottky barrier height of the metal–semiconductor interface, i.e. the intimate boundary of the metal layer with the n-type semiconductor, here GaAs. In magnitude, it equals the energy difference between two quantities, one on the metal side and the other on the semiconductor side. On the metal side, the Fermi level E_F of the metal is taken. On the semiconductor side, the energy of the conduction band edge E_C of the semiconductor is the relevant quantity. As usual, N_C is the effective density of states in the conduction band. It is related to temperature and is equal to 4.7×10^{17} $(T/300)^{1.5}$. Also, N_D is the donor concentration, and ε_s is the relative permittivity of the semiconductor. The symbol R_g stands for the total resistance confronted by the gate current. It includes two components, one intrinsic and another extrinsic. The intrinsic component is determined by the width of the depletion region. The extrinsic component consists of the contact resistance. A^* is the effective Richardson constant, equal to 14.7×10^4 A m^{-2} K^{-2}.

In this equation, the basis of the third term is that the leakage current flowing through an SBD (area A) is

$$I_{D0} = A^* A T^2 \exp\{(-q\phi_{bn})/(k_B T)\}. \quad (7.17)$$

Further, in this equation, the first term due to the built-in voltage V_{bi} and the third term arising from voltage V_l generated by the leakage current depend on temperature whereas the middle term due to the pinch-off voltage V_{p0} is temperature-independent. An alternative equation for V_{bi} in place of that given in equation (7.16) is derived by assuming that the Fermi level E_F is situated midway between the conduction band edge E_C and the intrinsic energy level E_i. Then the built-in potential V_{bi} is approximated as (Shoucair and Ojala 1992)

$$\begin{aligned} V_{bi} &= \left(\frac{k_B T}{q}\right) \ln\left(\frac{N_D}{n_i}\right) - \frac{E_C - E_F}{q} = \left(\frac{k_B T}{q}\right) \ln\left(\frac{N_D}{n_i}\right) - \left(\frac{k_B T}{q}\right) \ln\left(\frac{N_C}{N_D}\right) \\ &= \left(\frac{k_B T}{q}\right) \ln\left(\frac{N_D}{n_i} \times \frac{N_D}{N_C}\right) = \left(\frac{k_B T}{q}\right) \ln\left(\frac{N_D^2}{n_i N_C}\right) \end{aligned} \quad (7.18)$$

$$\therefore V_{Th} = \left(\frac{k_B T}{q}\right) \ln\left(\frac{N_D^2}{n_i N_C}\right) - \left(\frac{q N_D a^2}{2\varepsilon_0 \varepsilon_s}\right) - \left\{ R_g WLA^* T^2 \exp\left(\frac{q\Phi_{bn}}{k_B T}\right) \right\} \quad (7.19)$$

where

$$N_C = 4.7 \times 10^{17} \left(\frac{T}{300}\right)^{1.5} \quad (7.20)$$

$$n_i = 2.51 \times 10^{19}\left\{\left(\frac{m_n^*}{m_0}\right)\left(\frac{m_p^*}{m_0}\right)\right\}^{0.75}\left(\frac{T}{300}\right)^{1.5}\exp\left(-\frac{E_g}{2k_BT}\right) \tag{7.21}$$

$$\begin{aligned} &= 2.51 \times 10^{19}[\{1.028 + (6.11 \times 10^{-4})T - (3.09 \times 10^{-7})T^2\} \\ &\quad \times\{0.61 + (7.83 \times 10^{-4})T - (4.46 \times 10^{-7})T^2\}]^{0.75} \\ &\quad \times\left(\frac{T}{300}\right)^{1.5}\exp\left(-\frac{E_g}{2k_BT}\right) \end{aligned} \tag{7.22}$$

obtained by substituting the temperature-dependent expressions for m_n^*/m_0 and m_p^*/m_0. The model is applied to the calculation of the drain current in the sub-threshold and saturation operational regimes of a MESFET in the temperature range 273–673 K (Won *et al* 1999). It may be mentioned here that this model takes into consideration the changes in electron and hole effective masses with temperature but overlooks the thermally induced energy bandgap variation.

7.10 The high-temperature electronic technique for enhancing the performance of MESFETs up to 300 °C

From the discussion in preceding sections, it appears as if the problems associated with operation at high temperatures can only be solved by cleverly improvised device process techniques. Process modification is a complex approach. For simplification, one must look for low-cost, process-free alternative solutions. One such technique, the high-temperature electronic technique, is based on electrical biasing of the substrate with a correct polarity voltage of the required magnitude (Narasimhan *et al* 1999). The I_{DS}–V_{DS} characteristics of the MESFET at room temperature (25 °C) and 300 °C showed that: (i) the transconductance g_m decreased by a factor of 56% from 160 mS mm^{-1} at 25 °C to 70 mS mm^{-1} at 300 °C; (ii) the leakage current increased by over five orders of magnitude from 25 °C to 275 °C. A careful inspection revealed that these temperature-induced variations in I_{DS}-V_{DS} characteristics were caused by an increase in the thermally generated leakage current. This leakage current altered the behavior of the GaAs substrate, which was semi-insulating at 25 °C. It made the substrate semiconducting at 300 °C. Consequently, the output resistance in the saturation region had a lower finite value at 300 °C instead of the desired infinite value.

The device characteristics at 300 °C can be restored to the 25 °C values by applying a 6 V bias to the substrate (Narasimhan *et al* 1999). Upon re-measuring and re-plotting the I_{DS}–V_{DS} characteristics at 300 °C, it was found that: (i) the transconductance g_m became comparable with the values at 25 °C; (ii) the leakage current decreased to a value which was lower than at 25 °C. Overall, the drain current at 300 °C was higher than its value at 25 °C and gate-controllable too. Thus stable operation of the device at high temperature was achieved without introducing any process complications.

7.11 The operation of GaAs complementary heterojunction FETs from 25 °C to 500 °C

A complementary heterostructure FET (CHFET; Wilson *et al* 1996) in GaAs parlance is the analogue of CMOS phraseology in silicon. As the CMOS structure consists of n- and p-channel MOSFETs, the CHFET structure comprises n- and p-channel GaAs FETs (figure 7.8). Looking at the layers in this diagram, the heterostructure is: $Al_{0.75}Ga_{0.25}As/In_{0.25}Ga_{0.75}As/GaAs$. In this heterostructure, the $Al_{0.75}Ga_{0.25}As$ layer with a bandgap 2 eV plays the same role in the MESFET device as the silicon dioxide in a MOSFET. Due to its insulating nature, the gate leakage current of a CHFET is substantially reduced in comparison to that of bulk GaAs MESFETs. The $In_{0.25}Ga_{0.75}As$ layer constitutes the channel of the CHFET. The layer offers the advantage that the mobility of carriers, both electrons and holes, is 30% higher in this layer than in bulk GaAs. Underneath the $In_{0.25}Ga_{0.75}As$ layer is a delta-doped layer with silicon as the dopant. It is meant to control the threshold voltage. Ohmic contacts are made of InGe alloy with stability established up to 400 °C. At 500 °C, the contact resistance began to increase but stayed below 2 Ωmm. Gate contacts are based on refractory metallization (WSi). The n-channel CHFETs performed sufficiently well digitally at 400 °C. The same is not

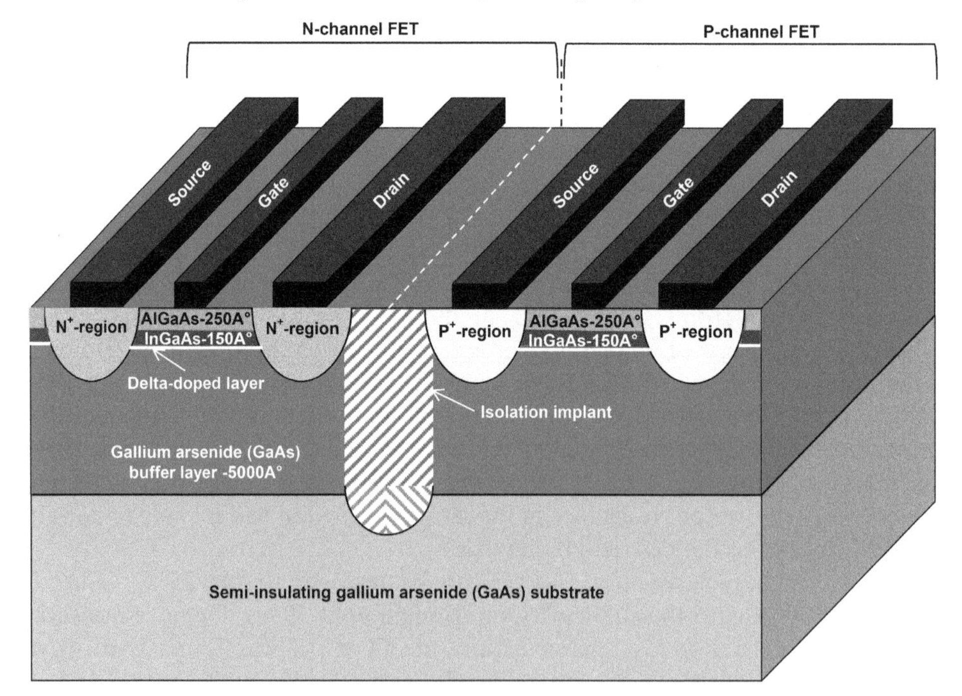

Figure 7.8. The $Al_{0.75}Ga_{0.25}As/In_{0.25}Ga_{0.75}As/GaAs$ CHFET.

true for the p-channel device. For both n- and p-channel CHFETs, the leakage current was an issue with the p-channel device being inferior to the n-channel type. Nonetheless, the forward gate leakage current of an n-channel CHFET was less than that of an equivalent MESFET. Unmodified ring oscillators based on CHFETs were recorded as working up to 420 °C. The power consumption of the ring oscillator leaped to 10.9 mW at 350 °C from its room-temperature (25 °C) value of 4.3 mW (Wilson *et al* 1996).

The CHFET provides a fast digital technology for high temperatures. Simulations showed that the CHFETs displayed a drain leakage current, which was limited by the gate due to effective Schottky barrier lowering under reverse bias conditions.

7.12 GaAs bipolar transistor operation up to 400 °C

For specific applications, bipolar transistors are favored in comparison to MESFETs. They have a lower 1/*f* noise. The intermodulation distortion introduced by them is also less troublesome. Moreover, they provide intrinsic current gain.

Bipolar transistors (working in up-mode with the n^+ epitaxial layer as the emitter and the surface n-layer as the collector) were fabricated by ion implantation into bulk n^+ GaAs through the following sequence of steps (Doerbeck *et al* 1982): Se implantation (shallow) for the n-layer → annealing → Be implantation (deep) for p-base layer formation → annealing → localized Be implantation for p^+ region formation to make contacts → annealing → localized Be implantation for isolation region formation. An n–p–n structure was thus obtained. The masking for ion implantation and the surface passivation layer is done through silicon nitride. For n-type regions, the contact material is alloyed Au–Ge–Ni. For p-type regions, it is alloyed Au–Zn. For interconnections, it is Ti–Au.

The current gain of the bipolar transistor exhibited a small increase with increasing temperatures up to 300 °C. Above this temperature, it started to decrease. However, the device showed current gain even beyond 400 °C. The collector–emitter leakage current with the base open (I_{CEO}) is temperature-independent up to 200 °C but increases above this temperature due to an increase in the leakage current of the collector–base diode. The device failed cataclysmically at around 400 °C due to melting and flowing of the gold layer on the contacts.

A 15-stage ring oscillator circuit, mounted in a ceramic package, was tested in the temperature range 25 °C to 390 °C. At a bias voltage of 1.75 V, the input current was 5 mA at 25 °C but increased to 7 mA at 385 °C. In the same temperature interval, the gate delay time stretched from 2.5 ns to 4.8 ns. The output signal increased three-fold. Circuit failure occurred at 390 °C due to metallization rupture (Doerbeck *et al* 1982).

7.13 A GaAs-based HBT for applications up to 350 °C

The performance characteristics of n–p–n HBTs with AlGaAs/GaAs heterojunctions were measured up to 350 °C (Fricke *et al* 1992). The Al mole fraction in the wide bandgap AlGaAs emitter layer was fixed at 0.45 to achieve a reasonable current gain value at the high operating temperature. The arguments for arriving at this value are as follows: if N_D and N_A are the doping concentrations of the emitter and base, respectively, v_{nB} is the velocity of electrons at the emitter end of the base, v_{pE} is the velocity of holes at the base end of the emitter and k_B stands for the Boltzmann constant, the maximum current gain β_{max} of an HBT at temperature T is expressed in terms of the valence band offset ΔE_V as

$$\beta_{max} = \left(\frac{N_D}{N_A}\right)\left(\frac{v_{nB}}{v_{pE}}\right)\exp\left(\frac{\Delta E_V}{k_B T}\right). \tag{7.23}$$

Since $\Delta E_V = 0.2$ eV for Al mole fraction 0.45, we have

$$\left[\exp\left(\frac{\Delta E_V}{k_B T}\right)\right]_{298\text{ K}} = \exp\left(\frac{0.2}{8.617 \times 10^{-5} \times 298}\right) = \exp(7.7886) = 2412.94 \tag{7.24}$$

$$\left[\exp\left(\frac{\Delta E_V}{k_B T}\right)\right]_{623\text{ K}} = \exp\left(\frac{0.2}{8.617 \times 10^{-5} \times 623}\right) = \exp(3.7255) = 41.49. \tag{7.25}$$

At room temperature, $T = 298$ K, the exponential term is 2413. At $T = 623$ K it is 42. The ratio of exponential terms in the two cases gives the factor by which the hole injection into the emitter is suppressed by the heterojunction. This factor is found to be 57.45. Considering this value as adequate for repressing the injection of holes, the above Al fraction was set at 0.45.

The emitter and collector contacts were made of Ni/Au/Ge/Ni (figure 7.9). The base contact was Ti–Pt–Au. The contact metallizations were subjected to RTA. $Si_xN_yO_z$ was used as a surface passivation to prevent As out-diffusion from GaAs. The DC characteristics of the HBTs were measured at 300 K, 423 K, 573 K and 623 K. The HBTs were functional up to 623 K. However, the characteristics were degraded as the temperature increased. Two distinct temperature ranges were identified. (i) Room temperature to 573 K: the hole current across the emitter–base heterojunction increases. Consequently, the ideality factor n_B of the base current decreases. The current gain β also diminishes as does the small signal current gain h_{FE}. (ii) 573 K to 623 K: the leakage current across the collector–base diode increases. As a result, n_B, β and h_{FE} decrease. All throughout the temperature range from room temperature to 673 K, a common-emitter small-signal current gain of 35 was maintained. Furthermore, the stability of DC characteristics was confirmed after performing a large number of heating/cooling cycles as well as storage at 573 K for 24 h (Fricke *et al* 1992).

7.14 Al_xGaAs_{1-x}/GaAs HBT

Basu and Sarkar (2011) proposed an analytical mode on similar lines to the SiGe HBT model presented by them and described in chapter 6. The main points of this

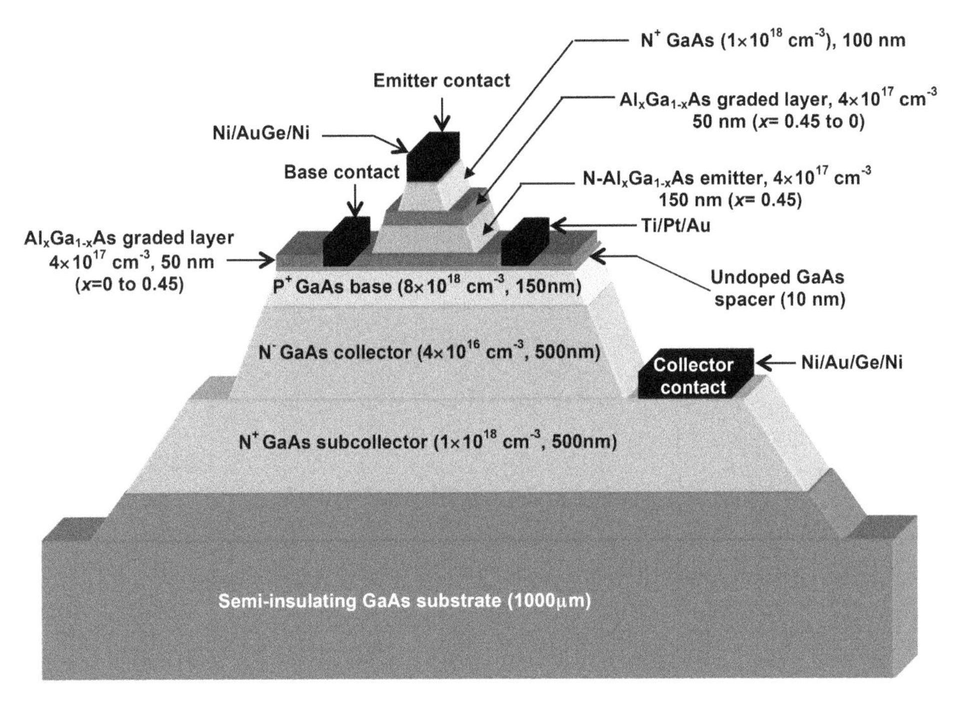

Figure 7.9. Layers in the AlGaAs/GaAs HBT.

model are given below: for Al_xGaAs_{1-x}/GaAs HBT, the equation for current gain $(\beta)_{\text{AlGaAs/GaAs}}$ is

$$(\beta)_{\text{AlGaAs/GaAs}} = \left(\frac{N_{\text{DE}}}{N_{\text{AB}}}\right)\left(\frac{W_{\text{E}}}{W_{\text{B}}}\right)\left\{\frac{D_{\text{n,GaAs}}(T)}{D_{\text{p,AlGaAs}}(T)}\right\}\exp\left(\frac{\Delta E_{\text{g}}}{k_{\text{B}}T}\right). \tag{7.26}$$

The electron mobility $\mu_{\text{n,GaAs}}$ in GaAs is

$$\mu_{\text{n,GaAs}}(T) \propto T^{-2.1} \tag{7.27}$$

or

$$\mu_{\text{n,GaAs}}(T) = K_1 T^{-2.1}, \tag{7.28}$$

where K_1 is a constant. To find K_1, we note that $\mu_{\text{n,GaAs}} = 8500\ \text{cm}^2\ \text{V}^{-1}\ \text{s}^{-1}$ at $T = 300$ K, so that

$$\begin{aligned} K_1 &= \mu_{\text{n,GaAs}}(T)/T^{-2.1} = \mu_{\text{n,GaAS}}(T) \times T^{2.1} \\ &= 8500 \times (300)^{2.1} = 1.353 \times 10^9\ \text{cm}^2\ \text{V}^{-1}\ \text{s}^{-1}\ \text{K}^{2.1} \end{aligned} \tag{7.29}$$

$$\therefore \mu_{\text{n,GaAs}}(T) = 1.353 \times 10^9 T^{-2.1} \tag{7.30}$$

$$\therefore D_{\mathrm{n,GaAs}}(T) = \left(\frac{k_{\mathrm{B}}T}{q}\right)1.353 \times 10^9 T^{-2.1}. \tag{7.31}$$

The hole mobility in $Al_xGa_{1-x}As$ is

$$\mu_{\mathrm{p,AlGaAs}}(T) \propto T^{-1} \tag{7.32}$$

or

$$\mu_{\mathrm{p,AlGaAs}}(T) = K_2 T^{-1}, \tag{7.33}$$

where K_2 is a constant determined from the condition that at $T = 300$ K, for AlGaAs (Shur 1995, Basu and Sarkar 2011)

$$\mu_{\mathrm{p,AlGaAs}}(300\ \mathrm{K}) = 370 - 970x + 740x^2, \tag{7.34}$$

where x is the mole fraction of aluminum in the alloy $Al_xGa_{1-x}As$

$$\therefore K_2 = \mu_{\mathrm{p,AlGaAs}}(T)/T^{-1} = \mu_{\mathrm{p,AlGaAS}}(T) \times T \tag{7.35}$$

$$= \mu_{\mathrm{p,AlGaAs}}(300) \times 300 = 300(370 - 970x + 740x^2)\ \mathrm{cm^2\ V^{-1}\ s^{-1}\ K} \tag{7.36}$$

$$\therefore \mu_{\mathrm{p,AlGaAs}}(T) = 300(370 - 970x + 740x^2)T^{-1} \tag{7.37}$$

$$\therefore D_{\mathrm{p,AlGaAs}}(T) = \left(\frac{k_{\mathrm{B}}T}{q}\right)300(370 - 970x + 740x^2)T^{-1}. \tag{7.38}$$

The difference between the energy bandgaps of the $Al_xGa_{1-x}As$ emitter and GaAs base is (Shur 1995, Basu and Sarkar 2011)

$$(\Delta E_{\mathrm{g}})_{\mathrm{AlGaAs/GaAs}} = 1.25x \text{ for } x < 0.4. \tag{7.39}$$

The emitter and base transit times $\{\tau_{\mathrm{E}}(T)\}_{\mathrm{AlGaAs/GaAs}}$ and $\{\tau_{\mathrm{B}}(T)\}_{\mathrm{AlGaAs/GaAs}}$ are:

$$\begin{aligned}\{\tau_{\mathrm{E}}(T)\}_{\mathrm{AlGaAs/GaAs}} &= W_{\mathrm{E}}^2/\left[2\beta D_{\mathrm{p,AlGaAs}}(T)\right]\\ &= W_{\mathrm{E}}^2/\left[2(\beta)_{\mathrm{AlGaAs/GaAs}} D_{\mathrm{p,AlGaAs}}(T)\right]\end{aligned} \tag{7.40}$$

$$\begin{aligned}&= W_{\mathrm{E}}^2/\left[2\left(\frac{N_{\mathrm{DE}}}{N_{\mathrm{AB}}}\right)\left(\frac{W_{\mathrm{E}}}{W_{\mathrm{B}}}\right)\left\{\frac{D_{\mathrm{n,GaAs}}(T)}{D_{\mathrm{p,AlGaAs}}(T)}\right\}\exp\left(\frac{\Delta E_{\mathrm{g}}}{k_{\mathrm{B}}T}\right)D_{\mathrm{p,AlGaAs}}(T)\right]\\ &= W_{\mathrm{E}}^2/\left[2\left(\frac{N_{\mathrm{DE}}}{N_{\mathrm{AB}}}\right)\left(\frac{W_{\mathrm{E}}}{W_{\mathrm{B}}}\right)D_{\mathrm{n,GaAs}}(T)\exp\left(\frac{\Delta E_{\mathrm{g}}}{k_{\mathrm{B}}T}\right)\right]\end{aligned} \tag{7.41}$$

$$\{\tau_{\mathrm{B}}(T)\}_{\mathrm{AlGaAs/GaAs}} = W_{\mathrm{B}}^{2}/\{2D_{\mathrm{n,GaAs}}(T)\}. \tag{7.42}$$

From the above analytical model, Basu and Sarkar (2011) found that high gain along with low forward transit time were achievable when a small fraction of Al was introduced in the emitter region of an AlGaAs/GaAs HBT.

7.15 Discussion and conclusions

Gallium arsenide is fundamentally different from silicon in terms of it's being a compound formed from two distinct elements, its crystal structure, the nature of its bandgap and the absence of a good quality insulating oxide layer. As an upshot of these basic differences, the devices which have generated the greatest interest in GaAs are MESFETs and HBTs. To prevent sublimation of the volatile arsenic, the LEC technique is used for growing single crystals of GaAs. Common ohmic and Schottky contact metallization schemes for GaAs are modified for high-temperature operation. Despite the higher intrinsic temperature of GaAs, initial experiments on commercial devices developed for room-temperature operation showed that these devices were even inferior to silicon MOSFETs at high temperatures. They exhibited high leakage currents, which originated from the gate and drain leakage components. Therefore, blocking these leakage current paths by using heterojunctions helped in raising the upper temperature limit at which GaAs MESFETs operated well. The high-temperature performance of GaAs MESFETs can be restored to be equivalent to room-temperature operation without any process iteration by taking advantage of high-temperature electronic techniques. The CHFET structure in GaAs is similar to the CMOS structure in silicon technology. The n-channel devices in the CHFET configuration show better performance than the p-channel devices. The GaAs bipolar transistor also functions well at high temperatures. GaAs HBTs worked, with degraded performance, up to 623 K.

Review exercises

7.1. Complete the following sentences for gallium arsenide. (a) Silicon is an elemental semiconductor but gallium arsenide is .... (b) Silicon crystallizes in a diamond structure but gallium arsenide crystallizes in .... (c) Silicon is an indirect bandgap semiconductor but gallium arsenide is an .... (d) The oxide grown on silicon is advantageous for microelectronics but the oxides on gallium arsenide are .... (e) The most popular device in silicon ICs is the MOSFET whereas in gallium arsenide ICs, it is the .... (f) In a silicon MOSFET, the gating action is provided through the silicon dioxide dielectric, while in a gallium arsenide MESFET, the gating action is provided by ....

7.2. (a) Which allows a high leakage current: an SBD in a GaAs MESFET or an insulating oxide film in a Si MOSFET? (b) Which has a high diffusion-dominated leakage current at high temperatures: a Si MOSFET or GaAs MESFET?

7.3. Is there any analogue of SOI technology in GaAs ICs? If not, how are the different components isolated (a) at low temperatures and (b) at high temperatures?
7.4. Can you name another important device used in gallium arsenide ICs besides the GaAs MESFET?
7.5. What is the intrinsic temperature for gallium arsenide? How high is it compared to that of silicon?
7.6. Why is it necessary to surround the GaAs melt with liquid boron trioxide during single crystal growth? What is this technique called? How are single crystals of GaAs grown?
7.7. What elements are used for p-type doping of gallium arsenide? What are the common n-type dopants in gallium arsenide? What is the doping technique used?
7.8. What happens if the sample is heated during ion implantation? How is damage introduced by implantation removed?
7.9. Discuss the contributions of the constituent layers in the Au/Ge/Ni/Ti metallization scheme for n-type GaAs? Why is RTA suitable for this scheme? Is this scheme thermally stable?
7.10. Describe a suitable metallization scheme for n-type GaAs which can sustain high-temperature operation. What is the maximum allowed temperature for this scheme? What happens if any residual gold is left unconsumed during annealing?
7.11. What is the barrier height for a Ti–Pt–Au Schottky contact to GaAs? Explain the roles of the three constituent layers. How can the thermal stability of this contact scheme be improved? What is the maximum temperature limit for the scheme?
7.12. What is the barrier height of a Au–LaB_6 Schottky contact to GaAs? How high a temperature can be allowed for this contact without damage?
7.13. Describe experiments carried out on commercial GaAs devices in the 25 °C to 400 °C temperature range. What are the ohmic contacts made of? What material is used for Schottky contacts? Write the equations describing the variation of mobility, saturation velocity and threshold voltage. Why did the GaAs MESFETs perform worse than Si MOSFETs despite the larger bandgap of gallium arsenide than silicon?
7.14. What are the two components of leakage current in a GaAs MESFET at high temperatures? What are the paths of flow for these components? Explain how it is possible to raise the operating temperature limit of GaAs MESFETs by taking advantage of heterojunction barriers to stop leakage current flows?
7.15. Write the equations for the drain–source current of a MESFET in the linear and saturation regions. Explain the symbols used. What parameters in these equations make the drain current temperature dependent?
7.16. Write the equation for the threshold voltage of a MESFET and explain the symbols in this equation. What is the physical explanation for the

three terms in this equation? Which of the terms in this equation depend on temperature?

7.17. Is it possible to improve the performance of a GaAs MESFET up to 300 °C without process modification? If yes, what is the technique used for this purpose called?

7.18. Describe the high-temperature electronic technique for restoring the 300 °C values of electrical parameters of a GaAs to room-temperature values.

7.19. What is the analog of CMOS structure of silicon technology in GaAs technology? What are the equivalents of NMOS and PMOS transistors in this analog?

7.20. Which layer in the $Al_{0.75}Ga_{0.25}As/In_{0.25}Ga_{0.75}As/GaAs$ heterostructure plays the same role in a GaAs MESFET as silicon dioxide in a MOSFET? What property of this layer helps in reducing the gate leakage?

7.21. Which layer in the $Al_{0.75}Ga_{0.25}As/In_{0.25}Ga_{0.75}As/GaAs$ heterostructure plays the role of a channel in a GaAs MESFET? What advantage is offered by this layer as compared to bulk GaAs?

7.22. How are the ohmic and Schottky contacts made for the $Al_{0.75}Ga_{0.25}As/In_{0.25}Ga_{0.75}As/GaAs$ heterostructure? How do the n- and p-channel CHFETs perform at a high temperature? How does the power consumption of ring oscillators based on CHFETs vary with temperature?

7.23. Describe the process sequence for fabrication of a GaAs bipolar transistor. How does this transistor work at 300 °C? What are the effects of temperature on current gain and leakage current? How does the ring oscillator circuit using this device function at high temperature? At what temperature does it fail?

7.24. Explain, with calculations, why the Al mole fraction in the AlGaAs emitter of an AlGaAs/GaAs HBT is fixed at 0.45? What are the emitter and collector contacts made of? What is the base contact made of? What is the passivation material used? Up to what temperature is the HBT device functional? How does the device function in the temperature ranges: room temperature to 573 K and 573 K to 623 K? How much is the performance degraded?

References

Basu S and Sarkar P 2011 Analytical modeling of AlGaAs/GaAs and Si/SiGe HBTs including the effect of temperature *J. Electron Devices* **9** 325–29

Doerbeck F H, Duncan W M, Mclevige W V and Yuan H-T 1982 Fabrication and high-temperature characteristics of ion-implanted GaAs bipolar transistors and ring-oscillators *IEEE Trans. Indust. Electron.* **29** 136–9

Eun J and Cooper J A Jr 1993 High-temperature ohmic contact technology to n-type GaAs *ECE Technical Reports, Purdue University* TR-EE 93-7, pp 14–18

Fricke K, Hartnagel H L, Lee W-Y and Würfl J 1992 AlGaAs/GaAs HBT for high-temperature applications *IEEE Trans. Electron Devices* **39** 1977–81

Fricke K, Hartnagel H L, Schütz R, Schweeger G and Würfl J 1989 A new GaAs technology for stable FETs at 300 °C *IEEE Electron Device Lett.* **10** 577–9

Kacprzak T and Materka A 1983 Compact dc model of GaAs FETs for large-signal computer calculations *IEEE J. Solid State Circuits* **18** 211

Narasimhan R, Sadwick L P and Hwu R J 1999 Enhancement of high-temperature high-frequency performance of GaAs-based FETs by the high-temperature electronic technique *IEEE Trans. Electron Devices* **46** 24–31

Shoucair F S and Ojala P K 1992 High-temperature electrical characteristics of GaAs MESFETs (25–400 °C) *IEEE Trans. Electron Devices* **39** 1551–7

Shur M 1987 *GaAs Devices and Circuits* (New York: Springer) pp 161–2

Shur M 1995 *Physics of Semiconductor Devices* (New Delhi: Prentice Hall India), 704 pages

Wilson C D and O'Neill A G 1995 High temperature operation of GaAs based FETs *Solid-State Electron.* **38** 339–43

Wilson C D, O'Neill A G, Baier S M and Nohava J C 1996 High temperature performance and operation of HFETs *IEEE Trans. Electron Devices* **43** 201–6

Won C-S, Ahn H K, Han D-Y and El Nokali M A 1999 DC characteristic of MESFETs at high temperatures *Solid-State Electron.* **43** 537–42

Würfi J, Singh J K and Hartnagel H L 1990 Reliability aspects of thermally stable LaB_6–Au Schottky contacts to GaAs *Reliability Physics Symposium* (*New Orleans, LA 26–29 March*), pp. 87–93

Chapter 8

Silicon carbide electronics for hot environments

Silicon carbide is endowed with unique material properties that can be utilized for the fabrication of high-temperature, high-power and fast switching devices, promising a greater impact on high-voltage devices than Si and GaAs due to its higher breakdown field. The electron mobility in the 4H-SiC polytype is double the mobility in the 6H-SiC polytype perpendicular to the *c*-axis, while parallel to the *c*-axis the ratio of mobility in 4H-SiC to that in 6H-SiC is 10. This difference makes 4H-SiC more attractive. Important SiC devices are p–n junction diodes, SBDs and JFETs. MOSFET development has slowed down due to the low channel mobilities achieved. BJTs are receiving increasing attention due to their low on-resistance, made possible by electron injection and resulting conductivity modulation. SiC technology is progressing from the stage of research in semiconductor laboratories to commercial-scale production. The supply of low-cost, defect-free large-area substrates will considerably boost this transition. This is a matter of the greatest concern in moving towards technological maturity.

8.1 Introduction

Silicon carbide electronics have progressed from the research phase to commercial manufacturing, mounted on the most appropriate polytype 4H-SiC for high-temperature circuits. Silicon carbide occurs in a large number of polytypes, around 150–250. The packing sequences of the close-crowded bi-atomic strata in these polytypes constitute the criteria for distinguishing them from each other. All these polytypes do not allow easy growth. Amongst those found, the 4H-SiC and 6H-SiC polytypes are available as substrates for device fabrication. Between 4H-SiC and 6H-SiC, the former is superior for electronic device fabrication due to the higher mobility of charge carriers in this polytype and its wider bandgap.

The bandgap of silicon carbide (3.23 eV for 4H-SiC) is larger than that for silicon (1.12 eV) by about 2.9 times. So, operation above 873 K is feasible. Its electrical breakdown field (3 MV cm^{-1}) is higher than that of silicon (0.3 MV cm^{-1}) by a

doi:10.1088/978-0-7503-1155-7ch8

factor of ten. Taking advantage of the higher field, the thickness of the uniformly doped conduction region is drastically reduced, enabling a sharp fall of the on-resistance of devices. Its thermal conductivity (4.9 W cm K^{-1}) is 3.27 times higher than that of silicon (1.5 W cm K^{-1}). Hence, a greater power density is obtainable from SiC, giving more power per unit area of the chip. On the opposite side, the lower electron mobility in SiC (800–900 cm^2 V s^{-1}) than silicon (1400 cm^2 V s^{-1}) is a major disadvantage of SiC. Similar remarks apply to the hole mobility. The low carrier mobilities are sufficient to provide RF performance in the frequency range 8×10^9–12×10^9 Hz but are inadequate beyond this limit. At microwave frequencies ($<10^{10}$ Hz), SiC devices compete well with Si and GaAs components. p–n junction diodes, JFETs and thyristors are fabricated using SiC, but MOSFETs have lagged behind, mainly due to the lower carrier mobilities. In table 8.1, the salient properties of silicon carbide related to device and circuit design are listed. Figures 8.1 and 8.2 show the crystalline structures of polytypes of silicon carbide.

8.2 Intrinsic temperature of silicon carbide

It is interesting to perform the intrinsic temperature calculations for silicon carbide in line with those for silicon, and compare the values. For 4H-SiC,

$$m_n^* = 0.39m_0,\ m_p^* = 0.82m_0,\ E_g = 3.23 \text{ eV}. \tag{8.1}$$

Hence,

$$m^* = \left(\frac{m_n^* m_p^*}{m_0^2}\right)^{0.75} = \left(\frac{0.39m_0 \times 0.82m_0}{m_0^2}\right)^{0.75} = 0.4253. \tag{8.2}$$

For 4H-SiC, the equation

$$\ln\{0.207/(m^* T^{1.5})\} = -5.8025 \times 10^3 E_g \tag{8.3}$$

takes the form

$$T\ln\{0.207/(0.4253T^{1.5})\} = -5.8025 \times 10^3 \times 3.23 = -18742.075 \tag{8.4}$$

or

$$\begin{aligned} T\ln 0.4867 - T\ln T^{1.5} &= -18742.075 \\ \therefore -0.72011T - T\ln T^{1.5} &= -18742.075. \end{aligned} \tag{8.5}$$

Suppose,

$$T = 1590 \text{ K} \tag{8.6}$$

$$\begin{aligned} \text{lhs} &= -0.72011T - T\ln T^{1.5} = -0.72011 \times 1590 - 1590\ln 1590^{1.5} \\ &= -1144.9749 - 17581.001969 = -18725.976869. \end{aligned} \tag{8.7}$$

If

$$T = 1591.5 \text{ K}$$

Table 8.1. Properties of silicon carbide (4H-SiC polytype).

Property	Value	Property	Value	Property	Value
Chemical formula	4H-SiC	Dielectric constant	9.7	Electron saturated velocity (cm s^{-1})	2×10^7
Classification	IV–VI compound semiconductor	Thermal conductivity (W cm K^{-1})	4.9	Donors	Nitrogen Phosphorous
Crystal structure	Wurtzite (hexagonal unit cell)	Energy bandgap E_g (eV) at 300 K	3.23	Donor ionization energies ΔE_D (meV)	50, 92 (nitrogen) 54, 93 (phosphorous)
Color	Light brown (low n-doping)	Electrical breakdown field (V cm^{-1})	3×10^6 parallel to c-axis	Acceptor impurities	Aluminum boron
Density at 300 K (g cm^{-3})	3.211	Intrinsic carrier concentration (cm^{-3})	5×10^{-9}	Acceptor ionization energies ΔE_A (meV)	200 (aluminum) 285 (boron)
Lattice constant (Å)	$a = 3.073$ $b = 10.053$	Electron mobility (cm^2 Vs^{-1})	900 parallel to c-axis, 800 perpendicular to c-axis		
Melting point (°C)	3103 ± 40 K at 35 atm	Hole mobility (cm^2 V s^{-1})	115		

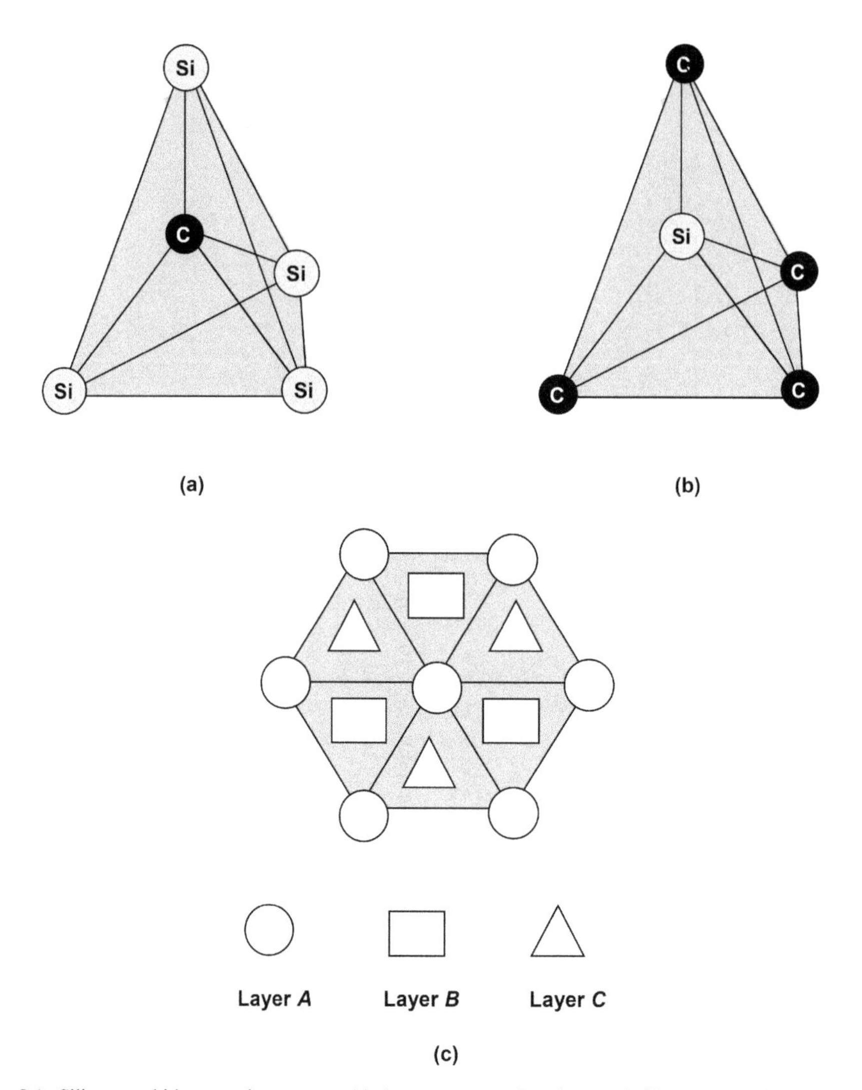

Figure 8.1. Silicon carbide crystal structure. (a) Arrangement of carbon and silicon atoms in a tetrahedron with one carbon atom surrounded by four silicon atoms, (b) a similar arrangement with one silicon atom surrounded by four carbon atoms and (c) naming of the layers as A, B and C according to the positions occupied by the atoms in the tetrahedron.

$$\begin{aligned}\text{lhs} &= -0.72011T - T\ln T^{1.5} = -0.72011 \times 1591.5 - 1591.5\ln 1591.5^{1.5}\\ &= -1146.055 - 17599.83888 = -18745.89388.\end{aligned} \tag{8.8}$$

When

$$T = 1592\ \text{K}$$

$$\begin{aligned}\text{lhs} &= -0.72011T - T\ln T^{1.5} = -0.72011 \times 1592 - 1592\ln 1592^{1.5}\\ &= -1146.415 - 17606.1183 = -18752.5333.\end{aligned} \tag{8.9}$$

$$\therefore T \approx 1591.5\ \text{K}.$$

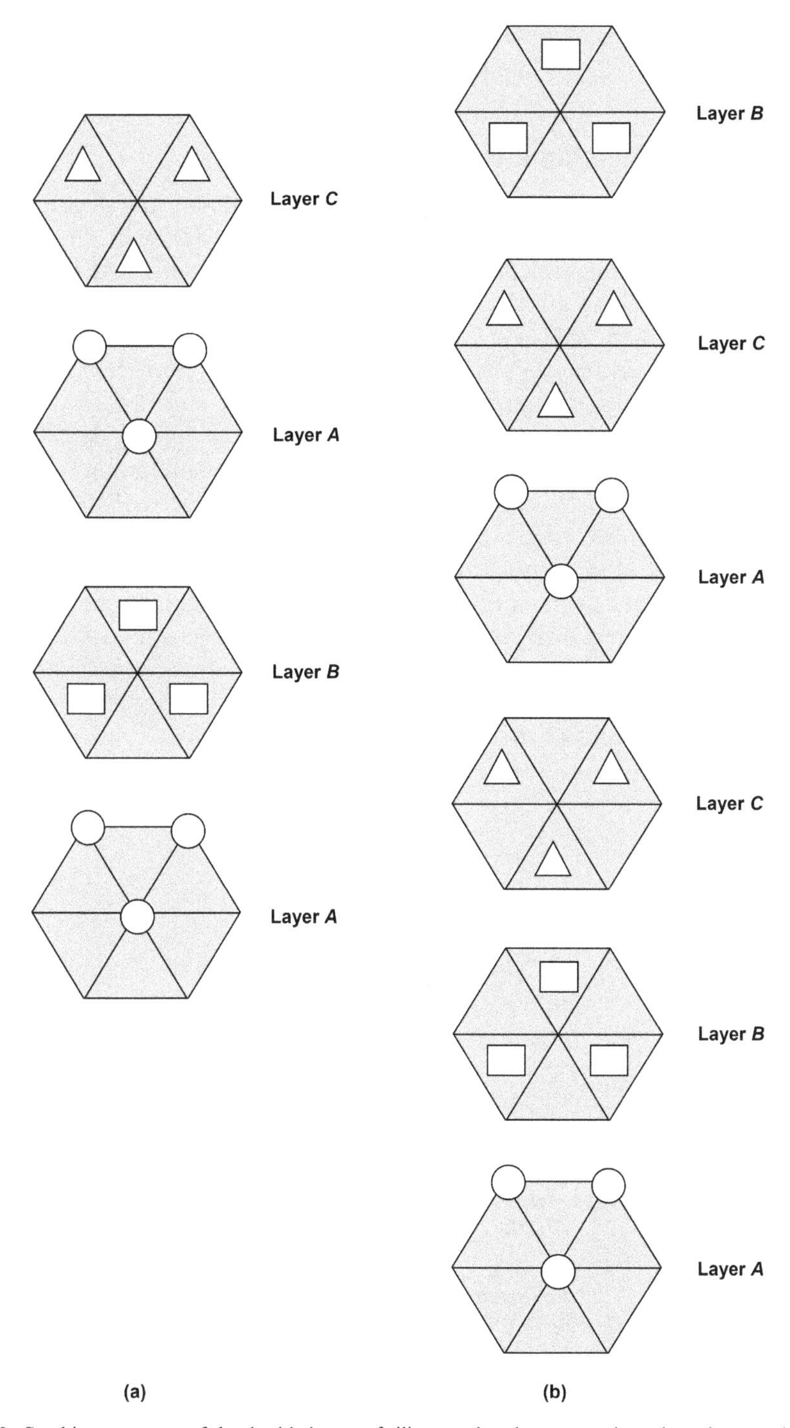

Figure 8.2. Stacking sequence of the double layers of silicon and carbon to produce the polytypes. (a) 4H-SiC (ABAC) and (b) 6H-SiC (ABCACB); H stands for hexagonal.

This temperature is 1591.5 K −588.63 K = 1002.87 K higher than that for silicon (588.63 K) and 1591.5 K − 756.6 K = 834.9 K higher than that for GaAs (756.6 K).

8.3 Silicon carbide single-crystal growth

The traditional method of growing large diameter single-crystal silicon by crystal pulling from the melt, e.g. the CZ process, supplies the bulk of crystals for semiconductor device manufacturing. Unfortunately, this method is not applicable to silicon carbide owing to the fact that instead of melting, it sublimes at a temperature lower than 2000 °C.

Silicon carbide single crystals are grown by a sublimation method referred to as physical vapor deposition (figure 8.3) in which the SiC formed by the reaction between molecular species containing silicon and carbon is directly deposited on the seed crystal. These molecular species are obtained from a subliming source of SiC lying in the vicinity of the seed crystal. For crystal growth, a seed of {0001} substrate is generally used. The SiC single crystals are grown along the direction of the [0001] *c*-axis. This method of growth is called *c*-face growth.

Of the well-known defects in SiC crystals the following can be mentioned:

(i) Open core screw dislocations termed micropipes, which are responsible for the failure of SiC diodes below their avalanche breakdown limit.
(ii) Domain walls or grain boundaries leading to leakage current flow and producing cracks in the wafer during epitaxy.

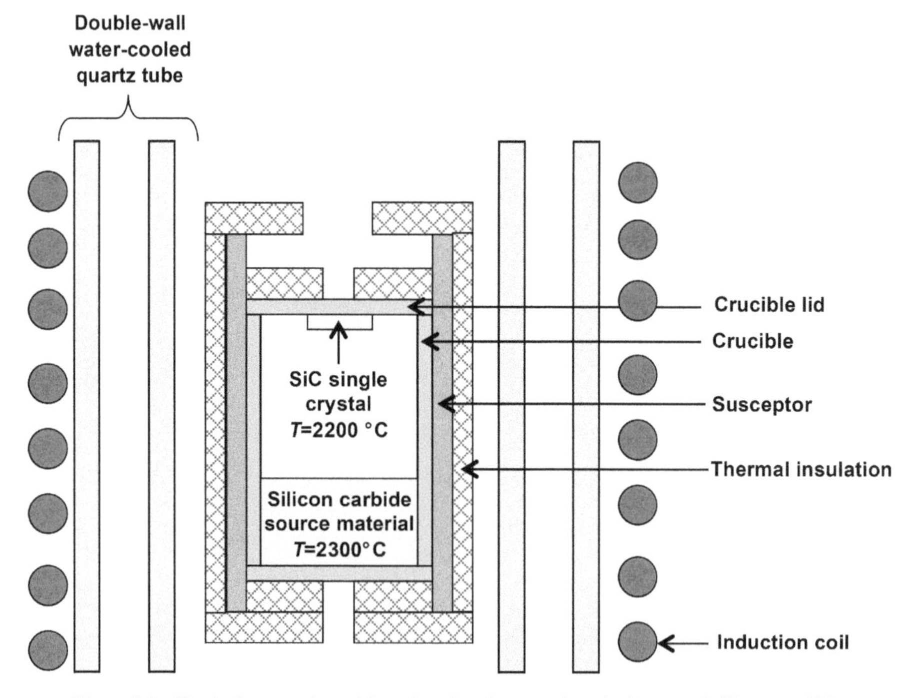

Figure 8.3. Physical vapor deposition chamber for growing single-crystal silicon carbide.

(iii) Threading screw dislocations affecting the leakage currents in Schottky diodes.
(iv) Basal plane dislocations, whose presence in the active layer of PIN devices degrades their forward voltage characteristics.
(v) Remnant nitrogen and boron contamination which is detrimental to highly pure undoped semi-insulating SiC substrates.

8.4 Doping of silicon carbide

An accepted n-type dopant for SiC is nitrogen. p-type doping is performed using aluminum. Even at very high temperatures, ~2000 °C, the diffusion coefficients of these dopants in SiC are extremely low. The energy levels of these impurities are shown in figure 8.4. Therefore, the popular technique of thermal diffusion of dopants into semiconductors is precluded outright. A practical option is ion implantation. During ion implantation, the SiC wafer may be kept at a temperature between normal room temperature to as high as 900 °C. The implantation step is invariably followed by annealing at a temperature higher than 1700 °C. The

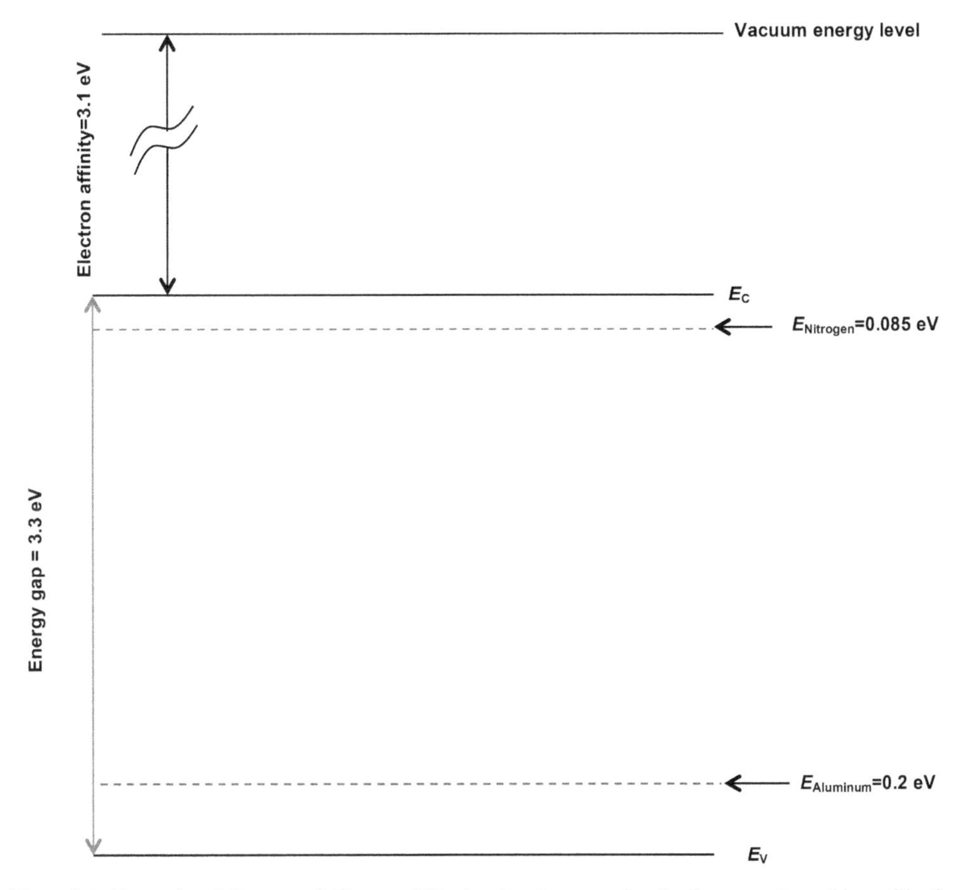

Figure 8.4. Energy band diagram of silicon carbide showing the energy levels of commonly used impurities for n- (nitrogen) and p-doping (Al).

temperature of the SiC wafer during ion implantation as well as in the annealing cycle is a critical determinant of the resultant doping profile.

The above techniques are applied when doping needs to be done selectively in defined areas of the SiC wafer for device fabrication. Otherwise, *in situ* doping is carried out during epitaxial growth of SiC.

8.5 Surface oxidation of silicon dioxide

An encouraging aspect of SiC wafer processing is that oxide films can be grown over SiC surfaces much like silicon, although the oxidation rates are comparatively slower than Si and also depend upon whether the silicon- or carbon-terminated face is oriented towards the growing oxide. Nevertheless, the growth of good-quality oxide films with low interface state and trap concentrations has been a major roadblock. The reliability of the oxide films, particularly at high electric fields and elevated temperatures, is questionable.

8.6 Schottky and ohmic contacts to silicon carbide

For Schottky contacts with n- and p-type SiC, various combinations of transition metals have been attempted. These metals have barrier heights in the range 0.9–1.7 eV. However, the Schottky behavior is either transformed to ohmic or serious deterioration of performance takes place at temperatures exceeding 873 K. An ohmic contact to n-type SiC is commonly established with Ni_2Si obtained by deposition of Ni followed by its silicidation by high-temperature annealing at >900 °C, providing a thermally stable contact. For an ohmic contact to p-type SiC, aluminum is still the preferred choice but its operation at high temperatures is not recommended because it has a low melting point. Low-resistance contact materials with high work functions are necessary to equipoise the large bandgap and electron affinity of SiC.

8.7 SiC p–n diodes

Using the edge-termination technique, planar p–n diodes with a blocking capability of 1400 V showed a forward drop of 6.2 V at 4000 A cm^{-2}. The differential on-resistance was <1 mΩ cm^{-2} (Peters *et al* 1997). 4.5 kV 4H-SiC diodes could switch waveforms up to 770 A. The diodes used a two-fold implanted junction termination extension (JTE) consisting of two p-rings surrounding the p^+-anode (Braun *et al* 2002).

8.7.1 SiC diode testing up to 498 K

A breakdown voltage of 12.9 kV was obtained on 100 μm thick epitaxial layers. These layers were grown on n^+ 4H-SiC substrates (Sundaresan *et al* 2012). Carrier lifetimes were measured using the open circuit voltage decay (OCVD) technique. They ranged from 2 to 4 μs at room temperature and increased to 14 μs at 498 K. To test the switching behavior at high currents, a module consisting of five 10 kV PIN rectifiers was connected as a free-wheeling diode with silicon insulated gate bipolar transistor (IGBT). 1100 V, 100 A switching was carried out through the IGBT-PIN rectifier module up to 498 K. The peak recovery current rose from −48 A at 298 K to −120 A at 498 K. The reverse recovery time increased from 168 ns at 298 K to 528 ns at 498 K. These effects were produced by the positive TC of the lifetime of carriers injected into the n-base.

8.7.2 SiC diode testing up to 873 K

A primary need to guarantee high-temperature operation of SiC diodes is provision of thermally stable ohmic contacts. This calls for proper selection of metals which can withstand the high-temperature conditions. It was found (Kakanakov *et al* 2002) that Ti/Al/p-SiC and Ni/n-SiC contacts were not degraded upon ageing at 773 K to 783 K in N_2 ambient for 100 h. On injecting high current density (10^3 A cm^{-2}) through these contacts and raising the temperature to 723 K, the contacts exhibited stability. In a related study (Chand *et al* 2014), Ni or Pt was sputtered on Ni/Si ohmic contacts on SiC diodes (figure 8.5). Ageing experiments were conducted in air at different temperatures: 673 K, 773 K and 873 K. Forward current–voltage characteristics of the diodes were recorded. The motive was to examine the changes in junction and series resistances of the diodes. Platinum metallized diodes showed stability of the current–voltage characteristics as well as invariability of series resistance at all the above temperatures. The finding that the diode ideality factor was 1.02 at 873 K evidenced that the diode was functioning properly at this temperature. However, the series resistance of the nickel-metallized diodes increased, although only slightly, due to oxidation of the nickel surface. This happened after 7 h at 673 K.

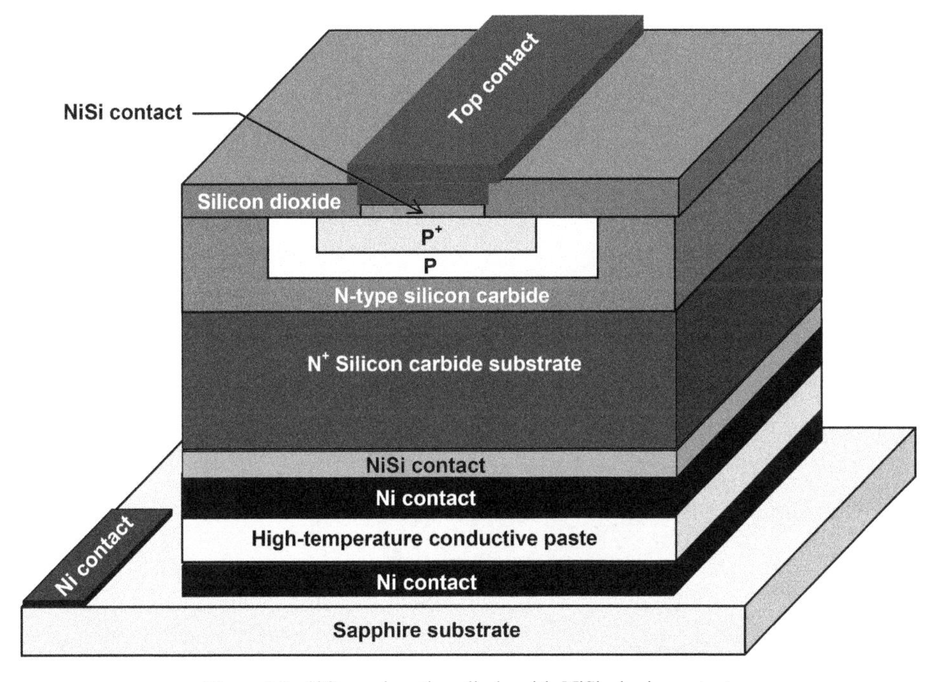

Figure 8.5. SiC p–n junction diode with NiSi ohmic contact.

8.7.3 Operation of SiC integrated bridge rectifier up to 773 K

Stable operation of an integrated bridge rectifier circuit was demonstrated up to 773 K (Shao *et al* 2014). The turn-on voltage of the diode decreased from 2.6 V at room temperature (290 K) to 1.4 V at 773 K. The average shift rate was 2.48 mV K^{-1}. The on-resistance decreased from 2.3 kΩ to 0.9 kΩ in the same temperature range with an average rate of 2.9 Ω K^{-1}. From room temperature to 773 K, the DC output voltage of the rectifier changed from 5.4 V to 6.4 V. An important figure-of-merit of the rectifier is the voltage conversion efficiency (V_{CE}) = output DC voltage/effective DC voltage of the input sine wave (0.707 V_{peak}). The V_{CE} was estimated as 76.4% at room temperature and 90.5% at 773 K. The success of the SiC rectifier circuit promises applications in extreme environments.

8.8 SiC Schottky-barrier diodes

Schottky diodes are closer to ideal switches than p–n diodes with reference to reverse recovery charge and recovery softness. SiC SBDs essentially resemble Si SBDs in construction (figure 8.6). They are majority carrier devices. They do not show any reverse recovery behavior like p–n diodes. However, some reverse recovery effects are observed due to the parasitic capacitances and inductances in the circuit, and also due to the p-rings, which are incorporated in the chip to decrease leakage current.

SBDs rank as the first commercialized SiC device. SiC SBDs expanded the blocking voltage range from 200 V for Si SBDs to over 1000 V for SiC SBDs. But the forward voltage of SiC SBDs, ~1–2 V, is greater than that of Si SBDs. The larger forward voltage arises from the wider bandgap of SiC than Si.

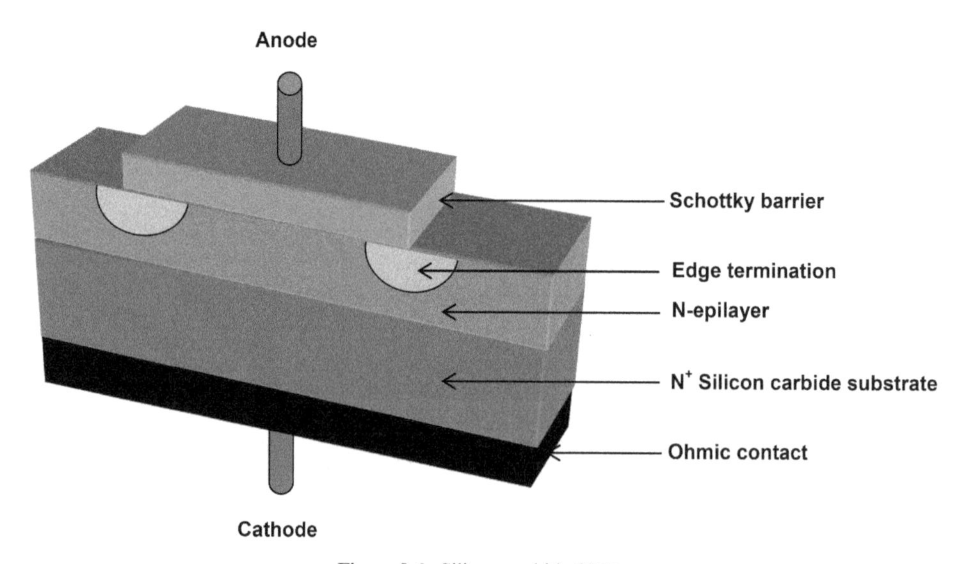

Figure 8.6. Silicon carbide SBD.

8.8.1 Temperature effects on Si and SiC Schottky diodes

The forward voltage of both Si and SiC diodes decreases with temperature. In one comparative study, through the characterization of commercial Si and SiC diodes it was found that the forward voltage V_D of Si and SiC diodes can be represented as a function of temperature (Ozpineci and Tolbert 2003)

$$V_D^{Si} = 0.3306\exp(-0.0103t° C) + 0.5724 \tag{8.10}$$

$$V_D^{SiC} = 0.2785\exp(-0.0046t° C) + 0.7042. \tag{8.11}$$

As can be seen, the forward voltages for both materials decrease as temperature increases. However, the same remarks are not true in both cases for the on-resistances R_D of Si and SiC diodes. The equations for Si and SiC on-resistance-temperature variation are (Ozpineci and Tolbert 2003)

$$R_D^{Si} = 0.2136\exp(-0.293t° C) + 0.0529 \tag{8.12}$$

$$R_D^{SiC} = -0.1108\exp(-0.0072t° C) + 0.2023. \tag{8.13}$$

In opposition to the negative TC of the on-resistance of Si SBDs, the TC of the on-resistance of SiC SBDs is positive, enabling easy paralleling of SiC SBDs.

The reverse recovery time (t_{rr}) of SiC SBDs (~20 ns) is shorter than that of Si SBDs (~40 ns). Therefore, they can be operated at higher switching frequencies. Moreover, the t_{rr} of SiC SBDs remains constant with increasing temperature. We know that power losses consist of two components: conduction and switching losses. Due to temperature-independent reverse recovery characteristics, SiC diodes can work at elevated temperatures with lower switching losses than Si diodes. In contrast, the t_{rr} of Si SBDs increases with an increase in temperature. The longer t_{rr} of Si SBDs at high temperatures leads to enormous switching losses.

6H-SiC SBDs with a forward voltage drop of 1.1 V between 298 K and 473 K at 100 A cm^{-2} current density and sharp breakdown voltage characteristics of 400 V at 298 K were reported (Bhatnagar *et al* 1992). These diodes showed improved reverse recovery behavior over fast silicon PIN diodes. Merged PIN/Schottky (MPS) diodes made from 4H-SiC with reverse voltage capability V_B = 4300 V showed a specific on-resistance = 20.9 mΩ cm (Wu *et al* 2006).

8.8.2 Schottky diode testing up to 623 K

The thermal stability of the ohmic contacts in two commercially available SiC Schottky diode dies was examined under 1000 h vacuum storage conditions at 623 K without applying any bias (O'Mahony *et al* 2011). These diodes had Al metallization as the Schottky anode for wiring. They had Ni/Ag metallization on the cathode bond pad for affixing the die. After high-temperature storage, the Al anode remained stable and showed <5% change in Schottky barrier height or diode ideality factor. But the die attach metal was seriously affected, resulting in an appreciable increase in series resistance to the extent of a hundred-fold increase in one type of diode.

8.8.3 Schottky diode testing up to 523 K

High-temperature (523 K) reverse-bias endurance tests were carried out on 600 V, 6 A 4H-SiC Schottky diodes (Testa *et al* 2011). The forward voltages at 100 μA to 6 A and the reverse currents of the diodes at 600 V changed insignificantly after 1 000 h. This experiment confirmed the stability of the electrical parameters of these devices.

8.9 SiC JFETs

The late 1980s and early 1990s witnessed efforts towards the development of SiC JFETs, hindered by the low mobilities, transconductance and SiC material quality issues. But a SiC JFET was easier to implement than the MOSFET because of the absence of oxide–semiconductor interface quality problems (Baliga 2006). The fruition of these efforts led to the emergence of two main designs (figure 8.7): the lateral channel JFET (LCJFET) and vertical trench JFET (VTJFET) structures. The structural differences between LCJFET and VTJFET are observed as differences in their performance characteristics. Again, two versions of LCJFET, both n-type and with a buried p-layer, were put forward to be used for applications in either high-speed switching or low on-state losses. (i) The fast switching version has a smaller gate–drain area producing lower gate–drain capacitance. The pre-channel region between the source junction and the channel is responsible for the high on-resistance. (ii) The low on-state resistance version has a structure such that the direct flow of current from the source through the channel to the drift region reduces the on-resistance, but the high gate resistance due to the high resistivity p-type SiC material at the gate junction slows down the switching rate.

Both versions (i) and (ii) are normally-on or depletion mode devices which require a negative gate-source voltage ~ -15 V less than the pinch-off voltage to be applied between the p^+ gate and n^+-source terminals to constrict the channel and keep the JFET turned off. The reverse breakdown voltage of the gate–source junction is > -35 V. For normally-off operation, a normally-on JFET is combined with a low-voltage enhancement mode silicon MOSFET (Siemieniec and Kirchner 2011). This is known as the direct-driven concept.

Power electronic applications need enhancement-mode devices. Applying a voltage of ~2–3 V can turn them on but these devices have larger specific on-resistance. In the VTJFET, both enhancement-mode VTJFET (EMVTJFET) and depletion-mode VTJFET (DMVTJFET) types are realized (Casady *et al* 2010). Cross-sectionally, EMVTJFET and DMVTJFET are identical. The thickness of the vertical channel of EMVTJFET is 10% narrower than that of DMVTJFET. Further, the channel doping of EMVTJFET is 10% lower than that of DMVTJFET. Consequently, the on-resistance of EMVTJFET is 15% higher than that of DMVTJFET. Also, the saturation current of EMVTJFET is 50% of that of DMVTJFET. These JFETs were made from 4H-SiC. They had equivalent die areas (4.5 mm^2) and were rated at 1200 V.

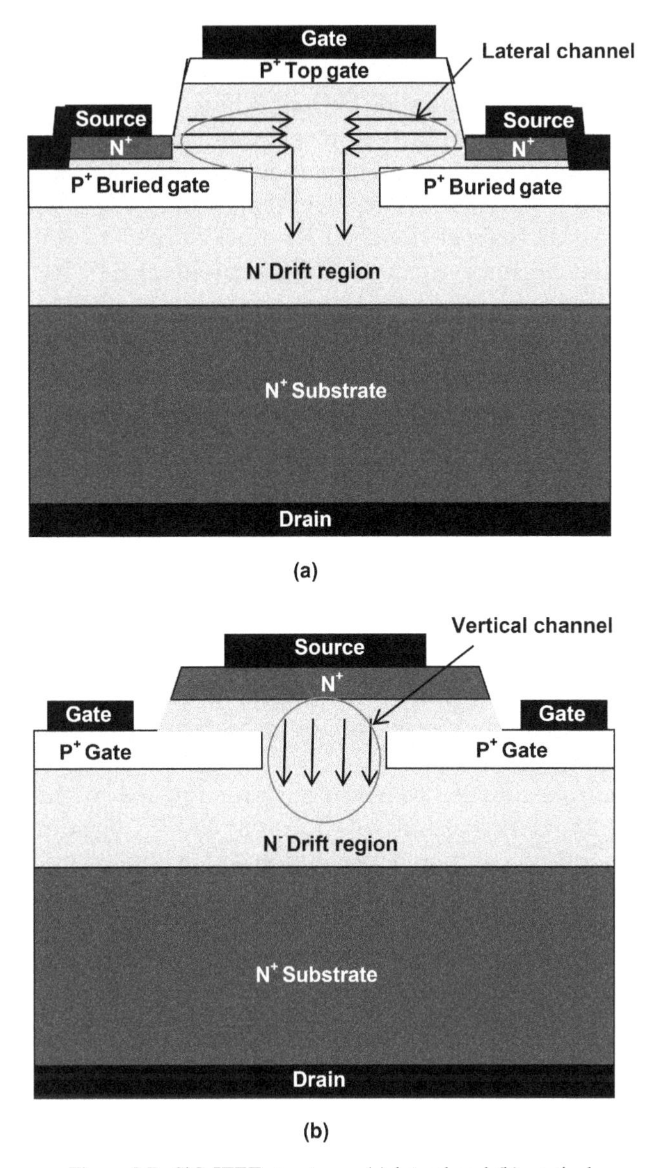

Figure 8.7. SiC JFET structures: (a) lateral and (b) vertical.

In a VTJFET, the anti-parallel body diode is absent. The presence of such a diode in LCJFET is a distinct advantage of this structure. An external anti-parallel diode must therefore be connected to the VTJFET. It is utilized during switching transitions.

An enhancement type 5.3 kV JFET was fabricated from 4H-SiC. It has a low specific on-resistance (69 mΩ cm^2). The turn-off time is 47 ns (Asano *et al* 2001, 2002). This JFET is called a static expansion channel JFET (SEJFET). An area-efficient cell structure is used and a suitable JTE technique is employed.

A buried-grid JFET (BGJFET) (Malhan *et al* 2006, Tanaka *et al* 2006, Malhan *et al* 2009, Lim *et al* 2010) consists of several cells at a small pitch. As a result, the on-resistance is low and the saturation current is high. Fabrication is more difficult than for a LCJFET. As decided by the channel width and channel doping, this type of JFET is configured as either normally-on or normally-off. The on-resistance of the normally-off BGJFET is 30% lower than that of the VTJFET (Lim *et al* 2010).

A dual-gate vertical channel trench JFET (DGVTJFET) (Malhan *et al* 2006, 2009) combines the attributes of a LCJFET and BGJFET. Its low gate–drain capacitance leads to fast switching while the decrease in cell pitch together with gate control minimizes the on-resistance. It is available as both types, normally-on as well as normally-off, with the latter type providing a much larger saturation current. However, the fabrication process of these JFETs is complex.

8.9.1 Characterization of SiC JFETs from 25 °C to 450 °C

Several 2.5 A, 1200 V SiC JFETs (figure 8.8) have been comprehensively characterized (Funaki *et al* 2006) from room temperature to 450 °C by packaging the devices in dedicated TO-258 packages, that were able to withstand high temperatures, and by using specially fabricated fixtures. These studies were conducted for modeling and the extraction of parameters which can be used for circuit simulations. DC current–voltage characteristics were measured on a curve tracer while AC capacitance–voltage and impedance–voltage measurements were performed on an impedance analyzer.

From the forward conduction characteristics, for $V_{GS} = 0$ V, the saturated drain current is ~4 A at 25 °C. It decreases to ~0.75 A at 450 °C. The threshold voltage is −15.8 V at 25 °C and becomes more negative at 450 °C; the new value is −18.2 V. The transconductance parameter is 2.4×10^{-2} at 25 °C. It decreases to 0.25×10^{-2} at 450 °C. From the reverse conduction characteristics, the saturation current increases from 10^{-10} A at 25 °C to 10^{-6} A at 450 °C. From the AC characteristics, the

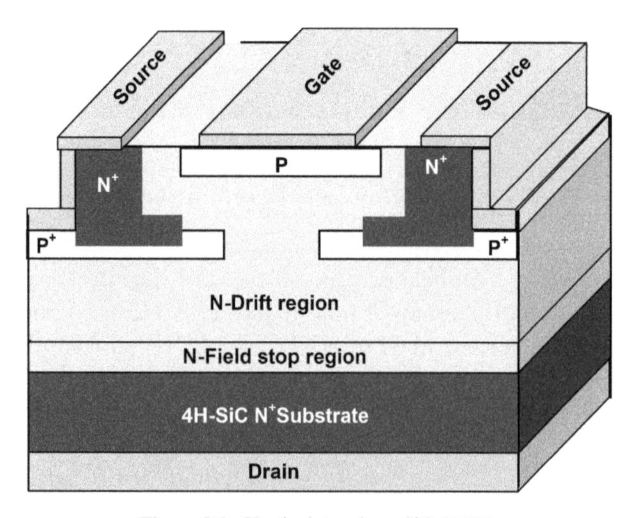

Figure 8.8. Vertical topology SiC JFET.

capacitance C_{GS} at $V_{GS} = 0$ V increases from 450 pF at 25 °C to 575 pF at 450 °C. The capacitance C_{GD} shows an identical trend. The capacitance C_{DS} decreases as V_{DS} increases but remains practically unaffected by temperature.

8.9.2 500 °C operational test of 6H-SiC JFETs and ICs

The vital electrical parameters of depletion-mode epitaxial JFETs with gate lengths of 10 µm showed changes smaller than 10% over 3007 h of operation at 500 °C in air ambient (Neudeck *et al* 2008). In these JFETs (figure 8.9), an n-type source and drain were formed by nitrogen implantation. The p-type epitaxial gate layer was formed by Al implantation. Multilayer Ti/$TaSi_2$/Pt metallization was used to make the three contacts.

The differential amplifier IC consists of three resistors and two JFETs. The JFETs were source-coupled and operated with a 40 V source.

The NOR gate IC consists of three JFETs and the resistors. +20 V and −24 V power supplies were used for the NOR gate.

The test run was carried out by Neudeck *et al* by keeping the JFETs and ICs in an oven containing ordinary room air. Measurement data were periodically recorded

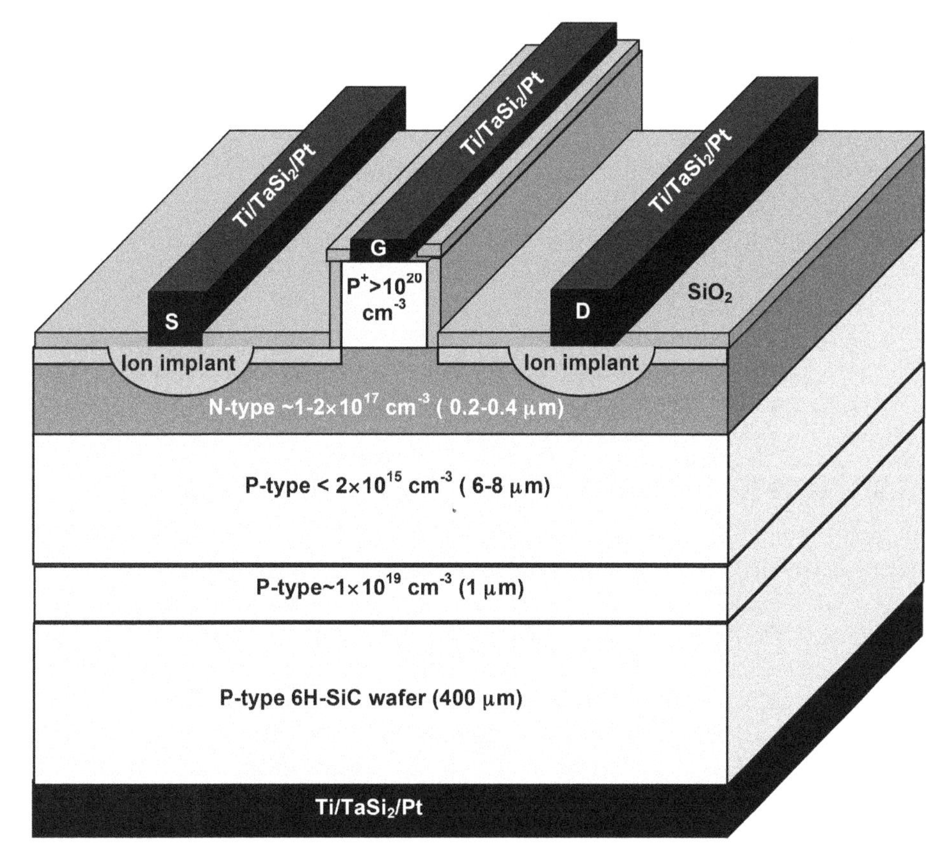

Figure 8.9. A 6H-SiC JFET without Si_3N_4 surface passivation.

during continuous operation under electrical bias at 500 °C. Typical magnitudes of changes in JFET parameters at 500 °C storage can be visualized from the following readings (Neudeck *et al* 2008):

(i) the capacitor leakage current density J_R under 50 V stress decreased from 604 μA cm^{-2} after 100 h to 104 μA cm^{-2} after 3007 h;
(ii) the threshold voltage remained at −11.8 V throughout;
(iii) the drain current I_{DSS} with $V_G = 0$ V, $V_D = 20$ V decreased from 1.36 mA after 100 h to 1.31 mA after 3007 h;
(iv) the transconductance g_{m0} at $V_D = 20$ V decreased from 214 μS after 100 h to 205 μS after 3007 h;
(v) the drain–source resistance R_{DS} for $V_G = 0$ V increased from 4.64 kΩ after 100 h to 4.83 kΩ after 3 007 h;
(vi) the drain leakage current I_{OFF} at $V_D = 50$ V, $V_G = -15$ V decreased from 37.1 μA after 100 h to 0.19 μA after 3007 h.

For the differential amplifier IC:

(i) the voltage gain A_V at 100 Hz was 2.99 and 2.91 after 100 h and 3007 h, respectively;
(ii) voltage gain A_V at 10 kHz was 2.18 after 100 h and 2.19 after 3007 h;
(iii) the unity voltage gain frequency f_T was 32 kHz after 100 h and rose to 33 kHz after 3007 h.

For the NOR-gate IC:

(i) the output high-voltage V_{OH} with both inputs at low voltage (= −7.5 V) was −1.47 V after 100 h and −1.67 V after 3007 h;
(ii) the output low voltage V_{OL} keeping one input at high voltage (= −2.5 V) and the other input at low voltage (= −7.5 V) was −8.12 V after 100 h and −8.14 V after 3007 h.

The above data of Neudeck *et al* amply demonstrate the feasibility of JFETs and ICs for long-duration functioning capability at 500 °C.

8.9.3 6H-SiC JFET-based logic circuits for the 25 °C to 550 °C range

Depletion-mode JFETs were used to fabricate inverter, NAND gate and NOR gate circuits to provide a DC transfer characteristic with a broad noise margin at 550 °C (Soong *et al* 2012). In the JFET structure, the top p^+ layer serving as the gate, controls the depletion of the underlying n-type epitaxial layer acting as the channel layer. Below this n-type epi-layer is a p^+ epi-layer on a p-type SiC substrate. Nitrogen ion implantation is used to form the n^+ source and drain. Following deposition of a 20 nm SiO_2 layer, a stack of Ti, $TaSi_2$ and Pt is sputtered and protected by a Si_3N_4 layer. JFETs and logic circuits are completed, and interconnections are formed. The circuits are encapsulated in dual-in-line packages using gold bonding wires.

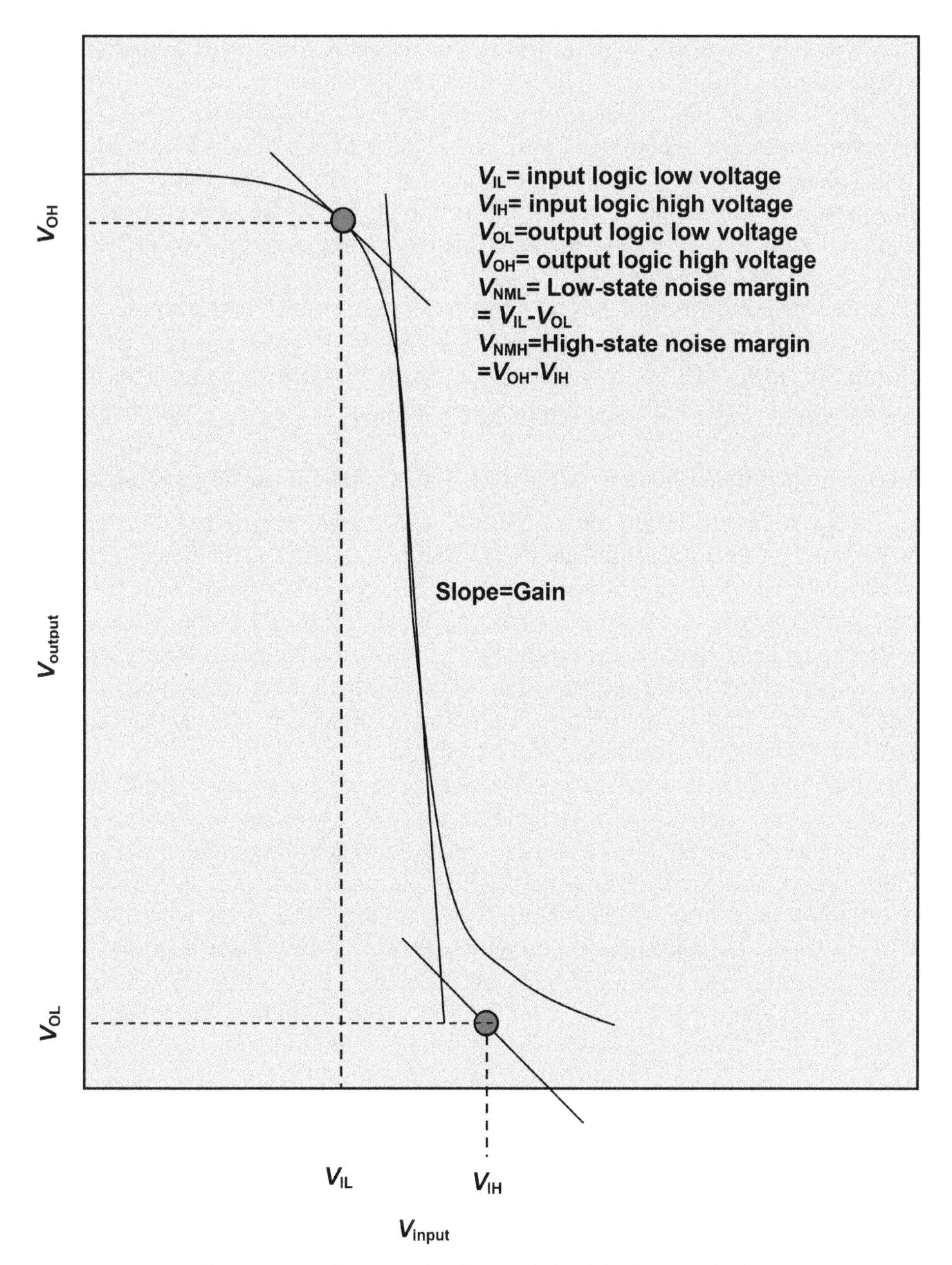

Figure 8.10. Voltage transfer characteristic of the inverter circuit.

The performance of the inverter is evaluated from its voltage transfer characteristic (VTC). Figure 8.10 shows the typical shape of the VTC of the inverter. For the temperature interval 25 °C to 550 °C, Soong *et al* noticed: (i) a gradient of VTC, falling sharply in an almost perpendicular manner; (ii) up to a temperature of 550 °C, the voltage gain exceeds −20; and (iii) the logic threshold is located in the middle of the logic swing.

To get an idea of the influence of temperature on a SiC inverter: (i) V_{OL} = low-level output voltage = −8.00 V at 25 °C and −8.69 V at 550 °C; (ii) V_{OH} = high-level output voltage = −0.50 V at 25 °C and −3.97 V at 550 °C; (iii) V_{NML} = low-level noise margin = 1.80 at 25 °C and 1.52 at 550 °C; (iv) V_{NMH} = high-level noise margin = 4.91 at 25 °C and 2.17 at 550 °C; and (v) voltage gain = −26 at 25 °C, −20 at 500 °C and −12 at 550 °C.

The DC characteristics of NAND and NOR gates were similar to those of the inverter circuit. Dynamic testing of NAND and NOR gates showed satisfactory performance up to 550 °C. Thus a technological platform for the fabrication of reliable digital circuits with operational capability up to 550 °C was upheld.

8.9.4 Long operational lifetime (10 000 h), 500 °C, 6H-SiC analog and digital ICs

The JFETs were fabricated on a 6H-SiC wafer over which a p^-SiC epi-layer (2×10^{15} cm^{-3}) was grown followed by an n-SiC epi-layer (1×10^{17} cm^{-3}) (Neudeck *et al* 2009). The devices employed a mesa-etched heavily Al-doped p^+ (2×10^{19} cm^{-3}) epi-gate structure with nitrogen-doped n-type implantations for the source and drain. The thickness and doping concentration of the sub-channel p-type layer were carefully controlled to reduce parasitic substrate-to-channel capacitance. A self-aligned nitrogen implant minimized the parasitic resistance between the gate and source/drain. Contacts were made of $Ti/TaSi_2/Pt$.

The diced chips were encapsulated by Neudeck *et al* in custom ceramic packages with 96% ceramic substrate and Au metallization. Wire bonding was performed with 1 mil diameter Au wires. The packages were fixed to ceramic printed circuit boards (PCBs). Device and circuit testing was carried out under electrical bias at 500 °C in an oven filled with room air of uncontrolled humidity. The JFET with W/L = 100 μm/10 μm was tested under drain voltage V_D = 50 V sweep. Gate bias step V_G was stepped from 0 to −16 V in −2 V intervals. The 200 μm/10 μm JFET was operated under V_D = 50 V and V_G = −5 V. After the first, 100th, 1000th and 10 000th hours at 500 °C, the I_D–V_D characteristics were recorded. The drain saturation current at zero gate bias (I_{DSS}), transconductance g_m and threshold voltage V_{Th} of the JFETs were normalized with respect to their corresponding values I_{DSS0}, g_{m0} and V_{Th0} at the 100th hour of 500 °C testing. The current I_{DSS} changed by less than 10% over 10 000 h at 500 °C. The transconductance g_m showed a similar trend. V_{Th} varied by less than 1% over this time span (Neudeck *et al* 2009).

8.9.5 Characterization of 6H-SiC JFETs and differential amplifiers up to 450 °C

JFETs were fabricated on Al-doped 6H-SiC wafers by growing an Al-doped p^- epi SiC layer, followed by an n^- epi SiC layer and a p^+ epi-SiC layer (Patil *et al* 2007), see figure 8.11. Nitrogen ion implantation at a high dose was performed for the source and drain. The contact metal for the source, drain and gate was Al deposited by sputtering. Up to 300 °C, the chips were probed by mounting on a hot chuck. For electrical characterization at temperatures above 300 °C, the diced JFET chips were

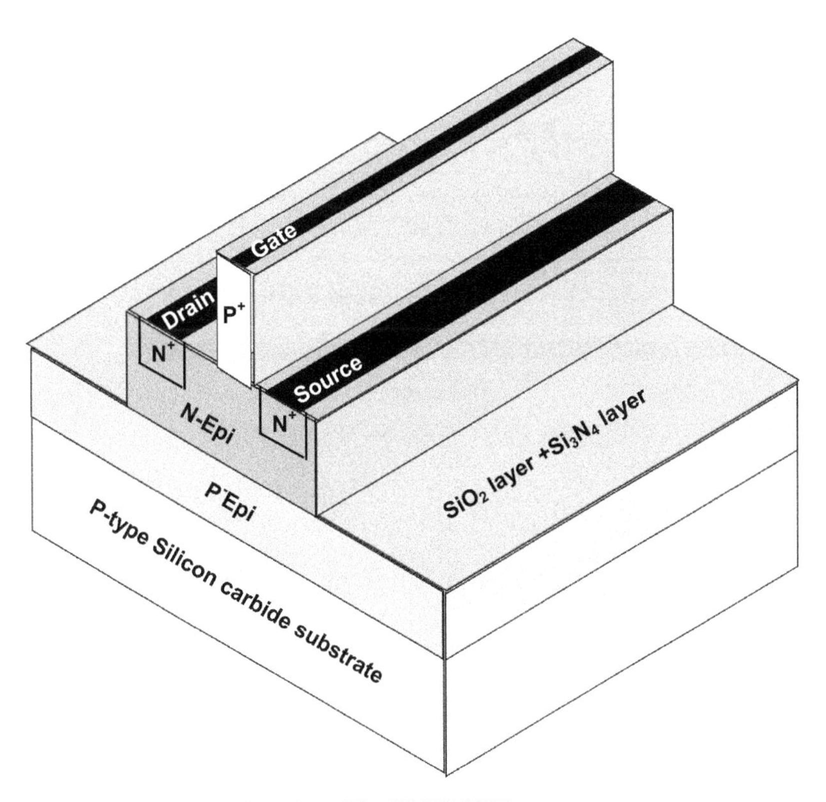

Figure 8.11. The 6H-SiC JFET structure.

encapsulated in ceramic dual-in-line packages (DIPs). Measurement of the I_{DS}–V_{DS} and I_{DS}–V_{GS} characteristics of the W/L = 100 μm/100 μm JFETs from 25 °C to 450 °C showed that the characteristics were well-behaved throughout, i.e. the nature of characteristics was preserved in the whole range. However, Al contacts created some instability near the upper limit of temperature. The experimental characteristics fully conformed to the 3/2 power JFET model. This allowed the extraction and computation of many JFET parameters. The pinch-off current (I_p) is 0.28 at 25 °C. It decreases to one-half at 450 °C. The pinch-off voltage (V_{po}) is 11.90 V at 25 °C, changing almost negligibly to 11.13 V at 450 °C. The channel length modulation parameter (λ) also varies insignificantly. Up to 300 °C, the threshold voltage V_{Th} follows the −2.3 mV °C^{-1} rule. The source/drain resistance $R_{S,D}$ increases from 4.54 kΩ at 25 °C to 13.39 kΩ at 450 °C.

Differential amplifiers fabricated with SiC JFETs were operated at high temperature with off-chip components at room temperature (Patil *et al* 2009). They worked successfully up to 450 °C albeit with altered parametric values. The DC transfer characteristics of a three-stage differential amplifier with external biasing and passive loads were measured at room temperature and 450 °C. These 6H-SiC JFETs had W/L = 110 μm/10 μm. The JFET-based differential amplifiers gave a low-frequency voltage gain of 87 dB at 25 °C, falling to 50 dB at 450 °C. The unity gain bandwidth product is 335 kHz at 25 °C, decreasing to 201 kHz at 450 °C.

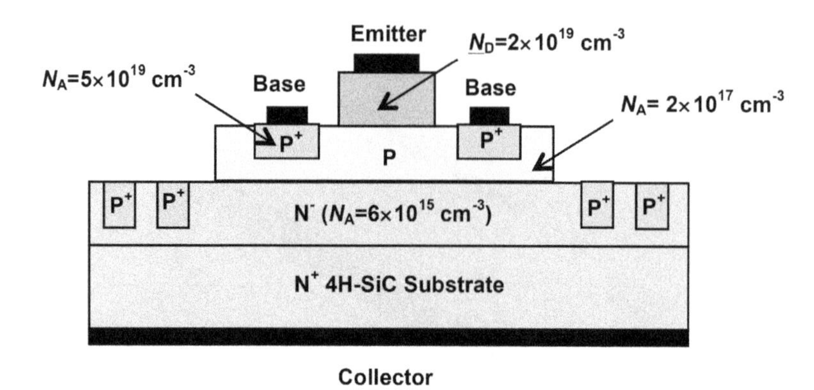

Figure 8.12. Cross-section of a SiC BJT.

Further, the input offset voltage at 450 °C is larger than the room-temperature value. For comparison, an externally biased SiC multistage amplifier with active loads gave a low-frequency voltage gain of 70 dB and a gain bandwidth product of 1300 kHz at 450 °C.

8.10 SiC bipolar junction transistors

SiC BJTs (figure 8.12) can provide a low on-resistance feature along with a high breakdown voltage. Freedom from the gate oxide problem permits their operation at a high temperature, unlike SiC MOSFETs.

4H-SiC BJTs with a voltage V_{CEO} above 1000 V and a DC current gain of 32 at collector current density $J_C = 319$ A cm^{-2} showed specific on-resistance 17 mΩ cm^2 at $J_C = 289$ A cm^{-2} (Luo *et al* 2003). These BJTs used an Al-free ohmic contact to the base.

4H-SiC BJTs blocked more than 480 V and conducted at $J_C = 239$ A cm^{-2} with specific on-resistance of 14 mΩ cm^2. A current gain of ⩾35 was obtained at $J_C = 40$–239 A cm^{-2} (Zhang *et al* 2003). The peak gain value was 38 at $J_C = 114$ A cm^{-2}.

4H-SiC BJTs with a blocking voltage of 9.2 kV at a leakage current of 4.6 μA showed a DC common-emitter current gain of 7 and a specific on-resistance of 33 mΩ cm^2 without considering current spreading (Zhang *et al* 2004).

4H-SiC BJTs (blocking voltage 1836 V) were tested for power switching (300 V, >7 A) and found to switch at 20–80 ns at par with Si unipolar devices. The switching speed remained unaffected up to a high temperature of 275 °C. The high-frequency, high-temperature capability of these BJTs was proved (Sheng *et al* 2005).

4H-SiC BJTs having a collector–emitter voltage with the base open ($V_{CEO} = 757$ V) and a current gain of 18.8, conducted up to 5.24 A at a forward voltage $V_{CE} = 2.5$ with a specific on-resistance of 2.9 mΩ cm^2 up to $J_C = 859$ A cm^{-2} (Zhang *et al* 2006).

4H-SiC BJTs were demonstrated using a graded base profile (Zhang *et al* 2009). They had a current gain of ~33 and a collector–emitter breakdown voltage (V_{CEO}) > 1000 V. The specific on-resistance was 2.9 mΩ cm^2.

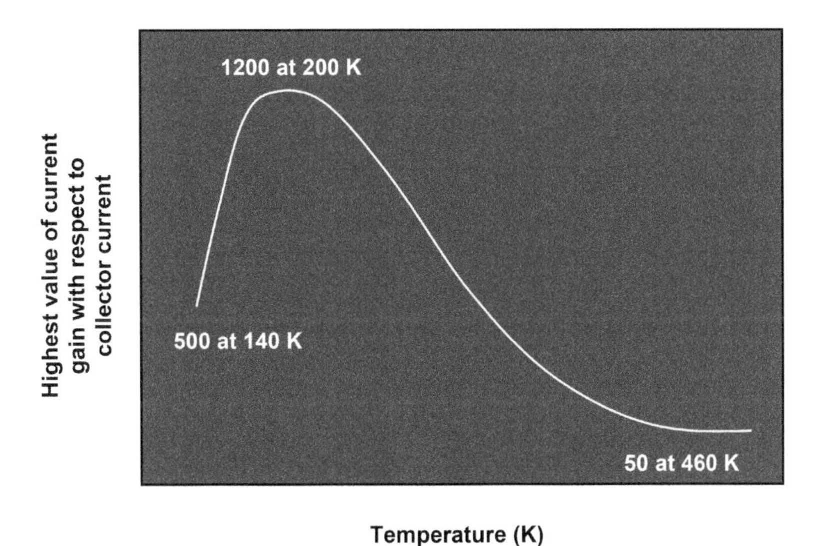

Figure 8.13. Current gain of silicon carbide bipolar junction transistor versus temperature.

8.10.1 Characterization of SiC BJTs from 140 K to 460 K

4H-SiC BJTs were characterized for temperature dependence in the range 140–460 K (Asada *et al* 2015). Surprisingly, the temperature behavior of SiC BJTs is strikingly different from that of Si BJTs. Whereas the current gain of Si BJTs decreases monotonically as temperature falls, SiC BJTs show two types of behavior, namely the reverse behavior of an increase in current gain up to a certain temperature followed by similar behavior of a decrease in current gain with further decrease of temperature, as shown in figure 8.13.

(i) When the temperature descended from 460 to 200 K, the current gain rose from 110 to 1200. This rise of current gain with decreasing temperature was interpreted in terms of an uplift of emitter injection efficiency. The efficiency was augmented because the aluminum acceptors in the base region of the transistor were only partially ionized at the low temperature; a higher temperature was necessary for the completion of ionization. So, the free carrier concentration in the base was low.

(ii) Subsequently, when the temperature was brought down from 200 to 140 K, the trend was opposite. The current gain started to decrease. It dropped down from 1200 at 200 K to 515 at 140 K. Interpretation for this change of trend is provided on the premise that the injected carrier concentration from the emitter into the base surpasses the low hole concentration in the base leading to the creation of a high injection condition at low collector currents. This high injection condition is held responsible for the decline of current gain.

8.10.2 Performance assessment of SiC BJT from −86 °C to 550 °C

In another investigation (Nawaz *et al* 2009), the current gain of SiC n–p–n BJTs was reported to decrease from 50 at room temperature to half the value (25) at 548 K. On decreasing the temperature below room temperature, the current gain attained a peak value of 111 at 187 K. Below 187 K, the current gain fell abruptly through the carrier freeze-out effect. The on-resistance measurements for the SiC BJT were also performed. Starting from a room-temperature value of 7 mΩ cm^2, the on-resistance increased to 28 mΩ cm^2 at 548 K. However, the on-resistance remained practically constant when the temperature was decreased to 187 K. At temperatures lower than

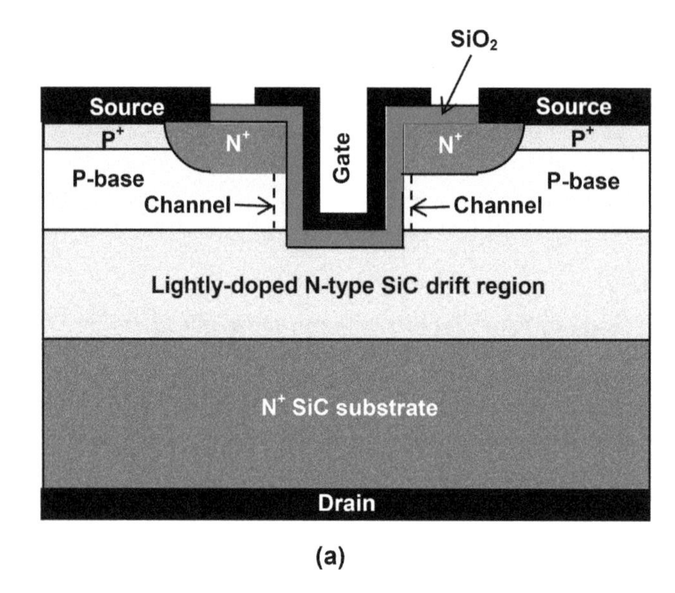

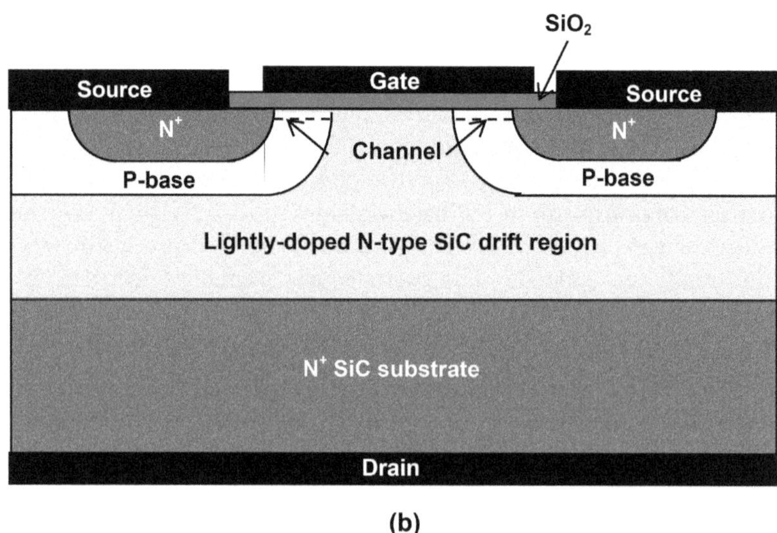

Figure 8.14. SiC MOSFETs: (a) UMOSFET and (b) DMOSFET.

187 K, the on-resistance climbed up suddenly, which is caused by carrier freeze-out. Packaged BJTs displayed satisfactory performance up to 823 K.

8.11 SiC MOSFETs

The voltage-controlled, normally-off devices are favorites of power electronic circuit designers. In the vertical trench U-shaped MOSFET (UMOSFET), figure 8.14(a), the channel is formed on the sidewalls of trenches excavated by reactive ion etching. The structure suffers from the problem of oxide breakdown at the corners of the trench (Cooper *et al* 2002). When close to avalanche breakdown, the peak electric field in SiC is 3×10^6 V cm^{-1}. Then the electric field in the oxide is higher by a factor equal to the ratio of the dielectric constant of SiC against the dielectric constant of SiO_2 (= 9.7/3.9 = 2.49). Hence, this electric field = $2.49 \times 3 \times 10^6$ V cm^{-1} = 7.46×10^6 V cm^{-1}. This high field value is very near to the dielectric strength of SiO_2 (10^7 V cm^{-1}). Current crowding at the trench corner further intensifies the field. At high temperatures or during prolonged operation, the situation is even more serious. SiC UMOSFETs with a blocking capability ~260 V and on-resistance equal to 10–50 mΩ cm^2 were fabricated in the 1990s. The channel mobility of these devices is low and the gate oxide quality needs improvement. In 1996, a planar DMOSFET structure (figure 8.14(b)) was advanced as a step to circumvent some of the oxide problems faced with UMOSFET. It was fabricated by successive implantation of aluminum or boron ions for the p-type base and nitrogen ions for the n^+ source. A three-fold improvement in blocking voltage up to 760 V could be achieved. Several variants of DMOSFET design were developed. A 6.1 kV static induction injected accumulated FET (SIAFET) with specific on-resistance of 732 mΩ cm^2 was developed (Takayama *et al* 2001).

Regarding the temperature effects on silicon dioxide on SiC and SiC MOSFETs, reliability studies have shown that satisfactory performance is assured if the electric field is restricted below 4×10^6 V cm^{-1}, and the temperature is below 150 °C. Long-term operation of SiC MOSFETs at temperatures higher than 200–250 °C seems to be unviable.

8.12 Discussion and conclusions

Amongst the two popular polytypes of SiC, namely 4H-SiC and 6H-SiC, the former reigns supreme by virtue of its mobility and bandgap advantages. Owing to the inapplicability of the conventional CZ process for single-crystal growth of silicon carbide, physical vapor deposition is widely used. Nitrogen and aluminum are used as n- and p-type dopants, respectively, and dopants are introduced by ion implantation followed by thermal annealing. Silicon carbide allows surface oxidation. Nonetheless, the oxide quality needs appreciable improvement. In SiC technology, different contact metallizations are available for making reliable ohmic and Schottky contacts. In several experimental studies conducted at temperatures up to 783 K, SiC p–n diodes as well as Schottky diodes have been found to function in a satisfactory manner, establishing their credentials for high-temperature survivability. Concerning SiC JFETs and MOSFETs, the former could easily be implemented in two versions, namely lateral channel and VTJFETs. JFETs and JFET-based analog and digital circuits were subjected to several trials at temperatures up to 550 °C. Their performance was encouraging. Silicon carbide BJTs have been fabricated providing low on-resistance with high breakdown voltage, but they behave differently from Si BJTs with regard to the variation of current gain with temperature, namely the opposite behavior in a certain temperature range and similar behavior in another range. SiC MOSFETs have hitherto shown a relatively poor performance than Si MOSFETs because of inferior oxide quality and the low mobility of carriers.

Review exercises

8.1. Mention two properties of 4H-SiC which make it superior to 6H-SiC.
8.2. Up to what frequencies are carrier mobilities in silicon carbide able to provide RF performance?
8.3. Why is it possible to provide a lower on-resistance with a higher breakdown voltage using silicon carbide in place of silicon?
8.4. How much broader is the bandgap of silicon carbide than silicon? How much higher is the intrinsic temperature of silicon carbide than silicon? Give answers as ratios for the two materials.
8.5. Why is the CZ process for growing single-crystal silicon from the melt not applicable to silicon carbide? What is the process by which silicon carbide single crystals are grown? What are the common defects in silicon carbide crystals?
8.6. What are the common dopants for producing n- and p-type silicon carbide? Is it possible to introduce impurities by thermal diffusion? If not, what is the technique used for doping silicon carbide? How is *in situ* doping carried out?
8.7. Is it possible to grow silicon dioxide on silicon carbide surfaces? Is the oxidation rate slower or faster than in silicon? Does it depend on whether the silicon or carbon atom is facing towards the growing oxide film? How reliable are these oxides grown over silicon carbide?
8.8. What metals are used for making ohmic contacts to: (a) n-type silicon carbide and (b) p-type silicon carbide?
8.9. What was the carrier lifetime for the 12.9 kV SiC p–n diode at room temperature and at 498 K? What were the peak recovery current values at the two temperatures?
8.10. Mention two types of contacts, which were not degraded upon ageing at 773–783 K. Discuss the relative performance of p–n diodes in which Ni or Pt was sputtered on Ni/Si ohmic contacts on SiC diodes, and the diodes were kept at 873 K.
8.11. Describe the performance of a SiC integrated bridge rectifier up to 773 K with reference to the effect of temperature on the following parameters as compared to their values at room temperature: (i) the turn-on voltage and

the on-resistance of a diode and (ii) the DC output voltage and voltage conversion efficiency of the rectifier.

8.12. In what respects is an SBD better than a p–n diode? How does a silicon carbide Schottky diode outperform a silicon p–n diode? In what way is it inferior?

8.13. How does temperature affect the forward voltages across SBDs made of silicon carbide and silicon? What is the effect on on-resistances of the two types of diodes? Which type of diodes can be easily paralleled without risk of thermal runaway and why?

8.14. Differentiate between silicon carbide and silicon SBD regarding their reverse recovery times. How does this difference impact the performances of the two types of diodes at high temperatures?

8.15. What were the results of 623 K testing on commercial Schottky diodes with Al metallization as the Schottky anode and Ni/Ag metallization on the cathode bond pad? How did the Schottky diodes perform in high-temperature (523 K) reverse-bias endurance tests?

8.16. Why did SiC JFET allow easier implementation than SiC MOSFET? What are the two main JFET designs pursued in SiC?

8.17. What are the main features of the two types of lateral channel JFET? Are these designs of normally-on or normally-off type? Can a normally-on JFET be operated as a normally-off type? If so, how?

8.18. What types of switching devices are required in power electronics: enhancement or depletion type? Does the VTJFET enable realization of both types? Which type of structure has lower saturation current? Which type has lower on-resistance?

8.19. An anti-parallel diode is present in which type: LCJFET or VTJFET? How is the deficiency of the diode overcome in the structure in which this diode is not present?

8.20. What are full forms of the acronyms of the following types of JFETs: (a) SEJFET, (b) BGJFET and (c) DGVTJFET? Describe the construction and salient features of each type of JFET.

8.21. Describe the performance parameter variation of SiC JFETs from 25 °C to 450 °C in respect of: (a) the saturated drain current, (b) threshold voltage, (c) transconductance, (d) reverse current and (f) gate-source and drain-source capacitances.

8.22. Describe the variations in vital parameters over the extended time period for the 500 °C operational test of: (a) 6H-SiC JFETs, (b) differential amplifier IC and (c) NOR gate IC in respect of drain current I_{DSS} and transconductance g_{m0} for JFET; voltage gain A_V at 10 kHz and unity voltage gain frequency f_T for a differential amplifier IC; and output high-voltage V_{OH} and output low voltage V_{OL} for a NOR-gate IC.

8.23. Describe the characteristics of a SiC JFET-based inverter with respect to V_{OH}, V_{OL} and V_{NML} variation in the 25 °C to 550 °C range. How did the NAND and NOR gates work in this temperature range?

8.24. How did the current I_{DSS}, the transconductance g_m, and threshold voltage V_{Th} of JFETs vary during the 10 000 h, 500 °C test on 6H-SiC analog and digital ICs?
8.25. What is the low-frequency voltage gain of the JFET-based differential amplifier at 25 °C and at 450 °C? What are the unity gain–bandwidth products at 25 °C and at 450 °C? How do the pinch-off current, pinch-off voltage and source/drain resistance of JFET change from 25 °C to 450 °C?
8.26. Give two examples of silicon carbide bipolar transistors mentioning the breakdown voltage and specific on-resistance in each case.
8.27. How is the trend of variation in current gain with temperature (140–460 K) different for Si BJT and SiC BJT? How does one account for this difference of behavior between the BJTs made from these two materials?
8.28. Explain the increase in current gain of a SiC BJT as the temperature is lowered from 460 K to 200 K.
8.29. Explain the decrease in current gain of a silicon carbide bipolar transistor when the temperature is decreased from 200 K to 140 K.
8.30. How does the on-resistance of a SiC BJT vary on decreasing the temperature from 548 K to 187 K and below 187 K? Why?
8.31. What is the problem of oxide breakdown in a vertical trench UMOSFET? Does the problem become more serious at high temperatures? Does SiC MOSFET operation at temperatures above 200 °C to 250 °C seem to be possible?

References

Asada S, Okuda T, Kimoto T and Suda J 2015 Temperature dependence of current gain in 4H-SiC bipolar junction transistors *Japan. J. Appl. Phys.* **54** 04DP13

Asano K, Sugawara Y, Hayashi T, Ryu S, Singh R, Palmour J and Takayama D 2002 5 kV 4H-SiC SEJFET with low RonS of 69 mΩcm^2 *Proc. 14th Int. Symp. on Power Semiconductor Devices and ICs* (Piscataway, NJ: IEEE) pp 61–4

Asano K, Sugawara Y, Ryu S, Singh R, Palmour J, Hayashi T and Takayama D 2001 5.5 kV normally-off low RonS 4H-SiC SEJFET *Proc. 13th Int. Symp. Power Semiconductor Devices and ICs (Osaka, 4–7 June)* (Piscataway, NJ: IEEE) pp 23–6

Baliga B J 2006 *Silicon Carbide Power Devices Devices* (Singapore: World Scientific) 528 pages

Bhatnagar M, McLarty P K and Baliga B J 1992 Silicon-carbide high-voltage (400 V) Schottky barrier diodes *IEEE Electron Device Lett* **13** 501–3

Braun M, Weis B, Bartsch W and Mitlehner H 2002 4.5 kV SiC pn-diodes with high current capability *10th Int. Conf. Power Electron. Motion Control (Cavtat and Dubrovnik)* pp 1–8

Casady J B, Sheridan D C, Kelley R L, Bondarenko V and Ritenour A 2010 A comparison of 1200 V normally-off and normally-on vertical trench SiC power JFET devices *Mater. Sci. Forum* **679–680** 641–4

Chand R, Esashi M and Tanaka S 2014 P–N junction and metal contact reliability of SiC diode in high temperature (873 K) environment *Solid-State Electron.* **94** 82–5

Cooper J A Jr, Melloch M R, Singh R, Agarwal A and Palmour J W 2002 Status and prospects for SiC power MOSFETs *IEEE Trans. Electron Devices* **49** 658–64

Funaki T, Kashyap A S, Mantooth H A, Balda J C, Barlow F D, Kimoto T and Hikihara T 2006 Characterization of SiC JFET for temperature dependent device modeling *37th IEEE Power Electronics Specialists Conference*

Kakanakov R, Kassamakova-Kolaklieva L, Hristeva N, Lepoeva G and Zekentes K 2002 Thermally stable low resistivity ohmic contacts for high power and high temperature SiC device applications *Proc. 23rd Int. Conf. Microelectronics* (*Niš, 12–15 May*) vol 1 (Piscataway, NJ: IEEE) pp 205–8

Lim J K, Bakowski M and Nee H P 2010 Design and gate drive considerations for epitaxial 1.2 kV buried grid N-on and N-off JFETs for operation at 250 °C *Mater. Sci. Forum* **645–648** 961–4

Luo Y, Zhang J and Alexandrov P 2003 High voltage (>1 kV) and high current gain (32) 4H-SiC power BJTs using Al-free ohmic contact to the base *IEEE Electron Device Lett.* **24** 695–7

Malhan R K, Bakowski M, Takeuchi Y, Sugiyama N and Schöner A 2009 Design, process, and performance of all-epitaxial normally-off SiC JFETs *Phys. Status Solidi* A **206** 2308–28

Malhan R K, Takeuchi Y, Kataoka M, Mihaila A P, Rashid S J, Udrea F and Amaratunga G A J 2006 Normally-off trench JFET technology in 4 H silicon carbide *Microelectron. Eng.* **83** 107–11

Nawaz M, Zaing C, Bource J, Schupbach M, Domeij M, Lee H-S and Östling M 2009 Assessment of high and low temperature performance of SiC BJTs *Mater. Sci. Forum* **615–617** 825–8

Neudeck P G, Garverick S L, Spry D J, Chen L-Y, Beheim G M, Krasowski M J and Mehregany M 2009 Extreme temperature 6H-SiC JFET integrated circuit technology *Phys. Status Solidi* A **206** 2329–45

Neudeck P G, Spry D J, Chen L-Y, Beheim G M and Okojie R S *et al* 2008 Stable electrical operation of 6H-SiC JFETs and ICs for thousands of hours at 500°C *IEEE Electron Device Lett.* **29** 456–9

O'Mahony D, Duane R, Campagno T, Lewis L, Cordero N, Maaskant P, Waldron F and Corbett B 2011 Thermal stability of SiC Schottky diode anode and cathode metalisations after 1000 h at 350 °C *Microelectron. Reliab.* **51** 904–8

Ozpineci B and Tolbert L M 2003 Characterization of SiC Schottky diodes at different temperatures *IEEE Power Electron. Lett.* **1** 54–7

Patil, Fu X-A, Anupongongarch C, Mehregany M and Garverick S L 2007 Characterization of silicon carbide differential amplifiers at high temperature CSIC 2007 *IEEE Compound Semiconductor Integrated Circuit Symposium (Portland, OR, 14–17 October)* pp 1–4

Patil A C, Fu X-A, Anupongongarch C, Mehregany M and Garverick S L 2009 6H-SiC JFETs for 450 °C differential sensing applications *J. Microelectromech. Syst.* **18** 950–61

Peters D, Schörner R, Hölzlein K-H and Friedrichs P 1997 Planar aluminum-implanted 1400 V 4 H silicon carbide p–n diodes with low on resistance *Appl. Phys. Lett.* **71** 2996–7

Shao S, Lien W-C, Maralani A and Pisano A P 2014 Integrated 4H-silicon carbide diode bridge rectifier for high temperature (773 K) environment *44th European Solid State Device Research Conf. (Venice, 22–26 September)* pp 138–41

Sheng K, Yu L C, Zhang J and Zhao J H 2005 High temperature characterization of SiC BJTs for power switching applications *Int. Semiconductor Device Research Symp. (Bethesda, MD, 7–9 December)* pp 168–9

Siemieniec R and Kirchner U 2011 The 1200 V direct-driven SiC JFET power switch *EPE 2011 (Birmingham)* pp 1–10

Soong C-W, Patil A C, Garverick S L, Fu X and Mehregany M 2012 550 °C integrated logic circuits using 6H-SiC JFETs *IEEE Electron Device Lett.* **33** 1369–71

Sundaresan S G, Sturdevant C, Marripelly M, Lieser E and Singh R 2012 12.9 kV SiC PiN diodes with low on-state drops and high carrier lifetimes *Mater. Sci. Forum* **717–720** 949–52

Takayama D, Sugawara Y, Hayashi T, Singh R, Palmour J, Ryu S and Asano K 2001 Static and dynamic characteristics of 4–6 kV 4H-SiC SIAFETs *Proc. 2001 Int. Symp. on Power Semiconductor Devices and ICs (Osaka)* pp 41–4

Tanaka Y, Okamoto M, Takatsuka A, Arai K, Yatsuo T, Yano K and Kasuga M 2006 700-V 1.0-mΩ-cm^2 buried gate SiC-SIT (SiC-BGSIT) *IEEE Electron Device Lett.* **27** 908–10

Testa A, De Caro S, Russo S, Patti D and Torrisi L 2011 High temperature long term stability of SiC Schottky diodes *Microelectron. Reliab.* **51** 1778–82

Wu J, Fursin L, Li Y, Alexandrov P, Weiner M and Zhao J H 2006 4.3 kV 4H-SiC merged PiN/Schottky diodes *Semicond. Sci. Technol.* **21** 987

Zhang J, Alexandrov P, Burke T and Zhao J H 2006 4H-SiC power bipolar junction transistor with a very low specific on-resistance of 2.9 mΩ.cm^2 *IEEE Electron Device Lett.* **27** 368–70

Zhang J, Luo Y, Alexandrov P, Fursin L and Zhao J H 2003 A high current gain 4H-SiC NPN power bipolar junction transistor *IEEE Electron Device Lett.* **24** 327–9

Zhang J H, Fursin L, Li X Q, Wang X H, Zhao J H, VanMil B L, Myers-Ward R L, Eddy C R and Gaskill D K 2009 4H-SiC bipolar junction transistors with graded base doping profile *Mater. Sci. Forum* **615–617** 829–32

Zhang J, Zhao J H, Alexandrov P and Burke T 2004 Demonstration of first 9.2 kV 4H-SiC bipolar junction transistor *Electron. Lett.* **40** 1381–82

Chapter 9

Gallium nitride electronics for very hot environments

Gallium nitride is of interest not only for optical devices covering the full visible spectrum and stretching far into the ultraviolet (UV) region, but also for microwave power devices capable of operating at much higher temperatures compared to traditional silicon and gallium arsenide components. In GaN electronics, the device that has drawn the most attention is the high electron mobility transistor (HEMT). Difficulty in the p-doping of GaN impedes bipolar device development. Therefore, investigations have primarily focused on MESFETs, MISFETs, HBTs and HEMTs. Amongst these, the HEMTs offer better carrier transport properties than MESFETs, and have consequently received more consideration. While AlGaN/GaN HEMTs could not withstand temperatures $\sim$500 °C, the use of lattice-matched InAlN/GaN heterojunctions was found to extend the capability up to 1000 °C owing to non-existence of mechanical strain. Polarization could be preserved in the heterostructures up to 1000 °C. Thus ferroelectric polarization instabilities were avoided, and thermal and chemical stabilities comparable to ceramic materials could be attained.

9.1 Introduction

A side-by-side comparison of the properties of gallium nitride with silicon and silicon carbide reveals that the bandgap of gallium nitride (3.39 eV) is 3 times that of silicon (1.12 eV) and 1.05 times that of silicon carbide (3.23 eV). Insofar as high-temperature operation is concerned, gallium nitride compares well with silicon carbide but is far ahead of silicon, giving temperatures above 600 °C. From the viewpoint of thermal conductivity, gallium nitride (1.3 W cmK^{-1}) is slightly inferior to silicon (1.5 W cmK^{-1}) but much inferior to silicon carbide (4.9 W cmK^{-1}). The electron mobility (1000 cm^2 Vs^{-1}) of GaN is of the same order as that of SiC (800–900 cm^2 Vs^{-1}) as is its hole mobility (200 cm^2 Vs^{-1} for GaN) against 115 cm^2 Vs^{-1} for SiC. High-speed ICs for high-temperature applications use AlGaN/GaN

doi:10.1088/978-0-7503-1155-7ch9

heterojunction FETs (HFETs), also called HEMTs. In table 9.1, the physical properties of wurtzite gallium nitride structure routinely required for design computations of GaN devices and circuits are compiled; the other structure of GaN is the zinc blende variety. The wurtzite crystalline structure is depicted in figure 9.1(a) and the zinc blende structure in figure 9.1(b).

9.2 Intrinsic temperature of gallium nitride

Following in the footsteps of the previously considered semiconductors Si, GaAs and 4H-SiC, it is worthwhile to calculate the intrinsic temperature of gallium nitride. For GaN,

$$m_{\mathrm{n}}^{*} = 0.20m_0,\ m_{\mathrm{p}}^{*} = 1.5m_0,\ E_{\mathrm{g}} = 3.39\ \mathrm{eV}$$

$$m^{*} = \left(\frac{m_{\mathrm{n}}^{*}m_{\mathrm{p}}^{*}}{m_0^2}\right)^{0.75} = \left(\frac{0.20m_0 \times 1.5m_0}{m_0^2}\right)^{0.75} = 0.4054. \tag{9.1}$$

For GaN, the equation

$$\ln\{0.207/(m^{*}T^{1.5})\} = -5.8025 \times 10^3 E_{\mathrm{g}} \tag{9.2}$$

is recast as

$$T\ln\{0.207/(0.4054T^{1.5})\} = -5.8025 \times 10^3 \times 3.39 = -19670.475$$

or

$$T\ln 0.51061 - T\ln T^{1.5} = -19670.475 \tag{9.3}$$

$$\therefore -0.672T - T\ln T^{1.5} = -19670.475. \tag{9.4}$$

Suppose,

$$T = 1660\ \mathrm{K}$$

$$\mathrm{lhs} = 0.672 \times 1660 - 1660\ln 1660^{1.5} = -1115.52 - 18462.286 = -19577.806. \tag{9.5}$$

If

$$T = 1665\ \mathrm{K}$$

$$\mathrm{lhs} = 0.672 \times 1665 - 1665\ln 1665^{1.5} = -1118.88 - 18525.407 = -19644.287. \tag{9.6}$$

When

$$T = 1667\ \mathrm{K}$$

$$\begin{aligned}\mathrm{lhs} &= 0.672 \times 1667 - 1667\ln 1667^{1.5} = -1120.22 - 18550.662 \\ &= -19670.882.\end{aligned} \tag{9.7}$$

$$\therefore T \approx 1667\ \mathrm{K}. \tag{9.8}$$

Table 9.1. Properties of gallium nitride (wurtzite).

Property	Value	Property	Value	Property	Value
Chemical formula	GaN	Lattice constant (Å)	A = 3.186 Å, c = 5.186 Å		
Molecular mass (g mol^{-1})	83.73	Melting point (°C)	2500	Electron mobility (cm^2 Vs^{-1})	1000
Classification	III–V compound semiconductor	Dielectric constant	9.5 (static) 5.35 (high frequency)	Hole mobility (cm^2 Vs^{-1})	200
Crystal structure	Wurtzite	Thermal conductivity (W cmK^{-1})	1.3	Electron diffusion coefficient (cm^2 s^{-1})	25
Color	Yellow	Energy bandgap E_g (eV) at 300 K	3.39 (direct)	Hole diffusion coefficient (cm^2 s^{-1})	5
Density at 300 K (g cm^{-3})	6.15	Electrical breakdown field (V cm^{-1})	5×10^6	Electron saturated velocity (cm s^{-1})	2×10^7
		Intrinsic carrier concentration (cm^{-3})	1.9×10^{-10}	Minority-carrier lifetime (s)	10^{-8}

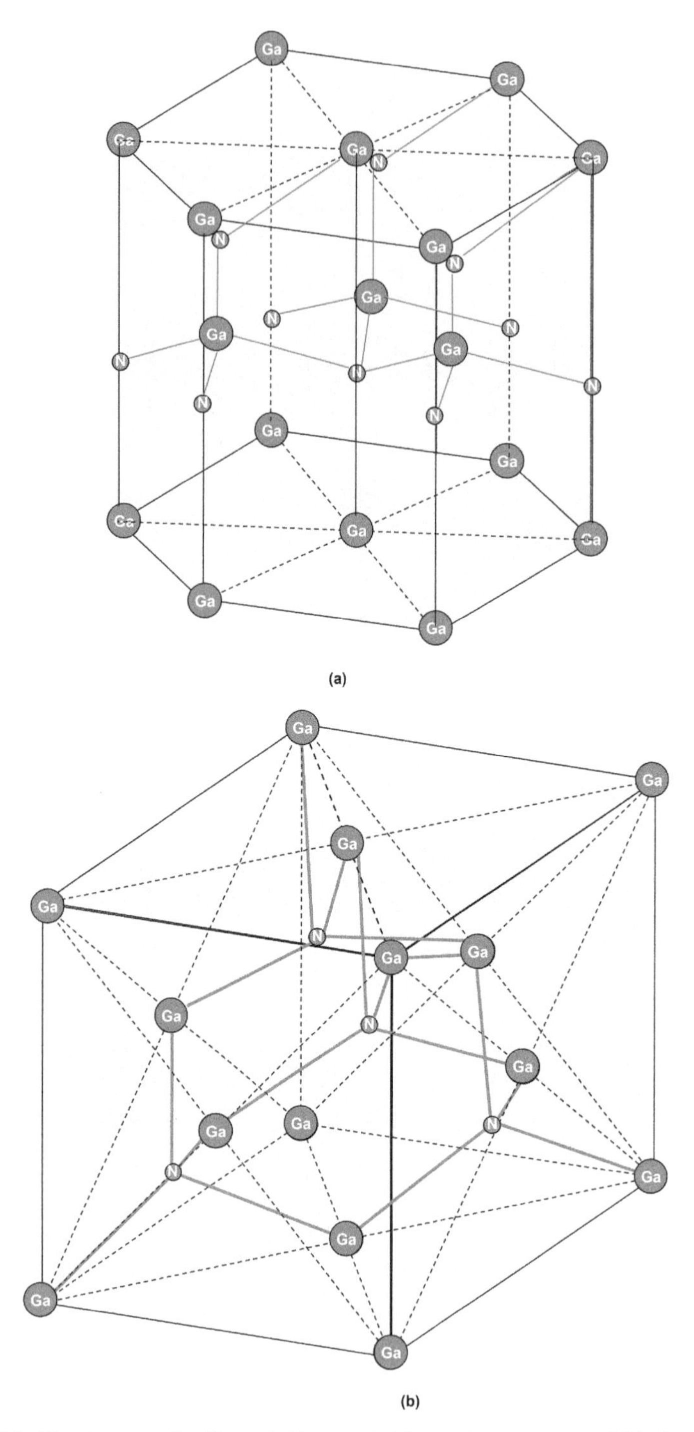

Figure 9.1. The structure of gallium nitride crystal: (a) wurtzite structure and (b) zinc blende.

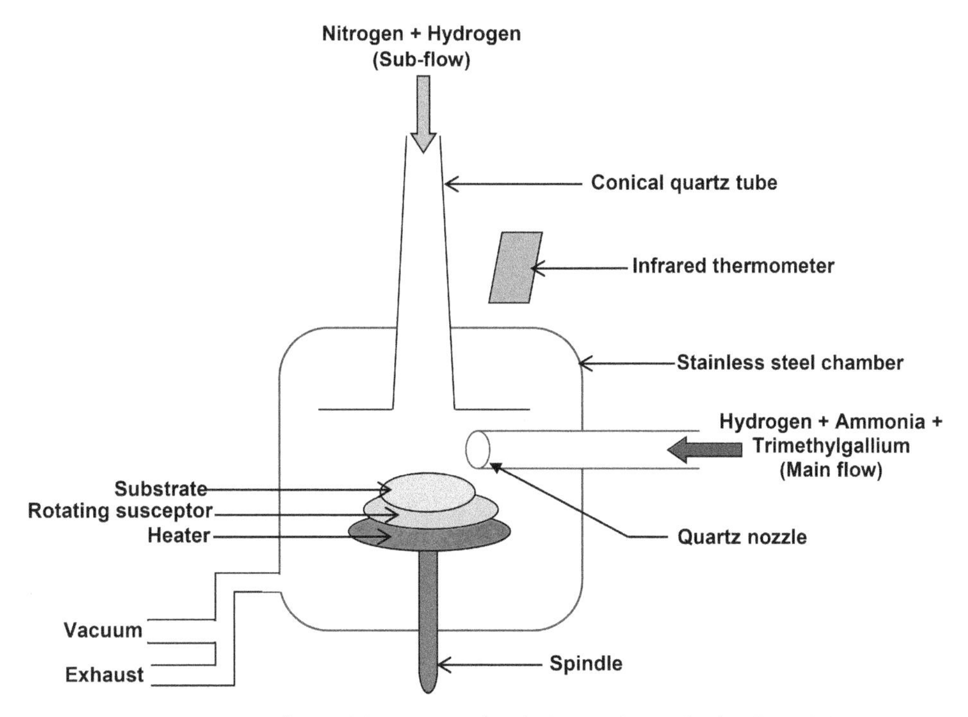

Figure 9.2. A two-flow MOCVD reactor for single-crystal growth of gallium nitride.

The intrinsic temperature for gallium nitride (1667 K) is slightly higher, by 1667 − 1591.5 = 75.5 K, than SiC (1591.5 K) but much higher, by 1667 − 588.63 = 1078.37 K, for Si (588.63 K).

9.3 Growth of the GaN epitaxial layer

GaN, AlN, AlGaN/GaN and InGaN heterostructures are grown epitaxially by the metal-organic CVD (MOCVD) technique (figure 9.2). Substrates used include silicon carbide, sapphire, silicon, etc. Silicon carbide is preferred due to its high thermal conductivity. Sapphire and silicon are low-cost materials. A resistive AlN nucleation layer is used to isolate the devices from silicon or silicon carbide. The non-uniformities on 100 mm diameter SiC substrates are <2%. The precursors for Ga, Al, In and N are trimethylyl gallium, trimethylaluminum, trimethylindium and ammonia, respectively. Growth is carried out at a typical pressure of 76 Torr. The growth temperature is ~1000 °C. The carrier gas for metal organic compounds is hydrogen. Flow rate of trimethylgallium is generally kept at 1–10 mmoles min^{-1}; that of trimythylaluminum at 0.6–1.5 mmoles min^{-1}.

9.4 Doping of GaN

The carrier concentration in the deposited films is $<10^{15}$ cm^{-3}. The resistive film can be doped as n-type with silicon as the donor impurity using disilane. Concentrations up to 10^{19} cm^{-3} are easily attainable. p-type doping is performed with magnesium as the acceptor impurity. As the energy level of magnesium is 160 meV above the

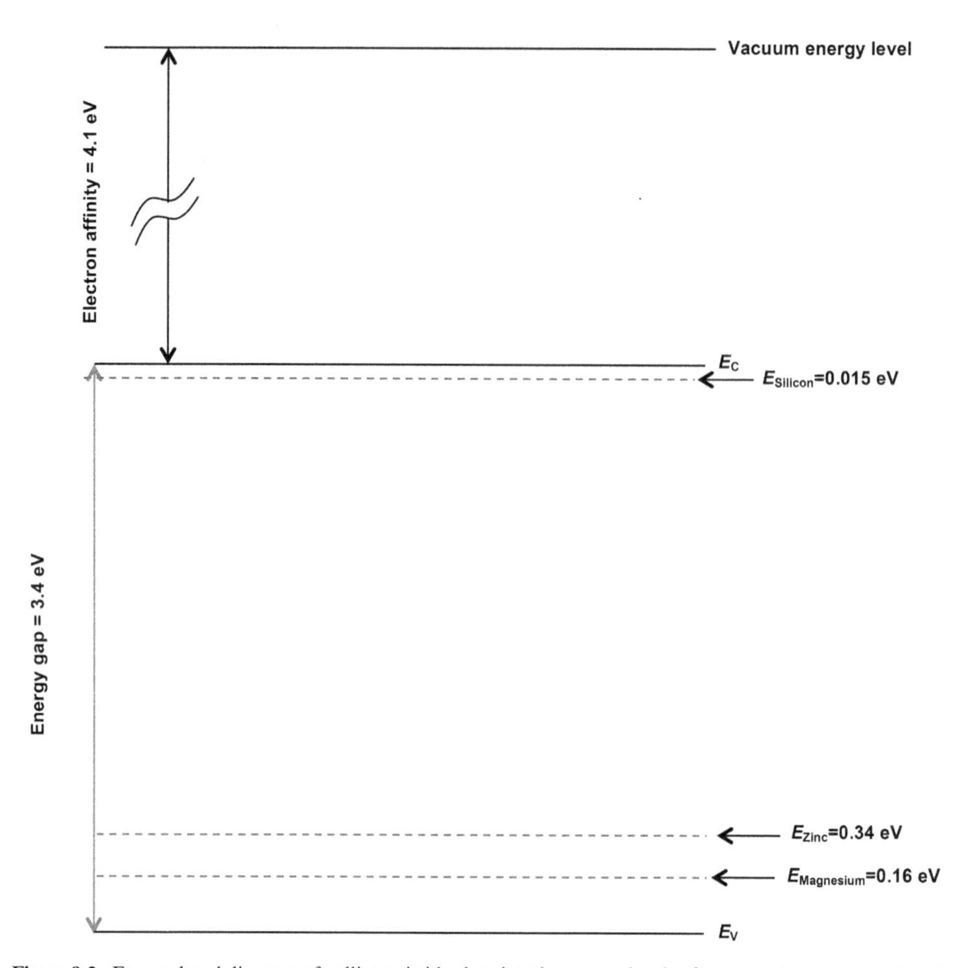

Figure 9.3. Energy band diagram of gallium nitride showing the energy levels of commonly used impurities for n- (Si) and p-doping (Zn, Mg).

valence band of GaN, only a small fraction <0.01 of the dopant is ionized at room temperature producing a maximum concentration up to 10^{18} cm^{-3}. The energy band diagram of gallium nitride is shown in figure 9.3.

9.5 Ohmic contacts to GaN

9.5.1 Ohmic contacts to n-type GaN

Ohmic contacts to n-type GaN are easily realized. Composite metal layer Ti/Al/Ni/Au (15 nm/220 nm/40 nm/50 nm) contact on reactive-ion etched n-type GaN (4×10^{17} cm^{-3}) surface, followed by RTA at 900 °C for 30 s in N_2 ambient, gives a low resistance (= 8.9×10^{-8} Ω cm^2). Here, Ti and Ni are e-beam evaporated; Al and Au are thermally evaporated (Fan *et al* 1996, Ruvimov *et al* 1996). The reaction of GaN with Ti forms TiN, which produces a large number of n-vacancies acting as donors in GaN. The resultant increase of donor density decreases the contact resistance. The Ni layer between Al and Au serves as a diffusion barrier between these layers, otherwise Al will react with Au to form $AlAu_4$. It is viscous at low temperatures and its lateral flow may lead to shorting.

A low contact resistance $\sim 10^{-4}\ \Omega\ cm^2$ was obtained on n^+ GaN with impurity concentration $\geqslant 10^{19}\ cm^{-3}$ by sputtering 500–1200 Å thick $WSi_{0.45}$ (Pearton *et al* 1998). The metallization was stable up to 1000 °C.

9.5.2 Ohmic contacts to p-type GaN

Low resistance ohmic contacts to p-type GaN are more difficult to accomplish, mainly due to the low acceptor concentrations available, the rarity of high work function metals, and the presence of compensating hydrogen interstitials and nitrogen vacancies. Commonly used metals are Pd, Pt, Ni or a high work function metal. This metal is covered with a gold layer to prevent oxidation. Pd/Ag/Au/Ti/Au (1 nm/50 nm/10 nm/30 nm/20 nm) contacts deposited by e-beam evaporation and subsequently annealed at 800 °C for 1 min give a resistance = $1 \times 10^{-6}\ \Omega\ cm^2$ due to formation of an alloy between Ag, Au and p-type GaN (Adivarahan *et al* 2001) producing a p^+ region at the metal–semiconductor interface.

9.6 Schottky contacts to GaN

Multilayer Au/Pt/Ti (50 Å/300 Å/1500 Å) gave a Schottky barrier height of 0.84 eV in the as-deposited condition on an n-GaN layer (Macherzyński *et al* 2009). RTA in a hydrogen/nitrogen mixture in the ratio 1:10 at 300 °C for 20 s decreased the barrier height to 0.61 eV. The barrier height decreased further to 0.48 eV when the annealing was performed at a temperature from 400 °C to 700 °C.

9.7 GaN MESFET model with hyperbolic tangent function

In this model, the drain–source current I_{DS} is approximately represented for large-signal devices by a simpler version of the Shockley equation in terms of a hyperbolic tangent function (Kacprzak and Materka 1983, Kabra *et al* 2008, Shashikala and Nagabhushana 2010); see figure 9.4. The equation given by Kacprzak and Materka (1983) is

$$I_{DS} = I_{DSS}\left(1 - \frac{V_{GS}}{V_{Th} + \gamma V_{DS}}\right)^2 \tanh\left(\frac{\alpha V_{DS}}{V_{Th} - \gamma V_{DS}}\right). \tag{9.9}$$

This equation was modified by Kabra *et al* (2008) by replacing V_{GS} by $(V_{GS}-V_{bi})$, and for temperature dependence as follows:

$$I_{DS} = I_{DSS}\left(\frac{T_0}{T}\right)\left(1 - \frac{V_{GS} - V_{bi}}{V_{Th} + \gamma V_{DS}}\right)^2 \tanh\left(\frac{\alpha V_{DS}}{V_{Th} - \gamma V_{DS}}\right), \tag{9.10}$$

where I_{DSS} is the saturated drain current at $V_{GS} = 0$, T_0 is room temperature (300 K), T is the temperature at which the calculation is performed, V_{Th} is the threshold voltage, $\alpha = -0.56$ is an empirical constant determining the saturation

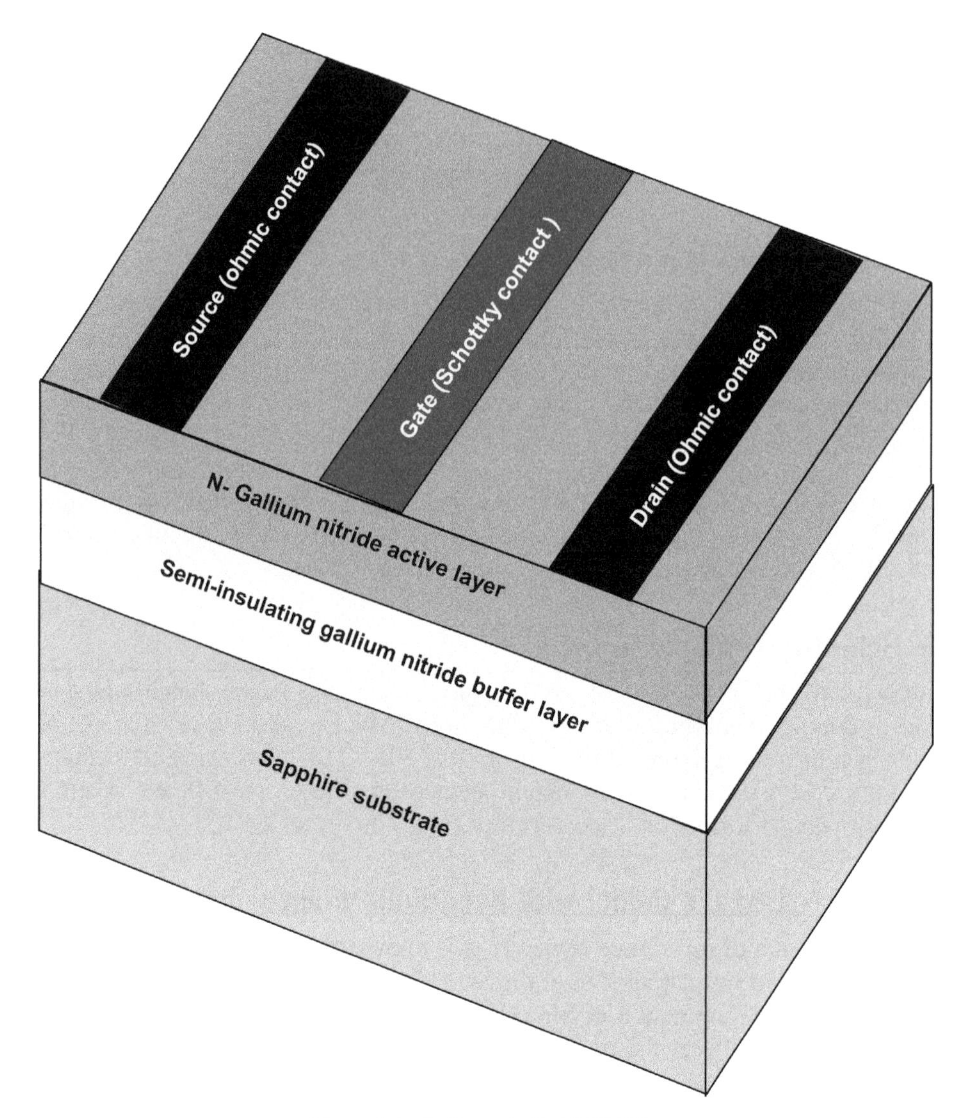

Figure 9.4. Gallium nitride MESFET.

voltage of the drain current and $\gamma = -0.0001$ represents the effective threshold voltage displacement with V_{DS} (Kabra *et al* 2008). The current I_{DSS} is given by

$$I_{DSS} = (I_{fc}/2)\left[\left(2 + \frac{R_s I_{fc}}{V_p}\right) - \sqrt{\left(2 + \frac{R_s I_{fc}}{V_p}\right)^2 - \left(\frac{4}{V_p}\right)\{V_p - (V_{bi} + V_{GS})\}}\right], \quad (9.11)$$

where $R_s = 160\ \Omega$ is the series resistance of the source, I_{fc} is the full saturation current expressed in terms of channel width W and the temperature-dependent variables namely the saturation velocity $v_{sat}(T)$ and the conducting channel charge $Q_a(T)$ as

$$I_{fc} = qWv_{sat}(T)Q_a(T) \quad (9.12)$$

with

$$v_{sat}(T) = v_{sat}(T_0) - \sigma_v T \tag{9.13}$$

$$Q_a(T) = N_D \sqrt{\frac{2\varepsilon_0 \varepsilon_s(T) V_p}{q N_D}}, \tag{9.14}$$

where σ_v is an empirical constant = 98 ms^{-1} K^{-1} and $v_{sat}(T_0)$ is the saturation velocity at a temperature T_0(= 300 K) = 2.8743 × 10^7 cm s^{-1}. The symbols V_{bi}, V_p and V_1 have the same meanings as for GaAs (see equations (7.15), (7.16), (7.18) and (7.19)) with proper modifications for the differences between GaN and GaAs, e.g. in the equation for V_p, the relative permittivity ε_s is replaced by a temperature-dependent dielectric constant defined by an identical equation as saturation velocity

$$\varepsilon_s(T) = \varepsilon_s(T_0) - \sigma_s T, \tag{9.15}$$

where σ_s is an empirical constant = 10^{-4} K^{-1}. The dielectric constant $\varepsilon_s(T_0)$ at T_0(= 300 K) is 9. Furthermore, the intrinsic carrier concentration $n_i(T)$ is written in terms of the effective density of states in the conduction band $N_c(T)$, the effective density of states in the valence band $N_v(T)$ and the energy gap $E_g(T)$ of the semiconductor as

$$n_i(T) = \sqrt{N_c(T) N_V(T)} \exp\left\{-\frac{q E_g(T)}{2 k_B T}\right\} \tag{9.16}$$

$$N_c(T) = N_c(T_0) T^{1.5} = 4.3 \times 10^{14} T^{1.5} \text{ cm}^{-3} \text{ K}^{-3/2} \tag{9.17}$$

$$N_V(T) = N_V(T_0) T^{1.5} = 8.9 \times 10^{15} T^{1.5} \text{ cm}^{-3} \text{ K}^{-3/2} \tag{9.18}$$

$$E_g(T) = E_g(T_0) - \frac{\eta_a T^2}{T + \eta_b} \text{ (Varshini expression)} \tag{9.19}$$

$N_c(T_0)$ = effective density of states in the conduction band of GaN at T_0(= 300 K)
$N_V(T_0)$ = effective density of states in the valence band of GaN at T_0(= 300 K)
$E_g(T_0)$ = energy bandgap of GaN at T_0(= 300 K) = 3.427 eV
η_a = an empirical constant = 9.39 × 10^{-4} eV K^{-1}
η_b = an empirical constant = 772 K.

Transconductance g_m is obtained by differentiating the I_{DS} expression with respect to V_{GS}

$$
\begin{aligned}
g_{\mathrm{m}} &= \left(\frac{\partial I_{\mathrm{DS}}}{\partial V_{\mathrm{GS}}}\right)_{V_{\mathrm{DS}}=\mathrm{constant}} \\
&= \frac{\partial}{\partial V_{\mathrm{GS}}}\left\{I_{\mathrm{DSS}}\left(\frac{T_0}{T}\right)\left(1-\frac{V_{\mathrm{GS}}-V_{\mathrm{bi}}}{V_{\mathrm{Th}}+\gamma V_{\mathrm{DS}}}\right)^2 \tanh\left(\frac{\alpha V_{\mathrm{DS}}}{V_{\mathrm{Th}}-\gamma V_{\mathrm{DS}}}\right)\right\} \\
&= \left(\frac{T_0}{T}\right)\tanh\left(\frac{\alpha V_{\mathrm{DS}}}{V_{\mathrm{Th}}-\gamma V_{\mathrm{DS}}}\right) \times \frac{\partial}{\partial V_{\mathrm{GS}}}\left\{I_{\mathrm{DSS}}\left(1-\frac{V_{\mathrm{GS}}-V_{\mathrm{bi}}}{V_{\mathrm{Th}}+\gamma V_{\mathrm{DS}}}\right)^2\right\} \\
&= \left(\frac{T_0}{T}\right)\tanh\left(\frac{\alpha V_{\mathrm{DS}}}{V_{\mathrm{Th}}-\gamma V_{\mathrm{DS}}}\right)\left\{\frac{\mathrm{d}I_{\mathrm{DSS}}}{\mathrm{d}V_{\mathrm{GS}}} \times \left(1-\frac{V_{\mathrm{GS}}-V_{\mathrm{bi}}}{V_{\mathrm{Th}}+\gamma V_{\mathrm{DS}}}\right)^2 + I_{\mathrm{DSS}}\right. \\
&\quad \left. \times 2\left(1-\frac{V_{\mathrm{GS}}-V_{\mathrm{bi}}}{V_{\mathrm{Th}}+\gamma V_{\mathrm{DS}}}\right)\left(0-\frac{1-0}{V_{\mathrm{Th}}+\gamma V_{\mathrm{DS}}}\right)\right\} \\
&= \left(\frac{T_0}{T}\right)\tanh\left(\frac{\alpha V_{\mathrm{DS}}}{V_{\mathrm{Th}}-\gamma V_{\mathrm{DS}}}\right)\left(1-\frac{V_{\mathrm{GS}}-V_{\mathrm{bi}}}{V_{\mathrm{Th}}+\gamma V_{\mathrm{DS}}}\right) \\
&\quad \times \left\{\frac{\mathrm{d}I_{\mathrm{DSS}}}{\mathrm{d}V_{\mathrm{GS}}} \times \left(1-\frac{V_{\mathrm{GS}}-V_{\mathrm{bi}}}{V_{\mathrm{Th}}+\gamma V_{\mathrm{DS}}}\right) + I_{\mathrm{DSS}} \times 2\left(-\frac{1}{V_{\mathrm{Th}}+\gamma V_{\mathrm{DS}}}\right)\right\} \\
&= \left(\frac{T_0}{T}\right)\tanh\left(\frac{\alpha V_{\mathrm{DS}}}{V_{\mathrm{Th}}-\gamma V_{\mathrm{DS}}}\right)\left(1-\frac{V_{\mathrm{GS}}-V_{\mathrm{bi}}}{V_{\mathrm{Th}}+\gamma V_{\mathrm{DS}}}\right) \\
&\quad \times \left\{\frac{\mathrm{d}I_{\mathrm{DSS}}}{\mathrm{d}V_{\mathrm{GS}}}\left(1-\frac{V_{\mathrm{GS}}-V_{\mathrm{bi}}}{V_{\mathrm{Th}}+\gamma V_{\mathrm{DS}}}\right) - \frac{2I_{\mathrm{DSS}}}{V_{\mathrm{Th}}+\gamma V_{\mathrm{DS}}}\right\},
\end{aligned}
\tag{9.20}
$$

where from equation (9.11),

$$
\begin{aligned}
\frac{\mathrm{d}I_{\mathrm{DSS}}}{\mathrm{d}V_{\mathrm{GS}}} &= (I_{\mathrm{fc}}/2) \times \frac{\mathrm{d}}{\mathrm{d}V_{\mathrm{GS}}}\left[\left(2+\frac{R_{\mathrm{s}}I_{\mathrm{fc}}}{V_{\mathrm{p}}}\right) - \sqrt{\left(2+\frac{R_{\mathrm{s}}I_{\mathrm{fc}}}{V_{\mathrm{p}}}\right)^2 - \left(\frac{4}{V_{\mathrm{p}}}\right)\{V_{\mathrm{p}}-(V_{\mathrm{bi}}+V_{\mathrm{GS}})\}}\right] \\
&= (I_{\mathrm{fc}}/2) \times \frac{\mathrm{d}}{\mathrm{d}V_{\mathrm{GS}}}\left(2+\frac{R_{\mathrm{s}}I_{\mathrm{fc}}}{V_{\mathrm{p}}}\right) - (I_{\mathrm{fc}}/2) \\
&\quad \times \frac{\mathrm{d}}{\mathrm{d}V_{\mathrm{GS}}}\left[\sqrt{\left(2+\frac{R_{\mathrm{s}}I_{\mathrm{fc}}}{V_{\mathrm{p}}}\right)^2 - \left(\frac{4}{V_{\mathrm{p}}}\right)\{V_{\mathrm{p}}-(V_{\mathrm{bi}}+V_{\mathrm{GS}})\}}\right] \\
&= 0 - (I_{\mathrm{fc}}/2) \times (1/2)\left[\left(2+\frac{R_{\mathrm{s}}I_{\mathrm{fc}}}{V_{\mathrm{p}}}\right)^2 - \left(\frac{4}{V_{\mathrm{p}}}\right)\{V_{\mathrm{p}}-(V_{\mathrm{bi}}+V_{\mathrm{GS}})\}\right]^{1/2-1} \\
&\quad \times -\left(\frac{4}{V_{\mathrm{p}}}\right) \times \frac{\mathrm{d}}{\mathrm{d}V_{\mathrm{GS}}}\{V_{\mathrm{p}}-(V_{\mathrm{bi}}+V_{\mathrm{GS}})\} \\
&= -(I_{\mathrm{fc}}/2) \times (1/2)\left[\left(2+\frac{R_{\mathrm{s}}I_{\mathrm{fc}}}{V_{\mathrm{p}}}\right)^2 - \left(\frac{4}{V_{\mathrm{p}}}\right)\{V_{\mathrm{p}}-(V_{\mathrm{bi}}+V_{\mathrm{GS}})\}\right]^{1/2-1} \times -\left(\frac{4}{V_{\mathrm{p}}}\right) \times -1 \\
&= -\frac{I_{\mathrm{fc}}/V_{\mathrm{p}}}{\sqrt{\left(2+\frac{R_{\mathrm{s}}I_{\mathrm{fc}}}{V_{\mathrm{p}}}\right)^2 - \left(\frac{4}{V_{\mathrm{p}}}\right)\{V_{\mathrm{p}}-(V_{\mathrm{bi}}+V_{\mathrm{GS}})\}}}.
\end{aligned}
\tag{9.21}
$$

Output conductance g_d is found by differentiation of the expression for I_{DS} with respect to V_{DS}

$$\begin{aligned}
\left(\frac{\partial I_{DS}}{\partial V_{DS}}\right)_{V_{GS}=\text{constant}} &= \frac{\partial}{\partial V_{DS}}\left\{I_{DSS}\left(\frac{T_0}{T}\right)\left(1-\frac{V_{GS}-V_{bi}}{V_{Th}+\gamma V_{DS}}\right)^2\tanh\left(\frac{\alpha V_{DS}}{V_{Th}-\gamma V_{DS}}\right)\right\} \\
&= I_{DSS}\left(\frac{T_0}{T}\right)\times\frac{d}{dV_{DS}}\left\{\left(1-\frac{V_{GS}-V_{bi}}{V_{Th}+\gamma V_{DS}}\right)^2\tanh\left(\frac{\alpha V_{DS}}{V_{Th}-\gamma V_{DS}}\right)\right\} \\
&= I_{DSS}\left(\frac{T_0}{T}\right)\left[2\left(1-\frac{V_{GS}-V_{bi}}{V_{Th}+\gamma V_{DS}}\right)^{2-1}\left\{0-\frac{0-\gamma(V_{GS}-V_{bi})}{(V_{Th}+\gamma V_{DS})^2}\right\}\right. \\
&\quad\times\tanh\left(\frac{\alpha V_{DS}}{V_{Th}-\gamma V_{DS}}\right)+\left(1-\frac{V_{GS}-V_{bi}}{V_{Th}+\gamma V_{DS}}\right)^2\operatorname{sech}^2\left(\frac{\alpha V_{DS}}{V_{Th}-\gamma V_{DS}}\right) \\
&\quad\left.\times\frac{\alpha(V_{Th}-\gamma V_{DS})+\gamma\alpha V_{DS}}{(V_{Th}-\gamma V_{DS})^2}\right] \qquad (9.22) \\
&= I_{DSS}\left(\frac{T_0}{T}\right)\left(1-\frac{V_{GS}-V_{bi}}{V_{Th}+\gamma V_{DS}}\right)\left[2\tanh\left(\frac{\alpha V_{DS}}{V_{Th}-\gamma V_{DS}}\right)\left\{\frac{\gamma(V_{GS}-V_{bi})}{(V_{Th}+\gamma V_{DS})^2}\right\}\right. \\
&\quad\left.+\left(1-\frac{V_{GS}-V_{bi}}{V_{Th}+\gamma V_{DS}}\right)\operatorname{sech}^2\left(\frac{\alpha V_{DS}}{V_{Th}-\gamma V_{DS}}\right)\times\frac{\alpha V_{Th}}{(V_{Th}-\gamma V_{DS})^2}\right] \\
&= I_{DSS}\left(\frac{T_0}{T}\right)\left(1-\frac{V_{GS}-V_{bi}}{V_{Th}+\gamma V_{DS}}\right)\left[2\tanh\left(\frac{\alpha V_{DS}}{V_{Th}-\gamma V_{DS}}\right)\left\{\frac{\gamma(V_{GS}-V_{bi})}{(V_{Th}+\gamma V_{DS})^2}\right\}\right. \\
&\quad\left.+\left(1-\frac{V_{GS}-V_{bi}}{V_{Th}+\gamma V_{DS}}\right)\left\{1-\tanh^2\left(\frac{\alpha V_{DS}}{V_{Th}-\gamma V_{DS}}\right)\right\}\frac{\alpha V_{Th}}{(V_{Th}-\gamma V_{DS})^2}\right].
\end{aligned}$$

The cut-off frequency f_T of the MESFET is

$$f_T = g_m/\{2\pi(C_{gs}+C_{gd})\}, \qquad (9.23)$$

where C_{gs} is the gate–source capacitance

$$C_{gs} = \{\varepsilon_0\varepsilon_s(T)WL\}/[a - I_{DS}/\{qN_D v_{sat}(T)W\}], \qquad (9.24)$$

where a is the thickness of the active n-GaN layer, and W and L stand for channel width and length, respectively. C_{gd} is the gate–drain capacitance

$$C_{gd} = \{\varepsilon_0\varepsilon_s(T)\pi W_S L\}/(2W_D), \qquad (9.25)$$

where W_S is the depletion region thickness at the source and W_D is the depletion region thickness at the drain. The variation of the relative permittivity of the semiconductor and the saturation velocity of the charge carriers with temperature are described by the aforementioned equations (9.15) and (9.13) respectively.

The maximum frequency f_{max} of oscillations is

$$f_{max} = (kf_T)/\sqrt{g_d(R_S + R_G)}, \tag{9.26}$$

where $k = 0.34$ is constant of proportionality, $R_S = 160\ \Omega$ is the source resistance and $R_G = 10\ \Omega$ is the gate resistance.

The I_{DS}–V_{GS} characteristics of the GaN MESFET simulated from this model, along with the various MESFET parameters such as transconductance, output conductance, cut-off frequency and maximum frequency of oscillation, were found to agree well with experimental results up to 200 °C (Kabra *et al* 2008).

9.8 AlGaN/GaN HEMTs

9.8.1 Operation of AlGaN/GaN HEMTs on 4H-SiC/sapphire substrates from 25 °C to 500 °C

A comparative experimental study was performed to understand the high-temperature performance of $Al_{0.26}Ga_{0.74}N$/GaN HEMTs fabricated on 4H-SiC and sapphire substrates using atmospheric pressure MOCVD (Arulkumaran *et al* 2002). HEMT fabricated on a SiC substrate is shown in figure 9.5. The devices comprised the

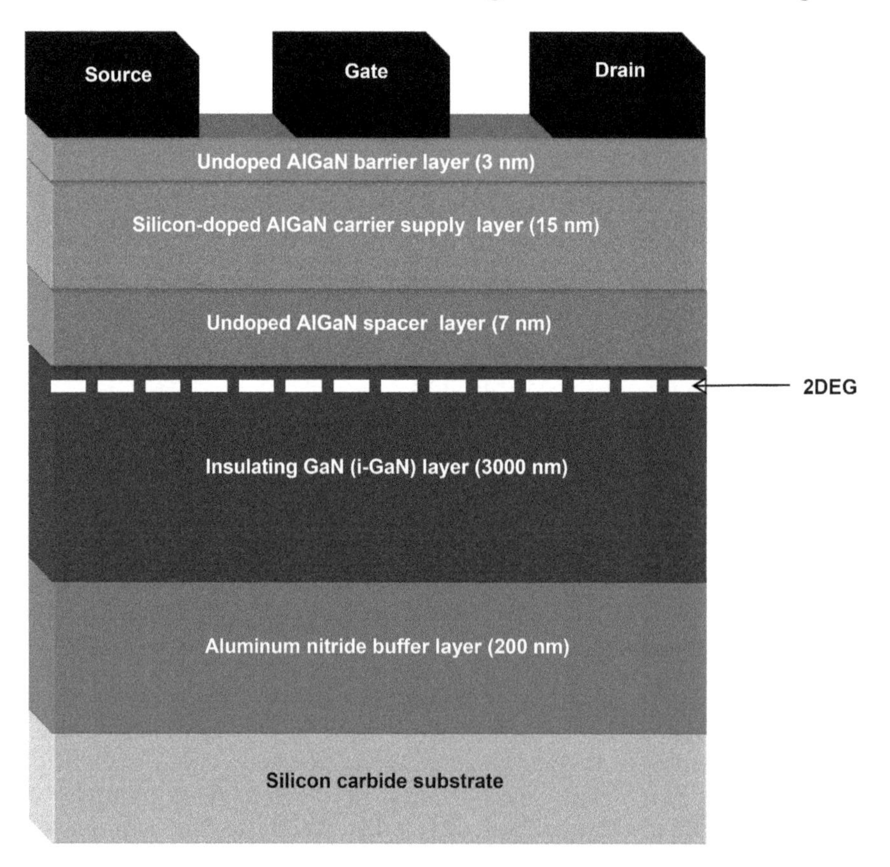

Figure 9.5. The structure of AlGaN/GaN HEMT on a SiC substrate showing the several constituent layers.

following successive layers: an AlGaN barrier layer (3 nm thick and undoped), an AlGaN carrier supply layer (15 nm thick, Si-doped), an AlGaN spacer layer (7 nm thick, undoped), an i-GaN layer (3000 nm thick, insulating), a buffer layer (200 nm thick AlN for the SiC substrate/30 nm thick i-GaN for the sapphire substrate) and the SiC/sapphire substrate. Gate metallization was Pd/Ti/Au (40 nm/40 nm/80 nm) of proven capability up to 500 °C. After being subjected to thermal stress at 500 °C, HEMTs fabricated on 4H-SiC substrates showed superior DC characteristics compared to those on sapphire substrates. For both types of HEMTs, the transconductance g_m and the drain current I_D decreased as the temperature was increased, mainly because of the reduction in two-dimensional electron gas (2DEG) mobility and carrier velocity. The general trends for HEMTs on SiC substrates are as follows:

(i) The maximum transconductance $g_{m(max)}$ decreased from 210 mS mm^{-1} at 25 °C to 33 mS mm^{-1} at 500 °C, but was 201 mS mm^{-1} when re-measured at 25 °C.
(ii) The maximum drain current I_{Dmax} was 510 mA mm^{-1} at 25 °C and declined to 110 mA mm^{-1} at 500 °C but shot up to 550 mA mm^{-1} when measured after restoring the temperature to 25 °C (figure 9.6). In this graph, the I_{DS} values were from 0–550 mA mm^{-1}, V_{DS} from 0 V to 20 V and the steps were at V_{GS} = 0.5 V (Arulkumaran *et al* 2002).
(iii) The source resistance R_s was 2.6 Ωmm and 12.8 Ωmm at 25 °C and 500 °C, respectively; it recovered to 3.5 Ωmm on returning to 25 °C.
(iv) The drain resistance R_d climbed from 8.4 Ωmm at 25 °C to 35 Ωmm at 500 °C, but became 6.5 Ωmm upon returning to 25 °C.

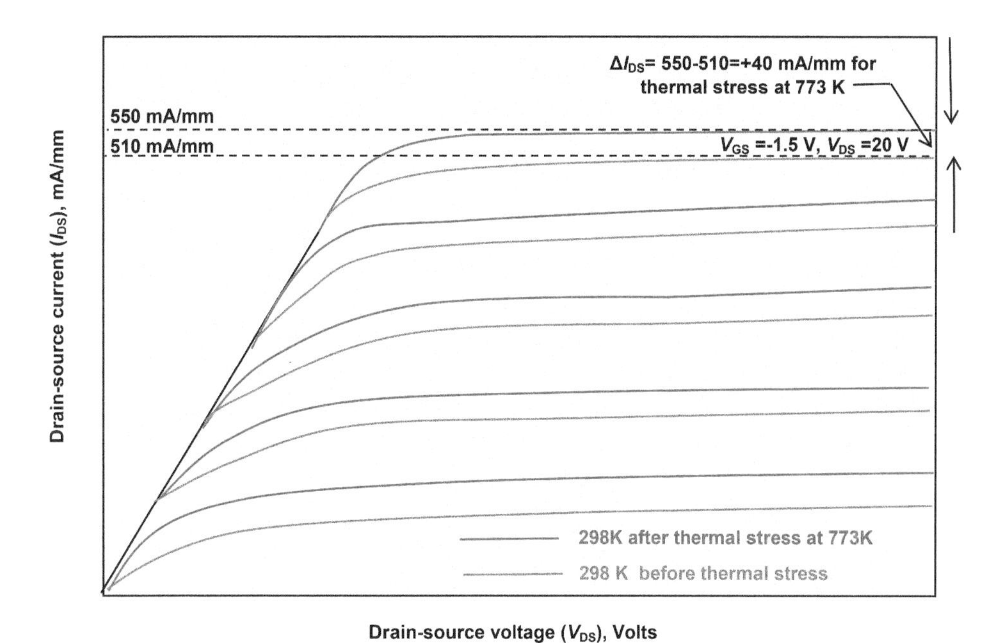

Figure 9.6. Increase in the room-temperature drain–source current of AlGaN/GaN HEMTs after thermal stressing at 773 K.

Thus the HEMTs operated with diminished capabilities at elevated temperatures. They resumed their normal functioning after cooling to 25 °C with practically the same I_{DS}–V_{DS} characteristics. However, they showed an increase in their room-temperature I_{DS} with respect to the values they had before thermal stress (Arulkumaran *et al* 2002).

9.8.2 Life testing of AlGaN/GaN HEMTs from 150 °C to 240 °C

MOCVD-grown 0.25 μm AlGaN/GaN HEMTs fabricated on SiC substrates and mounted on a dual-in-line package, were subjected to systematic life testing at high temperatures in nitrogen ambient (Chou *et al* 2004). In terms of their construction, these HEMTs contained the following layers: SiC substrate, AlN nucleation layer (100 nm), GaN buffer layer (1000 nm) and top $Al_{0.28}Ga_{0.72}N$ layer (25 nm). Ohmic contacts were made using Ti/Al/Pt/Au annealed at 860 °C in N_2. The Schottky contacts were made of Pt/Au metal stacks. The T-shaped gate was patterned and passivated with PECVD nitride.

Commencing from a temperature of 150 °C, the temperature was increased in steps of 15 °C up to a maximum limit of 240 °C. The HEMTs were kept for 48 h at each temperature step. An electrical stress was applied with V_{DS} = 10 V and I_{DS} = 500 mA mm^{-1}. An important result of this experiment was that the degradation of I_{DS}–V_{DS} characteristics starts at a temperature of 195 °C. The degradation is observed as a fall in transconductance. The drain current also decreases. The channel on-resistance registers an increase. Despite the variation of these electrical parameters of HEMTs, neither the gate area nor the ohmic contacts showed signs of any inter-diffusion of metal. There was no degradation of the gate diode as there was neither any modification of the ideality factor of the diode nor that of the Schottky barrier height of the gate. Gate leakage current also did not show any changes (Chou *et al* 2004). The quality of the AlGaN/GaN epitaxial layers was of utmost significance.

9.8.3 Power characteristics of AlGaN/GaN HEMTs up to 368 °C

The power characteristics of AlGaN/GaN HEMTs with a surface charge-controlled structure on SiC substrates, were examined at high temperatures (Adachi *et al* 2005). These HEMTs are provided with a capping n-GaN layer on the surface so that the constituent layers are: SiC, i-GaN, i-AlGaN, n-AlGaN and n-GaN; the 2DEG was formed between the i-GaN and i-AlGaN layers. They had Ti/Al ohmic contacts for source and drain connections, and Ni/Au gate electrodes. At a channel temperature of 269 °C, a frequency of 2.14 GHz and an operating voltage of 50 V, the linear gain was 12.3 dB and the maximum efficiency was 53.6%. When the channel temperature rose to 368 °C, the linear gain went down to 10.4 dB and the efficiency dropped to 43.9%. Further, from a channel temperature >100 °C onwards, the saturated output power begins to decrease. Its value at 368 °C is 2.1 dB less than at 59 °C. These changes were correlated with a decrease in drain current and hence with that in electron velocity. The drain current at 300 °C is half of its value at 30 °C. The runaway threat to AlGaN/GaN was very low due to the gate bias voltage remaining virtually constant (Adachi *et al* 2005).

9.8.4 Mechanisms of the failure of high-power AlGaN/GaN HEMTs at high temperatures

A review of the modes and mechanisms of the failure of high-power AlGaN/GaN HEMTs (Meneghesso *et al* 2008) showed that at frequencies <3.5 GHz, these HEMTs were highly reliable with a mean-time-to-failure (MTTF) $\geqslant 10^6$ h at a junction temperature of 180 °C. Good thermal stability was attained with Pt/Au, Ni/Au and Mo/Au Schottky contacts. The Mo/Au Schottky contacts along with Ti/Al/Ni/Au ohmic contacts, annealed at 900 °C for 30 s, showed a <2% change in ohmic contact resistivity after storage in nitrogen at 340 °C for 2 000 h. The Schottky barrier height increased from 0.8 eV to 0.95 eV and became stable at this value (Sozza *et al* 2005).

9.9 InAlN/GaN HEMTs

9.9.1 AlGaN/GaN versus InAlN/GaN HEMTs for high-temperature applications

The commonly used AlGaN/GaN structure in HEMT devices has the drawback that it is a highly strained and polarized heterojunction (Maier *et al* 2010). The mechanical strain affects its thermal stability, thereby deteriorating its performance at high temperatures. Therefore, AlGaN/GaN HEMTs have often failed in the temperature range 25 °C to 500 °C. For reliable operation, it is advisable to keep away from the prescribed failure limit.

When comparing the InAlN/GaN heterostructure to the usual AlGaN/GaN heterostructure, it was found that: (i) mechanical stress was absent in the lattice-matched InAlN/AlGaN structure; (ii) the quantum well polarization-induced charge obtained from the InAlN/GaN structure was higher than from the AlGaN/GaN structure by a factor of 2–3 giving a density of 2DEG sheet charge = 2.8×10^{-13} cm^{-2}. This translates to a possibility of a 205% increase in drain current with the InAlN/GaN structure (Kuzmik 2001).

9.9.2 InAlN/GaN HEMT behavior up to 1000 °C

The above facts have encouraged the adoption of the InAlN/GaN structure for designing robust HEMT devices. The electrical performance of InAlN/GaN HEMTs was assessed to stake its claim as an alternative choice to conventional AlGaN/GaN HEMTs for high-power and high-temperature applications. For sapphire-substrate HEMTs of gate length 0.25 μm, the maximum output current density was 3 A mm^{-1} at 77 K and 2 A mm^{-1} at room temperature (Medjdoub *et al* 2006). The transition frequency was 50 GHz and the maximum frequency was 60 GHz for a 0.15 μm gate length. The HEMT structure consists of a sapphire substrate–GaN buffer layer (2 μm), an AlN spacer (1 nm) and an $Al_{0.81}In_{0.19}N$ layer (13 nm, undoped). The ohmic contact was a Ti/Al/Ni/Au multilayer metallization annealed at 890 °C for 1 min. The Schottky contact was Ni/Au. On-chip in-vacuum testing of devices was carried out by increasing the temperature by intervals of 100 °C and maintaining the device at every temperature for 10 min. At 800 °C, the maximum drain current density was 850 mA mm^{-1}. A marginal decrease in pinch-off voltage was noticed. At 1000 °C, the maximum current density dropped to 600 mA mm^{-1}, but pinch-off voltage suffered serious degradation. However, the device was not irreversibly degraded. It recovered its original characteristics after cooling, indicating the temporary nature of the departure in device behavior (Medjdoub *et al* 2006).

9.9.3 Thermal stability of barrier layer in InAlN/GaN HEMTs up to 1000 °C

Barrier-layer scaling of InAlN/GaN HEMTs was studied by varying the $In_{0.17}Al_{0.83}N$ barrier layer thickness from 33 nm to 3 nm (Medjdoub *et al* 2008). The HEMT structures studied included the GaN buffer layer (2 μm), an AlN spacer layer (1.0 nm) and different thicknesses of the barrier layer, with Ti/Al/Ni/Au ohmic contacts and Ni/Au Schottky gates. In the thermal stress experiment, the temperature was incremented by 100 °C, kept at 100 °C for half an hour, and then lowered to room temperature to examine the electrical characteristics and to ascertain any possible damage. This sequence was continued up to 1000 °C. Above 1000 °C, melting of Au did not permit any further experimentation. Despite the rigorous thermal stressing undergone by the HEMTs, no detrimental effects on the gate forward current capability came to light. Neither were there any changes in the open channel current nor in the pinch-off voltage. The properties of the metallurgical interfaces remained unaffected. The integrity and characteristics of the InAlN/GaN barrier layer up to 3 nm thickness were preserved. The reverse current of the SBD was constant. This is a major advantage of the InAlN/GaN structure over $Al_{0.3}Ga_{0.7}N$/GaN devices, which showed degeneration of Schottky diode characteristics to ohmic behavior after a thermal stress trial, even with 25 nm thick barrier layers (Medjdoub *et al* 2008). These observations could vastly benefit ultra-high frequency HEMT design and fabrication.

9.9.4 Feasibility demonstration of HEMT operation at gigahertz frequency up to 1000 °C

The ruggedness and reliability of a HEMT device with respect to temperature is determined by the thermal stability of the different constituent layers and metal stacks (Maier *et al* 2010). $In_{0.185}Al_{0.815}N$/GaN HEMTs on sapphire substrates contained a 12 nm thick InAlN layer with a 1 nm thick AlN spacer layer separating the buffer and barrier layers; the buffer layer thickness was 3 μm (Maier *et al* 2012). The ohmic contact was Ti/Al/Ni/Au annealed at 800 °C in nitrogen. The thickness ratio of the different metal films in this stack was optimized to achieve stability at 1000 °C. The refractory metal molybdenum (Mo) was used as the gate metal. The passivation coating was a 30 nm thick silicon nitride film formed by the PECVD process at 340 °C. This process was ammonia-free. The surface was specially prepared prior to Si_3N_4 deposition. These measures assured avoidance of blistering and cracking issues on raising the temperature to 1100 °C. Operated under a 1 MHz large signal condition, the

HEMT devices were ramped up to 1000 °C in 48 h with intermediate stops at 500 °C, 700 °C and 900 °C before the final destination of 1000 °C. As the temperature was increased, the gate leakage current became obvious. At 1000 °C, the gate leakage current showed an incessant increase as testing time elapsed. The drain current increased at 1000 °C on account of the extra leakage produced in the buffer layer around 800 °C. Throughout the temperature range from room temperature up to 1000 °C, and over the total time period of testing, the threshold voltage was stable. The HEMTs were operated for 25 h at 1000 °C. Thus a feasibility demonstration of HEMT operation was successfully executed in the gigahertz frequency range on a temperature scale hitherto considered the solo province of ceramic materials and refractory metals. The study also showed that the polarization in HEMT structure was preserved up to this temperature (Maier *et al* 2012).

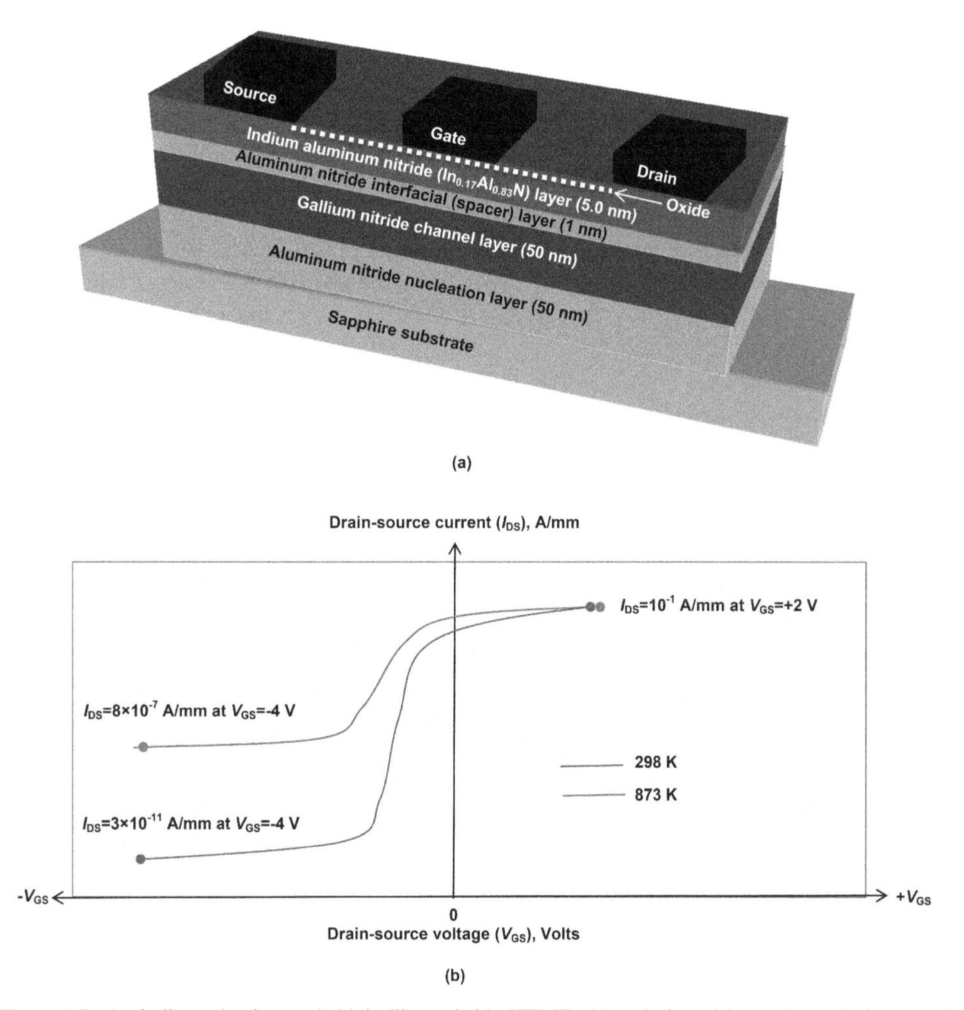

Figure 9.7. An indium aluminum nitride/gallium nitride HEMT: (a) a device with an ultra-thin body and (b) the transfer characteristics of the HEMT at 298 K and 873 K at $V_{DS} = 6$ V.

However, it was evident that the full capability of the HEMT structure remained unutilized. Parasitic effects influenced the operation of the device at high temperatures. One conspicuous effect was the increase in leakage current of the buffer layer in this temperature regime. This predicament is avoided by resorting to mesa technology using an ultra-thin HEMT body (Herfurth *et al* 2013), see figure 9.7. In this improved structure, the active region of the device is restricted only to the channel area. An exceedingly thin (50 nm) AlN nucleation layer was chosen. The buffer layer thickness was also reduced to 50 nm. Besides these layers, the structure included a 1 nm thick AlN interfacial layer and a 5 nm thick InAlN barrier layer. The carrier concentration in 2DEG was 1.4×10^{13} cm^{-2}. The ohmic contact stack was Ti/Al/Ni without any Au capping layer, annealed at 800 °C for 30 s. The Schottky contact for the gate was Cu overlaid with Pt. The HEMT devices were tested using tungsten carbide needles. Testing at 600 °C was carried out at a pressure of 10^{-6} mbar. At room temperature, the maximum drain current was 0.4 A mm^{-1} and the threshold voltage was –1.4 V. Further, at room temperature, the off-state current I_{off} was ~1 pA for gate width $W_G = 50$ μm with a current on/off ratio of $I_{on}/I_{off} > 10^{10}$ and subthreshold swing SS = 73 mV/dec. At 600 °C, the maximum drain current and threshold voltage values remained nearly unaltered. But I_{off} decreased to ~1 μA mm^{-1} and I_{on}/I_{off} slumped to a low but tolerable level of 10^6; the SS was 166 mV/dec. The increase in off-state gate leakage current at 600 °C was a major shortcoming. Room-temperature RF testing at 1 MHz gave a small-signal peak transconductance $g_m = 110$ mS mm^{-1}. The value of g_m was the same at 600 °C. No chemically or electrically induced material degradation was noticed during a 30 min trial experiment. The output power measurements in class A mode operation at 600 ° C gave an RF output power = 109 mW mm^{-1} at $V_{DS} = 8.75$ V. The transition frequency was 6.6 GHz and the maximum frequency was 30 GHz (Herfurth *et al* 2013).

9.10 Discussion and conclusions

The merits and demerits of GaN over SiC and Si were pointed out. The MOCVD technique for synthesis of GaN was described. n- and p-type dopants for GaN were mentioned and the difficulties in achieving high carrier concentrations in p-type doping were indicated. Various ohmic and Schottky contact schemes on GaN were mentioned, and the fabrication and performance characteristics of AlGaN/GaN HEMTs were discussed. This structure was crippled by the stress problem. The lattice-matched InGaN/GaN HEMT structure greatly helped in overcoming this issue. HEMT operation is feasible at gigahertz frequencies up to 1000 °C using this structure.

Review exercises

9.1. Compare GAN with SiC and Si with regard to thermal conductivity.
9.2. Compare GaN with SiC with respect to electron and hole mobilities.
9.3. The energy bandgap of GaN is slightly larger than that of SiC. What is the difference between their intrinsic temperatures?

9.4. What is the technique used for growing GaN-based heterostructures called? What are the substrates used? What precursors are used for Ga, Al, In and N? What is the typical pressure of growth? What is the temperature used?
9.5. What are the commonly used n-type and p-type dopants in gallium nitride? What are the maximum concentrations achieved in n- and p-type doped GaN?
9.6. What is the position of the energy level for p-type doping in gallium nitride? What practical difficulty is experienced in p-type doping due to this energy level position?
9.7. Discuss the roles of titanium and nickel in Ti/Al/Ni/Au multilevel metallization on n-type GaN. Give a metallization scheme which is stable up to 1000 °C.
9.8. Mention one metallization scheme for making an ohmic contact on p-type GaN. Give one example of Schottky contact metallization on n-type GaN.
9.9. What are the different structural layers used in the fabrication of AlGaN/GaN HEMTs on 4H-SiC/sapphire substrates? What was the gate metallization used? Which type of HEMTs showed superior DC characteristics? What were the values of the following parameters at 25 °C and 500 °C: (a) maximum transconductance, (b) maximum drain current, (c) source resistance and (d) drain resistance?
9.10. Describe the systematic life testing of AlGaN/GaN HEMTs from 150 °C to 240 °C. At what temperature does the degradation of I_{DS}–V_{DS} characteristics begin? In what form is it observed? Does the gate diode degrade in any way? Describe the variation in power characteristics of AlGaN/GaN HEMTs up to 368 °C.
9.11. How do the following contacts on HEMTs perform at high temperatures: (a) Mo/Au Schottky contacts and (b) Ti/Al/Ni/Au ohmic contacts?
9.12. What is the main shortcoming of the frequently used AlGaN/GaN structure in HEMT devices? In what respects is the InAlN/GaN heterostructure superior to AlGaN/GaN heterostructure?
9.13. Describe the structure of InAlN/GaN HEMT. What is the ohmic contact metallization? What is the Schottky contact metallization used? What is the maximum drain current density at 800 °C? What is its value at 1000 °C? Is the device permanently degraded at 1000 °C?
9.14. Compare the thermal stability of the barrier layer in InAlN/GaN HEMTs with AlGaN/GaN HEMTs up to 1000 °C.
9.15. Describe the performance of $In_{0.185}Al_{0.815}N$/GaN HEMTs with refractory metallization up to 1000 °C in the gigahertz frequency range. What was the passivation coating used? Did the threshold voltage remain stable?
9.16. What were the Schottky and ohmic contacts used in the mesa InAlN/GaN HEMTs? What was the small-signal peak transconductance g_m at room temperature and at 600 °C? What was the RF output power in class A operation at 600 °C?

References

Adachi N, Tateno Y, Mizuno S, Kawano A, Nikaido J and Sano S 2005 High temperature operation of AlGaN/GaN HEMT *2005 IEEE MTT-S Int. Microwave Symposium Digest* (12–17 June) pp 507–10

Adivarahan V, Lunev A, Asif Khan M, Yang J, Simin G, Shur M S and Gaska R 2001 Very-low-specific-resistance Pd/Ag/Au/Ti/Au alloyed ohmic contact to p GaN for high-current devices *Appl. Phys. Lett.* **78** 2781–3

Arulkumaran S, Egawa T, Ishikawa H and Jimbo T 2002 High-temperature effects of AlGaN/GaN high-electron-mobility transistors on sapphire and semi-insulating SiC substrates *Appl. Phys. Lett.* **80** 2186–8

Chou Y C *et al* 2004 Degradation of AlGaN/GaN HEMTs under elevated temperature life testing *Microelectron. Reliab.* **44** 1033–8

Fan S Z, Mohammad S N, Kim W, Aktas O, Botchkarev A E and Morkoc H 1996 Very low resistance multilayer ohmic contact to n-GaN *Appl. Phys. Lett.* **68** 1672–4

Herfurth P, Maier D, Lugani L, Carlin J-F, Rösch R, Men Y, Grandjean N and Kohn E 2013 Ultrathin body InAlN/GaN HEMTs for high-temperature (600 °C) electronics *IEEE Electron Device Lett.* **34** 496–8

Kabra S, Kaur H, Haldar S, Gupta M and Gupta R S 2008 Temperature dependent analytical model of sub-micron GaN MESFETs for microwave frequency applications *Solid-State Electron* **52** 25–30

Kacprzak T and Materka A 1983 Compact DC model of GaAs FETs for large-signal computer calculation *IEEE J. Solid State Circuits* **18** 211–3

Kuzmik J 2001 Power electronics on InAlN/(In)GaN: prospect for a record performance *IEEE Electron Device Lett.* **22** 510–2

Macherzyński W, Paszkiewicz B, Szyszka A, Paszkiewicz R and Tłaczała M 2009 Effect of annealing on electrical characteristics of platinum based Schottky contacts to n-GaN layers *J. Electr. Eng.* **60** 276–8

Maier D, Alomari M, Grandjean N, Carlin J-F and di Forte-Poisson M-A *et al* 2010 Testing the temperature limits of GaN-based HEMT devices *IEEE Trans. Device Mater. Reliab.* **10** 427–36

Maier D, Alomari M, Grandjean N, Carlin J-F, Diforte-Poisson M-A, Dua C, Delage S and Kohn E 2012 InAlN/GaN HEMTs for operation in the 1000 °C regime: a first experiment *IEEE Electron Device Lett.* **33** 985–7

Medjdoub F, Alomari M, Carlin J-F, Gonschorek M, Feltin E, Py M A, Grandjean N and Kohn E 2008 Barrier-layer scaling of InAlN/GaN HEMTs *IEEE Electron Device Lett.* **29** 422–5

Medjdoub F, Carlin J-F, Gonschorek M, Feltin E, Py M A, Ducatteau D, Gaquière C, Grandjean N and Kohn E 2006 Can InAlN/GaN be an alternative to high power/high temperature AlGaN/GaN devices? *IEDM '06 Int. Electron Devices Meeting* (*San Francisco, CA, 11–13 December*) pp 1–4

Meneghesso G, Verzellesi G, Danesin F, Rampazzo F, Zanon F, Tazzoli A, Meneghini M and Zanoni E 2008 Reliability of GaN high-electron-mobility transistors: state of the art and perspectives *IEEE Trans. Device Mater. Reliab.* **8** 332–43

Pearton S J, Donovan S M, Abernat C R, Ren F, Zolper J C, Cole M W and Shul R J 1998 High temperature stable WSi_x ohmic contacts on GaN *IEEE Fourth Int. High Temperature Electronics Conf.* 296–300

Ruvimov S, Liliental-Weber Z, Washburn J, Duxstad K J, Haller E E, Fan Z-F, Mohammad S N, Kim W, Botchkarev A E and Morkoc H 1996 Microstructure of Ti/Al and Ti/Al/Ni/Au ohmic contacts for n-GaN *Appl. Phys. Lett.* **69** 1556–8

Shashikala B N and Nagabhushana B S 2010 Modeling of GaN MESFETs at high temperature, *22nd Int. Conf. on Microelectronics* (*ICM 2010*)

Sozza A *et al* 2005 Evidence of traps creation in GaN/AlGaN/GaN HEMTs after a 3000 hour on-state and off-state hot-electron stress *IEEE Electron Device Meeting* pp 590–3

Chapter 10

Diamond electronics for ultra-hot environments

Ranked at the highest place for mechanical hardness and thermal conductivity among all known materials, diamond has the most interesting thermal properties of semiconductors, and is seen as the perfect material for power electronics, fulfilling the requirements of a niche market for robust high-power devices. From its calculated figures of merit based on intrinsic diamond of low defect density and with shallow donors/acceptors for doping, diamond may in future replace SiC and GaN by providing less lossy devices with a better trade-off between on-resistance and breakdown voltage. Practically, however, the achievements are too far from predicted values. Tremendous efforts will be necessary to overcome the technological difficulties in order to reach the goal of making this human ornamental gemstone the real jewel of microelectronics.

10.1 Introduction

Compared to the silicon bandgap of 1.12 eV, diamond has a bandgap of 5.5 eV, which is 4.9 times higher than that of silicon. In fact, the bandgap of diamond approaches that of an insulator. Operating temperatures of >1000 °C can be realized by using diamond as the fabrication material for devices. The breakdown electric field for diamond (10 MV cm^{-1}) is 33 times larger than that of silicon (0.3 MV cm^{-1}). It is 3.33 times that for silicon carbide (3 MV cm^{-1}). A much lower specific on-resistance is achievable using the thin conducting regions in diamond devices than is possible with silicon or silicon carbide devices. The thermal conductivity of diamond (15 W cm K^{-1}) is 10 times that of silicon (1.5 W cm K^{-1}) and 3 times that of silicon carbide (4.9 W cm K^{-1}). Therefore, diamond devices can provide substantially higher power per unit area than either silicon or silicon carbide components. Unlike silicon carbide, the carrier mobilities in diamond, ~2000 cm^2 V s^{-1}, are better than those available in silicon (electron mobility 1400 cm^2 V s^{-1}). So, the mobility disadvantage upsetting silicon carbide does not impose any encumbrance on the good features of diamond. In addition to the above good features, diamond displays exceptional

doi:10.1088/978-0-7503-1155-7ch10

hardness and outstanding wear resistance. It offers high stability against the degradation effects due to chemical or thermal effects or upon irradiation. Thus diamond displays a unique set of properties which are several-fold superior in every category when compared to any other wide bandgap semiconductor, making it the 'true dream material' for designers of microelectronic devices, sensors, microwave and acoustic wave filters, and microelectromechanical systems (MEMS). Table 10.1 lists the major properties of interest to engineers working on diamond device and circuit design. Diamond crystallizes in the structural form displayed in figure 10.1.

Diamond electronics is still in its preliminary phase. Many processing issues for the exploitation of diamond as a base material for semiconductor device fabrication are still technological challenges to be resolved. The present research outcomes trail far behind the idealistic prospects.

10.2 Intrinsic temperature of diamond

Before embarking on detailed discussions of diamond electronics, let us repeat the intrinsic temperature calculations for diamond and examine where diamond stands with respect to other semiconductors for high-temperature applications. This will help us in appreciating the significance of diamond in a proper perspective relative to competing materials. For diamond

$$m_n^* = 1.9m_0,\ m_p^* = 0.8m_0,\ E_g = 5.46\text{–}5.6\ \text{eV}$$

$$\therefore m^* = \left(\frac{m_n^* m_p^*}{m_0^2}\right)^{0.75} = \left(\frac{1.90m_0 \times 0.8m_0}{m_0^2}\right)^{0.75} = 1.3689. \tag{10.1}$$

The equation for intrinsic temperature calculation

$$\ln\{0.207/(m^* T^{1.5})\} = -5.8025 \times 10^3 E_g \tag{10.2}$$

is converted to the form

$$T \ln\{0.207/(1.37T^{1.5})\} = -5.8025 \times 10^3 \times 5.5 \tag{10.3}$$

or

$$T \ln 0.151 - T \ln T^{1.5} = -31913.75 \tag{10.4}$$

$$\therefore -1.89T - T \ln T^{1.5} = -31913.75. \tag{10.5}$$

Suppose,

$$T = 2357.3\ \text{K} \tag{10.6}$$

$$\begin{aligned} \text{lhs} &= -1.89T - T\ln T^{1.5} = -1.89 \times 2357.3 - 2357.3 \ln 2357.3^{1.5} \\ &= -4455.297 - 27457.6141 = -31912.9111. \end{aligned} \tag{10.7}$$

Table 10.1. Properties of diamond.

Property	Value	Property	Value	Property	Value
Chemical symbol	C (carbon)	Lattice constant (Å) at 300 K	3.5668	Intrinsic resistivity (Ωcm)	1×10^{42}
Refractive index	2.424 at 546.1 atm.	Melting point (°C)	3773	Electron mobility ($cm^2\ V\ s^{-1}$)	2400
Classification	Non-metal	Dielectric constant	5.7 (1–10 kHz)	Hole mobility ($cm^2\ V\ s^{-1}$)	2100
Crystal structure	Diamond cubic	Thermal conductivity ($W\ cm\ K^{-1}$)	10–20	Electron saturated velocity ($cm\ s^{-1}$)	2.7×10^7
Color	Colorless to yellow/brown	Energy bandgap E_g (eV) at 300 K	5.46–5.60	Donors	Phosphorous arsenic
Density at 300 K ($g\ cm^{-3}$)	3.516–3.525	Electrical breakdown field ($V\ cm^{-1}$)	1×10^7	Donor ionization energies ΔE_D (meV)	590 (phosphorous) 410 (arsenic)
Number of atoms/cm^3	1.763×10^{23}	Intrinsic carrier concentration (cm^{-3})	$\sim 1 \times 10^{-27}$	Acceptor and ionization energy ΔE_A (meV)	Boron 370

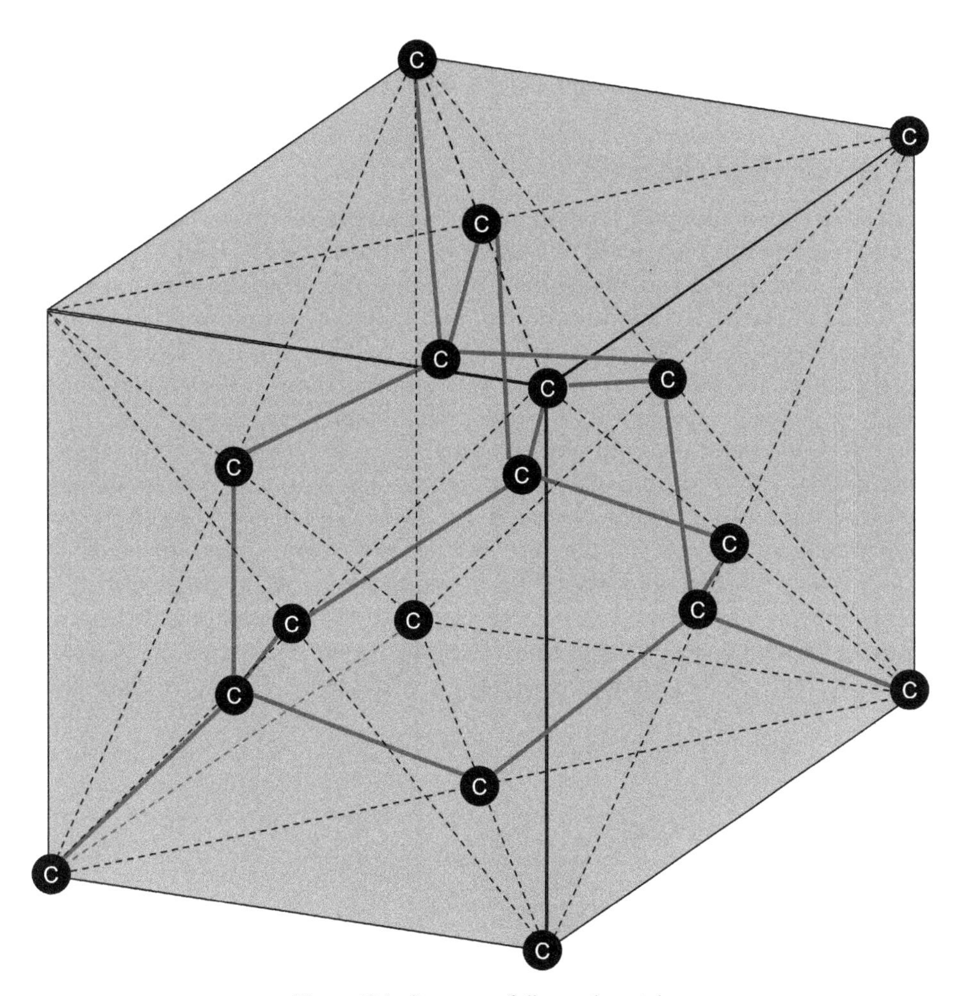

Figure 10.1. Structure of diamond crystal.

If

$$T = 2357.35\,\text{K} \tag{10.8}$$

$$\begin{aligned}\text{lhs} &= -1.89T - T\ln T^{1.5} = -1.89 \times 2357.35 - 2357.35\ln 2357.35^{1.5} \\ &= -4455.3915 - 27458.2715 = -31913.663.\end{aligned} \tag{10.9}$$

When

$$T = 2357.4\,\text{K} \tag{10.10}$$

$$\begin{aligned}\text{lhs} &= -1.89T - T\ln T^{1.5} = -1.89 \times 2357.4 - 2357.4\ln 2357.4^{1.5} \\ &= -4455.486 - 27458.9289 = -31914.4149.\end{aligned} \tag{10.11}$$

$$\therefore\ T = 2357.35\,\text{K}. \tag{10.12}$$

This temperature represents the highest of all the materials examined in the preceding chapters. It is much higher than for GaN (1667 K) or SiC (1591.5 K).

10.3 Synthesis of diamond

Diamond has been known to humankind for more than 200 years, but its identity as elemental carbon was recognized only in 1796. Various types of diamonds have been extensively studied and subdivided into four categories. This classification is based on the type of chemical impurities present and their concentration in the diamond. The categories of diamond are: type Ia, type Ib, type IIa and type IIb. The type Ia category comprises 98% of all diamonds obtained naturally. Their main impurity content is nitrogen, up to 0.1%. Type Ib diamonds represent about 0.1% of diamonds found naturally with nitrogen content up to 0.05%. Type IIa diamonds constitute 1%–2% of natural diamonds and are practically free from impurities. Type IIb diamonds are present in around 0.1% of natural diamonds with a very low impurity content like type IIa; they are the one of the rarest diamond varieties.

The artificial synthesis of diamond through a high-pressure and high-temperature (HPHT) process was reported in 1955 (Bundy *et al* 1955); hence it is called HPHT diamond. Diamond is synthesized by another method known as CVD, see figures 10.2 and 10.3. Hence, this diamond is termed CVD-grown diamond or

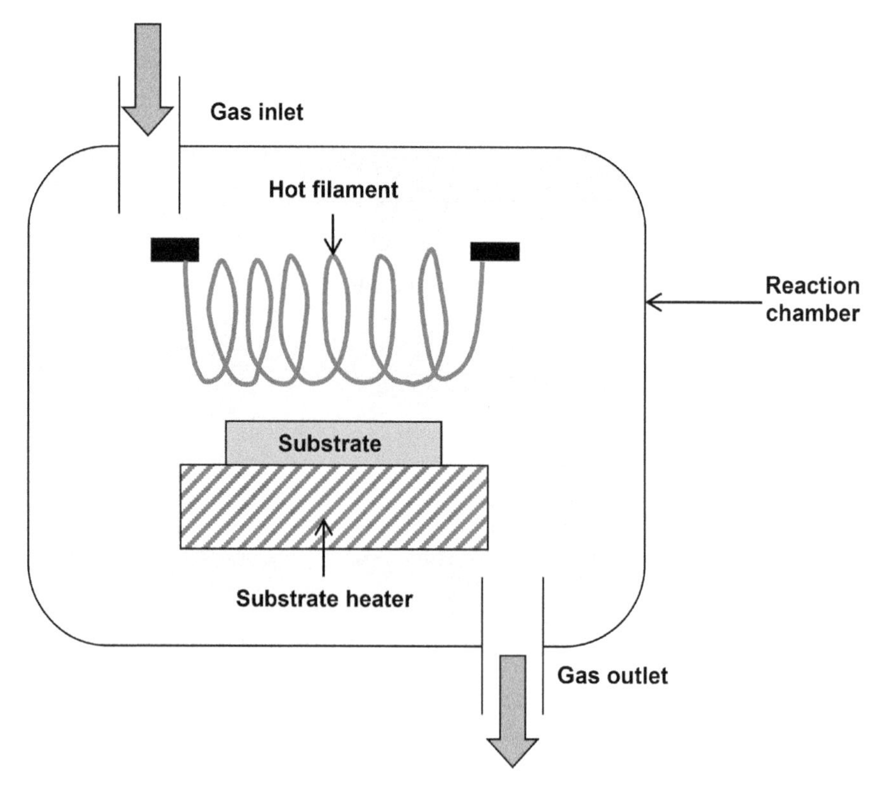

Figure 10.2. Hot-filament CVD of diamond.

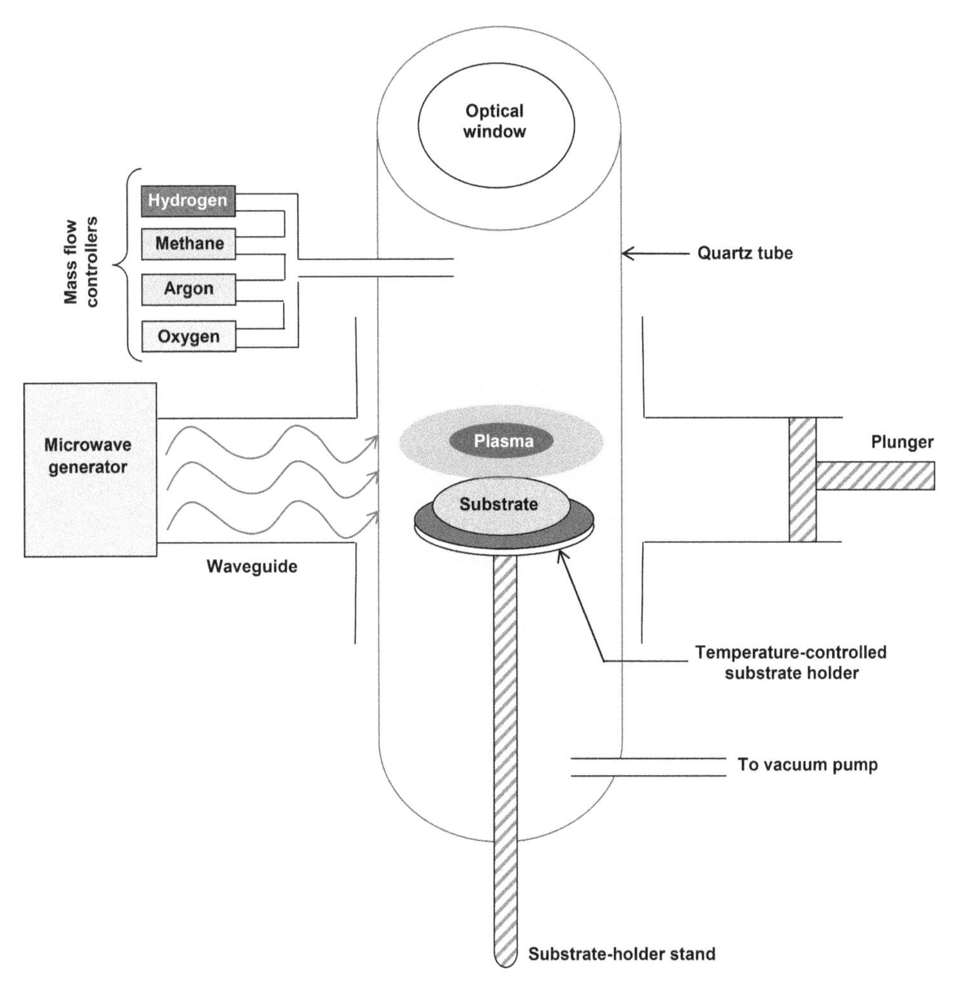

Figure 10.3. Microwave plasma reactor for diamond synthesis by CVD.

CVD diamond. In general, man-made diamond is called synthetic diamond, cultured diamond or cultivated diamond.

CVD is an established commercial technique. It is an atomistic process yielding highly dense, good-quality, adherent films. In this technique, a solid film is slowly deposited on a hot substrate kept inside a vacuum chamber at low pressure, and serving as a reactor. This deposition takes place by chemical reaction between gaseous carbon-containing precursor molecules (usually a hydrocarbon such as methane and hydrogen) introduced into the reactor. In the growth chamber, the gases are ionized into chemically reactive radicals by any of the following means: a hot filament, a DC or microwave plasma, an oxyacetylene flame, a welding torch, a laser or an e-beam. Typical growth conditions are: 160 Torr, 1000 °C to 1200 °C, 3% N_2/CH_4 and high methane proportion (12% CH_4/H_2) (Yan *et al* 2002). To produce gem-quality diamond, a seed crystal is inserted into the gas mixture. Then seeded growth of diamond occurs, replicating the structure of the seed crystal.

10.4 Doping of diamond

The absence of donor and acceptor impurities with shallow energy levels, i.e. at small distances below the conduction band and above the valence band, is the main bottleneck confronting diamond electronics. Substitutional doping by conventional methods such as thermal diffusion and ion implantation has not yet been able to provide high carrier concentrations and mobilities. The energy levels of the available impurities are appreciably deep. Hence, the ionization energies of these impurities are much larger than the thermal energy at room temperature.

10.4.1 n-type doping

Attempts to produce n-type diamond, either by diffusion, by impurity incorporation during CVD growth, or by ion implantation and high-temperature annealing, have so far met with limited success. Very often, nitrogen is used to dope diamond as n-type. But the energy level of the nitrogen impurity is very deep (1.7 eV below the conduction band), see figure 10.4. Hence, thermal energy is unable to cause ionization at room temperature. The fraction of impurities that are ionized at room temperature is very small. Nitrogen can be thermally activated only at temperatures >600 °C to 700 °C. As a result, nitrogen doping cannot provide the required high electron concentrations. Nitrogen-doped diamond is practically an insulator at room temperature.

Besides nitrogen, phosphorous has been used for n-type doping of diamond (Koizumi 2006, Kato *et al* 2007). The energy level of phosphorous, located at 0.59 eV, is much shallower than that for nitrogen but its depth is still too large to meet the requirement.

Phosphorous and nitrogen co-doped diamond epitaxial films have been grown with a phosphorous concentration of 2×10^{16}–3.6×10^{17} atoms cm^{-3} and a nitrogen concentration from 4×10^{17} to 3×10^{18} atoms cm^{-3} (Cao *et al* 1995).

Sulphur (0.32 eV) and lithium (0.16 eV) have also been tried. Sulphur was implanted with energies up to 400 keV (Hasegawa *et al* 1999).

10.4.2 p-type doping

p-type doping of diamond is done by ion implantation of boron. The activation energy of boron decreases as its concentration increases. It lies in the range 0–0.43 eV above the valence band. It becomes zero at boron concentrations $>1.7 \times 10^{20}$ cm^{-3} and is 0.35 eV at a boron concentration of 1×10^{18} cm^{-3}. Boron can also be incorporated during CVD growth of diamond. However, the presence of any hydrogen in the plasma inhibits the control of boron doping to obtain p-type diamond. Efforts to achieve high hole mobility have been made by several workers, e.g. by a cold-implantation–rapid-annealing process (CIRA; Prins 1988), with a final annealing temperature of 1723 K (Fontaine *et al* 1996) giving a hole mobility of 400 cm^2 V s^{-1}; by ion implantation at a high energy of ~MeV (Prawer *et al* 1997) yielding a mobility of 600 cm^2 V s^{-1} (Uzan-Saguy *et al* 1998).

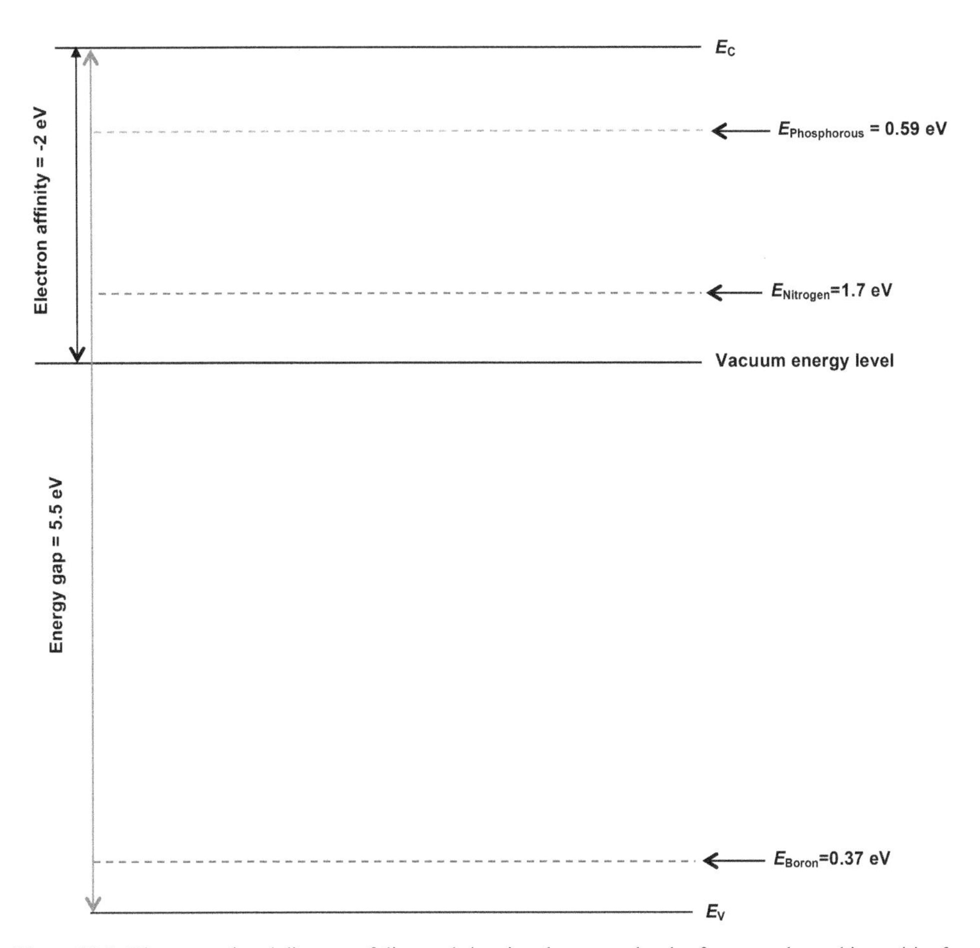

Figure 10.4. The energy band diagram of diamond showing the energy levels of commonly used impurities for n- (phosphorous, nitrogen) and p-doping (boron).

10.4.3 p-doping by hydrogenation termination of the diamond surface

p-type conduction in diamond can also be obtained by hydrogenation of the diamond surface in hydrogen plasma and cooling to room temperature in a hydrogen atmosphere (Kawarada 1996). More than 90% of charge carriers exist in a <10 nm thick conducting layer formed on the surface showing p-type conductivity $\sim 10^{-4}$–10^{-5} S, with a hole concentration of $\sim 10^{13}$ cm^{-2} and hole mobility $\sim$100–150 cm^2 V s^{-1}. Such a surface with a two-dimensional hole gas is said to be hydrogen-terminated. Thus an intrinsic conducting channel containing holes as charge carriers is formed in a sub-surface layer without any impurity doping, in sharp contrast to the extrinsic boron-doped acceptor layer, which is only partially activated and hence only incompletely conducting at room temperature.

The p-type conduction is chemically and thermally unstable. It is vulnerable to environmental effects. By exposure to oxygen ambient, by heating in air above

200 °C or annealing in vacuum, the p-type conduction is destroyed because of dehydrogenation of the surface. For utilization of this p-type layer as a part of an electronic device, it needs proper protection. Passivation of the hydrogenated surface by hydrogenated carbon films formed by an RF plasma CVD process was found to improve its thermal stability up to 200 °C (Yamada *et al* 2004). Surface passivation is crucial for robust device development.

The low density of surface states $<10^{11}$ cm^{-2} exhibited by the H-terminated surface makes it appropriate for an FET channel. Moreover, an ohmic contact on the p-type conducting layer can be formed using metals of high electronegativity. The contact resistance for Au or Pd is $<10^{-5}$ Ωcm^{-2}.

In opposition to the conducting nature of a H-terminated diamond surface, an oxygen-terminated diamond surface behaves as an insulator. Oxygen termination is accomplished by exposure of the diamond surface to oxygen plasma. A H-terminated surface can be rendered insulating by such treatment in which hydrogen is substituted by oxygen. This affords a method of making the active areas of a device conducting and the adjoining passive areas insulating by selective treatment with hydrogen and oxygen.

10.5 A diamond p–n junction diode

An UV light-emitting diode (LED) was fabricated on diamond (Koizumi *et al* 2001), see figure 10.5. The starting substrate was boron-doped single crystal diamond. The diode was formed on its mechanically polished {111} surface. Upon the surface, the boron-doped layer (1–2 × 10^{17} cm^{-3}, 1 μm) was grown using trimethylboron (TMB; $(CH_3)_3B$) as the source of boron impurity. Then a phosphorous-doped layer (7–8 × 10^{18} cm^{-3}, 2 μm) was grown with phosphine (PH_3), as the source of phosphorous impurity. Separate microwave plasma-enhanced CVD (MPCVD) systems were used for boron and phosphorous doping. The activation energies for B and P were determined as 0.37 eV and 0.59 eV, respectively. The hole mobility was 60 cm^2 V s^{-1} and the electron mobility was 300 cm^2 V s^{-1}. For the p-layer, Ti–Au contacts were made. For the n-layer, graphitic dots formed by Ar implantation through a mask were capped with Ti–Au. The turn-on voltage was 6–7 V. The reverse leakage current was 10^{-8} A at –20 V bias. UV light emission was observed at a wavelength of 235 nm on applying a forward voltage of 20 V.

10.6 Diamond Schottky diode

10.6.1 Diamond Schottky diode operation up to 1000 °C

The vertical-structure Schottky diodes (figure 10.6) were fabricated on synthetic boron-doped IIb substrates. The boron concentration was $>10^{20}$ cm^{-3} (Vescan *et al* 1997a). In the MPCVD-grown active layer, the surface doping concentration decreased sharply to 10^{16} cm^{-3} to achieve a breakdown voltage >50 V. The surface was exposed to oxygen plasma. For the Schottky contact, a nitrogen doped Si:W alloy was deposited by ion beam sputtering. It was covered with gold. Up to 773 K, temperature has little effect on the reverse leakage current. Above 773 K, the leakage

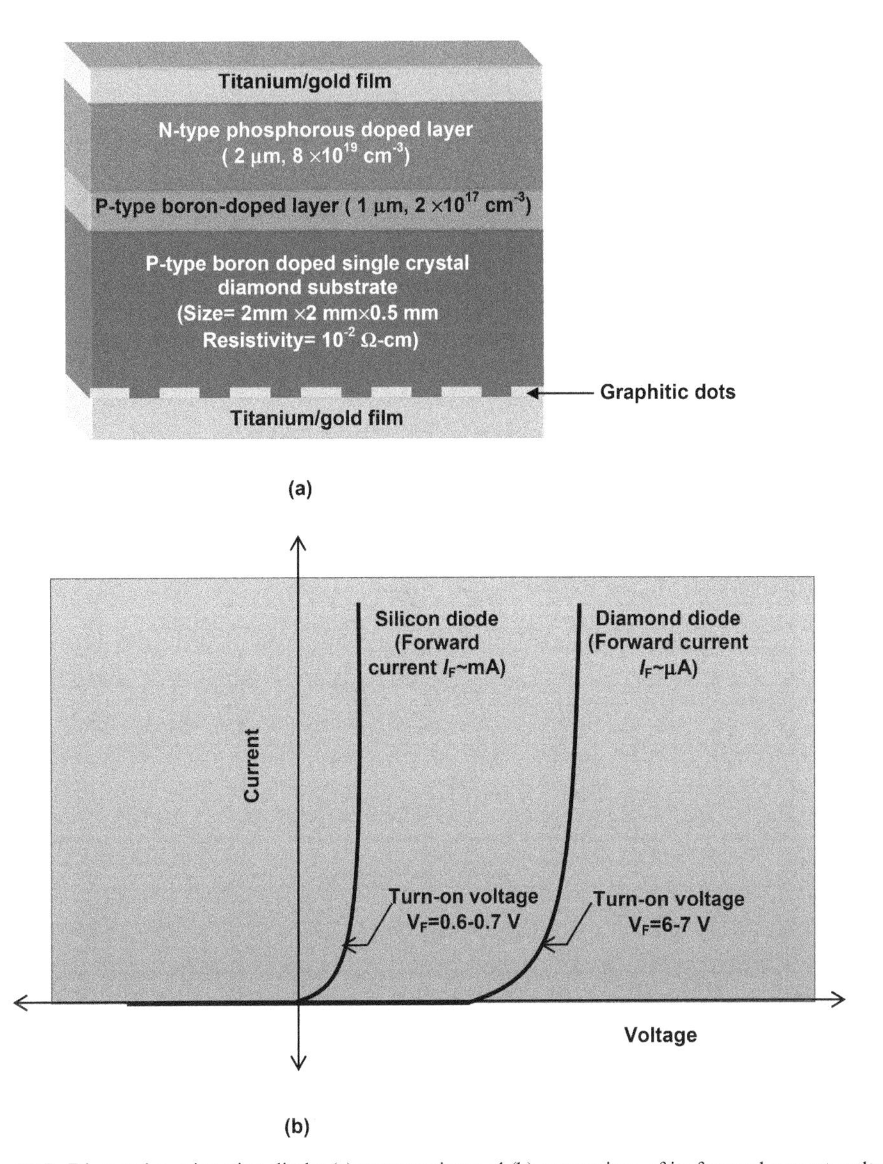

Figure 10.5. Diamond p–n junction diode: (a) construction and (b) comparison of its forward current–voltage characteristics with those of a silicon diode.

current increases. It happened because the amorphous Si contact layer re-crystallizes at ~873 K. At still higher temperatures, the SBD failed permanently as the W:Si diffusion barrier stopped working.

In order to operate SBDs at higher temperatures, the Au layer was not applied in some diodes. These diodes, fabricated without the Au film, were functional up to 1000 °C (Vescan *et al* 1997a). In these diodes as well, the reverse leakage current was boosted up above a temperature of 500 °C. Nonetheless, the slope of the reverse current was degraded along with decline of breakdown voltage to 30 V only when

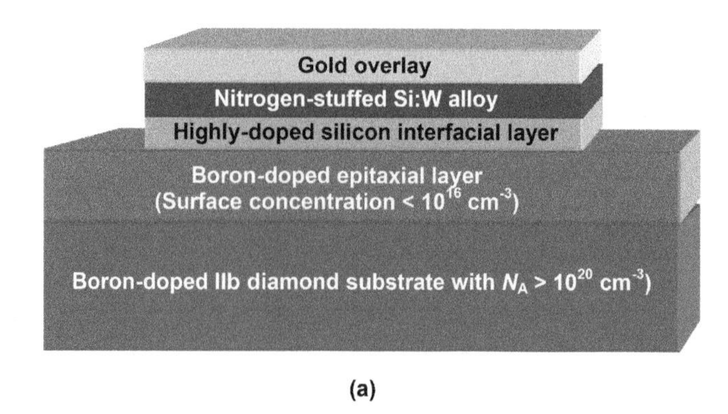

(a)

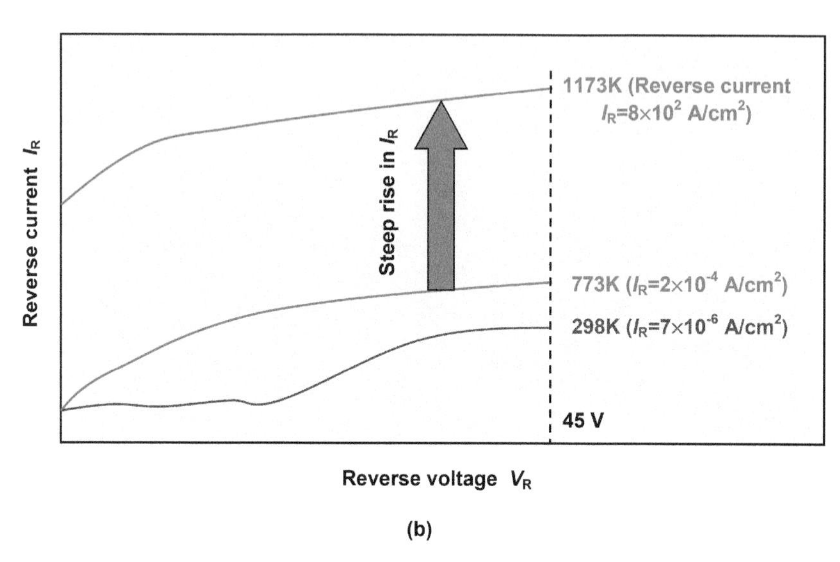

(b)

Figure 10.6. Diamond Schottky diode: (a) constituent layers, (b) satisfactory reverse operation up to 773 K and the sharp increase in leakage current above this temperature.

the diodes were stored at 1000 °C for 15 min before applying the voltage. The on-state current I_{on} was still 10 times the off-state current I_{off} at 1000 °C.

From the Richardson plot, the barrier height was estimated to be 1.9 eV. At low temperatures, the ideality factor departs from the unity value. This shows the presence of current generation mechanisms deviating from thermionic emission. As the temperature increases, the ideality factor approaches unity indicating that the thermionic emission current is the predominant component. Further, as the diodes were un-passivated, surface leakage was apparent above 800 °C.

Thus at high temperatures, the forward characteristics of the Schottky diode show perfect behavior with an ideality factor of 1.01, governed by thermionic emission through a 1.9 eV high barrier. The reverse characteristics responsible for the low I_{on}/I_{off} ratio point towards the presence of defects, which are activated by temperature (Vescan *et al* 1997a).

10.6.2 Long-term operation of diamond Schottky barrier diodes up to 400 °C

A methodical comparison of diamond SBDs against SiC and Si diodes was presented (Tatsumi *et al* 2009). Diamond SBDs were fabricated in a pseudo-vertical configuration. Starting with an insulating Ib (100) diamond substrate, p^+- and p-layers were formed by CVD. Both these layers were boron-doped. By increasing the microwave power during CVD, a large thickness of boron-doped p-layer (14 μm) was obtained. This enabled the realization of a high breakdown voltage of 2.8 kV.

Ti/Pt/Au ohmic electrodes were deposited to contact with the p^+-layer below the p-layer. The Pt Schottky electrodes with barrier height of 2 eV were deposited on the surface of the p-layer after subjecting it to oxidation treatment to make it insulating. This was a surface preparation step to form the proper interface between the surface and the contact. It was necessary because any hydrogen exposure had the adverse effect of making the surface conducting, thereby increasing the leakage current and reducing the Schottky barrier height. A breakdown field of 3.1×10^6 V cm^{-1} was attained, which was >2.4×10^6 V cm^{-1} for SiC p–n diode.

In comparison to Si or SiC, high temperature favorably affects the forward operation of a diamond Schottky diode. This difference originates from the position of impurity energy level in the bandgap. For Si, the energy level of the boron impurity is 44 meV above the valence band. The energy level for boron in SiC is 200 meV and in diamond it is 370 meV. Looking at these energy levels, it is evident that in silicon all impurities are ionized at room temperature and maximum conductivity is obtained at room temperature. As temperature increases, the mobility decreases and hence the conductivity decreases. In SiC, complete ionization of impurity atoms takes place at a higher temperature. Therefore, the temperature of maximum conductivity is higher than that for silicon. For the same reason, the temperature of the maximum conductivity of diamond is even higher than that for SiC. It is in the range of 100 °C to 200 °C. For a diamond SBD, the forward current density at 8 V is 3000 A cm^{-2}, which is threefold that available with SiC SBD.

Regarding the reverse characteristics, the leakage current of the SiC Schottky diode was very high at room temperature. It rose to 100 mA cm^{-2} at 162 °C at 1.5×10^6 V cm^{-1}. The leakage current of the diamond Schottky diode was reported to be less than the limit of measurement at room temperature. At 142 °C it was 0.1 mA cm^{-2} (Tatsumi *et al* 2009).

The long-term thermal stability of diodes with Ru Schottky electrodes was judged by keeping the diodes at 400 °C, and measuring the forward and reverse current–voltage characteristics after 100 h, 250 h, 500 h, 1000 h and 1500 h. The characteristics did not show any perceptible changes during this period, confirming that the diamond SBDs provide energy saving and high power capabilities at 400 °C (Tatsumi *et al* 2009).

10.7 Diamond BJT operating at <200 °C

Prins (1982) demonstrated bipolar transistor action in natural p-type diamond by implanting carbon to produce n-type regions. A low current gain was achieved. The prospects of bipolar diamond devices were examined by Aleksov and fellow workers

(Aleksov *et al* 2000). Heavily doped (1×10^{20} cm^{-3}) p-type boron-doped synthetic diamond crystals were used as the starting material. p^{+}-emitter (10^{20} cm^{-3}) and p^{-}-collector (10^{17} cm^{-3}) regions were grown using a solid boron source. The base region was formed by nitrogen doping (1.5×10^{18} cm^{-3}). Emitter contacts were made with Ti/Au while base contacts were formed with sputtered phosphorous-doped silicon capped with W–Si/Au. Measurements of the forward and reverse current–voltage characteristics of the diodes from 20 °C to 400 °C showed that the ideality factor was close to unity, pointing towards the domination of the diffusion current component, whereas above 10–15 V, the high value of the current was due to leakage currents arising from the strong electric field, estimated as 5×10^{5} $V^{-1}cm$, in the neutral base. The base has a high resistivity (10 Ωcm at 20 °C). Further, the reverse-biased junction showed large leakage currents, which together with base resistance limited the operation of diamond BJT to the nA range and temperatures <200 °C. The common-emitter current gain β was 1.1. The low gain was interpreted to be caused by high leakage current. It was shown theoretically that nitrogen-doped diamond BJTs can give a DC current gain of ~30 000 upon reduction of leakage current (Aleksov *et al* 2000).

10.8 Diamond MESFET

10.8.1 Hydrogen-terminated diamond MESFETs

Several research groups fabricated diamond MESFETs on hydrogen-terminated diamond with successively improved maximum drain current, transconductance, switching and other characteristics. Salient features of a few studies are outlined below.

An enhancement-mode metal semiconductor FET (MESFET) was fabricated using an aluminum Schottky contact on a p-type hydrogen-terminated homoepitaxial layer, and gold for the source and drain contacts (Kawarada *et al* 1994). The aluminum gate length was 10–40 μm. A transconductance g_m = 20–200 μS mm^{-1} was obtained.

Enhancement-mode MESFETs were fabricated by Gluche *et al* 1997 by growing 100 nm thick homoepitaxial diamond films on 1b (nitrogen doped) single crystal

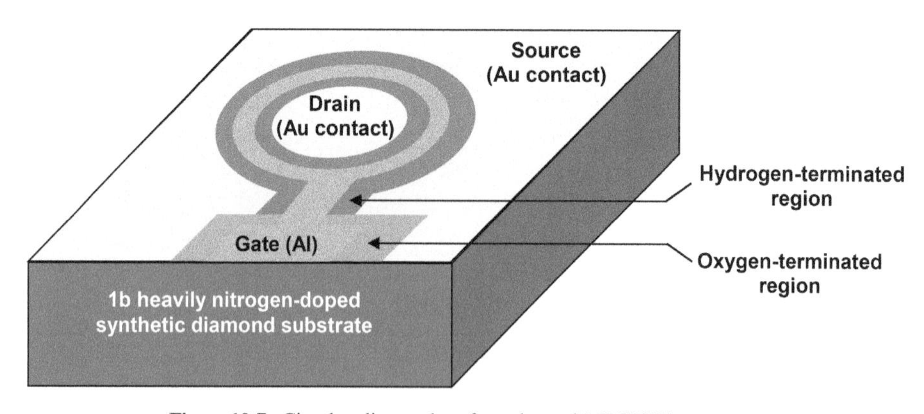

Figure 10.7. Circular diamond surface-channel MESFET structure.

diamond substrates by MPCVD at 595 °C using 1.5% CH_4 in H_2 (figure 10.7). As-grown surfaces were hydrogen-terminated showing p-type conductivity. Al Schottky and Au ohmic contacts were formed. A channel length of 3 μm gave a maximum drain current of 90 mA mm^{-1} and a 200 V gate–drain breakdown voltage (Gluche *et al* 1997).

Cu-gate diamond MESFETs of gate length 2–3 μm, fabricated on hydrogen-terminated MPCVD-grown diamond, gave a maximum transconductance = 70 mS mm^{-1} at V_{GS} = –1 5 V and V_{DS} = 5 V, with a cut-off frequency $f_T = 2 \times 10^9$ Hz and maximum frequency of oscillation f_{max} = 7 GHz (Taniuchi *et al* 2001).

Al-Schottky gated MESFETs with Au source/drain contacts and a channel length L_G = 0.2 μm gave a maximum drain current I_{Dmax} = 275 mA mm^{-1}, at gate voltage = –3.5 V, maximum transconductance g_m = 100 mS mm^{-1} and high cut-off frequencies: transition frequency $f_T = 24.6 \times 10^9$ Hz, maximum frequencies $f_{max(MAG)} = 63 \times 10^9$ Hz and $f_{max(U)} = 80 \times 10^9$ Hz, where MAG is the maximum available gain and U is unilateral gain (Kubovic *et al* 2004). The achieved saturated output power density was 0.35 W mm^{-1} at 1 GHz.

Kasu *et al* (2006) reported substantial improvement in RF output power. For a gate width of 1 mm and gate length of 0.4 μm, the maximum output power was 1.26 W. At an output power of 0.84 W, the device temperature rose by merely 0.6 °C. This was enabled by the extremely high thermal conductivity of diamond.

Al-Schottky gate diamond MESFETs of gate length 0.1 μm were fabricated on polycrystalline diamond films (Ueda *et al* 2006). This work deviated from the trend of using single crystal diamond to explore the polycrystalline material as an alternative (see figure 10.8). This was necessary because the commercial HPHT diamond substrate was 4 mm in size, whereas for mass manufacturing of diamond devices, a minimum of 4″ diameter diamond wafers are required. A quasi-2D hole channel was formed by H-passivation in an MPCVD system. For this device, $f_T \sim 45 \times 10^9$ Hz, $f_{max} \sim 129 \times 10^9$ Hz and g_m = 143 mS mm^{-1} at V_{DS} = –8 V (Ueda *et al* 2006).

For production of hole carriers and formation of a hole channel on the hydrogen-terminated diamond surface, the presence of some adsorbed molecules is necessary.

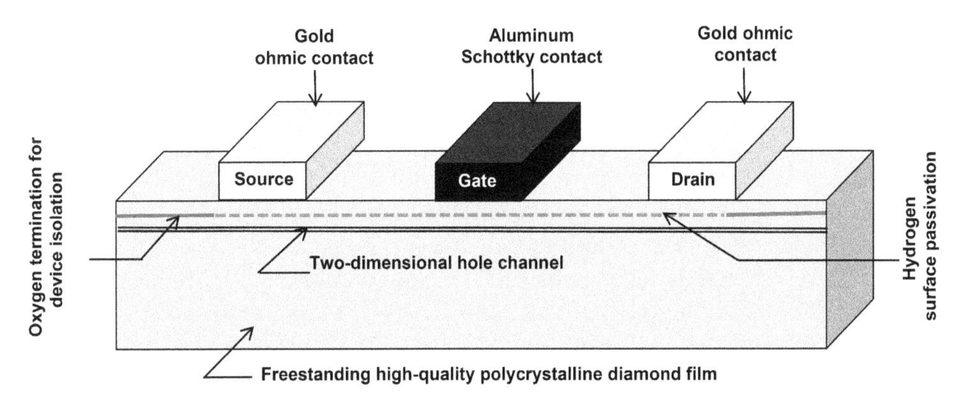

Figure 10.8. MESFET formed on polycrystalline diamond film.

Significant improvement in the characteristics of diamond transistors was observed on exposure to nitrogen dioxide (NO_2) gas (Kubovic and Kasu 2009). By this exposure, the hole sheet charge concentration increased to 1.3×10^{14} cm^{-2}. Maximum I_{DS} increased by a multiplication factor of 1.8, g_m by a factor of 1.5 and f_T by 1.6.

Hydrogen-terminated diamond MESFETs with gate lengths of 250, 120 and 50 nm were fabricated on a homoepitaxial diamond substrate (Russell *et al* 2015). For hydrogen termination, the substrate was exposed to hydrogen plasma for 30 min at 580 °C. The gate metal was Al/Au (25 nm/25 nm). The device had Au ohmic contacts on the source and drain. Extrinsic peak transconductance for 50 nm, 120 nm and 250 nm MESFETs were 78, 137 and 92 mS mm^{-1}, respectively, while the extrinsic f_T of respective devices was 53×10^9 Hz, 45×10^9 Hz and 19×10^9 Hz.

10.8.2 Electrical characteristics of diamond MESFETs in 20 °C to 100 °C temperature range

The above efforts concentrated on studying diamond MESFETs as a microwave or power device. A microwave power device needs scrupulous attention to thermal stability. The heat produced during device operation may considerably degrade its performance. Therefore, an extensive investigation of temperature effects on DC and RF characteristics of diamond MESFETs was called for (Ye *et al* 2006). This study was performed between 20 °C and 100 °C. At 20 °C, the maximum I_{DS} is 160 mA mm^{-1} which decreased to 120 mA mm^{-1} at 100 °C. The on-resistance was virtually temperature-independent, but the threshold voltage moved towards the negative side. Its value was 0.4 V at 20 °C and 0.18 V at 100 °C. The maximum transconductance showed a decline from 55 mS mm^{-1} at 20 °C to 48 mS mm^{-1} at 100 °C. The transition frequency f_T was 9×10^9 Hz at 20 °C. It decreased marginally to 8.5×10^9 Hz at 100 °C. The sheet resistance of the hydrogen-terminated surface maintained a constant value of 5 kΩsq in the range 20 °C to 150 °C, showing good thermal stability. On the whole, the device performance was satisfactory between 20 °C and 100 °C with the cut-off frequency nearly constant at 8–9 × 10^9 Hz (Ye *et al* 2006).

10.8.3 Hydrogen-terminated diamond MESFETs with a passivation layer

Passivation of the hole channel by an Al_2O_3 layer was found to provide a stable hole channel on the hydrogen-terminated surface up to 200 °C. NO_2 exposure was employed in conjunction with the Al_2O_3 passivation layer to increase the maximum drain current I_{DSmax} of the hydrogen-terminated MESFET up to 1.3 A mm^{-1} for a 1 μm gate length MESFET (Hirama *et al* 2012). The f_T was 10×10^9 Hz and f_{max} was 20×10^9 Hz over a V_{GS} range of 10 V. For the fabrication of Al_2O_3-passivated MESFETs, the H-terminated diamond surface between the source and drain contact regions was stabilized by exposure to NO_2 gas (2% in nitrogen) at atmospheric pressure. The exposure was performed for 35 min. Thereafter, the drain–source gap was covered with the 17 nm thick Al_2O_3 layer formed at 150 °C by ALD.

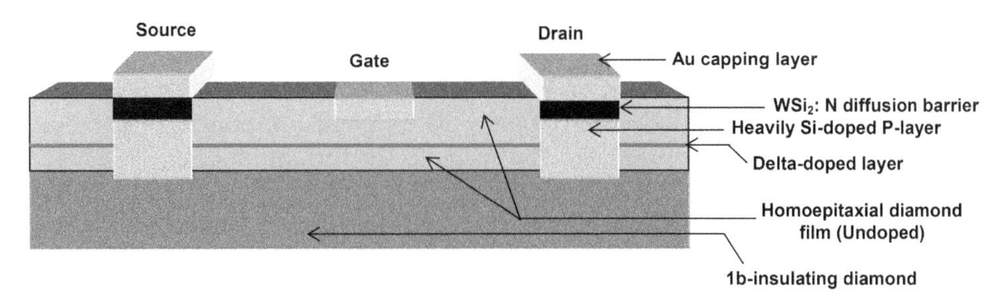

Figure 10.9. A delta-doped diamond MESFET with selectively grown ohmic contact zones.

In addition to Al_2O_3, several other dielectric materials have been used for passivation of diamond FETs, notable examples being AlN, HfO_2, $LaAlO_3$, Ta_2O_5, ZrO_2 and SiN_x (Wang *et al* 2015). The techniques used for this purpose include thermal evaporation, ALD and MOCVD.

10.8.4 Operation of pulse or delta boron-doped diamond MESFETs up to 350 °C

Boron dopant in diamond can be fully activated at room temperature if the peak doping concentration is in the range $\sim 10^{-20}$ cm^{-3}. Based on this condition for full activation, it is possible to fabricate an FET working on full charge activation. To achieve full modulation of the channel charge by the gate diode without exceeding the breakdown limit of the material, $\sim 3 \times 10^6$ to 1×10^7 V cm^{-1}, the total channel sheet charge must be $< 10^{13}$ cm^{-2}. The restriction applies when the thickness of the doped region is around 1–2 nm. This means that the doping profile required has a high peak concentration and a well-defined, narrow distribution. Such a distribution of dopant is defined by the delta (δ) function, and the semiconductor is said to be δ-doped or pulse-doped. The resultant spiky profile is a δ-doping profile or pulse profile.

For fabrication of a pulse-doped diamond MESFET (figure 10.9), a homoepitaxial diamond film was formed by MPCVD in CH_4 and H_2 plasma at 700 °C on a 1b insulating diamond substrate with nitrogen doping, followed by growth of an undoped 1 μm thick buffer layer (Vescan *et al* 1997b). A boron rod was inserted into the plasma for 5 s to obtain the pulse doping profile with peak concentration 10^{19} cm^{-3}. Selectively grown p^+-regions were formed for source and drain. Silicon-based metallization, metallurgically stable up to 750 °C, was employed both for Schottky and ohmic contacts. It consisted of a 50 nm thick contact layer made of highly doped sputter-deposited silicon, a 30 nm thick WSi_2:N diffusion barrier, and finally a 250 nm thick gold capping layer. Operation of the pulse-doped diamond MESFET was demonstrated up to 350 °C. The maximum drain–source voltage was 70 V. The maximum drain current was 35 μA mm^{-1} at room temperature. At 350 °C, it was 5 mA mm^{-1}. The maximum channel conductance was 0.22 mS mm^{-1} at 350 °C (Vescan *et al* 1997b).

10.8.5 Alternative approach to boron δ-doping profile

In another investigation (Aleksov *et al* 1999), a boron δ-doping profile was formed by eliminating any parasitic boron doping tails by compensation of p-type boron doping with n-type nitrogen doping. This leads to the formation of a p–n junction. In this junction, the n-type nitrogen doped portion is unactivated at room temperature. It behaves as a semi-insulating or lossy dielectric when the temperature is low and the frequency is high. But when the temperature is high and the frequency is low, the nitrogen dopant is activated, rendering the n-type portion conducting. Then this portion acts as a series resistance connected with the p–n junction. Utilizing this concept, two FET structures were studied. In the first structure, the nitrogen-doped diamond substrate served as a back gate. The second structure consisted of a nitrogen-doped gate layer formed on top of the δ-channel. These FETs showed drain currents up to 100 mA mm^{-1}. The channel was fully modulated at temperatures ~200 °C to 250 °C.

10.9 Diamond JFET

10.9.1 Operation of diamond JFETs with lateral p–n junctions up to 723 K

A boron-doped p-type channel is sandwiched between two selectively grown highly phosphorous-doped n$^+$ sidewall gates (Iwasaki *et al* 2013), see figure 10.10. Control of the depletion region thickness in the channel region is accomplished by applying the same voltage on both gates. The contact metallization scheme is Ti (30 nm)/Pt (30 nm)/Au (100 nm). By controlling the depletion regions from two sides, the channel width and therefore the drain current can be modulated more effectively than from a single side. Measurement of the I_{DS}–V_{DS} characteristics at 300 K and 673 K showed that:

(i) clearly distinguishable linear and saturation regions are obtained at both temperatures;
(ii) the current density obtained at 673 K is 1300 A cm^{-2}. It is 50 times the current density at 300 K, which is 25 A cm^{-2}; and
(iii) the specific on-resistance is 52.2 × 10^{-3} Ω cm^2 at 300 K and plummets to 1.8 × 10^{-3} Ω cm^2 at 723 K. Saturation of specific on-resistance was noticed above 500 K.

From the I_{DS}–V_{GS} transfer characteristics (Iwasaki *et al* 2013):

(i) the leakage current was found to be extremely low ~10^{-14} A up to 673 K;
(ii) the maximum transconductance at 300 K is 1.6 μS mm^{-1}. At 723 K, it rises to 34.6 μS mm^{-1}, increasing 22 times.

A high ratio between on-state and off-state currents $I_{on}/I_{off} > 10^6$ K is maintained from 300 K to 673 K with a peak value of 4 × 10^7 at 623 K. A very steep subthreshold swing, SS, is observed. It is nearly ideal, and this is particularly true for high temperatures. An interesting achievement is the attainment of high-voltage operation at high temperature with exceedingly low leakage current from

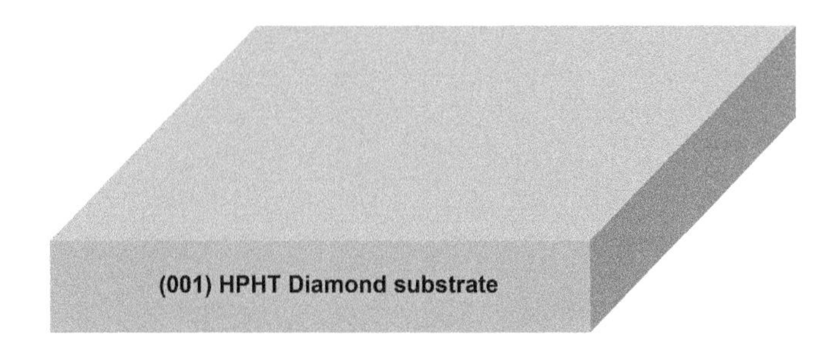

(a) Starting high-pressure high-temperature (HPHT) diamond substrate

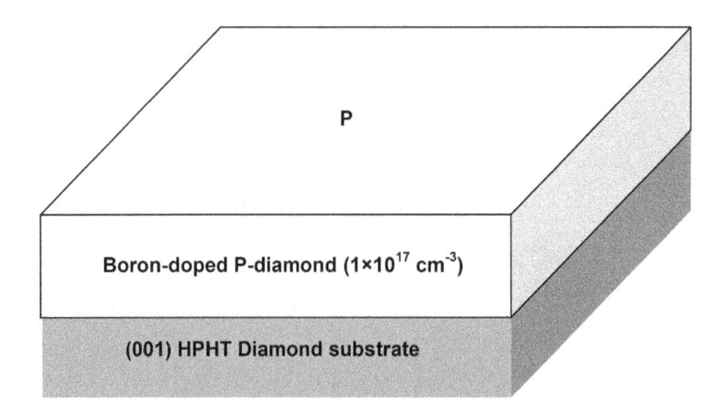

(b) P-layer formation by microwave plasma chemical vapor deposition (MPCVD).

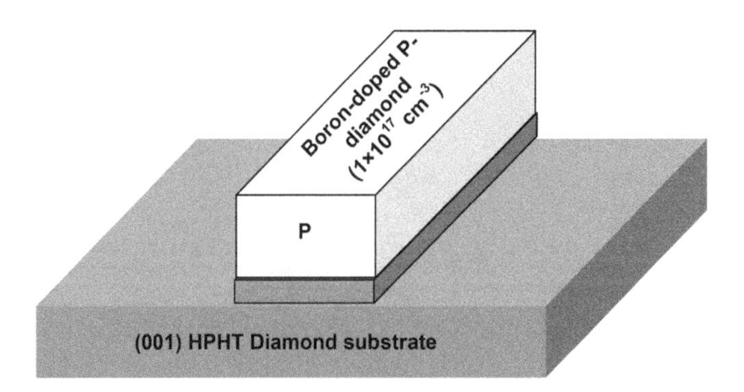

(c) Patterning of the P-layer by electron beam (e-beam) lithography followed by inductively-coupled plasma etching (ICP).

Figure 10.10. Fabrication process steps of diamond JFET with selectively deposited n^+ side gates.

the lateral p–n junctions between the drain and gate. From the transfer curves of the JFET (figure 10.11) plotted at 423 K, the leakage current was found to be $\sim 10^{-14}$ A at $V_{DS} = -10$ V; the same is true even for $V_{DS} = -100$ V (Iwasaki *et al* 2013). The low leakage current confirms its berth as a futuristic power device.

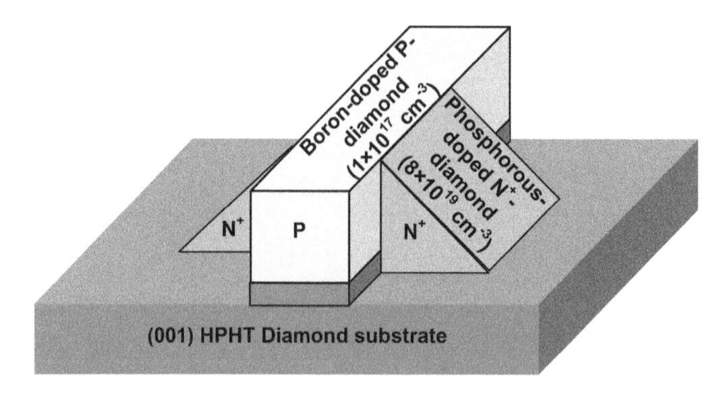

(d) Selectively-grown N^+ diamond forming a lateral P-N junction with the P-layer. The P-channel is sandwiched between N^+ side gates from its two sides.

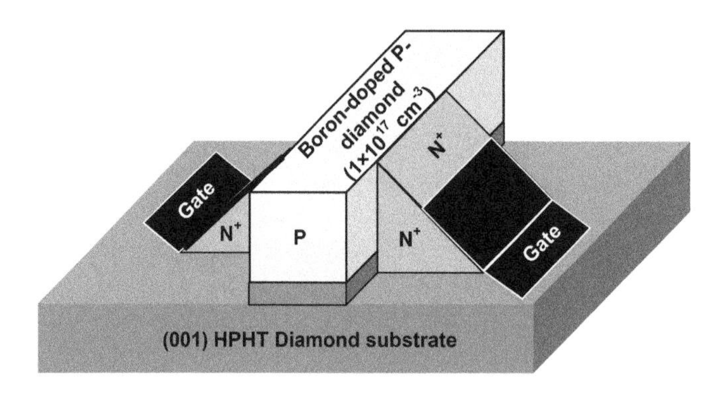

(e) Electrode deposition on the walls of side gates.

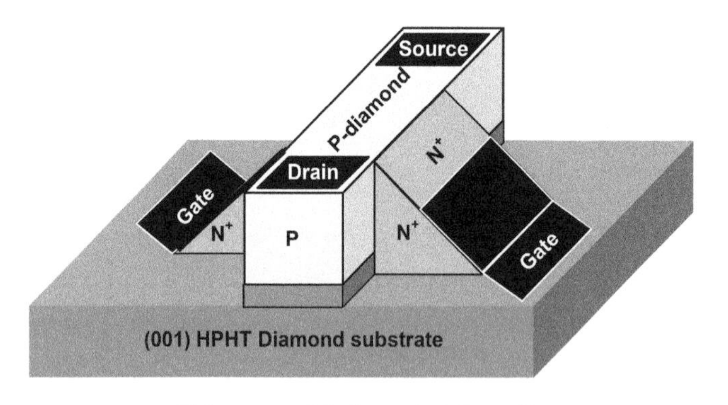

(f) Source and drain contact deposition

Figure 10.10. (Continued.)

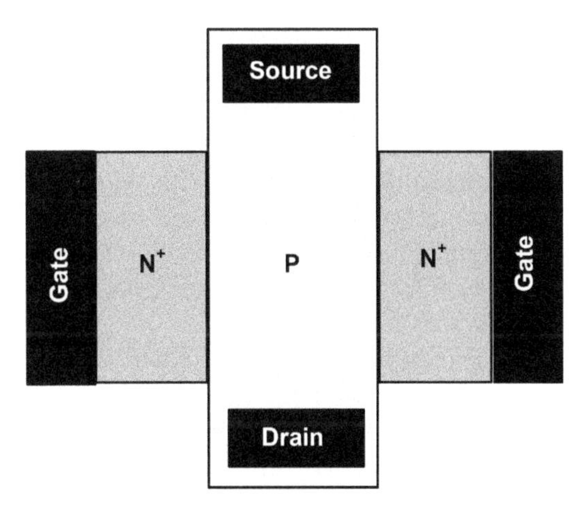

(g) Top view of the fabricated JFET structure showing the P-channel and N$^+$ side gates with contact metallization.

Figure 10.10. (Continued.)

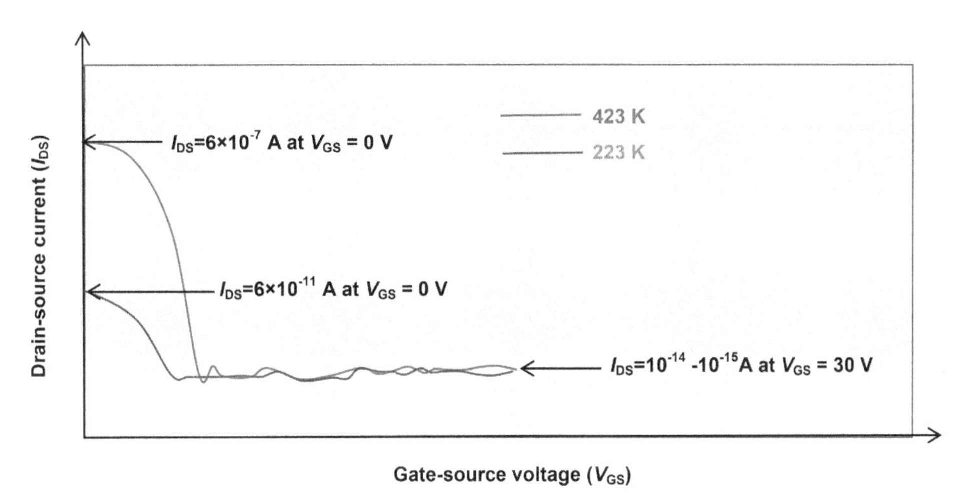

Figure 10.11. Transfer characteristics of diamond JFET at two temperatures at $V_{DS} = -10$ V.

10.10 Diamond MISFET

Matsudaira *et al* (2004) fabricated 0.2–0.5 μm gate length Cu/CaF_2/diamond MISFETs with evaporated CAF_2 as the gate insulator on p-type hydrogen-terminated homoepitaxial diamond films deposited on synthetic 1b (001) substrates by MPCVD using H_2-diluted CH_4 source gas (4%). With a 0.2 μm gate length, the highest f_T was 23×10^9 Hz and f_{max} was 25×10^9 Hz.

High-quality IIa-type polycrystalline CVD diamond has been used as a substrate for fabrication of diamond MISFETs (Hirama *et al* 2007a). A hole accumulation layer was used as the p-channel. Naturally oxidized evaporated Al formed the

Al_2O_3 gate insulator. On a 0.1 μm gate length device, f_T was 42 × 10^9 Hz. The maximum drain current was 650 mA mm^{-1}.

p-channel MOSFETs were fabricated with an Al_2O_3 gate insulator on (001) homoepitaxial and (110) preferentially oriented CVD diamond films of large grain size (figure 10.12). The gate electrode was formed with Al_2O_3/Al by oxidation of 3 nm thick Al film. Ohmic contacts of the drain and source were formed with gold. The devices showed a channel mobility of 20 cm^2 V s^{-1}, a drain–source current I_{DS} = −790 mA mm^{-1}, f_T = 45 × 10^9 Hz and a power density of 2.14 W mm^{-1} at 1 GHz (Hirama *et al* 2007b).

Let us now look at the studies conducted regarding MISFET operation in the 10–673 K range. Thermally stable two-dimensional hole gas (2DHG) up to 800 K was obtained at Al_2O_3/H-terminated C-H diamond interface using Al_2O_3 formed by high-temperature ALD process. MISFET performance was preserved in the temperature range 10–673 K (Kawarada *et al* 2014). The substrate used is 1b (100) synthetic diamond (figure 10.13). It is nitrogen doped at ~10^{17} cm^{-3}. On this diamond

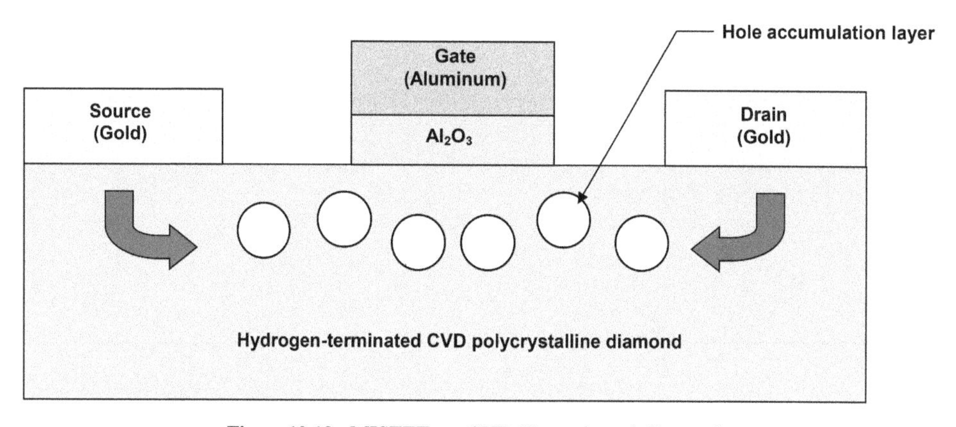

Figure 10.12. MISFET on CVD H-terminated diamond.

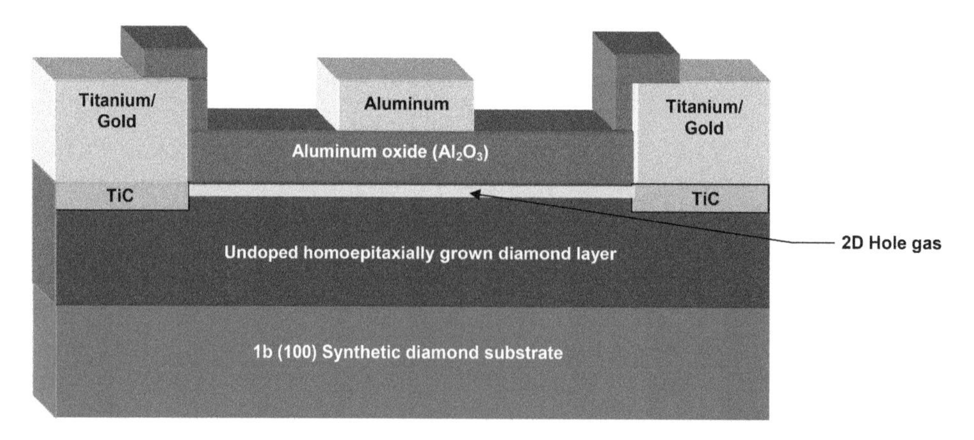

Figure 10.13. Diamond MOSFET with alumina gate dielectric and H-terminated channel.

substrate, a 0.5 μm thick undoped diamond layer is homoepitaxially grown by MPCVD. Excluding the regions occupied by source and drain metallization, the top surface is hydrogenated. Oxidation of the surface serves to isolate the different regions. The gate is insulated and passivated with a 10–200 nm thick alumina layer formed by ALD. This alumina layer also serves as the gate dielectric. Ohmic contacts to the source and drain are formed with Ti/Au layers and the gate metallization is Al film.

The I_{DS}–V_{GS} characteristics of a diamond MISFET with a gate insulator thickness of 32 nm were plotted at room temperature, 100 °C, 200 °C, 300 °C and 400 °C. For each case, the on-state current was ~30 A. At room temperature, the off-state current was 8×10^{-8} A, giving a ratio $I_{on}/I_{off} = 30/(8 \times 10^{-8}) = 3.75 \times 10^{8}$. At 300 °C, the off-state current increased to 6×10^{-4} A. Then the ratio $I_{on}/I_{off} = 30/(6 \times 10^{-4}) = 5 \times 10^{4}$. At 400 °C, $I_{off} = 2 \times 10^{-2}$ A, so that $I_{on}/I_{off} = 30/(2 \times 10^{-2}) = 1.5 \times 10^{3}$. For this MISFET, the maximum drain current I_{DSmax} was −23 mA mm^{-1} at 10 K. It began at −38 mA mm^{-1} at room temperature and changed to −35 mA mm^{-1} at 400 °C. Hence, the I_{DSmax} at a particular bias varies by <50% from room temperature to 400 °C.

The breakdown voltage was 996 V without any field plate. The maximum breakdown field was 3.6×10^{6} V cm^{-1}. In further improvement, the drain current was 100 mA mm^{-1} with a breakdown voltage >1600 V (Kawarada *et al* 2015). The values achieved are superior to those of diamond MESFETs and SiC planar MOSFETs.

10.11 Discussion and conclusions

Diamond provides a multitude of useful physical and chemical properties of which only a small percentage have been utilized. Diamond can be synthesized by CVD. It can be doped as n-type by nitrogen and phosphorous. p-type doping is performed through boron and hydrogen termination. But there are many practical problems. The energy levels of these dopants lie deep inside the bandgap, so they are not activated at room temperature as in silicon. Hydrogen termination is unstable and needs to be properly stabilized and protected. Diamond Schottky diodes have performed well up to 1000 °C. The current gain of BJT must be increased. Boron delta-doped MESFETs have been demonstrated to function up to 350 °C while MESFETs fabricated by hydrogen termination with a passivation layer worked at lower temperatures. Functional JFETs up to 723 K and MISFETs in the temperature range 10 K to 673 K have been reported, asserting their capabilities for high-temperature applications. Some technological challenges have been successfully dealt with, while many others are waiting to be solved if the gigantic opportunities offered by diamond are to be tapped.

Review exercises

10.1. Compare diamond with Si, GaAs, SiC and GaN regarding its bandgap, carrier mobility, breakdown field, thermal conductivity and fissure strength. Explain with proper justification why diamond is said to be a 'true dream material' of microelectronic device design engineers.

10.2. If silicon electronics is described as attaining maturity and touching the basic physical limits, how would you describe diamond electronics? Calculate the intrinsic temperature of diamond.

10.3. What are the four classes of diamond called? Mention the features of each class.

10.4. What do the terms HPHT diamond and CVD diamond stand for? How is diamond synthesized by CVD? How is gem-quality diamond obtained?

10.5. Why have conventional doping techniques such as thermal diffusion and ion implantation not been able to provide high carrier concentrations in diamond? Name two n-type doping elements that have been used for diamond. What are their energy level depths below the conduction band?

10.6. Discuss the use of boron as a p-type doping impurity in diamond. What is the effect of boron concentration on its activation energy? How does the presence of hydrogen in the plasma affect *in situ* boron doping during CVD?

10.7. How is p-type doping accomplished in diamond by surface hydrogenation? What is the typical depth below the diamond surface within which most of the charge carriers lie? How does this channel containing holes differ from a boron-doped layer?

10.8. Is the p-type diamond surface formed by hydrogenation stable from chemical and temperature viewpoints? How can it be destroyed? How can it be protected?

10.9. Can a hydrogen-terminated diamond surface be used as the channel of an FET?

10.10. How does an oxygen-terminated diamond surface behave? How is oxygen termination of a diamond surface achieved? Can a hydrogen-terminated diamond surface be modified in this way? What is the result of this modification?

10.11. Describe the fabrication of an UV emitting diamond p–n junction diode using boron and phosphorous as doping impurities. Discuss its optical characteristics.

10.12. Describe one method of fabrication of a diamond Schottky diode using a nitrogen doped Si:W alloy as the Schottky metal. What is the Schottky barrier height? What is the ideality factor at low temperatures? What is its value at high temperatures? What do these values signify?

10.13. Why is oxygen treatment necessary before deposition of a Schottky contact? What happens if it this step is omitted? Up to what temperature does the Schottky diode perform satisfactorily?

10.14. The energy level for boron impurity in silicon is 44 meV above the valence band. In SiC the energy position for boron is at 200 meV and in diamond it is at 370 meV. Explain how high temperature affects the forward operation of a diamond Schottky diode more favorably than for a SiC diode. Do the same remarks hold true for a comparison between a SiC diode with a Si diode? If so, how?

10.15. Compare the relative magnitudes of reverse leakage current at room temperature and at high temperature in a SiC Schottky diode with a diamond Schottky diode.
10.16. How are emitter, base and collector regions of a diamond BJT formed? What contact materials are used for these regions? What is the common-emitter current gain achieved? Why is it so low? How can the current gain be improved? How much current gain is expected theoretically after these improvements are made?
10.17. Describe the structure and fabrication of a hydrogen-terminated diamond MESFET without any passivation layer. Give typical experimental results on electrical characteristics. How does the MESFET perform in the 20–100 °C temperature range?
10.18. What is the effect of the Al_2O_3 passivation layer deposition in a hydrogen-terminated MESFET on the hole carrier concentration? How are the RF small-signal characteristics of the MESFET improved by this passivation layer? What other materials are used for passivation apart from Al_2O_3? What deposition techniques are used for preparing these layers?
10.19. What is meant by a delta doping profile? How does it provide full charge activation for the boron dopant at room temperature without exceeding the critical breakdown field of boron?
10.20. How is a delta doping profile of boron achieved by inserting a boron rod into the plasma? How is a diamond MESFET fabricated by this method? What is the maximum drain current at room temperature? What is the drain current value at 350 °C?
10.21. How is the delta doping profile formed by elimination of boron tails by compensating with opposite type of impurity? How is the MESFET fabricated based on this principle? At what temperatures is the channel fully modulated?
10.22. How is the channel doping performed in a diamond JFET? How are the sidewall gates doped? What is the contact metallization scheme used? How does this JFET perform electrically at 300 K and 673 K? What electrical features assure its bright prospects as a diamond microelectronic device for high-temperature operation?
10.23. Describe the fabrication of a diamond MISFET with thermally stable 2DHG up to 800 K. What is the ratio I_{on}/I_{off} at room temperature and at 400 °C? What is the maximum drain current at 10 K and at 400 °C? How does its performance compare with that of a MESFET?

References

Aleksov A, Denisenko A and Kohn E 2000 Prospects of bipolar diamond devices *Solid-State Electron.* **44** 369–75

Aleksov A, Vescan A, Kunze M, Gluche P, Ebert W, Kohn E, Bergmeier A and Dollinger G 1999 Diamond junction FETs based on δ-doped channels *Diam. Relat. Mater.* **8** 941–5

Bundy F P, Hall H T, Strong H M and Wentorf Jun R H 1955 Man-made diamonds *Nature* **176** 51–5

Cao G Z, Giling L J and Alkemade P F A 1995 Growth of phosphorus and nitrogen co-doped diamond films *Diam. Relat. Mater.* **4** 775–9

Fontaine F, Uzan-Saguy C, Philosoph B and Kalish R 1996 Boron implantation *in situ* annealing procedure for optimal p-type properties of diamond *Appl. Phys. Lett.* **68** 2264–6

Gluche P, Aleksov A, Vescan A, Ebert W and Kohn E 1997 Diamond surface-channel FET structure with 200 V breakdown voltage *IEEE Electron Device Lett.* **18** 547–9

Hasegawa M, Takeuchi D, Yamanaka S, Ogura M, Watanabe H, Kobayashi N, Okushi H and Kajimura K 1999 n-type control by sulfur ion implantation in homoepitaxial diamond films grown by chemical vapor deposition *Jpn. J. Appl. Phys.* 38, Part 2, No. 12B, 15 December 1999, pp L 1519–L 1522

Hirama K, Sato H, Harada Y, Yamamoto H and Kasu M 2012 Diamond field-effect transistors with 1.3A/mm drain current density by Al_2O_3 passivation layer *Japan J. Appl. Phys.* **51** 090112

Hirama K, Takayanagi H, Yamauchi S, Jingu Y, Umezawa H and Kawarada H 2007a Diamond MISFETs fabricated on high quality polycrystalline CVD diamond *Proc. 19th Int. Symp. on Power Semiconductor Devices and ICs (Jeju, Korea, 27–30 May)* pp 269–72

Hirama K, Takayanagi H, Yamauchi S, Jingu Y, Umezawa H and Kawarada H 2007b High-performance p-channel diamond MOSFETs with alumina gate insulator *IEEE Int. Electron Devices Meeting (Washington, DC, 10–12 December)* pp 873–6

Iwasaki T, Hoshino Y, Tsuzuki K, Kato H, Makino T, Ogura M, Takeuchi D, Okushi H, Yamasaki S and Hatano M 2013 High-temperature operation of diamond junction field-effect transistors with lateral p–n junctions *IEEE Electron Device Lett.* **34** 1175–7

Kasu M, Ueda K, Ye H, Yamauchi Y, Sasaki S and Makimoto T 2006 High RF output power for H-terminated diamond FETs *Diam. Relat. Mater.* **15** 783–6

Kato H, Makino T, Yamasaki S and Okushi H 2007 n-type diamond growth by phosphorus doping *MRS Proc.* **1039** 1039-P05-01

Kawarada H 1996 Hydrogen-terminated diamond surface and interface *Surf. Sci. Rep.* **26** 205–59

Kawarada H, Aoki M and Ito M 1994 Enhancement mode metal semiconductor field effect transistors using homoepitaxial diamonds *Appl. Phys. Lett.* **65** 1563–5

Kawarada H, Yamada T, Xu D, Tsuboi H, Saito T and Hiraiwa A 2014 Wide temperature (10 K–700 K) and high voltage (1000 V) operation of C–H diamond MOSFETs for power electronics application *IEEE Int. Electron Devices Meeting (San Francisco, CA, 15–17 December)* pp 11.2.1–4

Kawarada H, Yamada T, Xu D, Tsuboi H, Saito T, Kitabayashi Y and Hiraiwa A 2015 Diamond power MOSFETs using 2D hole gas with >1600 V breakdown *6th NIMS/MANA-Waseda University International Symposium (Tokyo, 29 July)*, 1 page

Koizumi S 2006 n-type doping of diamond *MRS Proc.* **956** 0956-J04-01

Koizumi S, Watanabe K, Hasegawa M and Kanda H 2001 Ultraviolet emission from a diamond p–n junction *Science* **292** 1899–901

Kubovic M and Kasu M 2009 Improvement of hydrogen-terminated diamond field effect transistors in nitrogen dioxide atmosphere *Appl. Phys. Express* **2** 086502

Kubovic M, Kasu M, Kallfass I, Neuburger M, Aleksov A, Koley G, Spencer M G and Kohn E 2004 Microwave performance evaluation of diamond surface channel FETs *Diam. Relat. Mater.* **13** 802–7

Matsudaira H, Miyamoto S, Ishizaka H, Umezawa H and Kaw H 2004 Over 20-GHz cutoff frequency submicrometer-gate diamond MISFETs *IEEE Electron Device Lett.* **25** 480–2

Prawer S, Nugent K W and Jamieson D N 1997 The Raman spectrum of amorphous diamond *Diam. Relat. Mater.* **7** 106–10

Prins J F 1982 Bipolar transistor action in ion implanted diamond *Appl. Phys. Lett.* **41** 950–2

Prins J F 1988 Activation of boron-dopant atoms in ion-implanted diamonds *Phys. Rev. B* **38** 5576–84

Russell S, Sharabi S, Tallaire A and Moran D A J 2015 RF operation of hydrogen-terminated diamond field effect transistors: a comparative study *IEEE Trans. Electron Devices* **62** 751–6

Saada D 2000 *p-type Diamond* http://phycomp.technion.ac.il/~david/thesis/node16.html

Taniuchi H, Umezawa H, Arima T, Tachiki M and Kawarada H 2001 High-frequency performance of diamond field-effect transistor *IEEE Electron Device Lett.* **22** 390–2

Tatsumi N, Ikeda K, Umezawa H and Shikata S 2009 Development of diamond Schottky barrier diode *SEI Tech. Rev.* **68** 54–61

Ueda K, Kasu M, Yamauchi Y, Makimoto T, Schwitters M, Twitchen D J, Scarsbrook G A and Coe S E 2006 Diamond FET using high-quality polycrystalline diamond with f_T of 45 GHz and f_{max} of 120 GHz *IEEE Electron Device Lett.* **27** 570–2

Uzan-Saguy C, Kalish R, Walker R, Jamieson D N and Prawer S 1998 Formation of delta-doped, buried conducting layers in diamond, by high-energy, B-ion implantation *Diam. Relat. Mater.* **7** 1429–32

Vescan A, Daumiller I, Gluche P, Ebert W and Kohn E 1997a Very high temperature operation of diamond Schottky diode *IEEE Electron Device Lett.* **18** 556–8

Vescan A, Gluche P, Ebert W and Kohn E 1997b High-temperature, high-voltage operation of pulse-doped diamond MESFET *IEEE Electron Device Lett.* **18** 222–4

Wang W, Hu C, Li S Y, Li F N, Liu Z C, Wang F, Fu J and Wang H X 2015 Diamond based field-effect transistors of a Zr gate with SiN_x dielectric layers *J. Nanomater.* **2015** 124640, 5 pages

Yamada T, Kojima A, Sawabe A and Suzuki K 2004 Passivation of hydrogen terminated diamond surface conductive layer using hydrogenated amorphous carbon *Diam. Relat. Mater.* **13** 776–9

Yan C-S, Vohra Y K, Mao H-K and Hemley R J 2002 Very high growth rate chemical vapor deposition of single-crystal diamond *Proc. Natl Acad. Sci.* **99** 12523–5

Ye H, Kasu M, Ueda K, Yamauchi Y, Maeda N, Sasaki S and Makimoto T 2006 Temperature dependent DC and RF performance of diamond MESFET *Diam. Relat. Mater.* **15** 787–91

Chapter 11

High-temperature passive components, interconnections and packaging

Progress in passive electronic devices for HTE must proceed hand-in-hand with the active devices using wide bandgap semiconductors. A passive component failure is no less catastrophic than that of an active one. Passive components cannot be ignored because their role is equally important for circuit operation as active components. However, they have hitherto received much less attention. This chapter surveys the passive component aspects of HTE, including resistors, capacitors and inductors. Several new metallization schemes proposed for interconnections are also examined. The packaging and housing needs of HTE are addressed. Most low-cost plastic packages are unable to cope with the thermal challenges, leaving the scenario in favor of metallic hermetic seals.

11.1 Introduction

To assemble an electronic circuit, active devices are the vital components. But together with active devices, several passive devices are required. These include resistors, capacitors, inductors and conductors/interconnections. This chapter will deal with these circuit components, and will also be concerned with packaging of the assembly.

11.2 High-temperature resistors

11.2.1 Metal foil resistors

A metal foil resistor (figure 11.1) consists of a metal alloy, e.g. nichrome with additives cemented to a ceramic substrate by an adhesive. The foil is photo-etched to form a metal pattern. Foil resistors can survive temperatures up to 240 °C (Hernik 2012). A low TCR is achieved by the balancing effect of the increase in resistance with rising temperature and the decrease in resistance by compressive forces generated in the foil bound to the substrate by its thermal expansion.

doi:10.1088/978-0-7503-1155-7ch11

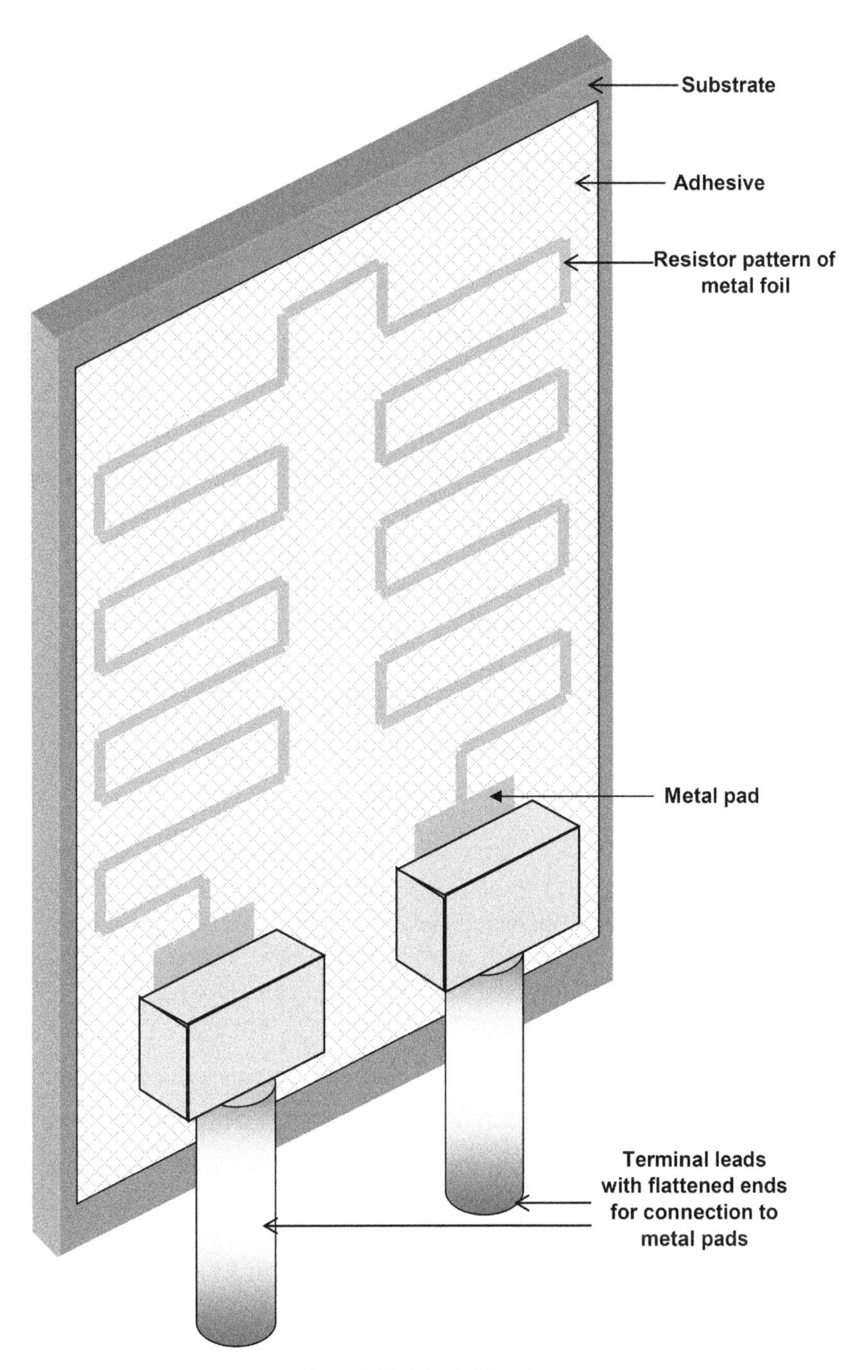

Figure 11.1. Metal foil resistor.

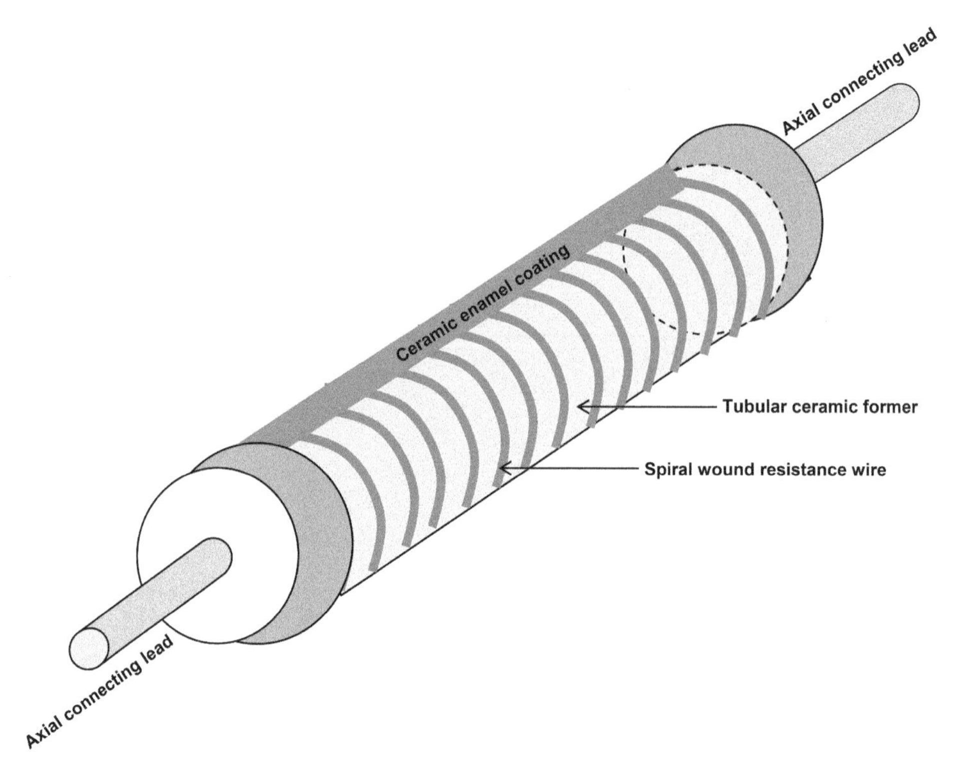

Figure 11.2. Wire wound resistor.

11.2.2 Wirewound resistors

A wirewound resistor (figure 11.2) is formed by winding long strands of insulated resistive wire, e.g. 20Cr–80Ni (nichrome) or 86Cu–2Ni–12Mn (manganin) or 72Fe–20Cr–5Al–3Co (kanthal A) or W (tungsten) around a non-conducting core such as a ceramic bobbin. Wirewound resistors have been reported to work from −55 °C to 275 °C (Ebbert 2014). Applying derating, they can operate up to still higher temperatures.

11.2.3 Thin-film resistors

A thin-film resistor (figure 11.3) is fabricated by thermal evaporation or sputtering of a metallic film 0.1 μm thick, e.g. NiCr or TaN (tantalum nitride), on a substrate, and then photolithographically etching the film into a resistive metal pattern. A typical example is as follows: film material: nichrome (NiCr); substrate: alumina (Al_2O_3); passivation: silicon nitride (Si_3N_4); protection: epoxy + silicone; terminations: gold <1 μm over nickel; operating temperature range: −55 °C to 215 °C; TCR: 15 ppm; structure: wrap-around chip resistor (Vishay 2015).

11.2.4 Thick-film resistors

A thick-film resistor (figure 11.4) is made by screen and stencil printing of a mixture of a glassy frit binder with a carrier of organic solvents and plasticizers, and metal

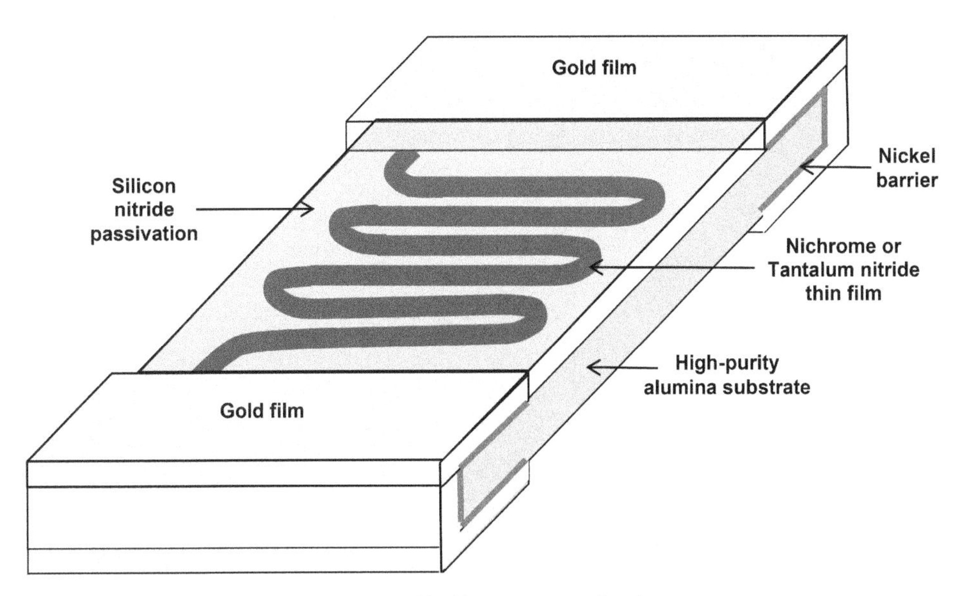

Figure 11.3. Thin-film wrap-around resistor.

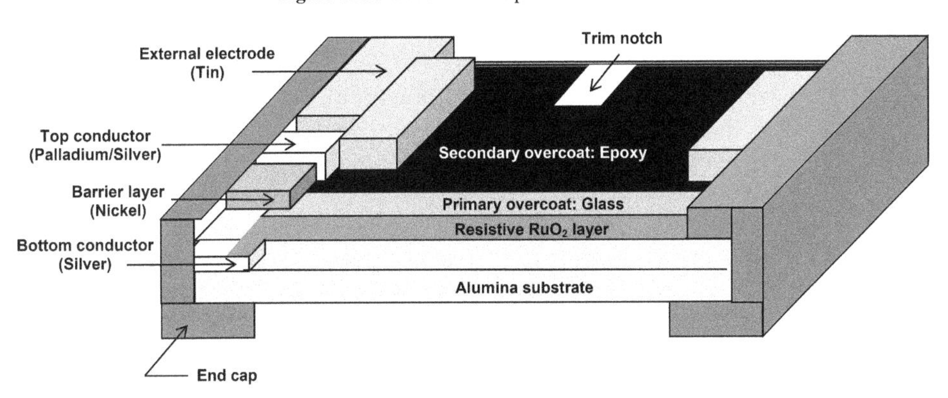

Figure 11.4. Anatomy of a thick-film chip resistor.

oxides of ruthenium, iridium and rhuthenium. The resistor film is ~100 μm thick. An example is as follows: film material: ruthenium oxide (RuO_2); substrate: alumina (Al_2O_3); protection: organic; terminations: gold <1 μm over nickel; operating temperature range: −55 °C to 215 °C; structure: chip resistor (Vishay 2014).

Thick-film-on-steel resistors for operation up to 400 °C are made by binding a thick ceramic dielectric glaze on a substrate of stainless steel (Morrison 2000). The steel substrate offers the advantages of bendability into desired shapes, and welding. Hence, the resistors can be shaped and fitted to the dimensions required by the application.

11.3 High-temperature capacitors

11.3.1 Ceramic capacitors

These capacitors (figure 11.5) comprise a ceramic material as the dielectric interposed between two conducting layers, e.g. a silver-plated ceramic disc.

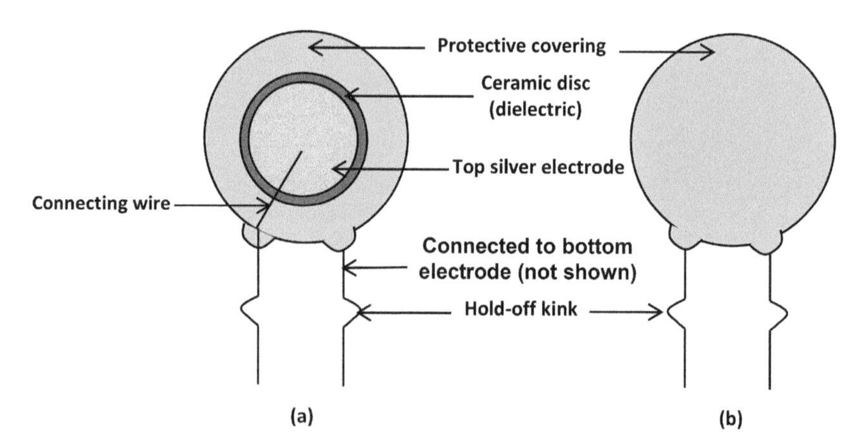

Figure 11.5. Ceramic capacitor: (a) without protective covering and (b) with protective covering.

The ceramic dielectric belongs to class 1 of the classification by the Electronics Industries Association (EIA). Class 1 dielectrics are of the negative positive zero (NP0) type, also referred to as C0G. They are the most stable formulations with respect to temperature. They are based on para-electric, i.e. non-ferroelectric, materials, e.g. TiO_2. They have a TC of 0 ± 30 ppm from −55 °C to +125 °C. From the thermal stability viewpoint, they are the most stable type of ceramic dielectrics. From the volumetric efficiency standpoint, they have the lowest efficiency due to which they are physically larger than other capacitors.

For compactness of size, a high dielectric constant is required. A dielectric based on $(1-x)(0.6Bi_{1/2}Na_{1/2}TiO_3-0.4Bi_{1/2}K_{1/2}TiO_3)-xK_{0.5}Na_{0.5}NbO_3$ has been reported (Dittmer *et al* 2012) to show a deviation of ±10% in relative permittivity of 2167 in the temperature range 54 °C to 400 °C for $x = 0.15$. The temperature insensitivity of this dielectric was ascribed to the existence of two types of polar nanoregions providing dissimilar relaxation mechanisms.

11.3.2 Solid and wet tantalum capacitors

Tantalum capacitors are of two types: solid and wet. These capacitors are electrolytic capacitors made with a tantalum anode.

In a solid tantalum capacitor (figure 11.6), the Ta anode is formed by compressing Ta powder around a Ta wire called the riser wire and sintering it at 1500°C to 1800 °C for pellet formation. The tantalum pellet is oxidized by anodization to tantalum pentoxide:

$$2Ta \rightarrow 2Ta^{5+} + 10e^-; 2Ta^{5+} + 10\,OH^- \rightarrow Ta_2O_5 + 5H_2O. \tag{11.1}$$

The tantalum pentoxide, thus electrochemically formed, has a dielectric constant of 26, providing high capacitance per unit volume. By immersion of the Ta_2O_5 grown

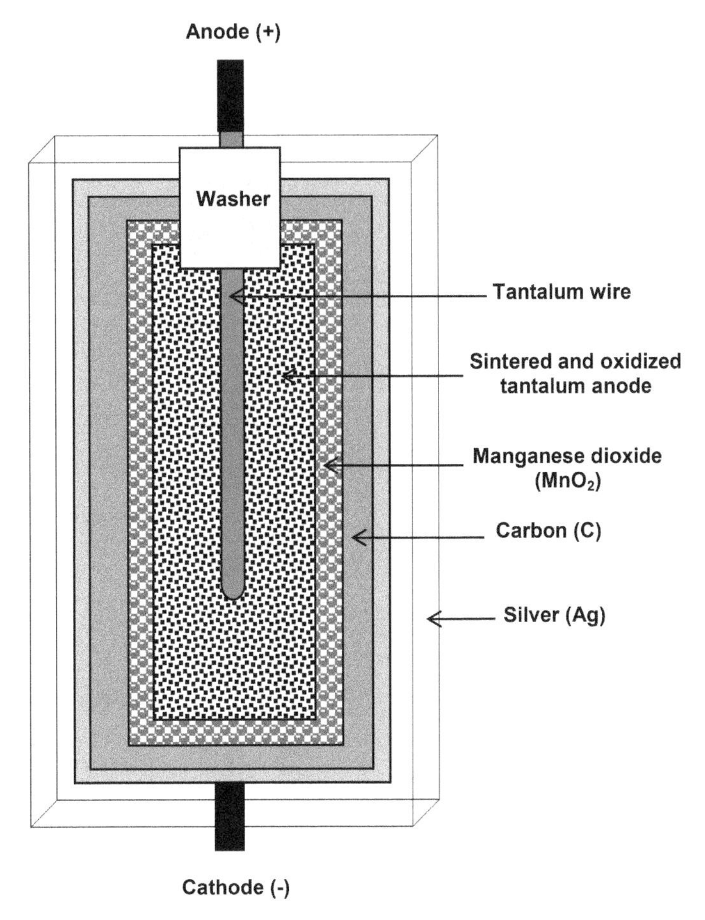

Figure 11.6. Solid tantalum capacitor.

pellet into manganese nitrate aqueous solution, and baking it at 250 °C, it is coated with manganese dioxide electrolyte via a pyrolysis reaction:

$$Mn(NO_3)_2 \rightarrow MnO_2 + 2NO_2. \quad (11.2)$$

Dipping it successively into graphite and silver forms the silver cathode termination while a tinned nickel anode lead is attached to the Ta riser wire for the external anode termination. For high-temperature operation >200 °C, the silver in the top coating is substituted by electroplated nickel (Freeman *et al* 2013).

In a wet tantalum capacitor (figure 11.7), the electrolyte is non-solid. The advantage derived by keeping the electrolyte in the wet condition is that it becomes possible to supply oxygen in weaker portions of the dielectric film. Thus an improved oxide layer can be formed in these regions. This facility helps in safely forming very thin tantalum pentoxide films, enabling the realization of very high capacitance values per unit volume with extremely low leakage current, permitting high-voltage operation. Additionally, the silver migration problems affecting the

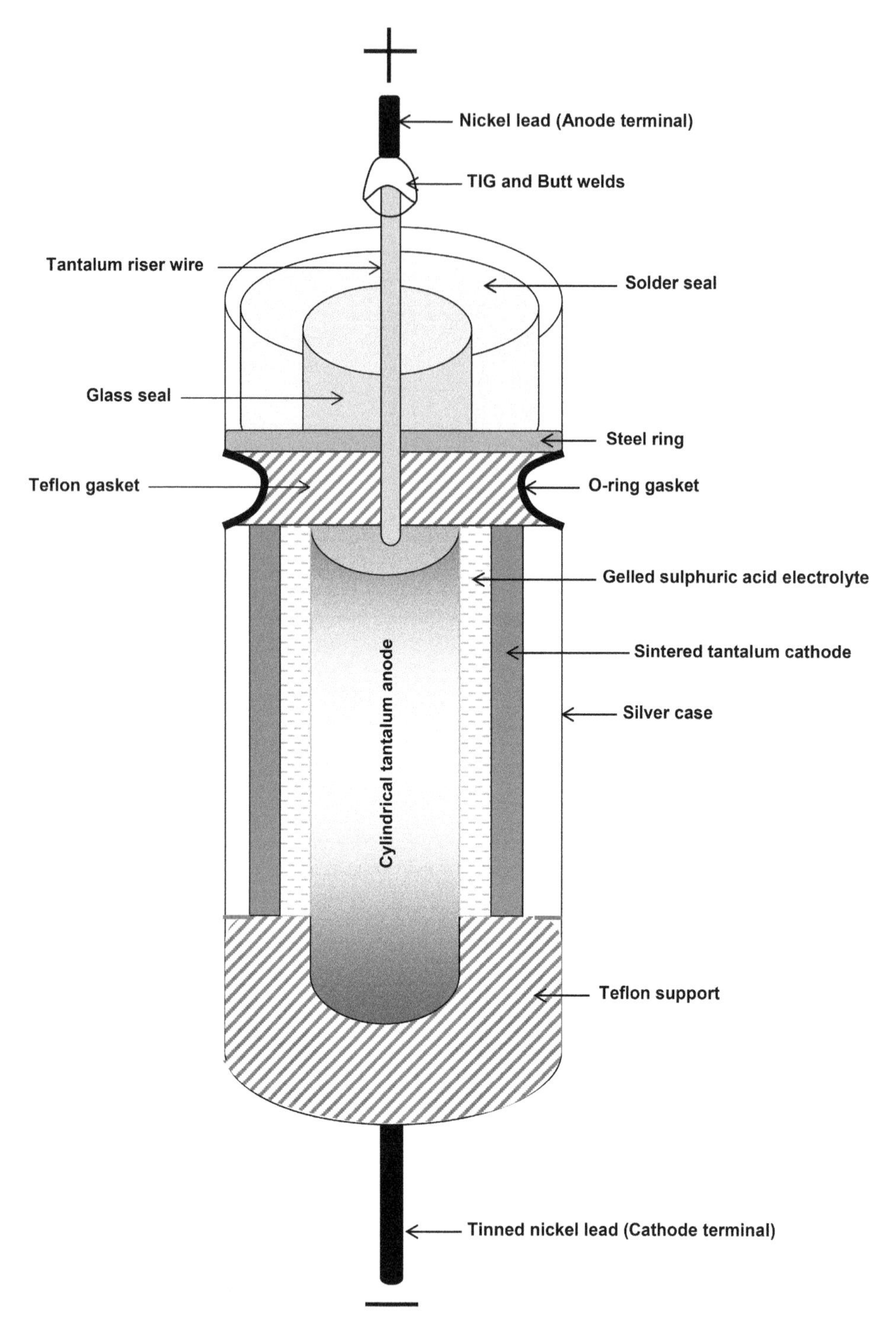

Figure 11.7. Wet tantalum capacitor.

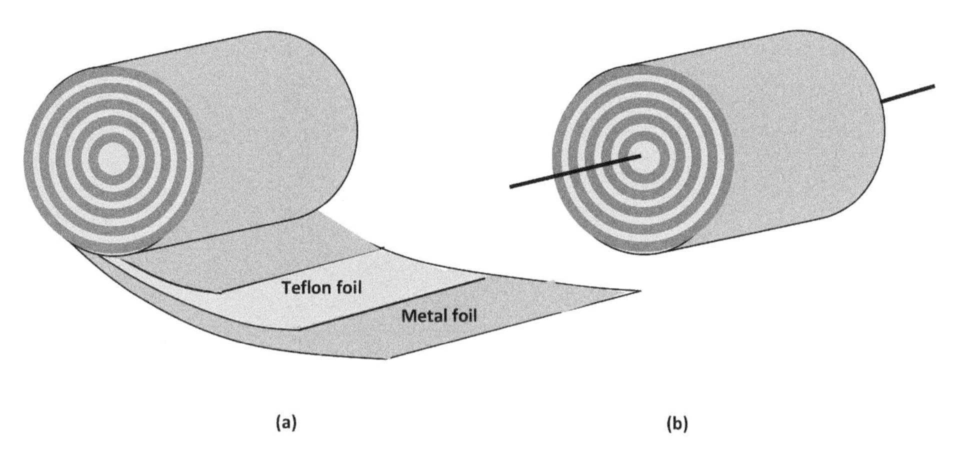

Figure 11.8. Teflon capacitor: (a) winding of the teflon/metal foils and (b) lead fixation.

silver case led to the use of the sintered tantalum encasing. The electrolyte is gelled sulfuric acid. It is filled in a case of pure tantalum.

11.3.3 Teflon capacitors

These capacitors (figure 11.8) are film/foil capacitors manufactured by covering a polytetrafluoroethylene (PTFE) film/foil with Al foil/evaporated film on both sides. PTFE is a chemically inert fluoropolymer of tetrafluoroethylene with a dielectric constant of 2.1, low dielectric loss and high insulation resistance, usable up to 250 °C (Bauchman 2012). Teflon capacitors are used in critical mission mode applications.

11.4 High-temperature magnetic cores and inductors

11.4.1 Magnetic cores

Powder cores for elevated temperature operation are made from alloys of Ni and Fe in various compositions (Magnetics 2016). After grinding the alloys into small particles, they are coated with a non-magnetic, non-organic, and non-conducting material. The powder thus obtained is annealed at 500 °C. The annealed powder is pressed to form toroids of the desired size (figure 11.9). The stresses in the toroids are relieved by annealing at temperatures >500 °C. The annealed toroids are coated with epoxy paint. During operation, the maximum temperature should not exceed 200 °C.

11.4.2 Inductors

Wirewound inductors

Surface mount device (SMD) power inductors in a wirewound configuration for operation from −40 °C to +150 °C are made by winding a coil around a ferrite drum core (TDK 2014); see figure 11.10. Magnetic coupling with neighboring electronic components is secured by an octagonal ferrite core. Interference and power losses are thereby prevented.

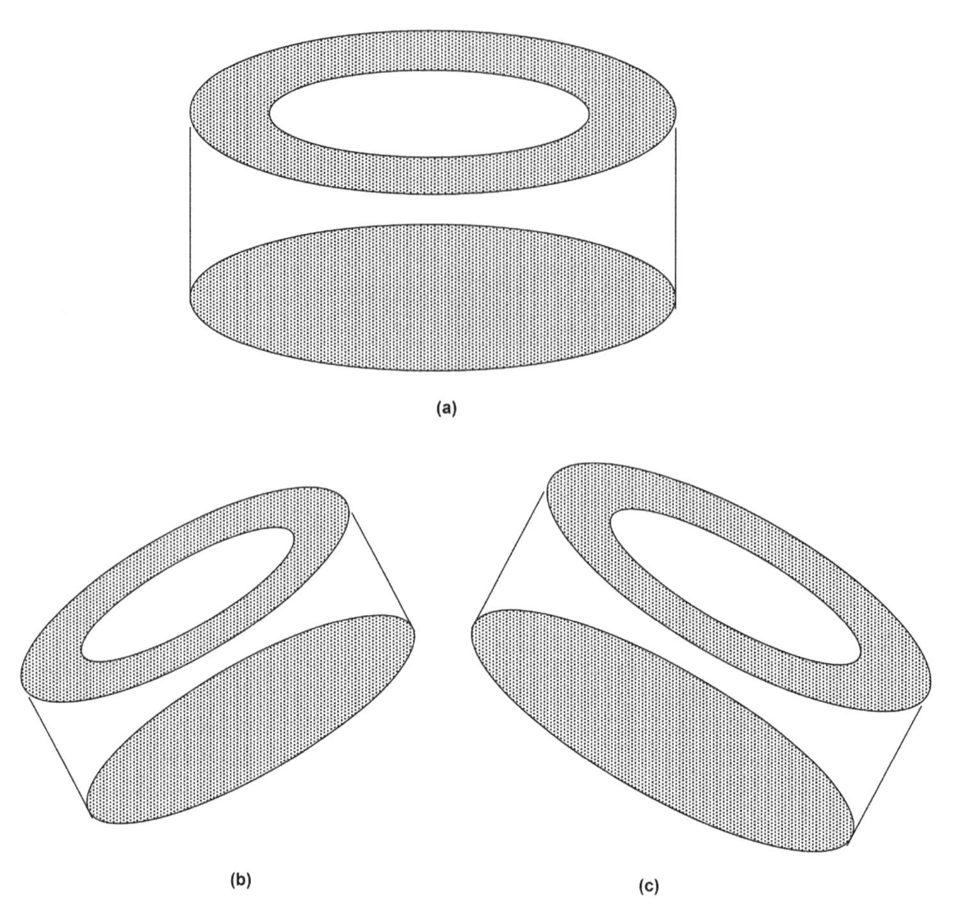

Figure 11.9. Toroids of different sizes: (a) large, (b) small and (c) medium.

Micro-inductors

Planar micro-inductors (figure 11.11) for operation up to 200 °C are fabricated using yttrium iron garnet (YIG) magnetic layers (Haddad *et al* 2013) by the following sequence of steps. (i) Piercing 500 μm diameter holes in a 1 mm thick YIG layer in alignment with the terminals of the inductor. (ii) Stuffing the holes with copper and removing the superfluous copper by polishing to achieve surface flatness. (iii) Depositing a Ti/Cu seed layer on the substrate by e-beam evaporation. (iv) Laminating and patterning a thick layer of dry film photoresist. (v) Electroplating the copper windings using the photoresist as a mold. (vi) Removing the dry film photoresist with NaOH and wet etching the Ti–Cu seed layer for separation of copper turns. (vii) Polishing the copper windings and covering them with a thick bismaleimide (BMI) resin. (viii) Adding another YIG magnetic layer and clasping the inductor to bond the two YIG layers with BMI resin.

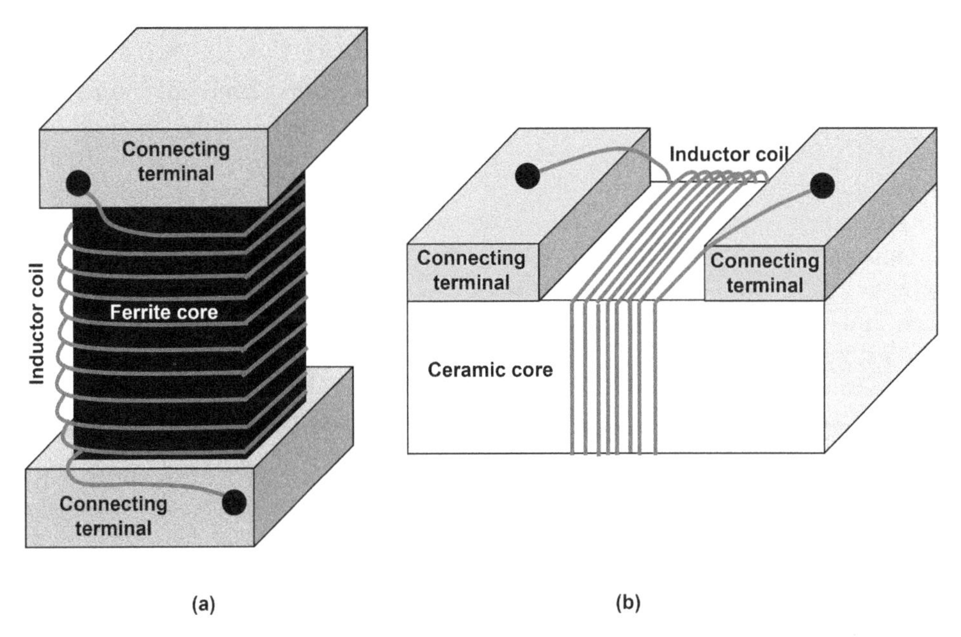

Figure 11.10. Wirewound SMD chip inductor with: (a) a ferrite core and (b) ceramic core.

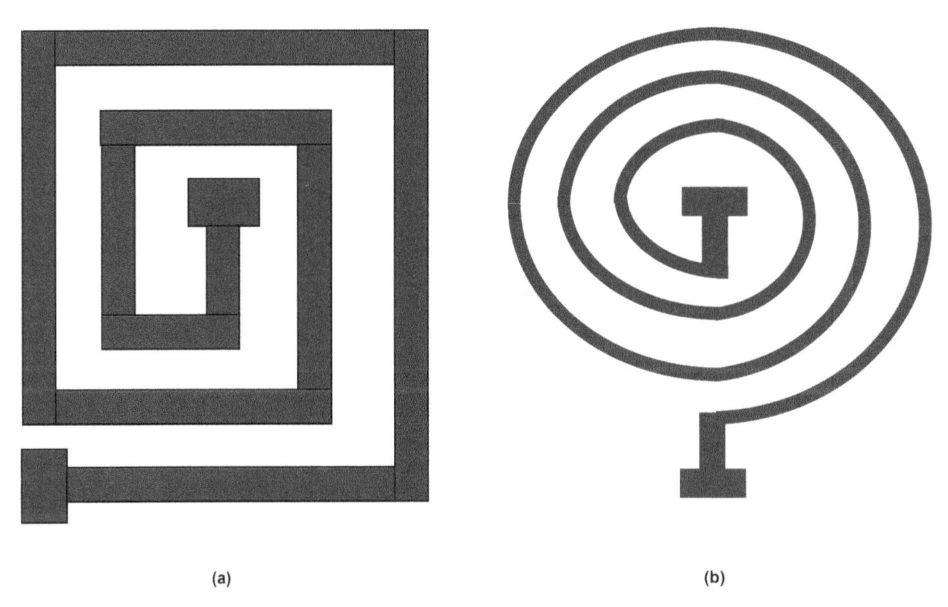

Figure 11.11. Planar micro-inductor layouts: (a) rectangular spiral and (b) circular spiral.

11.5 High-temperature metallization

The adjunct metallization schemes for high-temperature electronic devices should be on par with the thermal capabilities of the devices.

11.5.1 Tungsten metallization on silicon

To raise the high-temperature limit of metallization on silicon to temperatures beyond 450 °C, the refractory metal tungsten is a favorable choice because of its matching coefficient of thermal expansion (CTE) with silicon. The obstacle to be overcome is the formation of tungsten discilicide (WSi_2) by reaction between tungsten and silicon at 600 °C. The formation of WSi_2 layer can be prevented by providing a diffusion barrier on Si. A $TiSi_2$/TiN layer, obtained by RTA in nitrogen, serves as an effective diffusion barrier and adhesion layer (Chen and Collinge 1995, Chen and Collinge 1996, Madou 2002). The $TiSi_2$ layer is formed at the Ti–Si interface and TiN layer on the exposed surface of the Ti film.

11.5.2 Tungsten: nickel metallization on nitrogen-doped homoepitaxial layers on p-type 4H- and 6H-SiC substrates

An n-type ohmic contact was formed on silicon carbide by sputtering a tungsten: nickel layer containing 75 atomic % tungsten and 25 atomic % nickel (Evans *et al* 2012). The thickness of the W:Ni layer was 100 nm. The composite layer was covered with 20 nm thick silicon film. The role of silicon film was to prevent premature oxidation of tungsten, which occurs during venting of the chamber and annealing. The W:Ni metallization performed well at high temperatures. It survived at 1000 °C in an argon atmosphere for 15 hours. This metallization scheme is expected to provide reliable operation at 600 °C.

11.5.3 Nickel metallization on n-type 4H-SiC and Ni/Ti/Al metallization on p-type 4H-SiC

The metal deposition was performed by e-beam evaporation and sputtering (Smedfors *et al* 2014, Smedfors 2014). RTA of the contacts was performed in Ar or N_2 ambient. The metallizations were tested in the temperature range from −40 °C to +500 °C. The Ni/Ti/Al ohmic contacts on p-type material showed an increase in specific contact resistivity at −40 °C by a factor of 5 with respect to the room-temperature value (6.75×10^{-4} Ωcm^2 at 25 °C). They exhibited a tenfold decrease in the specific contact resistivity at 500 °C with reference to the 25 °C value. The Ni ohmic contacts on n-type material showed improvement with an increase in temperature. The variation in specific contact resistivity was comparatively smaller.

11.5.4 A thick-film Au interconnection system on alumina and aluminum nitride ceramic substrates

Au thick-film was screen printed on a ceramic substrate. The thick Au film proved to be a low resistance and stable substrate metallization through a 1500 h exposure to atmospheric oxygen at 500 °C (Chen *et al* 2001). The metal film was tested in both unbiased and biased electrical conditions (50 mA DC). The resistance varied by ~0.1% in 1000 h. The shear strength of Au film on an Al_2O_3 substrate at 500 °C was reported to decrease by a fraction 0.8 of its value at 350 °C. But the shear strength value at 350 °C was approximately the same as at room temperature. Thermal cycling tests were performed on 44 gold thick-film to gold wire bonds at a temperature rate of 32 °C/min for 120 cycles and at 53 °C/min for 100 more cycles in an oxidizing ambient under an electrical bias condition. The survival of the bonds indicated good bond strength.

11.6 High-temperature packaging

Electronic packaging is a functional link between two vital segments of an electronic system. On one side are the delicate electronic devices and on the other side is the remaining system incorporating a variety of operations such as die attachment, wire bonding, passivation, interconnections between the components and housing of the complete assembly inside a safe enclosure that protects it from mechanical disturbances and the surroundings. High-temperature electronic packaging differs from the conventional moderate temperature packaging catering to room-temperature conditions with small excursions, primarily in the exclusion of materials and related processes, producing components that cannot withstand the high temperatures, and inclusion of any new materials with the demonstrated capability to tolerate the same. From this consideration, it is necessary to re-examine the list of commonly used materials starting from substrate selection, and then moving to die attachment, wire bonding and further, and discard those which do not fulfill the expectations, and add any novel material that promises high-temperature forbearance.

11.6.1 Substrates

Ceramic substrates are ideal for HTE. Among the competing materials—alumina (Al_2O_3), aluminum nitride (AlN), boron nitride (BN) and silicon nitride (Si_3N_4)—aluminum nitride stands out supreme (Chasserio *et al* 2009). In addition to very high thermal conductivity (175 W mK^{-1} against 28.1 W mK^{-1} for alumina), it offers good chemical stability, dielectric strength and stability, flexural strength, a matching CTE with Si and SiC, and good thermal shock resistance. The matching of the thermal expansion coefficient assures less stress creation, both on the device and at the solder joints during thermal cycling. However, it suffers from a cost disadvantage, where alumina is distinctively better. Alumina is inferior to aluminum nitride from the point of view of thermal expansion coefficient mismatch, thermal shock resistance and thermal conductivity. Silicon nitride provides very high flexural strength and a good thermal coefficient matching with that of Si and SiC. Together with these qualities, it has good thermal shock resistance and chemical stability, but it suffers from being unacceptably costly. Boron nitride fails on many parameters, particularly on flexural strength, and is eliminated from the contest.

11.6.2 Die-attach materials

Examples of lead-based materials are (Manikam and Cheong 2011) Pb95–Sn5 and Pb97.5–Ag1.5–Sn1 with liquidus temperatures of 312 °C and 309 °C, respectively. Some lead-free gold-based solutions are Au100, AuNi18, Au thick-film paste and AuGe12, with maximum operating temperatures of 1063 °C, 950 °C, >600 °C and 356 °C. Silver nanoparticle paste was sintered at a temperature <350 °C (Guo *et al*

2015). The sintering joint showed good electrical and thermal characteristics with a melting point near that of pure silver (960 °C), and is useful for high-temperature operation >350 °C.

11.6.3 Wire bonding

Due to the intermetallic compound formation between Al–Au and Al–Cu with the resultant decrease in bond strength at temperatures <200 °C, aluminum wires bonded to nickel surfaces serve as an alternative to the above combinations (Barlow and Elshabini 2006). They do not degrade below 350 °C by intermetallic formation.

A comprehensive investigation of nickel wire (25 μm) bonding to nickel metallization pads (750 nm) on a 3C–SiC substrate is described by Burla *et al* (2009) using the ultrasonic technique. The wire bonds provided a pull strength of 13.1 gf (gram-force), about 4.4 times stronger than for Au wire bonds and 5.2 times stronger than for Al wire bonds. The wire bonds were mechanically and electrically tested up to 550 °C. Nickel wire bonds exhibited good performance on a chemical sensor (up to 280 °C) and a resonant device (up to 950 °C).

Problems of weak wire bonds and mismatching of the CTEs between the bond wires and devices are avoidable by attaching a CTE-matched cover as the top contact over the semiconductor device using spherical bumps of a high-temperature soft solder in place of the wires (Barlow and Elshabini 2006).

11.6.4 Hermetic packaging

The hermetic package provides an airtight connection between the semiconductor device(s) inside and the wiring outside. Hermetic packages are based on glass-to-metal and ceramic-to-metal seal technologies.

Glass-to-metal seal packages

Glass-to-metal seals are of two types:

(i) *Matched seal.* Here, the CTEs of the metal and glass are identical. The sealing takes place through a chemical reaction between the oxide film on the metal surface and the glass. This type of seal is weak.

(ii) *Compression seal.* Here the metal has a higher CTE than glass. The resulting seal is much stronger than the matched seal. To create feed-throughs for passing electric current from a part inside the package to the external circuit, a metal body with the required number of holes is formed. Through the holes pass conducting metal wires surrounded by glass. The CTE of the metal body is higher than that of glass and the conducting metal wires. The parts are placed inside a fixture and transferred to a furnace at a high temperature. On cooling, the metal body contracts around the solidifying glass. Due to the concentric compressive force exerted by the metal body on the solidifying glass, the conducting metal wires are strongly fixed in place, surrounded by and insulated from the metal body by the intervening glass layer.

Figure 11.12 illustrates the use of glass-to-metal sealing in the common three-pin headers used for semiconductor device packaging.

Ceramic-to-metal seal packages
A metal film is deposited on the ceramic. The ceramic is sealed with the metal body by brazing the metal film on ceramic with the metal body (figure 11.13). In the case that a metal with a high CTE is to be sealed with ceramic, a transition piece of metal with a low CTE is included between the metal piece with a high CTE and the ceramic to avoid direct interfacing between the ceramic and the metal piece with a high CTE.

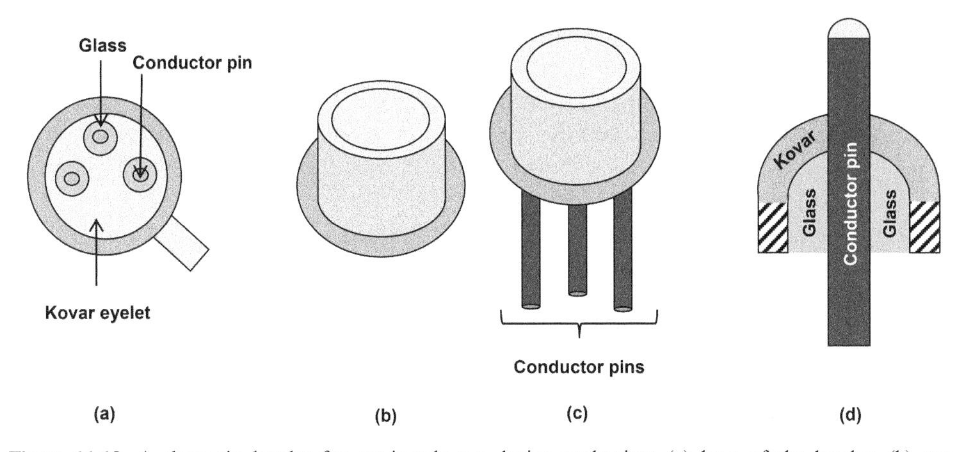

Figure 11.12. A three-pin header for semiconductor device packaging: (a) base of the header, (b) cap, (c) header fitted with cap and (d) diagram showing the glass-to-metal seal at the conductor pin and the hole in the kovar eyelet.

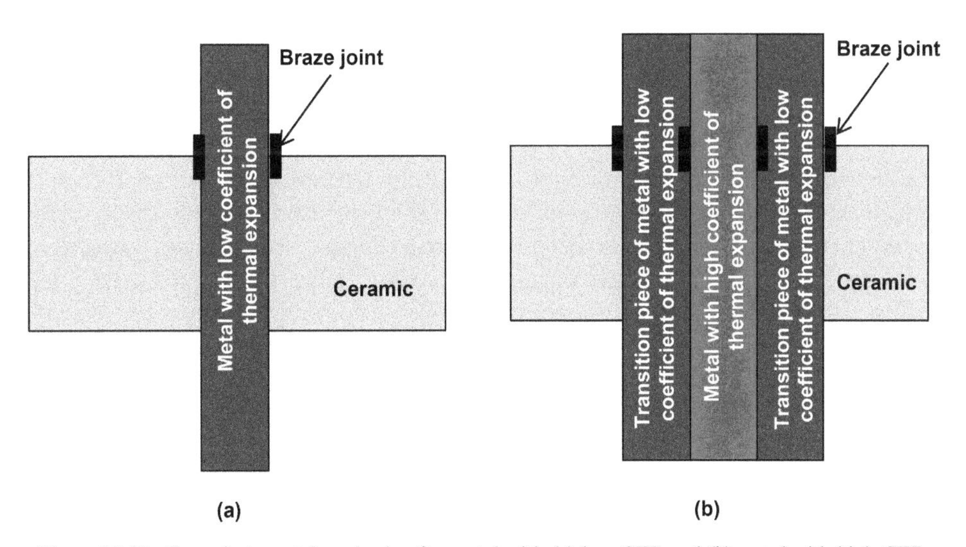

Figure 11.13. Ceramic-to-metal packaging for metal with (a) low CTE and (b) metal with high CTE.

11.6.5 Joining the two parts of hermetic packages

Hermetic packages consist of a base plate and an exactly fitting cap. The two parts are joined together to create a cavity in which the electronic die is placed by one of following processes (Schott 2009):

(i) *Resistance welding.* Electrodes are fixed on opposite sides to pass an electric current through the base plate and the cap which are mounted by aligning them in correct positions for welding. When the current flows, the heat produced by the electro-thermal process due to the electrical resistance of the base plate/cap melts them at the welding ridges to form a bond on cooling.

(ii) *Rolled seam welding.* This is a variant of the resistance welding in which the parts to be welded are pressed between rolling electrodes through which current is made to flow. Thus the parts are welded point by point using rollers in place of static electrodes. This process is applicable only when the parts to be welded together are made of conducting materials. In the case of ceramic packages, metallic frames are soldered to the ceramic parts. Electric current flowing between the rollers through the metallic frames joins the ceramic parts together.

(iii) *Cold welding.* Pressure applied to the surfaces of package parts causes ring welding when the surfaces are flat, parallel and clean, and the tooling is accurately designed.

(iv) *Laser welding.* The laser beam is focused at different points of the parts to be welded, fusing and connecting them together. However, the welding becomes difficult for surfaces coated with gold, silver or nickel due to reflection of the beam instead of it being absorbed.

(v) *Soldering.* Here a solder is used as the joining glue. The surfaces of the two parts must be wettable by the solder for soldering to take place.

11.7 Discussion and conclusions

Neither conventional materials nor the processes followed in the electronics industry can be used to fabricate systems capable of operating reliably at high temperatures. An all-round effort is necessary to choose the appropriate materials and processes at each step of system development to comply with temperature specifications. HTE involves mastering and leveraging the aforesaid technological advancements to fabricate rugged versions of previous medium-temperature designs or to evolve altogether new designs. To achieve this goal, diverse technologies capable of providing high-temperature sustaining components have to be interwoven with existing fabrication facilities to supply robust electronic systems for elevated temperature operation. This will necessitate the re-arrangement, rejuvenation and restructuring of the entire infrastructure.

Review exercises

11.1. What is a metal foil resistor? Describe the mechanism by which low TCR is achieved in this resistor structure.

11.2. What is a wirewound resistor? What is its typical upper temperature limit?

11.3. How are the following types of resistors fabricated: (a) thin-film resistor and (b) thick-film resistor? Give an idea of the resistive film thicknesses in the two types. How high a temperature can they sustain?
11.4. How do thick-film-on-steel resistors differ from conventional thick-film resistors? What is the advantage of using steel as a substrate material?
11.5. What is the composition of a class 1 dielectric used in ceramic capacitors? What are the virtues of this class? What is its limitation?
11.6. Give an example of a dielectric composition providing a high dielectric constant with low temperature sensitivity. What is the reason for the temperature insensitivity?
11.7. What are the materials used for the anode, cathode and dielectric layers in a wet tantalum capacitor? How is the dielectric film formed? What is the relative permittivity of the dielectric? How is the cathode film formed and how is the cathode metal termination deposited? What is the cathode metal termination used for high-temperature operation?
11.8. What is the dielectric constant of teflon? How high are the temperatures that a teflon capacitor can withstand?
11.9. How is a magnetic core for high-temperature operation made? How is a high-temperature sustaining wirewound inductor assembled?
11.10. Name the magnetic layer material used in a micro-inductor. Describe the process sequence for micro-inductor fabrication.
11.11. What difficulty is faced in using tungsten metallization over silicon at high temperatures? How is the difficulty overcome?
11.12. Up to what operating temperatures can tungsten:nickel metallization be used over silicon carbide? What is the reason for covering it with a silicon layer?
11.13. Compare the performance of Ni/Ti/Al metallization on p-type SiC with that of Ni metallization on n-type SiC with respect to the variation of specific contact resistivity of the metal layer with temperature.
11.14. How is high-temperature electronic packaging fundamentally different from low- and moderate-temperature packaging?
11.15. Prepare a comparative performance analysis of four substrate materials for HTE: aluminum nitride, aluminum oxide, silicon nitride and boron nitride.
11.16. Which has the higher thermal conductivity: aluminum nitride or aluminum oxide?
11.17. Which has higher flexural strength: silicon nitride or boron nitride?
11.18. Give one example each of lead-based, gold-based and silver-based die attach materials and mention the temperature limit for each case.
11.19. Give two examples of the materials of bonding wire-metal pad combinations used in microelectronics, which are prone to failure by intermetallic compound formation.
11.20. Describe the high-temperature performance of nickel wire bonding on nickel pads.

11.21. For reliable operation at high temperatures, it is necessary to eliminate the wire bonding step from the process. Discuss.
11.22. What are the two types of glass-to-metal seals? How are they made? How do they differ in sealing strengths?
11.23. How is a ceramic-to-metal seal made?
11.24. What are the main techniques used for joining together the base plate and cap of a hermetic package?
11.25. Discuss with reference to the following, what steps you would take to fabricate an electronic system working at high temperatures: substrates, die-attach materials, wire bonding, and packages.

References

Barlow F D and Elshabini A 2006 High-temperature high-power packaging techniques for HEV traction applications *Technical Report* ORNL/TM-2006/515 http://info.ornl.gov/sites/publications/files/Pub2285.pdf, pp 1–20

Bauchman M E 2011 A look at film capacitors *TTI* www.ttiinc.com/object/me-tti-20120111.html

Burla R K, Chen L, Zorman C A and Mehregany M 2009 Development of nickel wire bonding for high-temperature packaging of SiC devices *IEEE Trans. Adv. Packag.* **32** 564–74

Chasserio N, Guillemet-Fritsch S, Lebey T and Dagdag S 2009 Ceramic substrates for high temperature electronic integration *J. Electron. Mater.* **38** 164–74

Chen J and Colinge J P 1995 Tungsten metallization for high-temperature SOI devices *Mater. Sci. Eng.* B **29** 18–20

Chen J and Colinge J P 1996 Tungsten metallization system with TiN/$TiSi_2$ contact structure for thin film SOI devices *Trans. Third Int. High Temperature Electronics Conf.* (*Albuquerque, NM, June*) vol 1 pp V.27–V.31

Chen L-Y, Okojie R S, Neudeck P G, Hunter G W and Lin S-T 2001 Material system for packaging 500 °C SiC microsystems, in microelectronic, optoelectronic and MEMS packaging *Materials Research Society Symp. Proc., MRS Spring Meeting, (April 16–20)* ed J C Boudreaux, R H Dauskardt, H R Last and F P McCluskey vol 682 pp 79–90

Dittmer R, Anton E-M, Jo W, Simons H, Daniels J E, Hoffman M, Pokorny J, Reaney I M and Rödel J 2012 A high-temperature-capacitor dielectric based on $K_{0.5}Na_{0.5}NbO_3$-modified $Bi_{1/2}Na_{1/2}TiO_3$–$Bi_{1/2}K_{1/2}TiO_3$ *J. Am. Ceram. Soc.* **95** 3519–24

Ebbert P 2014 The wirebound resistor: 'the report of my death was an exaggeration *Riedon White paper.* https://riedon.com/media/pdf-tech/Pulse_Wirewound_Resistors.pdf

Evans L J, Okojie R S and Lukco D 2012 Development of an extreme high temperature n-type ohmic contact to silicon carbide *Mater. Sci. Forum* **717-720** 841–4

Freeman Y, Chacko A, Lessner P, Hessey S, Marques R and Moncada J 2013 High-temperature Ta/MnO_2 capacitors *CARTS Int. Proc.* (Houston, TX, 25–28 March) pp 1–5

Guo W, Zeng Z, Zhang X, Peng P and Tang S 2015 Low-temperature sintering bonding using silver nanoparticle paste for electronics packaging *J. Nanomater.* **2015** 897142

Haddad E, Martin C, Buttay C, Joubert C and Allard Bergogne D B 2013 High temperature, high frequency micro-inductors for low power DC–DC converters *EPE'13-ECCE Europe* (*Lille, France, September*) paper 390, hal-00874475

Hernik Y 2012 Precision resistors for energy, transportation, and high-temperature applications *Vishay Precision Groups* www.digikey.com/Web%20Export/Supplier%20Content/VishayPrecisionGroup_804/PDF/vishay-glance-precision-resistors-for-energy.pdf

Madou M J 2002 *Fundamentals of Microfabrication: The Science of Miniaturization* (Boca Raton, FL: CRC Press) 281

Magnetics © 2016 Using magnetic cores at high temperatures *Technical Bulletin* No. CG-06 http://wenku.baidu.com/view/74a86e170b4e767f5acfcee2.html

Manikam V R and Cheong K Y 2011 Die attach materials for high temperature applications: a review *IEEE Trans. Compon. Packag. Manuf. Technol.* **1** 457–78

Morrison D G 2000 Thick-film-on-steel resistors prove worthy in high-temperature, high-power applications *Electronic Design* http://electronicdesign.com/energy/thick-film-steel-resistors-prove-worthy-high-temperature-high-power-applications

Schott 2009 *New Hermetic Packaging and Sealing Technology Handbook* http://www.schott.com/english/news/press.html?NID=com2516

Smedfors K 2014 Ohmic contacts for high temperature integrated circuits in silicon carbide *Licentiate Thesis* Royal Institute of Technology, Stockhlom, pp 1–39

Smedfors K, Lanni L, Östling M and Zetterling C M 2014 Characterization of ohmic Ni/Ti/Al and Ni contacts to 4H-SiC from –40 °C to 500 °C *Mater. Sci. Forum* **778-780** 681–4

TDK 2014 Power inductors for advanced engine management: built tough for high temperatures *TDK* http://en.tdk.eu/tdk-en/373562/tech-library/articles/applications---cases/applications---cases/built-tough-for-high-temperatures/790022

Vishay 2014 High temperature (245 °C) thick film chip resistor *Vishay* www.vishay.com/docs/52032/chpht.pdf

Vishay 2015 High stability–high temperature (230 °C) thin film wraparound chip resistors, sulfur resistant *Vishay* www.vishay.com/docs/53050/pht.pdf

Chapter 12

Superconductive electronics for ultra-cool environments

Superconductive electronics is an extensive field covering both low- and high-temperature superconductors (HTS), and their applications in quantum interference devices, power transmission, microwave filters and logic circuits. The microscopic origins of superconductivity are explained by Bardeen–Cooper–Schieffer (BCS) theory whereas a macroscopic treatment is provided by the Ginzburg-Landau theory. The London equations can explain the well-known Meissner effect of exclusion of magnetic induction by superconductors. A widely used device in superconductive electronics is the JJ. The DC and AC Josephson effects as well as the inverse AC Josephson effects are described. The application of DC and AC SQUIDs in magnetometry are discussed. RFSQ logic, touted as a competitive beyond-CMOS technology in nanoelectronics, is briefly surveyed.

12.1 Introduction

As the name insinuates, superconductive electronics means electronics using superconducting materials. These are the materials which exhibit zero electrical resistance when cooled below a particular temperature known as the critical or transition temperature (T_C). Superconductive electronics has a broad scope covering (Superconductivite 2015):

(i) *SQUID magnetometers.* These are instruments used for accurate measurement of the strength and direction of a magnetic field at a point in space.
(ii) *Current limiters.* These are safety devices used for protecting an electrical plant from accidental faults such as short circuits or voltage overshoots.
(iii) *Efficient electronic filters.* These high-efficiency filters are made of superconductors.
(iv) *RSFQ logic.* This is a logic system based on flux quantum.

The above topics will now be elaborated.

doi:10.1088/978-0-7503-1155-7ch12

12.2 Superconductivity basics

12.2.1 Low-temperature superconductors

A low-temperature superconductor is a material, element or alloy which offers practically no resistance to the flow of electricity at a temperature near absolute zero (−273.15 °C). In such a material, the current can flow for an indefinitely long time, even for years without any reduction.

The resistivity of any normal conductor decreases with cooling but levels off at a certain value beyond which it does not decrease with temperature. Interestingly, the best conductors, such as copper, silver and gold, belong to this class. The resistivity of a superconductor also decreases up to a certain value, as does that of a normal conductor, but in contrast it suddenly drops to zero when a particular temperature is reached; this is the critical temperature, as indicated earlier. Many materials behave as superconductors below 30 K (−243.15 °C), e.g. Ga (1.1 K), Al (1.2 K), Hg (4.2 K), Nb (9.3 K) and La–Ba–Cu oxide (17.9 K). Superconductivity is the property by which these materials show extremely small electrical resistance/resistivity below their respective critical temperatures.

12.2.2 Meissner effect

A superconductor opposes any magnetic field applied by an external magnet (figure 12.1). This expulsion of the magnetic field by a superconductor from its interior is known as the Meissner effect (figures 12.2 and 12.3). Thus a superconductor is said to exclude any magnetic field, or superconductivity opposes magnetism.

The Meissner effect is not an abrupt transition at the superconductor surface. The distance up to which the magnetic field makes an incursion into the superconductor has a finite non-zero value. The characteristic depth ~10–100 nm of a superconductor up to which the external magnetic field can penetrate inside the material, as determined from the decay of the magnetic field to $1/e$ times its value at the surface, is referred to as its London penetration length/depth, denoted by λ or λ_L. Thus the variation of magnetic induction $B(x)$ along the x-direction is given by

$$B(x) = B(0)\exp(-x/\lambda_L), \tag{12.1}$$

where $B(0)$ is magnetic induction at the surface. The penetration depth is

$$\lambda_L = \sqrt{\frac{m}{\mu_0 n q^2}}, \tag{12.2}$$

where m, n, q are the mass, density and charge of superconducting electrons, and μ_0 is the vacuum permeability ($= 1.257 \times 10^{-6}$ Vs A m^{-1}). Taking $n = 1 \times 10^{29}$ m^{-3}, $m = 9.11 \times 10^{-31}$ kg, $q = 1.6 \times 10^{-19}$ C,

$$\lambda_L = \sqrt{\frac{9.11 \times 10^{-31}}{1.257 \times 10^{-6} \times 1 \times 10^{29} \times (1.6 \times 10^{-19})^2}} = 1.683 \times 10^{-8}\ \text{m} = 16.83\ \text{nm} \tag{12.3}$$

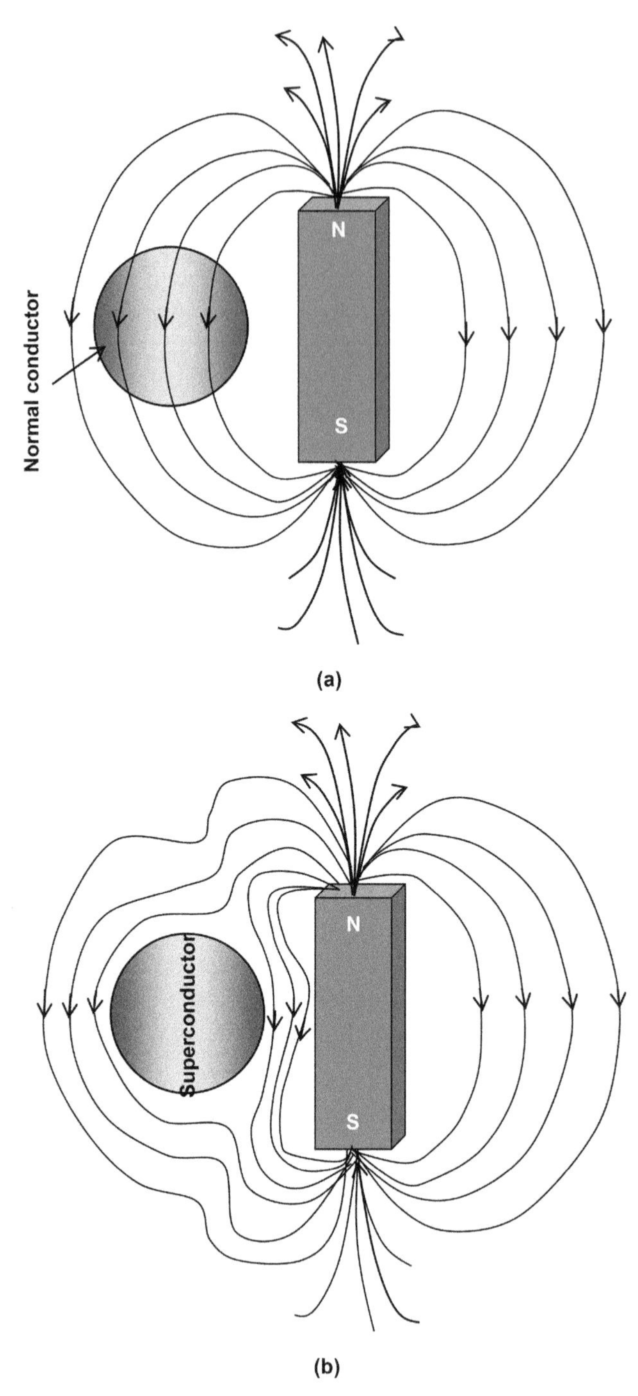

Figure 12.1. Difference in the behavior of superconductor from a normal electrical conductor when placed in an external magnetic field: (a) penetration of the magnetic lines of force through a normal conductor and (b) their exclusion by a superconductor.

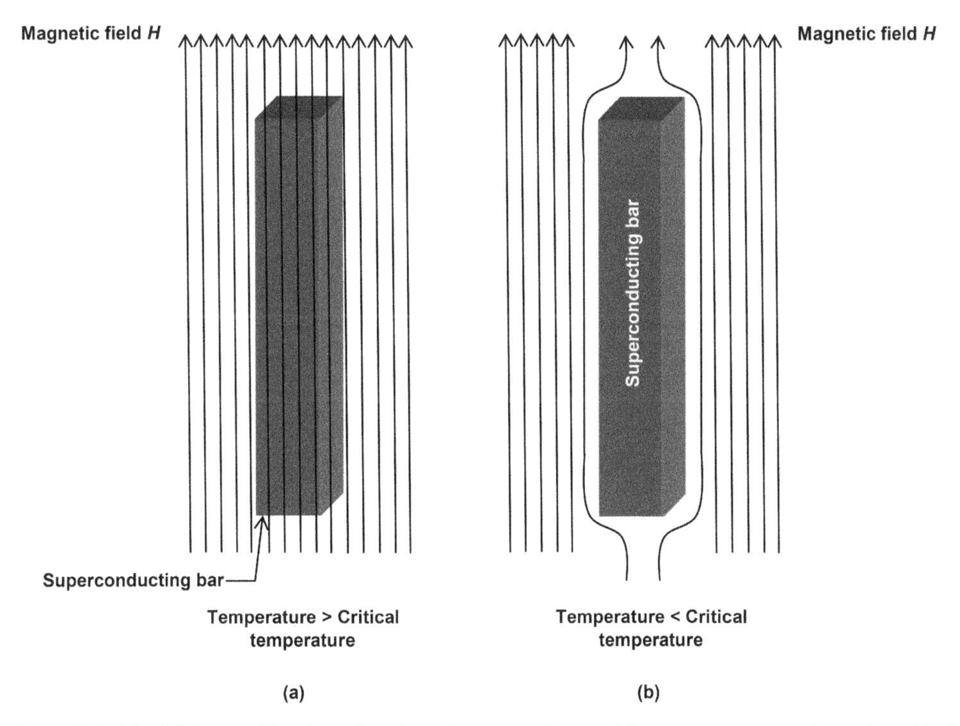

Figure 12.2. The Meissner effect for a bar-shaped superconductor: (a) at a temperature above the critical temperature and (b) at a temperature below the critical temperature.

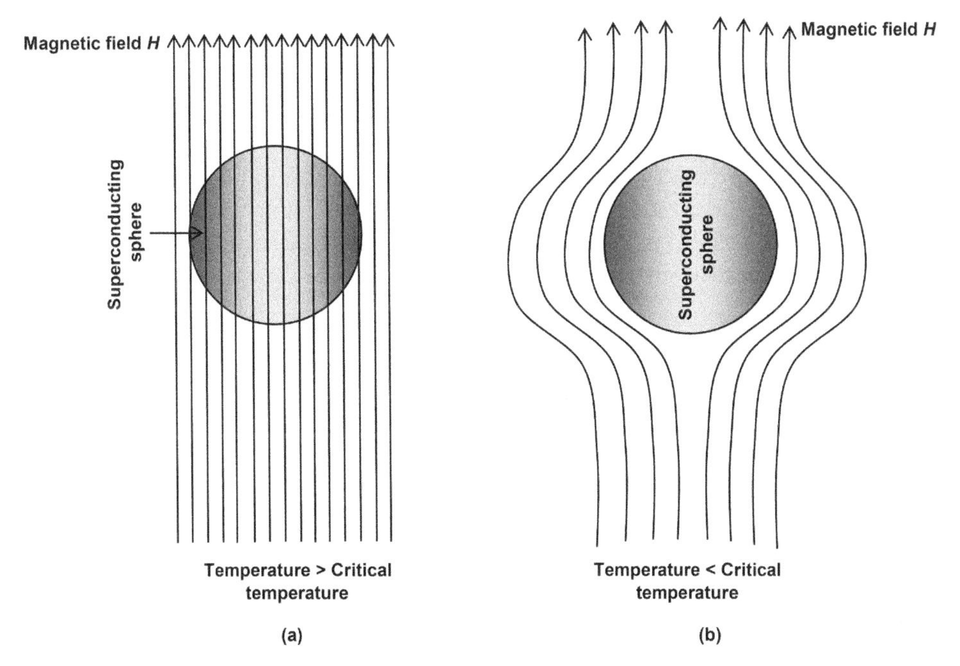

Figure 12.3. The Meissner effect for a spherical superconductor: (a) at a temperature above the critical temperature and (b) at a temperature below the critical temperature.

12.2.3 Critical magnetic field (H_C) and critical current density (J_C)

When maintaining the temperature and the current flowing through a semiconductor as constant, if the external magnetic field applied on the superconductor becomes greater than a particular value known as the critical magnetic field H_C, the superconductivity behavior of the material is lost. Similarly, keeping the temperature and the magnetic field acting on a superconductor unchanged, when the current density J flowing through a superconductor exceeds a critical value J_C, superconductivity is disrupted. Thus, a superconductor loses its superconducting property when any one of the group of three parameters is surpassed. These vital parameters are: critical temperature, critical magnetic field or critical current density value.

12.2.4 Superconductor classification: type I and type II

Superconductors are classified into two types (figure 12.4):

(i) *Type I or soft.* These are the materials which completely pass into the superconducting state through an abrupt transition when the temperature is brought down below the critical temperature, and vice versa. This implies that all the material under investigation is transformed into a superconductor. No chunk of the material is left untransformed. Examples of pure metals are aluminum (Al), lead (Pb), mercury (Hg), etc. The alloy $TaSi_2$ also behaves in this manner.

(ii) *Type II or hard.* The conversion of these materials into the superconducting state does not occur altogether at once in completeness. It is preceded by an intermediate stage in which some fraction of the material is superconducting while the remaining portion is conducting. An example of such materials is the alloy NbTi which is step-by-step, little-by-little changed into a superconductor. The mixed state comprising superconducting and normal conducting materials is known as the vortex state. It is so-called because in this state vortices (whirlwinds or eddies) of superconducting current encircle normal conductors. Slowly, the amount of superconducting material increases while that of conducting material decreases until all the material becomes a superconductor.

Type I superconductors retain their superconducting nature up to very low critical temperatures and critical magnetic fields. Hence, their practical utility is low. In contrast, type II superconductors are able to maintain their superconducting characteristics up to very high critical temperatures and critical magnetic fields. This capability considerably enhances their usefulness.

Type I superconductors display a single critical magnetic field strength H_C below which superconductivity begins or above which it ends (figure 12.5(a)). Type II superconductors show two critical magnetic field strengths (figure 12.5(b)): H_{C2} where superconductivity starts to produce a mixed material and H_{C1} where the whole mass of material becomes superconducting. On reversal, at H_{C1} there is a change from pure superconductor to mixed superconductor while at H_{C2} the superconductivity ceases.

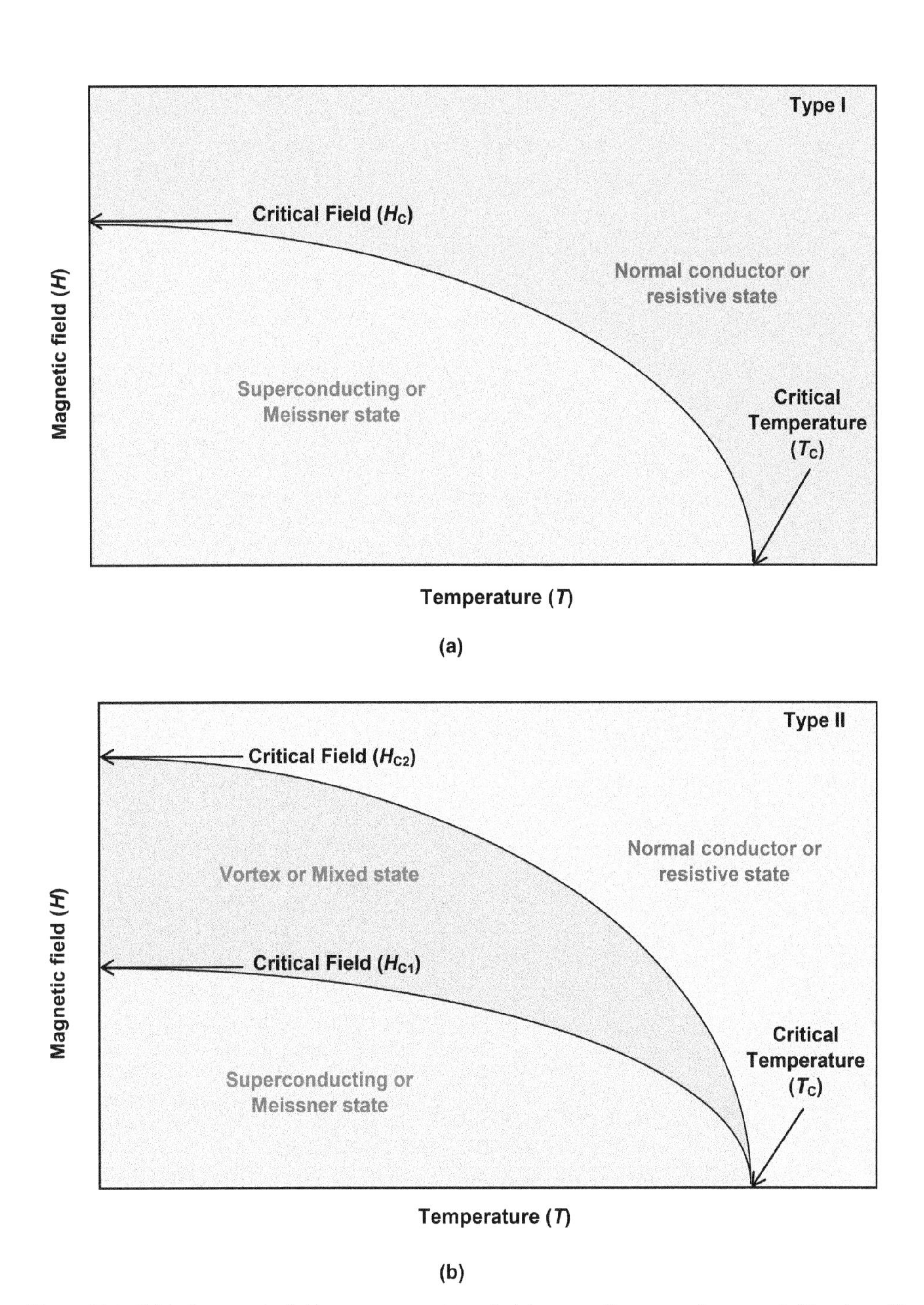

Figure 12.4. Critical magnetic field-temperature plots of: (a) a type I superconductor and (b) a type II superconductor.

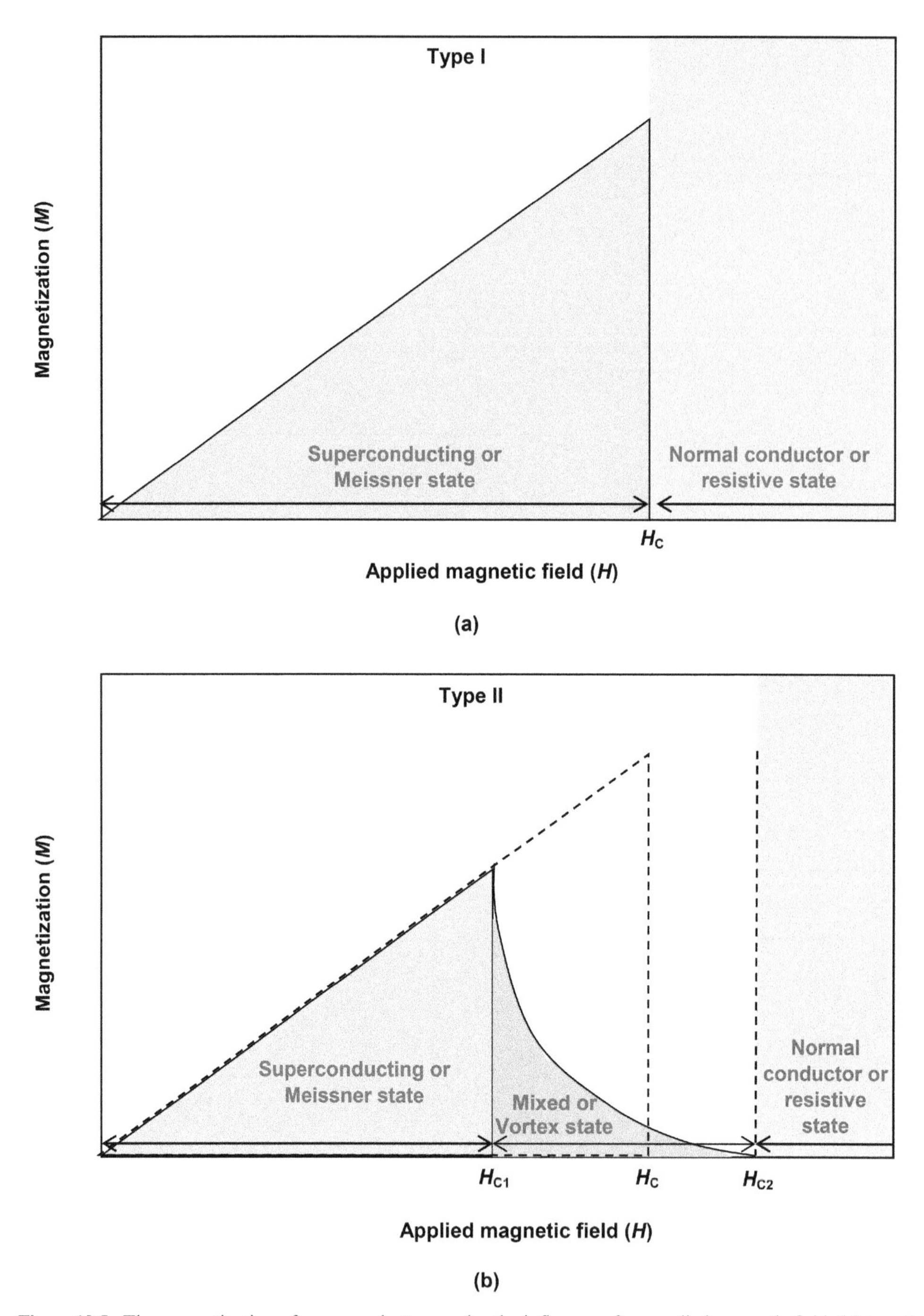

Figure 12.5. The magnetization of superconductors under the influence of an applied magnetic field: (a) type I and (b) type II.

Type I superconductors show a complete Meissner effect due to the thorough or out-and-out exclusion of the magnetic field. Type II superconductors do not show a complete Meissner effect. Rather, they show a mixed Meissner effect because the magnetic field is restricted to portions, which are in the normal conductor state, while super currents flow in remaining portions.

12.2.5 The BCS theory of superconductivity

This theory was propounded by John Bardeen, Leon Cooper and Robert Schrieffer (1959). It provides a microscopic interpretation of superconductivity. The theory is based on a clue from the observation of the behavior of a superconductor in a magnetic field. It is a well-known fact that a superconductor does not allow a magnetic field to enter inside its body. This affirmation implies that superconductivity has some relationship to the magnetic properties of materials. As we know, the magnetism results from the alignment of spins of electrons. Electrons are particles obeying Fermi–Dirac statistics; hence called fermions. They have two permissible spin values, $\pm\frac{1}{2}$. During the magnetization of a material, the randomly oriented spins of electrons are re-arranged to form an aligned configuration. In a superconductor, the electrons are arranged into pairs called Cooper pairs. The electrons in a Cooper pair work in a coordinated fashion. The two electrons in a Cooper pair may be located far apart, up to a distance of hundreds of nanometers. Thus they can coordinate from a distance. The distance between two electrons comprising the Cooper pair or the distance up to which the Cooper pair extends is known as the coherence length ξ. This length ranges from nanometers to micrometers. Each pair contains electrons of oppositely directed spins. Thus the spin of a Cooper pair becomes zero. As a result, the electron spin changes from a fractional number to an integral value. Particles with integral spins obey Bose–Einstein statistics and are called bosons. So, Cooper pairs are bosons like phonons or photons, and no longer fermions. Unlike fermions, which obey Pauli's exclusion principle, Cooper pairs, which are bosons, can condense into their lowest energy level below the critical temperature, resulting in loss of electrical resistance.

Let us see how a phonon interaction mediates weakly bound electrons in a Cooper pair (figure 12.6):

- (i) At a temperature below critical temperature, the lattice vibrations are minimum. So, the phonons are called virtual phonons.
- (ii) Contemplate the motion of an electron in a material. As the electron moves past the lattice ions, it disturbs the lattice ions from their normal positions, not destructively by collision but constructively by interaction of the negative charge on the electron with the positive charge of the nuclear cores of the ions in the material. Creating this disturbance, the electron continues its motion and moves past the lattice ions.
- (iii) The lattice ions momentarily shift from their normal positions in a delayed action after the electron has passed by. Due to this distortion, a region of net positive charge density is created in the vicinity of the electron.

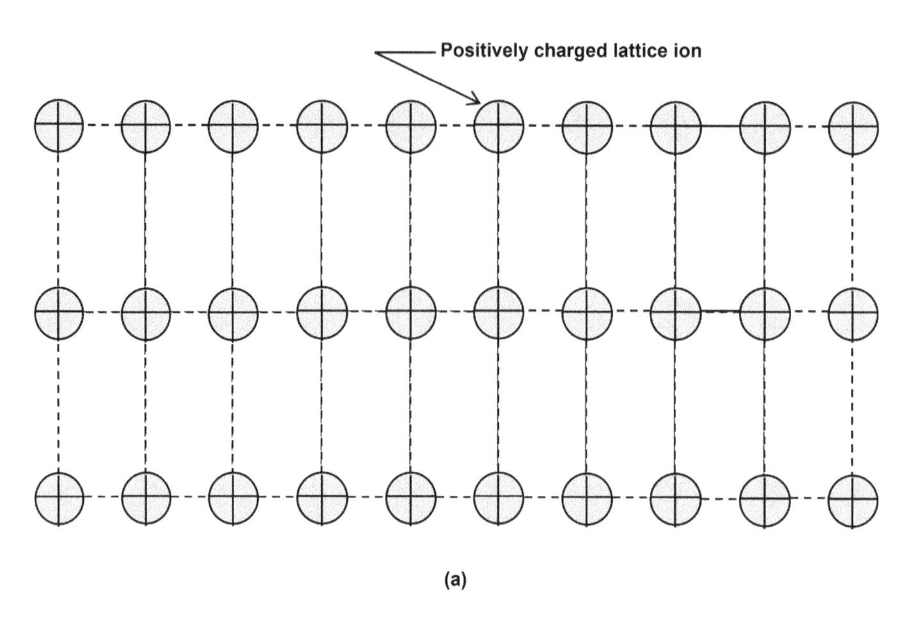

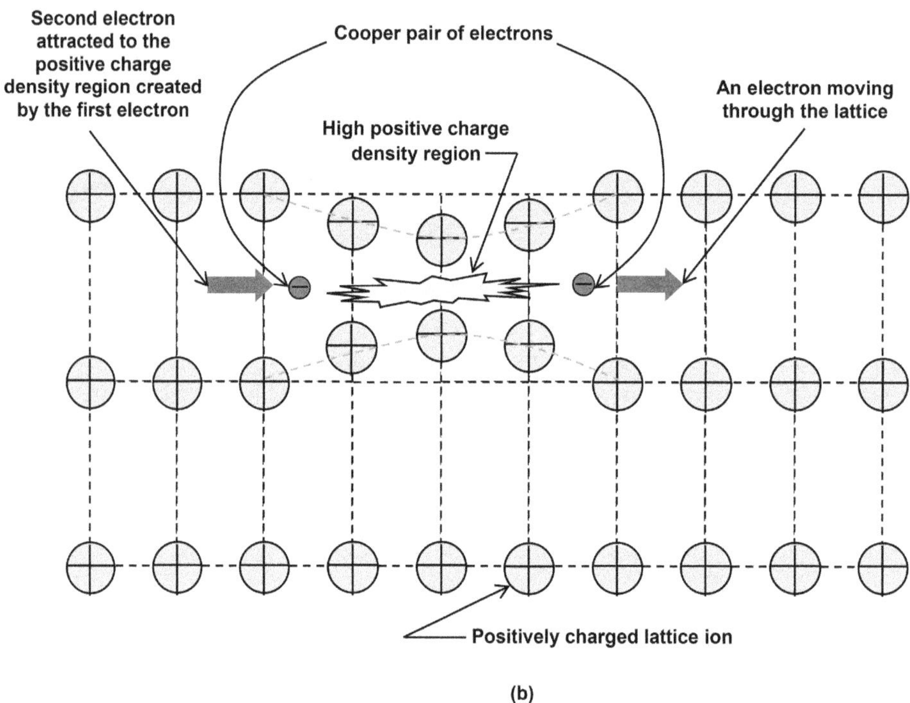

Figure 12.6. Illustration of BCS theory: (a) an undisturbed lattice of positive ions and (b) a distorted region of increased positive charge density created by a passing electron towards which another electron is attracted, resulting in the formation of a Cooper pair. This region disseminates through the crystal as a phonon.

(iv) A second electron having opposite momentum and spin to the first electron trails behind it. This second electron is attracted to the above region of positive charge created by the passage of the first electron. This trailing electron pairs up with the leading electron. Thus the attraction between the two electrons is mediated by the distortion of the lattice. Hence, it is an electron–phonon interaction because the phonon is a quantum of vibrational energy. The attractive force produced by lattice deformation is greater than the Coulomb repulsive force that will tend to push the electrons apart.

(v) After passage of the second electron, the lattice ions bounce back from their shifted positions to their original positions.

(vi) Another Cooper pair of electrons then undergoes an identical sequence of steps, causing electrical conduction.

(vii) The sequence is repeated for more Cooper pairs. Thus electric current constantly flows through the agency of Cooper pairs in the lattice. Cooper pairs serve as the carriers of current. The temperature-induced lattice vibrations cannot impede their movements in the manner they disturb the normal single electrons. This insusceptibility to lattice vibrations arises from the fact that the lattice itself mediates the attraction of one electron of the pair to its partner. Thus they are more stable than a single electron. This leads to decrease in the electrical resistance to negligible values.

To restate, at temperatures lower than the critical temperature, the electrons are kept in a paired state under the influence of a small attractive force between them. The attractive force originates from interaction of the electron with the lattice. In this interaction, the electrons do not undergo any scattering from lattice ions. There is no loss of energy as the supercurrent flows. In the absence of scattering from the ions, there is no electrical resistance. Above this temperature, the repulsive force between electrons causes them to become unpaired. These single electrons suffer scattering from ions as they traverse through the lattice. Due to this scattering, the material shows resistance.

12.2.6 Ginzburg–Landau theory

This is a macroscopic theory of superconductors based on physical intuition. It was proposed in 1950 before the microscopic BCS theory. It begins with proposition of a wavefunction Ψ. This wavefunction is a complex order parameter characterizing the superconducting state. The parameter $|\Psi|^2$ represents the density of superconducting electrons. Assuming that the free energy density of a superconductor can be expanded in powers of supercarrier density, the free energy density is minimized with respect to two parameters: the order parameter Ψ and the vector potential $\mathbf{A}$. This is done for stabilizing the energy of the superconducting state. The calculations are complex and merely numerical.

The free energy function is expanded as

$$F = \alpha\,|\Psi|^2 + \frac{\beta}{2}\,|\Psi|^4 + \gamma\,\left|\left(\nabla + \frac{2iq}{\hbar c}\mathbf{A}\right)\Psi\right|^2 + \frac{B^2}{8\pi}, \tag{12.4}$$

where $\mathbf{A}$ is the magnetic vector potential, q is the electronic charge and c is the velocity of light. Ψ is a complex-valued order parameter. The symbols α and β represent phenomenological parameters dependent on temperature. The derivative of the free energy function with respect to Ψ is equated to zero to minimize free energy with respect to Ψ

$$\frac{\mathrm{d}F}{\mathrm{d}\Psi} = 0 = 2\alpha|\Psi| + \frac{4\beta}{2}\,|\Psi|^2\Psi + 2\gamma\left(\nabla + \frac{2iq}{\hbar c}\mathbf{A}\right)^2\Psi. \tag{12.5}$$

Normalizing ψ such that

$$\gamma = \hbar^2/(4m) \tag{12.6}$$

$$\begin{aligned}
\frac{\mathrm{d}F}{d\Psi} &= 2\alpha\Psi + 2\beta\,|\Psi|^2\Psi + 2\left(\frac{\hbar^2}{4m}\right)\left(\nabla + \frac{2iq}{\hbar c}\mathbf{A}\right)^2\Psi \\
&= 2\alpha\Psi + 2\beta\,|\Psi|^2\Psi + 2\left(\frac{1}{4m}\right)\left\{\hbar\left(\nabla + \frac{2iq}{\hbar c}\mathbf{A}\right)\right\}^2\Psi \\
&= 2\alpha\Psi + 2\beta\,|\Psi|^2\Psi + \left(\frac{1}{2m}\right)\times\left\{\hbar^2\nabla^2 + 2\times\nabla\times\left(\frac{2iq}{\hbar c}\mathbf{A}\right)\times\hbar^2 + \left(\frac{2iq}{\hbar c}\mathbf{A}\right)^2\hbar^2\right\}\Psi \\
&= 2\alpha\Psi + 2\beta\,|\Psi|^2\Psi + \left(\frac{1}{2m}\right)\Bigg[-(-i\hbar\nabla)^2 - 2\times(-i\hbar\nabla) \\
&\qquad \times(-i\hbar)\left(\frac{2iq}{\hbar c}\mathbf{A}\right) - \left\{(-i\hbar)\left(\frac{2iq}{\hbar c}\mathbf{A}\right)\right\}^2\Bigg]\Psi \\
&= 2\alpha\Psi + 2\beta\,|\Psi|^2\Psi + \left(\frac{1}{2m}\right)\left[-(-i\hbar\nabla)^2 - 2\times(-i\hbar\nabla)\times\left(\frac{2q}{c}\mathbf{A}\right) - \left(\frac{2q}{c}\mathbf{A}\right)^2\right]\Psi \\
&= 2\alpha\Psi + 2\beta\,|\Psi|^2\Psi + \left(\frac{1}{2m}\right)\left\{-i\hbar\nabla - \left(\frac{2q}{c}\mathbf{A}\right)\right\}^2\Psi
\end{aligned} \tag{12.7}$$

$$\therefore\ \alpha\Psi + \beta|\Psi|^2\Psi + \left(\frac{1}{4m}\right)\left\{-i\hbar\nabla - \left(\frac{2q}{c}\mathbf{A}\right)\right\}^2\Psi = 0 \tag{12.8}$$

or

$$\{\alpha + \beta|\Psi|^2\}\Psi + \left(\frac{1}{4m}\right)\left\{-i\hbar\nabla - \left(\frac{2q}{c}\mathbf{A}\right)\right\}^2\Psi = 0. \qquad (12.9)$$

This is the first equation of Ginzburg–Landau.

The derivative of the free energy function with respect to **A** is equated to zero to minimize it with respect to A. Rewriting F by putting $\mathbf{B} = \nabla \times \mathbf{A}$

$$F = \alpha\,|\Psi|^2 + \frac{\beta}{2}\,|\Psi|^4 + \gamma\left\{\left(\nabla + \frac{2iq}{\hbar c}\mathbf{A}\right)\Psi \times \left(\nabla - \frac{2iq}{\hbar c}\mathbf{A}\right)\Psi^*\right\} + \frac{(\nabla \times \mathbf{A})^2}{8\pi}$$

$$\begin{aligned}\frac{dF}{d\mathbf{A}} &= 0 = 0 + 0 + \gamma\left(\frac{2iq}{\hbar c}\right)\left\{\Psi\left(\nabla - \frac{2iq}{\hbar c}\mathbf{A}\right)\Psi^* - \Psi^*\left(\nabla + \frac{2iq}{\hbar c}\mathbf{A}\right)\Psi\right\}\\ &\quad + \frac{2\{\nabla \times (\nabla \times \mathbf{A})\}}{8\pi} \qquad (12.10)\\ &= 0 + 0 + \gamma\left(\frac{2iq}{\hbar c}\right)\left\{\Psi\nabla\Psi^* - \Psi\frac{2iq}{\hbar c}\mathbf{A}\Psi^* - \Psi^*\nabla\Psi - \Psi^*\frac{2iq}{\hbar c}\mathbf{A}\Psi\right\} + \frac{\nabla \times \mathbf{B}}{4\pi}\\ &= \gamma\left(\frac{2iq}{\hbar c}\right)\left\{\left(\Psi\nabla\Psi^* - \Psi^*\nabla\Psi\right) - \frac{4iq}{\hbar c}\mathbf{A}\Psi^*\Psi\right\} + \frac{4\pi j}{4\pi c},\end{aligned}$$

since from Maxwell's fourth equation

$$\nabla \times \mathbf{B} = \frac{4\pi j}{c}, \qquad (12.11)$$

where the displacement current has been neglected,

$$\therefore \left(\frac{2\gamma iq}{\hbar c}\right)\left\{(\Psi\nabla\Psi^* - \Psi^*\nabla\Psi) - \frac{4iq}{\hbar c}\mathbf{A}\Psi^*\Psi\right\} + \frac{4\pi j}{4\pi c} = 0 \qquad (12.12)$$

or

$$\frac{j}{c} = -\left(\frac{2\hbar^2 iq}{4m\hbar c}\right)\left\{(\Psi\nabla\Psi^* - \Psi^*\nabla\Psi) - \frac{4iq}{\hbar c}\mathbf{A}\Psi^*\Psi\right\} \qquad (12.13)$$

or

$$\frac{j}{c} = -\left(\frac{\hbar iq}{2mc}\right)(\Psi\nabla\Psi^* - \Psi^*\nabla\Psi) - \left(\frac{2q^2}{mc^2}\right)\mathbf{A}\Psi^*\Psi \qquad (12.14)$$

$$\therefore j = -\left(\frac{\hbar iq}{2m}\right)(\Psi\nabla\Psi^* - \Psi^*\nabla\Psi) - \left(\frac{2q^2}{mc}\right)\mathbf{A}\Psi^*\Psi. \qquad (12.15)$$

This is the second equation of Ginzburg–Landau.

The Ginzburg–Landau theory does not treat the mechanisms responsible for superconductivity. Instead, the existence of superconductivity is postulated, and equations are derived to model its properties.

12.2.7 London equations

These equations were proposed by two brothers, Fritz and Heinz London, in the year 1935 as restrictions on classical electromagnetism for interpretation of the Meissner effect.

In a normal conductor, the acceleration force F_1 acting on an electron of charge q moving in an electric field E is

$$F_1 = qE. \tag{12.16}$$

The electron is accelerated by this force to a velocity v. On its way, it suffers collisions with metal ions which retard its motion. If τ is the time in which the electron comes to rest, its deceleration is given by

$$a = v/\tau. \tag{12.17}$$

If m is the mass of the electron, the deceleration force F_2 is

$$F_2 = -\ (mv)/\tau. \tag{12.18}$$

The equation of motion of the electron is

$$m\frac{\mathrm{d}v}{\mathrm{d}t} = F_1 + F_2 = qE - (mv)/\tau. \tag{12.19}$$

In a superconductor, there are no collisions causing scattering of an electron. Therefore,

$$F_2 = -\ (mv)/\tau = 0 \tag{12.20}$$

and the equation of motion reduces to

$$m\frac{\mathrm{d}v}{\mathrm{d}t} = qE \tag{12.21}$$

or

$$\frac{\mathrm{d}v}{\mathrm{d}t} = \frac{qE}{m}. \tag{12.22}$$

The current I is

$$I = \frac{\text{total charge/cross-sectional area}}{\text{time}}. \tag{12.23}$$

The current density is

$$J = \frac{\text{(total charge/volume)} \times \text{path length traversed}}{\text{time}} = nqv, \tag{12.24}$$

where nq = (number of electrons/volume) × charge of one electron, and v = path length traversed/time.

Taking derivative of both sides with respect to time

$$\frac{\mathrm{d}J}{\mathrm{d}t} = nq\frac{\mathrm{d}v}{\mathrm{d}t}. \tag{12.25}$$

Substituting the value of $\mathrm{d}v/\mathrm{d}t$ from equation (12.22) into equation (12.25)

$$\frac{\mathrm{d}J}{\mathrm{d}t} = nq\left(\frac{qE}{m}\right) = \frac{nq^2E}{m}. \tag{12.26}$$

This is the first equation of London.

Taking the curl of London's first equation

$$\nabla \times \frac{\mathrm{d}\mathbf{J}}{\mathrm{d}t} = \nabla \times \left(\frac{nq^2\mathbf{E}}{m}\right) = \left(\frac{nq^2}{m}\right)(\nabla \times \mathbf{E}) = \left(\frac{nq^2}{m}\right)\left(-\frac{\mathrm{d}\mathbf{B}}{\mathrm{d}t}\right) \tag{12.27}$$

since from Maxwell's third equation

$$\nabla \times \mathbf{E} = -\frac{\mathrm{d}\mathbf{B}}{\mathrm{d}t}. \tag{12.28}$$

Integrating both sides of equation (12.27) with respect to time

$$\nabla \times \mathbf{J} = -\frac{nq^2\mathbf{B}}{m}. \tag{12.29}$$

This is the second equation of London.

12.2.8 Explanation of Meissner's effect from London equations

Maxwell's fourth equation is differential form of Ampere's circuital law:

$$\nabla \times \mathbf{B} = \mu_0\mathbf{J} \tag{12.30}$$

excluding the displacement current term.

Taking the curl of both sides,

$$\nabla \times \nabla \times \mathbf{B} = \mu_0(\nabla \times \mathbf{J}) \tag{12.31}$$

or

$$\nabla \times \nabla \times \mathbf{B} = \mu_0\left(-\frac{nq^2\mathbf{B}}{m}\right), \tag{12.32}$$

where London's second equation has been used.

Applying the formula

$$\nabla \times \nabla \times \mathbf{B} = \nabla(\nabla \cdot \mathbf{B}) - \nabla^2 B, \tag{12.33}$$

equation (12.32) can be expressed as

$$\nabla(\nabla \cdot \mathbf{B}) - \nabla^2 B = \mu_0\left(-\frac{nq^2\mathbf{B}}{m}\right). \tag{12.34}$$

From Maxwell's second equation (Gauss law of magnetostatics)

$$\nabla \cdot \mathbf{B} = 0 \tag{12.35}$$

Equation (12.34) simplifies to

$$\nabla^2 B = \mu_0\left(\frac{nq^2\mathbf{B}}{m}\right) = \frac{B}{\left\{m/\left(\mu_0 nq^2\right)\right\}} = \frac{B}{\lambda_L^2}, \tag{12.36}$$

where

$$\lambda_L = \sqrt{\left\{m/\left(\mu_0 nq^2\right)\right\}}, \tag{12.37}$$

having the dimensions of length, is called the penetration depth; see equation (12.2). Equation (12.36) clearly shows that B is spatially uniform or

$$\nabla^2 B = 0 \tag{12.38}$$

only if $B = 0$. Hence, it is possible for B to be a uniform field inside the superconductor only if it is zero everywhere. So, a non-zero field cannot exist uniformly inside the superconductor. In other words, the field must be non-uniform or dependent on position.

Writing equation (12.36) in the x-direction,

$$\frac{d^2B}{dx^2} = \left(\frac{1}{\lambda_L^2}\right)B. \tag{12.39}$$

Let

$$B = \exp(px). \tag{12.40}$$

Then

$$p^2 \exp(px) - \left(\frac{1}{\lambda_L^2}\right)\exp(px) = 0 \tag{12.41}$$

or

$$\exp(px)\left\{p^2 - \left(\frac{1}{\lambda_L^2}\right)\right\} = 0 \tag{12.42}$$

$$\therefore p^2 - \left(\frac{1}{\lambda_L^2}\right) = 0. \quad (12.43)$$

Hence,

$$p = \pm\left(\frac{1}{\lambda_L}\right). \quad (12.44)$$

The general solution of the differential equation is

$$B = C_1 \exp\left(+\frac{1}{\lambda_L}\right)x + C_2 \exp\left(-\frac{1}{\lambda_L}\right)x, \quad (12.45)$$

$$\text{At } x = \infty,\ B = 0. \quad (12.46)$$

As the first term shows opposite behavior, the physically reasonable solution is

$$B = C_2 \exp\left(-\frac{1}{\lambda_L}\right)x. \quad (12.47)$$

$$\text{At } x = 0,\ B = B(0) \quad (12.48)$$

$$\therefore C_2 = B(0) \quad (12.49)$$

And

$$\therefore B = B(0)\exp\left(-\frac{1}{\lambda_L}\right)x. \quad (12.50)$$

Compare this to equation (12.1). Equations (12.1) and (12.50) are the same. The equation predicts an exponential decay of magnetic induction with distance from the external surface of the superconductor. So, λ_L is the distance over which the magnetic induction falls from the value $B(0)$ at the surface to e^{-1} times inside the superconductor. It is the penetration length. This exclusion of magnetic induction from the superconductor is the Meissner effect. The expulsion of magnetic field is achieved by the shielding supercurrents flowing in a sheath of thickness λ_L at the surface without any ohmic losses.

12.2.9 Practical applications

As superconductivity phenomena take place at very low temperatures, a primary requirement for utilizing these phenomena practically is to reach these low temperatures and maintain the temperatures, both of which are uphill tasks. So, these phenomena have not been put to use in everyday life but have found specific applications. One such application is MRI equipment, in which large magnetic fields are generated by electromagnets. If common conductors are used, the high currents flowing through the electromagnets will cause enormous heating. Coils made of superconducting materials are used in these gigantic electromagnets to solve the heating problem. Another application area of superconducting magnets is in particle accelerators. These are large devices in which elementary particles are paced to high speeds to smash against other particles in order to produce new particles.

12.2.10 High-temperature superconductor

In the context of superconductivity, the term ‘high temperature’ has different connotation than its common meaning because superconductivity was first observed close to absolute zero. A HTS (high-T_C) simply means one which shows superconducting behavior farther from absolute zero than a traditional low-temperature superconductor. In this context, a temperature around 90–100 K being close to liquid-nitrogen temperature (77.2 K) is sufficiently high for a material to qualify as a HTS. The HTS is generally made of a ceramic material. It behaves in the same way as an older low-temperature superconductor but at somewhat elevated temperatures with respect to its predecessors. The ultimate dream is to obtain a superconductor which works at room temperature, so that no extensive cooling arrangements are required for its practical utilization.

12.3 Josephson junction

The JJ is an important active device in superconductive electronics (Scientific American 1997). A JJ, also called a superconducting tunnel junction (STJ), is a sandwich structure consisting of two superconducting electrodes separated by a weak link such as a normal metal or a thin insulating tunnel barrier (figure 12.7). A JJ is artificially fabricated by using the following structures: (i) superconductor–non-superconducting-metal–superconductor or (ii) superconductor–insulator–superconductor. In structure (i), the metal thickness can be several microns, but in structure (ii), the insulator thickness must be very small ~3 nm. Several other ways of forming weak links have also been used, such as narrow constrictions, grain boundaries, etc.

12.3.1 The DC Josephson effect

Even when no DC voltage is applied across the junction, a DC current flows through the junction as long as the current is below a critical level (figure 12.8(a)). This DC current symbolizes the maximum value of supercurrent that can pass through the junction. The Cooper pairs on the opposite sides of the junction can be represented by wavefunctions ψ_1 and ψ_2. The current spontaneously flows by quantum-mechanical tunneling of Cooper pairs. In this tunneling, the collective quantum wave of Cooper pairs spills over to the other side of the junction causing current conduction. Tunneling of Cooper pairs is dissimilar to the tunneling of electrons in a normal tunnel junction because:

(i) The tunneling of Cooper pairs does not need any excitation or voltage whereas electrons require a finite voltage for tunneling.
(ii) Unlike a normal tunnel junction, the tunneling current of Cooper pairs does not encounter any resistance. So, if a current source is connected across the junction there is no voltage drop or power dissipation.

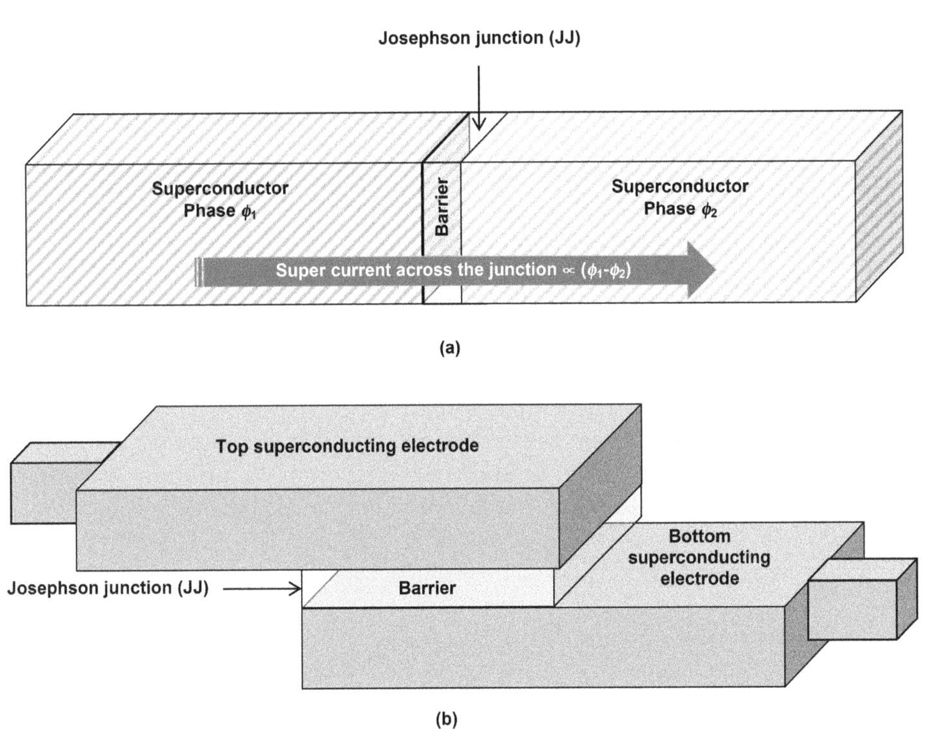

Figure 12.7. A JJ: (a) structure of the junction and (b) junction with electrodes.

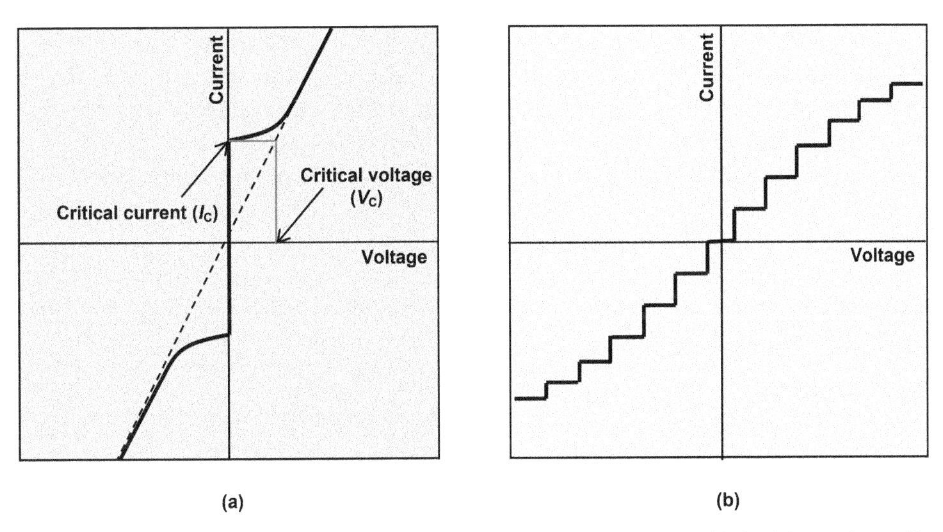

Figure 12.8. Current voltage characteristics for: (a) the DC Josephson effect and (b) the AC Josephson effect.

The current at zero DC voltage is called the Josephson current, and the phenomenon is known as the DC Josephson effect. The Josephson current is proportional to the trigonometric sine function of the phase difference between the wavefunctions ψ_1 and ψ_2. The phenomenon of DC Josephson effect was a surprise because it means that current can flow without any electric field. It is opposite to the commonly taught principle in physics that voltage = current × resistance. So, if voltage = 0, current = 0.

12.3.2 The AC Josephson effect

This effect is observed when a DC voltage is applied across the junction and the DC current exceeds the critical current. Then an oscillatory voltage is developed across the junction in reactionary response to phase variations (figure 12.8(b)). The characteristic frequency of this oscillation depends neither on the size of the superconductors on the two sides of the junction nor on their properties such as chemical composition or critical temperature. It depends only on the applied voltage and on the fundamental physical constants such as elementary electronic charge and Planck's constant. It is proportional to the voltage across the junction. Oscillations are generated at ~500 GHz mV^{-1} across the junction. This is the AC Josephson effect. It is also strange because it means that on applying a DC voltage, an AC current flows across the JJ.

12.3.3 Theory

For superconductivity, the system containing the JJ is assumed to be immersed inside a liquid-helium bath. The temperature of the bath is −269 °C. Further, the system is completely isolated from magnetic fields.

When the wavefunctions ψ_1 and ψ_2 of the Cooper pairs in the two superconducting electrodes of a JJ come closer together, they penetrate the barrier and try to couple with each other. As soon as the energy associated with coupling of wavefunctions becomes greater than the thermal fluctuation energy, the phases of the wavefunctions are locked. This enables the passage of Cooper pairs from one superconductor to another without incurring any energy loss. A simple derivation of the governing equations of the Josephson effect will be presented for a random location in the plane of the junction.

The temporal evolution of the wavefunctions ψ_1 and ψ_2 on the opposite sides of the tunnel barrier is described by the coupled time-dependent Schrodinger equations

$$i\hbar\frac{\partial\psi_1}{\partial t} = U_1\psi_1 + K\psi_2 \tag{12.51}$$

$$i\hbar\frac{\partial\psi_2}{\partial t} = U_2\psi_2 + K\psi_1, \tag{12.52}$$

where K is a constant representing the coupling through the tunnel barrier. It is a measure of the interaction between the wavefunctions. Its value depends on barrier width and composition. It also varies with temperature. It is called the barrier opacity constant. U_1, U_2 denote the states of lowest energy on the opposite sides of the barrier resulting from application of a voltage source $V = (V_1 - V_2)$ between the two sides. The difference of energies is

$$\Delta U = U_2 - U_1 = -2q(V_1 - V_2) = -2qV, \tag{12.53}$$

where $2q$ is the charge on the Cooper pair, q is the electronic charge and V is the potential difference across the junction.

The wavefunctions ψ_1 and ψ_2 can be conveniently expressed as

$$\psi_1 = \sqrt{\rho_1}\exp(\mathrm{i}\phi_1) \tag{12.54}$$

$$\psi_2 = \sqrt{\rho_2}\exp(\mathrm{i}\phi_2), \tag{12.55}$$

where ρ_1, ρ_2 are the densities of the Cooper pairs and ϕ_1, ϕ_2 are the phases of the wavefunctions.

Substituting for ψ_1, ψ_2 from equations (12.54) and (12.55) into equation (12.51), we obtain

$$\mathrm{i}\hbar\frac{\partial\left\{\sqrt{\rho_1}\exp(\mathrm{i}\phi_1)\right\}}{\partial t} = U_1\left\{\sqrt{\rho_1}\exp(\mathrm{i}\phi_1)\right\} + K\left\{\sqrt{\rho_2}\exp(\mathrm{i}\phi_2)\right\} \tag{12.56}$$

or

$$\begin{aligned} &\mathrm{i}\hbar \times (1/2)\rho_1^{-1/2}\left(\frac{\partial\rho_1}{\partial t}\right)\exp(\mathrm{i}\phi_1) + \left\{\mathrm{i}\hbar\sqrt{\rho_1} \times \exp(\mathrm{i}\phi_1) \times \mathrm{i}\left(\frac{\partial\phi_1}{\partial t}\right)\right\} \\ &= U_1\sqrt{\rho_1}\exp(\mathrm{i}\phi_1) + K\sqrt{\rho_2}\exp(\mathrm{i}\phi_2). \end{aligned} \tag{12.57}$$

Multiplying both sides by $\sqrt{\rho_1}/\{\exp(\mathrm{i}\phi_1)\}$, we have

$$\begin{aligned} &\mathrm{i}\hbar \times (1/2)\left(\frac{\partial\rho_1}{\partial t}\right) - \left\{\hbar\rho_1\left(\frac{\partial\phi_1}{\partial t}\right) \times \exp(\mathrm{i}\phi_1)\right\}\Big/\left\{\exp(\mathrm{i}\phi_1)\right\} \\ &= U_1\rho_1 + K\sqrt{\rho_1\rho_2}\left\{\exp(\mathrm{i}\phi_2)/\exp(\mathrm{i}\phi_1)\right\} \end{aligned} \tag{12.58}$$

or

$$\begin{aligned} &\mathrm{i}\hbar\left(\frac{\partial\rho_1}{\partial t}\right) - 2\hbar\rho_1\left(\frac{\partial\phi_1}{\partial t}\right) = 2U_1\rho_1 + 2K\sqrt{\rho_1\rho_2}\left[\exp\left\{\mathrm{i}(\phi_2 - \phi_1)\right\}\right] \\ &= 2U_1\rho_1 + 2K\sqrt{\rho_1\rho_2}\left\{\cos(\phi_2 - \phi_1) + \mathrm{i}\sin(\phi_2 - \phi_1)\right\}. \end{aligned} \tag{12.59}$$

Equating real and imaginary parts,

$$-2\hbar\rho_1\left(\frac{\partial\phi_1}{\partial t}\right) = 2U_1\rho_1 + 2K\sqrt{\rho_1\rho_2}\cos(\phi_2 - \phi_1) \tag{12.60}$$

or

$$\frac{\partial\phi_1}{\partial t} = -\frac{U_1}{\hbar} - \frac{K}{\hbar}\sqrt{\frac{\rho_2}{\rho_1}}\cos(\phi_2 - \phi_1) \tag{12.61}$$

and

$$i\hbar\left(\frac{\partial \rho_1}{\partial t}\right) = 2K\sqrt{\rho_1\rho_2} \times \mathrm{i}\sin(\phi_2 - \phi_1) \tag{12.62}$$

or

$$\frac{\partial \rho_1}{\partial t} = \frac{2K}{\hbar}\sqrt{\rho_1\rho_2}\,\sin(\phi_2 - \phi_1). \tag{12.63}$$

Substituting for ψ_1, ψ_2 from equations (12.54) and (12.55) into equation (12.52), we obtain

$$i\hbar\frac{\partial\left\{\sqrt{\rho_2}\exp(\mathrm{i}\phi_2)\right\}}{\partial t} = U_2\left\{\sqrt{\rho_2}\exp(\mathrm{i}\phi_2)\right\} + K\left\{\sqrt{\rho_1}\exp(\mathrm{i}\phi_1)\right\} \tag{12.64}$$

or

$$\begin{aligned} &i\hbar \times (1/2)\rho_2^{-1/2}\left(\frac{\partial \rho_2}{\partial t}\right)\exp(\mathrm{i}\phi_2) + \left\{i\hbar\sqrt{\rho_2} \times \exp(\mathrm{i}\phi_2) \times \mathrm{i}\left(\frac{\partial \phi_2}{\partial t}\right)\right\} \\ &= U_2\sqrt{\rho_2}\exp(\mathrm{i}\phi_2) + K\left\{\sqrt{\rho_1}\exp(\mathrm{i}\phi_1)\right\}. \end{aligned} \tag{12.65}$$

Multiplying both sides by $\sqrt{\rho_2}/\{\exp(\mathrm{i}\phi_2)\}$,

$$\begin{aligned} &i\hbar \times (1/2)\left(\frac{\partial \rho_2}{\partial t}\right) - \left\{\hbar\rho_2\left(\frac{\partial \phi_2}{\partial t}\right) \times \exp(\mathrm{i}\phi_2)\right\}\Big/\left\{\exp(\mathrm{i}\phi_2)\right\} \\ &= U_2\rho_2 + K\sqrt{\rho_1\rho_2}\left\{\exp(\mathrm{i}\phi_1)/\exp(\mathrm{i}\phi_2)\right\} \end{aligned} \tag{12.66}$$

or

$$\begin{aligned} &i\hbar\left(\frac{\partial \rho_2}{\partial t}\right) - 2\hbar\rho_2\left(\frac{\partial \phi_2}{\partial t}\right) = 2U_2\rho_2 + 2K\sqrt{\rho_1\rho_2}\left[\exp\left\{\mathrm{i}(\phi_1 - \phi_2)\right\}\right] \\ &= 2U_2\rho_2 + 2K\sqrt{\rho_1\rho_2}\left\{\cos(\phi_1 - \phi_2) + \mathrm{i}\sin(\phi_1 - \phi_2)\right\}. \end{aligned} \tag{12.67}$$

Equating real and imaginary parts,

$$-2\hbar\rho_2\left(\frac{\partial \phi_2}{\partial t}\right) = 2U_2\rho_2 + 2K\sqrt{\rho_1\rho_2}\cos(\phi_1 - \phi_2) \tag{12.68}$$

or

$$-\frac{\partial \phi_2}{\partial t} = \frac{U_2}{\hbar} + \frac{K}{\hbar}\sqrt{\frac{\rho_1}{\rho_2}}\cos(\phi_1 - \phi_2)$$

or

$$\frac{\partial \phi_2}{\partial t} = -\frac{U_2}{\hbar} - \frac{K}{\hbar}\sqrt{\frac{\rho_1}{\rho_2}}\cos(\phi_2 - \phi_1) \tag{12.69}$$

and

$$\frac{\partial \rho_2}{\partial t} = \frac{2K}{\hbar}\sqrt{\rho_1\rho_2} \times \sin(\phi_1 - \phi_2) = -\frac{2K}{\hbar}\sqrt{\rho_1\rho_2} \times \sin(\phi_2 - \phi_1). \tag{12.70}$$

Comparison of equations (12.63) and (12.70) shows that

$$\begin{aligned}\frac{\partial \rho_1}{\partial t} &= \text{rate of decrease of Cooper pair density } \rho_1 \text{ in superconductor 1} \\ &= -\frac{\partial \rho_2}{\partial t} = -\text{rate of decrease of Cooper pair density } \rho_2 \text{ in superconductor 2.}\end{aligned} \tag{12.71}$$

As these changes may bring about a charge imbalance between the electrons and background ions, currents flowing in the circuit connected to the JJ oppose them. The current density flowing from one superconductor to another is given by the time derivative of the density of Cooper pairs. The sign of the current density is obtained by comparing the JJ with a bulk superconductor. The direction of current density is opposite to that of the gradient of phase. In the JJ, the positive gradient of phase from superconductor 1 to superconductor 2 is represented by $\phi_2 - \phi_1 > 0$. For this polarity of current, transference of electrons in Cooper pairs must take place from superconductor 1 to superconductor 2. Hence, $\frac{\partial \rho_2}{\partial t} > 0$ and K is negative in sign, so that from equation (12.63), the current density is

$$j = \frac{\partial \rho_1}{\partial t} = 2\frac{K}{\hbar}\sqrt{\rho_1\rho_2}\sin(\phi_2 - \phi_1), \tag{12.72}$$

which can be recast in the form

$$j = j_C \sin\phi, \tag{12.73}$$

where

$$j_C = 2\frac{K}{\hbar}\sqrt{\rho_1\rho_2} \tag{12.74}$$

is the critical current density, and

$$\phi_2 - \phi_1 = \phi. \tag{12.75}$$

The critical current density j_C depends on K and the external magnetic field which is assumed to be absent.

For simplification, the superconducting metallic layers of the JJ are made of the same element niobium (Nb). Then $\rho_1 = \rho_2$, by taking identical superconductors and subtracting equation (12.61) from equation (12.69),

$$\frac{\partial \phi_2}{\partial t} - \frac{\partial \phi_1}{\partial t} = \frac{U_1}{\hbar} - \frac{U_2}{\hbar} \tag{12.76}$$

or

$$\frac{\partial (\phi_2 - \phi_1)}{\partial t} = \frac{U_1 - U_2}{\hbar}$$

or

$$\frac{\partial \phi}{\partial t} = \frac{2qV}{\hbar} \tag{12.77}$$

from equation (12.53).

Equations (12.71) and (12.77) constitute the basic governing equations of the Josephson effect. Let us see how they lead to the DC and AC Josephson effects.

The DC effect

Integrating equation (12.77) over time,

$$\int_{\phi_0}^{\phi(t)} \left(\frac{\partial \phi}{\partial t}\right) \mathrm{d}t = \frac{2q}{\hbar} \int_0^t V \mathrm{d}t \tag{12.78}$$

or

$$\phi(t) - \phi_0 = \frac{2q}{\hbar} \int_0^t V \mathrm{d}t. \tag{12.79}$$

If $V = 0$, $\phi(t) - \phi_0 = 0$ or, $\phi(t) = \phi_0$, hence

$$j = j_C \sin \phi_0, \tag{12.80}$$

which is a constant. Therefore, for $V = 0$, a constant DC current flows. The maximum value of this current is attained when $\sin \phi_0$ has peak value = 1. This maximum value is

$$j = j_C \sin \phi_0 = j_C \times 1 = j_C = 2\frac{K}{\hbar}\sqrt{\rho_1 \rho_2} \tag{12.81}$$

from equation (12.74).

This is exactly the DC Josephson effect, namely the appearance of a constant current without applied voltage and sustenance of this current up to a peak value j_C.

The AC effect

Now, if V is non-zero but has a magnitude = V, we have

$$\phi(t) - \phi_0 = \phi = \frac{2qVt}{\hbar} \tag{12.82}$$

so that

$$j = j_C \sin\phi = j_C \sin\left(\frac{2qVt}{\hbar}\right), \tag{12.83}$$

which is an oscillating current with a frequency

$$f = \frac{1}{2\pi}\left(\frac{2qV}{\hbar}\right) = \frac{1}{2\pi}\left\{\frac{2qV}{h/(2\pi)}\right\} = \frac{2qV}{h}. \tag{12.84}$$

For $V = 1\ \text{mV} = 10^{-3}\ \text{V}$,

$$f = \frac{2q}{h} = \frac{2 \times 1.6 \times 10^{-19} \times 10^{-3}}{6.63 \times 10^{-34}} = 4.827 \times 10^{11}\ \text{Hz}. \tag{12.85}$$

This is the AC Josephson effect in which application of a DC voltage V produces high-frequency AC with a frequency of $4.827 \times 10^{11}\ \text{Hz mV}^{-1}$.

12.3.4 Gauge-invariant phase difference

The foregoing treatment was based on the assumption that the JJ was not exposed to any magnetic field. The phase difference was taken as $\phi = \phi_2 - \phi_1$. This is not a gauge-invariant quantity. But it did not lead to any error because the effect of magnetic field on the JJ was not considered. But this ϕ value cannot represent a general situation wherein a magnetic field is present. Therefore the current density J thus determined is not true in a generalized perspective. In order to remedy this non-inclusion, in the presence of a magnetic vector potential **A**, defined in terms of the magnetic field **B** as

$$\mathbf{B} = \nabla \times \mathbf{A}, \tag{12.86}$$

the phase difference ϕ will be replaced by the gauge-invariant phase difference θ defined as

$$\theta = \phi_2 - \phi_1 - (2\pi/\Phi_0) \oint_1^2 \mathbf{A} \cdot \mathrm{d}\mathbf{s}, \tag{12.87}$$

where Φ_0 is the magnetic flux quantum given by

$$\Phi_0 = h/(2q) = 6.63 \times 10^{-34}/(2 \times 1.6 \times 10^{-19}) = 2.072 \times 10^{-15}\ \text{Wb}. \tag{12.88}$$

It has the same value for all superconductors. The quantization arises from the requirement for the macroscopic wave function to be single-valued. Due to this restriction, the total magnetic flux trapped by a superconducting ring can acquire only quantized values, which are integral multiples of the flux quantum. The magnetic flux Φ threading a loop is

$$\mathbf{\Phi} = \mathbf{B} \cdot \mathbf{s}, \tag{12.89}$$

where **s** is the area of the loop.

Let the Cooper pair concentration in the superconductor be $\rho(\mathbf{r})$ where $\mathbf{r}$ is the position vector and the wavefunction be $\psi(\mathbf{r})$. Then (Tsang 1997)

$$\rho(\mathbf{r}) = \Psi^*(\mathbf{r})\Psi(\mathbf{r}) = |\Psi(\mathbf{r})|^2 \tag{12.90}$$

$$\therefore \Psi(\mathbf{r}) = \sqrt{\rho(\mathbf{r})}\exp\{+\mathrm{i}\theta(\mathbf{r})\} \tag{12.91}$$

and

$$\Psi^*(\mathbf{r}) = \sqrt{\rho(\mathbf{r})}\exp\{-\,\mathrm{i}\theta(\mathbf{r})\}, \tag{12.92}$$

where $\theta(\mathbf{r})$ is the phase of the wavefunction.

The canonical momentum $\boldsymbol{p}$ of a classical particle of mass m moving with a velocity $\mathbf{v}$ is

$$\boldsymbol{p} = m\boldsymbol{v} + (q/c)\boldsymbol{A}, \tag{12.93}$$

where q is the electronic charge, c is the velocity of light and $\mathbf{A}$ is the magnetic vector potential.

Replacing $\mathbf{p}$ by the quantum-mechanical momentum operator $-\mathrm{i}\hbar\nabla$, we can write

$$m\boldsymbol{v} = -\mathrm{i}\hbar\nabla - (q/c)\boldsymbol{A} \tag{12.94}$$

$$\boldsymbol{v} = (1/m)\{-\,\mathrm{i}\hbar\nabla - (q/c)\boldsymbol{A}\}. \tag{12.95}$$

Let us recall the derivation of the relationship between the current density J and velocity v of free electrons. Consider a conductor of cross-sectional area A and length L. Then the volume of the conductor is

$$V = AL. \tag{12.96}$$

If n is the number of electrons/volume of the conductor, then the total number of electrons in the conductor is

$$N = ALn. \tag{12.97}$$

If q is the charge of a single electron, the total charge on all the electrons in the conductor is

$$Q = qALn. \tag{12.98}$$

Time taken by charge carriers to traverse the length L of the conductor

$$t = L/v. \tag{12.99}$$

Current I is defined as

$$I = Q/t. \tag{12.100}$$

Substituting for Q and t from equations (12.98) and (12.99) into equation (12.100), we obtain

$$I = Q/t = (qALn)/(L/v) = qAnv. \tag{12.101}$$

The current density is

$$J = I/A = qAnv/A = qnv. \tag{12.102}$$

In quantum mechanics

$$nv = \Psi^* \boldsymbol{v} \Psi. \tag{12.103}$$

Hence,

$$\mathbf{J} = q\Psi^* \mathbf{v} \Psi \tag{12.104}$$

Putting the values of Ψ from equation (12.91), Ψ^* from equation (12.92) and v from equations (12.95) into (12.104) for $\mathbf{J}$, we have

$$\begin{aligned}
\mathbf{J} &= q\Psi^* \mathbf{v}\, \Psi = q\sqrt{\rho}\, \exp(-\mathrm{i}\theta)(1/m)\{-\mathrm{i}\hbar\nabla - (q/c)\boldsymbol{A}\}\sqrt{\rho}\, \exp(+\mathrm{i}\theta) \\
&= q\sqrt{\rho}\, \exp(-\mathrm{i}\theta)(1/m)\left[\{-\mathrm{i}\hbar\nabla - (q/c)\boldsymbol{A}\}\sqrt{\rho}\, \exp(+\mathrm{i}\theta)\right] \\
&= q\sqrt{\rho}\, \exp(-\mathrm{i}\theta)(1/m) \\
&\quad \times \left[\left\{-\mathrm{i}\hbar\sqrt{\rho}\, \exp(+\mathrm{i}\theta) \times (+\mathrm{i})\nabla\theta - (q/c)\boldsymbol{A}\sqrt{\rho}\, \exp(+\mathrm{i}\theta)\right\}\right] \\
&= q\sqrt{\rho}\, \exp(-\mathrm{i}\theta)(1/m)\sqrt{\rho}\, \exp(+\mathrm{i}\theta)[\{-\mathrm{i}\hbar \times (+\mathrm{i})\nabla\theta - (q/c)\boldsymbol{A}\}] \\
&= (q\rho/m)[\{-\mathrm{i}\hbar \times (+\mathrm{i}\theta)\nabla\theta - (q/c)\boldsymbol{A}\}] = (q\rho/m)\{\hbar\nabla\theta - (q/c)\boldsymbol{A}\}.
\end{aligned} \tag{12.105}$$

But from Maxwell's fourth equation

$$\mathbf{J} = (c/4\pi)(\nabla \times \mathbf{B}). \tag{12.106}$$

Inside the superconductor, $B = 0$ by Meissner's effect. Hence, $\mathbf{J} = 0$, and therefore

$$(q\rho/m)\{\hbar\nabla\theta - (q/c)\boldsymbol{A}\} = 0 \tag{12.107}$$

or

$$\hbar\nabla\theta - (q/c)\boldsymbol{A} = 0$$

or

$$\hbar\nabla\theta = (q/c)\boldsymbol{A}. \tag{12.108}$$

Performing line integration of both sides along a closed path C inside the superconducting ring, we obtain

$$\oint \hbar\nabla\theta \cdot \mathrm{d}\boldsymbol{l} = \oint (q/c)\boldsymbol{A} \cdot \mathrm{d}\boldsymbol{l} \tag{12.109}$$

or

$$(\hbar c)\oint \nabla\theta \cdot \mathrm{d}\boldsymbol{l} = q\oint \boldsymbol{A} \cdot \mathrm{d}\boldsymbol{l} \tag{12.110}$$

On the left-hand side,

$$\oint \nabla\theta \cdot \mathrm{d}\boldsymbol{l} = \int_{\theta_1}^{\theta_2} \nabla\theta \cdot \mathrm{d}\boldsymbol{l} = (\theta_2 - \theta_1). \tag{12.111}$$

This is the phase difference on travelling once around the loop. The uniqueness of the Cooper-pair wavefunction imposes the restriction that the integral of the phase difference taken once around a closed loop can take on values equal to integer multiples of 2π only. Hence,

$$\theta_2 - \theta_1 = 2\pi\Xi, \tag{12.112}$$

where Ξ is an integer. So,

$$(\hbar c)\oint \nabla\theta \cdot \mathrm{d}\boldsymbol{l} = (\hbar c)2\pi\Xi. \tag{12.113}$$

On the right-hand side of equation (12.110), Stoke's theorem is applied to obtain

$$q\oint \boldsymbol{A} \cdot \mathrm{d}\boldsymbol{l} = q\iint (\nabla \times \mathbf{A}) \cdot \mathrm{d}\boldsymbol{s}. \tag{12.114}$$

Note that $\oint \boldsymbol{A}.\mathrm{d}\boldsymbol{l}$ is a line integral over the closed contour C; $\mathrm{d}\boldsymbol{l}$ is a linear element whereas

$$\iint (\nabla \times \mathbf{A}) \cdot \mathrm{d}\boldsymbol{s} \tag{12.115}$$

is a surface integral over the surface covering the interior of the closed path C; $\mathrm{d}\boldsymbol{s}$ is an areal element.

But

$$\nabla \times \mathbf{A} = \mathbf{B} \tag{12.116}$$

so that

$$q\oint \boldsymbol{A} \cdot \mathrm{d}\boldsymbol{l} = q\iint \mathbf{B} \cdot \mathrm{d}\boldsymbol{s} = q\mathbf{B} \cdot \mathbf{s} = q\mathbf{\Phi}, \tag{12.117}$$

where equation (12.89) has been used.

Combining equations (12.110), (12.113) and (12.117),

$$(\hbar c)2\pi\Xi = q\mathbf{\Phi} \tag{12.118}$$

$$\therefore \mathbf{\Phi} = \frac{(\hbar c)2\pi\Xi}{q} = \frac{hc}{2\pi q} \times 2\pi\Xi = \left(\frac{hc}{q}\right)\Xi = \Phi_0\Xi. \tag{12.119}$$

The line integral of **A** around a contour passing through the superconducting electrodes and the tunnel barrier yields the enclosed flux Φ. The path of integration of magnetic vector potential is taken from superconductor 2 to superconductor 1 across the tunnel barrier. Thus the modified super current density is

$$J = J_C \sin\theta = J_C \sin\left\{\phi_2 - \phi_1 - (2\pi/\Phi_0)\oint_1^2 \mathbf{A}\cdot \mathrm{d}\mathbf{s}\right\} \tag{12.120}$$

from equation (12.87). J_C is the modified critical current density. Its relationship with previous critical current density j_C will be derived. In terms of θ, the modified form of equation (12.77) is

$$\frac{2q(V_1 - V_2)}{\hbar} = \frac{\partial\theta}{\partial t} = \frac{\partial}{\partial t}\left\{\phi_2 - \phi_1 - (2\pi/\Phi_0)\oint_1^2 \mathbf{A}\cdot \mathrm{d}\mathbf{s}\right\}. \tag{12.121}$$

If, however, no magnetic field is present, one can take $\mathbf{A} = 0$. Then the equality $\theta = \phi$ is valid.

After knowing about this gauge-invariant form of current, let us apply a magnetic induction **B** across the junction along the $-\hat{y}$-direction. Then

$$\mathbf{A}\cdot \mathrm{d}\mathbf{s} = -Bx\mathrm{d}z \tag{12.122}$$

and

$$\oint_1^2 \mathbf{A}\cdot \mathrm{d}\mathbf{s} = \oint_0^d -Bx\mathrm{d}z = -Bx[z]_0^d = -Bxd \tag{12.123}$$

where x, $\mathrm{d}z$ are the dimensions of the JJ element in the x, z directions respectively, and d is the thickness of the tunnel barrier.

Substituting for $\oint_1^2 \mathbf{A}\cdot \mathrm{d}\mathbf{s}$ in the equation for current density, i.e. equation (12.120), equation (12.73) is rewritten as

$$j = j_C \sin\{\phi - (2\pi/\Phi_0)(-Bxd)\} = j_C \sin\{\phi + (2\pi/\Phi_0)(Bxd)\}. \tag{12.124}$$

Integrating from 0 to L, the length over which JJ extends along the x-direction

$$\begin{aligned} J &= \int_0^L j(x)\mathrm{d}x = \int_0^L j_C \sin\{\phi + (2\pi/\Phi_0)(Bxd)\}\mathrm{d}x \\ &= j_C \int_0^L \sin\{\phi + (2\pi/\Phi_0)(Bxd)\}\mathrm{d}x. \end{aligned} \tag{12.125}$$

Putting

$$(2\pi/\Phi_0)(Bd) = \eta \tag{12.126}$$

$$J = j_C \int_0^L \sin(\phi + \eta x)\,\mathrm{d}x, \tag{12.127}$$

let

$$u = \phi + \eta x, \tag{12.128}$$

then

$$du = \eta dx$$

or

$$dx = du/\eta. \tag{12.129}$$

Further,

$$\begin{aligned} J &= j_C \int_0^L (\sin u du)/\eta = (j_C/\eta) \int_0^L \sin u du = \left[(j_C/\eta) \times (-\cos u)\right]_0^L \\ &= (j_C/\eta) \times [-\cos\{\phi + (2\pi/\Phi_0)(Bxd)\}]_0^L \\ &= \left[j_C/\{(2\pi/\Phi_0)(Bd)\}\right] \times [-\cos\{\phi + (2\pi/\Phi_0)(Bxd)\}]_0^L \\ &= \left[Lj_C/\{(2\pi/\Phi_0)(BdL)\}\right] \\ &\quad \times [-\cos\{\phi + (2\pi/\Phi_0)(BLd)\} + \cos\{\phi + (2\pi/\Phi_0)(B \times 0 \times d)\}] \\ &= \left[Lj_C/(2\pi\Phi/\Phi_0)\right] \times [-\cos\{\phi + (2\pi/\Phi_0)(\Phi)\} + \cos(\phi)] \\ &= \left[Lj_C/(2\pi\Phi/\Phi_0)\right][\cos\phi - \cos\{\phi + (2\pi\Phi/\Phi_0)\}]. \end{aligned} \tag{12.130}$$

To find the maximum current flowing through the junction for all possible values of ϕ, let us find

$$\begin{aligned} dJ/d\phi &= \left[Lj_C/(2\pi\Phi/\Phi_0)\right] \times (d/d\phi)[\cos\phi - \cos\{\phi + (2\pi\Phi/\Phi_0)\}] \\ &= \left[Lj_C/(2\pi\Phi/\Phi_0)\right] \times [-\sin\phi + \sin\{\phi + (2\pi\Phi/\Phi_0)\}]. \end{aligned} \tag{12.131}$$

Setting

$$dJ/d\phi = 0 \tag{12.132}$$

$$-\sin\phi + \sin\{\phi + (2\pi\Phi/\Phi_0)\} = 0 \tag{12.133}$$

Or

$$\begin{aligned} &\sin\{\phi + (2\pi\Phi/\Phi_0)\} - \sin\phi = 0 \\ &2\sin[\{(\phi + 2\pi\Phi/\Phi_0) - \phi\}/2]\cos[\{(\phi + 2\pi\Phi/\Phi_0) + \phi\}/2] = 0 \\ &2\sin(\pi\Phi/\Phi_0)\cos\{\phi + (\pi\Phi/\Phi_0)\} = 0 \\ &2\sin(\pi\Phi/\Phi_0)\{\cos\phi\cos(\pi\Phi/\Phi_0) - \sin\phi\sin(\pi\Phi/\Phi_0)\} = 0 \\ &\cos\phi\cos(\pi\Phi/\Phi_0) - \sin\phi\sin(\pi\Phi/\Phi_0) = 0 \\ &\cos\phi\cos(\pi\Phi/\Phi_0) = \sin\phi\sin(\pi\Phi/\Phi_0) \\ &\cos(\pi\Phi/\Phi_0)/\sin(\pi\Phi/\Phi_0) = \sin\phi/\cos\phi \\ &\cot(\pi\Phi/\Phi_0) = \tan\phi. \end{aligned} \tag{12.134}$$

Therefore, from equations (12.130) and (12.134), the critical current density J_C corresponding to the gauge-invariant phase difference is obtained as follows

$$\begin{aligned} J_C &= \left\{Lj_C/(2\pi\Phi/\Phi_0)\right\}2\sin[\{2\phi + (2\pi\Phi/\Phi_0)\}/2] \\ &\quad \times \sin[\{(\phi + 2\pi\Phi/\Phi_0) - \phi\}/2] \\ &= \left\{Lj_C/(\pi\Phi/\Phi_0)\right\}\sin\{\phi + (\pi\Phi/\Phi_0)\}\sin(\pi\Phi/\Phi_0) \\ &= \left\{Lj_C/(\pi\Phi/\Phi_0)\right\}\{\sin\phi\cos(\pi\Phi/\Phi_0) + \cos\phi\sin(\pi\Phi/\Phi_0)\}\sin(\pi\Phi/\Phi_0) \\ &= \left[\left\{Lj_C/(\pi\Phi/\Phi_0)\right\}\sin(\pi\Phi/\Phi_0)\right] \\ &\quad \times [\cos\phi\sin(\pi\Phi/\Phi_0)\{\tan\phi\cot(\pi\Phi/\Phi_0) + 1\}] \\ &= \left[\left\{Lj_C/(\pi\Phi/\Phi_0)\right\}\sin(\pi\Phi/\Phi_0)\right]\left\{\cos\phi\sin(\pi\Phi/\Phi_0)(\tan^2\phi + 1)\right\} \end{aligned} \tag{12.135}$$

by applying equation (12.134). Hence,

$$\begin{aligned} J_C &= \left[\left\{Lj_C/(\pi\Phi/\Phi_0)\right\}\sin(\pi\Phi/\Phi_0)\right]\left\{\cos\phi\sin(\pi\Phi/\Phi_0)(\sec^2\phi)\right\} \\ &= \left[\left\{Lj_C/(\pi\Phi/\Phi_0)\right\}\sin(\pi\Phi/\Phi_0)\right]\{\sin(\pi\Phi/\Phi_0)/\cos\phi\}. \end{aligned} \tag{12.136}$$

But

$$\begin{aligned} \cos\phi &= 1/\sqrt{1 + \cot^2(\pi\Phi/\Phi_0)} = 1/\sqrt{1 + 1/\{\tan^2(\pi\Phi/\Phi_0)\}} \\ &= \tan(\pi\Phi/\Phi_0)/\sqrt{\tan^2(\pi\Phi/\Phi_0) + 1} \\ &= \tan(\pi\Phi/\Phi_0)/\sqrt{\sec^2(\pi\Phi/\Phi_0)} = \tan(\pi\Phi/\Phi_0)/\sec(\pi\Phi/\Phi_0) \\ &= \{\sin(\pi\Phi/\Phi_0)/\cos(\pi\Phi/\Phi_0)\}/\{1/\cos(\pi\Phi/\Phi_0)\} = \sin(\pi\Phi/\Phi_0). \end{aligned} \tag{12.137}$$

Therefore,

$$J_C = \left[\left\{Lj_C/(\pi\Phi/\Phi_0)\right\}\sin(\pi\Phi/\Phi_0)\right]\{\sin(\pi\Phi/\Phi_0)/\sin(\pi\Phi/\Phi_0)\} \tag{12.138}$$

or

$$J_C = Lj_C|\sin(\pi\Phi/\Phi_0)|/|(\pi\Phi/\Phi_0)|. \tag{12.139}$$

12.4 Inverse AC Josephson effect: Shapiro steps

Let us apply a combined (DC + AC) voltage $V(t)$. Let DC voltage = V_{DC}, AC voltage = $V_0 \cos \omega t$ where V_0 is the amplitude and ω is the angular frequency. Here, $V_0 \ll V_{DC}$ and ω is very high in the RF or microwave range. Thus a high-frequency (ω), small amplitude (V_0) AC voltage is superimposed on a large DC voltage (V_{DC}):

$$V(t) = V_{DC} + V_0 \cos \omega t. \tag{12.140}$$

Writing equation (12.77) for the AC Josephson effect

$$\frac{\partial\delta}{\partial t} = \frac{2q\{V(t)\}}{\hbar} = \frac{2q(V_{\mathrm{DC}} + V_0 \cos\omega t)}{\hbar}, \tag{12.141}$$

where δ is the phase difference.

Integrating both sides with respect to time,

$$\int_0^t \left(\frac{\partial\delta}{\partial t}\right) \mathrm{d}t = \left(\frac{2q}{\hbar}\right) \int V_{\mathrm{DC}} \mathrm{d}t + \left(\frac{2q}{\hbar}\right) \int V_0 \cos\omega t \,\mathrm{d}t \tag{12.142}$$

or

$$\int_0^t \partial\delta = \left(\frac{2q}{\hbar}\right) V_{\mathrm{DC}} t + \left(\frac{2qV_0}{\hbar\omega}\right) \sin\omega t$$

or

$$\begin{aligned} \delta(t) - \delta(0) &= \left(\frac{2q}{\hbar}\right) V_{\mathrm{DC}} t + \left(\frac{2qV_0}{\hbar\omega}\right) \sin\omega t \\ \therefore \delta(t) &= \delta(0) + \left(\frac{2q}{\hbar}\right) V_{\mathrm{DC}} t + \left(\frac{2qV_0}{\hbar\omega}\right) \sin\omega t. \end{aligned} \tag{12.143}$$

From equations (12.143) and (12.73), the current density j through the junction is

$$j = j_{\mathrm{C}} \sin\left\{\delta(0) + \left(\frac{2q}{\hbar}\right) V_{\mathrm{DC}} t + \left(\frac{2qV_0}{\hbar\omega}\right) \sin\omega t\right\}, \tag{12.144}$$

where j is a frequency-modulated current that can be analyzed using the following approximation:

$$\sin(x + \delta x) \approx \sin x + \delta x \cos x \tag{12.145}$$

giving

$$j = j_{\mathrm{C}}\left[\sin\left\{\delta(0) + \left(\frac{2q}{\hbar}\right) V_{\mathrm{DC}} t\right\} + \left(\frac{2qV_0}{\hbar\omega}\right) \sin\omega t \cos\left\{\delta(0) + \left(\frac{2q}{\hbar}\right) V_{\mathrm{DC}}\right\}\right], \tag{12.146}$$

where we have put

$$x = \delta(0) + \left(\frac{2q}{\hbar}\right) V_{\mathrm{DC}} t \tag{12.147}$$

$$\delta x = \left(\frac{2qV_0}{\hbar\omega}\right) \sin\omega t. \tag{12.148}$$

In the first term within square brackets,

$$\sin\left\{\delta(0) + \left(\frac{2q}{\hbar}\right) V_{\mathrm{DC}} t\right\} \tag{12.149}$$

$\hbar \rightarrow 0$ because $\hbar$ is very small, hence,

$$\left\{\delta(0) + \left(\frac{2q}{\hbar}\right)V_{\mathrm{DC}}t\right\} \rightarrow \infty. \tag{12.150}$$

sin(∞) oscillates between −1 and +1; therefore, its average value over time is zero. So, this term is zero.

The second term within square brackets

$$\left(\frac{2qV_0}{\hbar\omega}\right)\sin\omega t\cos\left\{\delta(0) + \left(\frac{2q}{\hbar}\right)V_{\mathrm{DC}}t\right\} \tag{12.151}$$

will also time-average to zero since cos(∞) behaves like sin(∞). But it can be made non-zero by choosing the frequency ω of the applied AC field such that

$$\omega = 2qV_{\mathrm{DC}}/\hbar \tag{12.152}$$

because then this term becomes

$$\left(\frac{2qV_0}{\hbar\omega}\right)\sin\{(2qV_{\mathrm{DC}}/\hbar)t\}\cos\left\{\delta(0) + \left(\frac{2q}{\hbar}\right)V_{\mathrm{DC}}t\right\}. \tag{12.153}$$

Using the trigonometric identity

$$\sin a\cos b=(1/2)\{\sin(a + b) + \sin(a - b)\}. \tag{12.154}$$

Equation (12.153) can be rewritten as

$$\begin{aligned}(1/2)&\left[\sin\left\{(2qV_{\mathrm{DC}}/\hbar)t + \delta(0) + \left(\frac{2q}{\hbar}\right)V_{\mathrm{DC}}t\right\}\right.\\ &\left.+ \sin\left\{(2qV_{\mathrm{DC}}/\hbar)t - \delta(0) - \left(\frac{2q}{\hbar}\right)V_{\mathrm{DC}}t\right\}\right]\\ &= (1/2)\left[\sin\left\{\left(\delta(0) + \left(\frac{4q}{\hbar}\right)V_{\mathrm{DC}}t\right)\right\} + \sin\{-\delta(0)\}\right].\end{aligned} \tag{12.155}$$

As argued previously, the term

$$\sin\left\{\left(\delta(0) + \left(\frac{4q}{\hbar}\right)V_{\mathrm{DC}}t\right)\right\} \tag{12.156}$$

will time-average to zero because $\hbar \rightarrow 0$, so that

$$j = j_{\mathrm{C}}\left(\frac{2qV_0}{\hbar\omega}\right)(1/2)\sin\{-\delta(0)\} = -j_{\mathrm{C}}\left(\frac{qV_0}{\hbar\omega}\right)\sin\{\delta(0)\}. \tag{12.157}$$

This is a constant DC. This phenomenon of obtaining zero-frequency or DC super current from combined (DC + AC) excitation is called the inverse AC effect.

Furthermore, the above analysis shows that at multiples of frequency ω, i.e. whenever

$$n\omega = 2qV_{DC}/\hbar \tag{12.158}$$

or

$$2qV_{DC} = n\hbar\omega, \tag{12.159}$$

where $n = 0, 1, 2, 3, \ldots$, the current density j has zero frequency. Hence, the DC current–voltage (I–V) characteristics of the junction contain a series of discrete steps having width = $2q/\hbar$. The well-defined constant-voltage DC spikes generated in the I–V characteristics in response to distinct frequencies of external RF or microwave signals are known as Shapiro spikes or steps. For these distinct frequency values, the JJ acts as a frequency-to-voltage converter. A precise determination of the ratio $q/\hbar$ can be done from this experiment.

In general, the current through the JJ can be written as (Grosso and Parravicini 2014)

$$j = j_C \sum_{+\infty}^{n=-\infty} J_n\left(\frac{2qV_0}{\hbar\omega}\right)\sin\left\{\delta(0) + \left(\frac{2q}{\hbar}\right)V_{DC}t + n\omega t\right\}, \tag{12.160}$$

where J_n are Bessel functions of the first kind and order n obtained from an infinite power series expansion:

$$J_n(x) = \sum_{k=0}^{\infty} \frac{(-1)^k}{k!\Gamma(k + n + 1)}\left(\frac{x}{2}\right)^{n+2k}. \tag{12.161}$$

12.5 Superconducting quantum interference devices

A SQUID works as an extremely sensitive magnetometer using one or more JJs and based on the principle of quantum interference. It is used for measuring very weak magnetic fields in microtesla or nanotesla (10^{-6}–10^{-9} T) ranges (Mehta 2011), and up to 5×10^{-18} T. A magnetometer is an instrument used to measure the magnitude and direction of a magnetic field at a point in space or the magnetization of a material. SQUIDs are of two types: (i) DC SQUID and (ii) AC or RF SQUID.

12.5.1 DC SQUID

A DC SQUID consists of two or more JJs connected in parallel (figure 12.9). Essentially, it is a dual-junction superconducting loop. It has a high sensitivity for magnetic field detection. Its fabrication is difficult and expensive.

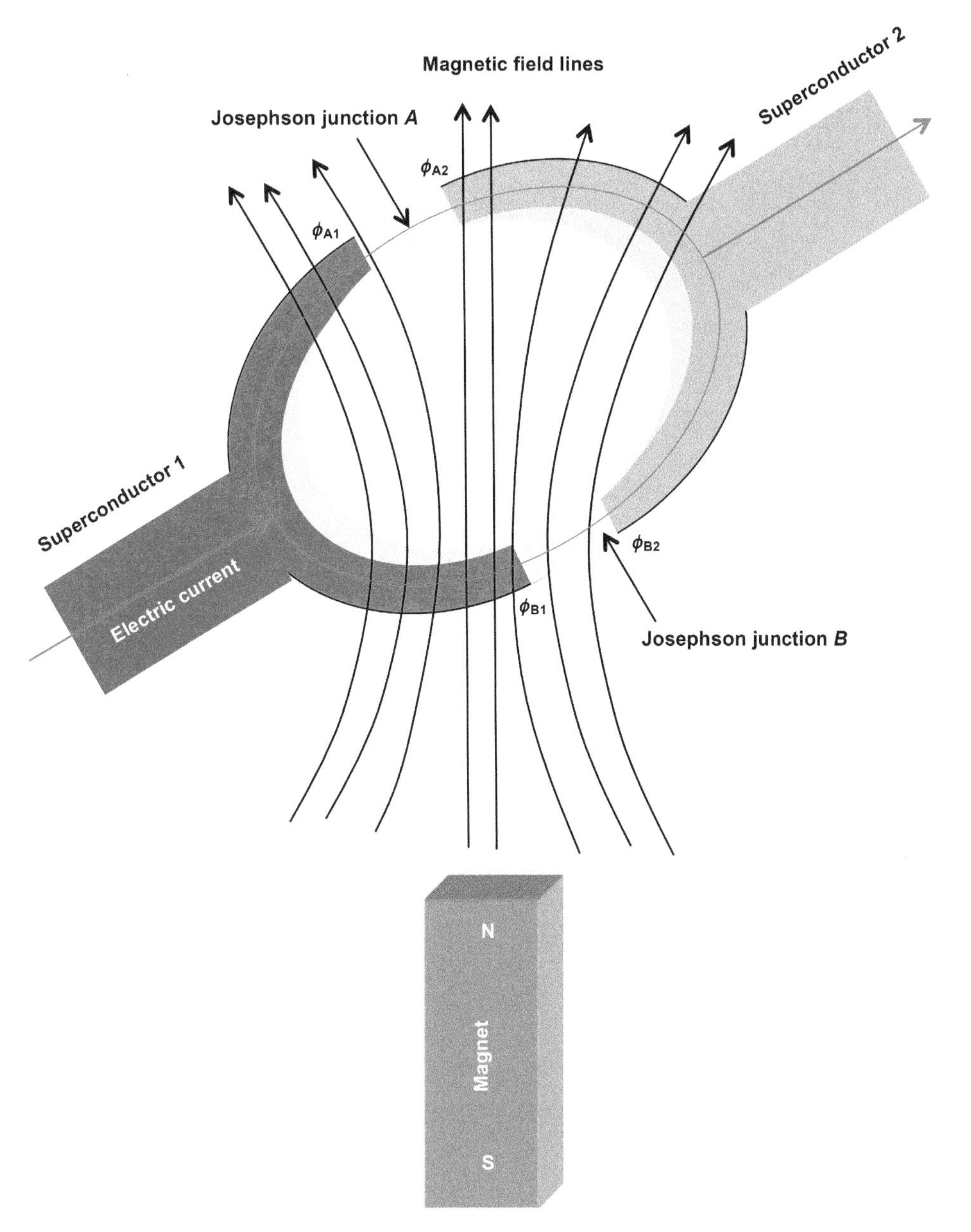

Figure 12.9. A DC SQUID.

Let us consider two weak links arranged in parallel connection as shown in figure 12.9. Let ϕ_{A1}, ϕ_{A2} be the phases adjacent to the junction A, and ϕ_{B1}, ϕ_{B2} be the phases adjacent to the junction B:

$$\phi_A = \phi_{A2} - \phi_{A1} \tag{12.162}$$

$$\phi_B = \phi_{B2} - \phi_{B1}. \tag{12.163}$$

Let ϕ_A, ϕ_B be the phase differences across the Josephson junctions A and B. Then the phase difference $\Delta\phi$ between junctions A, B is

$$\begin{aligned}\Delta\phi &= \phi_A - \phi_B = \phi_{A2} - \phi_{A1} - (\phi_{B2} - \phi_{B1}) = \phi_{A2} - \phi_{A1} - \phi_{B2} + \phi_{B1} \\ &= (\phi_{B1} - \phi_{A1}) + (\phi_{A2} - \phi_{B2}) = -\frac{2q}{\hbar}\int_{A1}^{B1} \mathbf{A}.\, ds - \frac{2q}{\hbar}\int_{B2}^{A2} \mathbf{A}.\, ds \\ &= -\frac{2q}{\hbar}\oint \mathbf{A}.\, ds = -\frac{2q}{\hbar}\Phi \\ &= -2\pi\frac{1}{h/(2q)}\Phi = -2\pi\frac{\Phi}{\Phi_0}\end{aligned} \tag{12.164}$$

since

$$h/(2q) = \Phi_0 \tag{12.165}$$

Φ is the magnetic flux penetrating the loop. Total current density flowing through the SQUID is

$$\begin{aligned} J &= J_C(\sin\phi_A + \sin\phi_B) \\ &= 2J_C[\sin\{(\phi_A + \phi_B)/2\}\cos\{(\phi_A - \phi_B)/2\}] \\ &= 2J_C[\sin\{(\phi_A + \phi_B)/2\}\cos\{(\Delta\Phi)/2\}] \end{aligned} \tag{12.166}$$

where the formula

$$\sin a + \sin b = 2\sin\{(a+b)/2\}\cos\{(a-b)/2\} \tag{12.167}$$

has been applied.

Substituting for $\Delta\phi$ from eq. (12.164) into eq. (12.166)

$$\begin{aligned} J &= 2J_C\left[\cos\left\{\left(-2\pi\frac{\Phi}{\Phi_0}\right)/2\right\}\sin\{(\phi_A + \phi_B)/2\}\right] \\ &= 2J_C\cos\left(\pi\frac{\Phi}{\Phi_0}\right)\sin\{(\phi_A + \phi_B)/2\} \end{aligned} \tag{12.168}$$

The current density has maximum value when

$$\sin\{(\phi_A + \phi_B)/2\} = 1 \tag{12.169}$$

and the maximum value is

$$J = 2J_C \cos\left(\pi\frac{\Phi}{\Phi_0}\right). \tag{12.170}$$

Consider a symmetric DC SQUID. A biasing current I is supplied. The current in each branch $= I/2$. If I_C is the critical current of one JJ, the critical current of the SQUID is $2I_C$. When an external magnetic flux $\Phi_{External}$ is superimposed perpendicular to the plane of the loop, a screening current I_S is produced. Flux quantization requires that

$$\Phi_{Total} = \Phi_{External} + LI_S = n\Phi_0, \tag{12.171}$$

where L is inductance of the loop, n is an integer and Φ_0 is a flux quantum. When $\Phi_{External} = n\Phi_0$, then $I_S = 0$. But when $\Phi_{External} = (n + ½)\Phi_0$, then $I_S = \pm(\Phi_0/2L)$.

Thus I_S varies periodically with $\Phi_{External}$. Due to the flow of screening current I_S around the SQUID loop, the critical current of the SQUID decreases from $2I_C$ to $(2I_C - 2I_S)$. Hence, the critical current is a periodic function of $\Phi_{External}$. If the SQUID is biased with a current a little above $2I_C$, the output voltage is a periodic function of $\Phi_{External}$. Thus the SQUID transduces the magnetic flux variations into voltage changes. An electronic circuit linearizes the periodic response of the SQUID.

An optical analogy for the DC SQUID is provided by the famous two-slit interference experiment performed by physicist Thomas Young (figure 12.10). In this

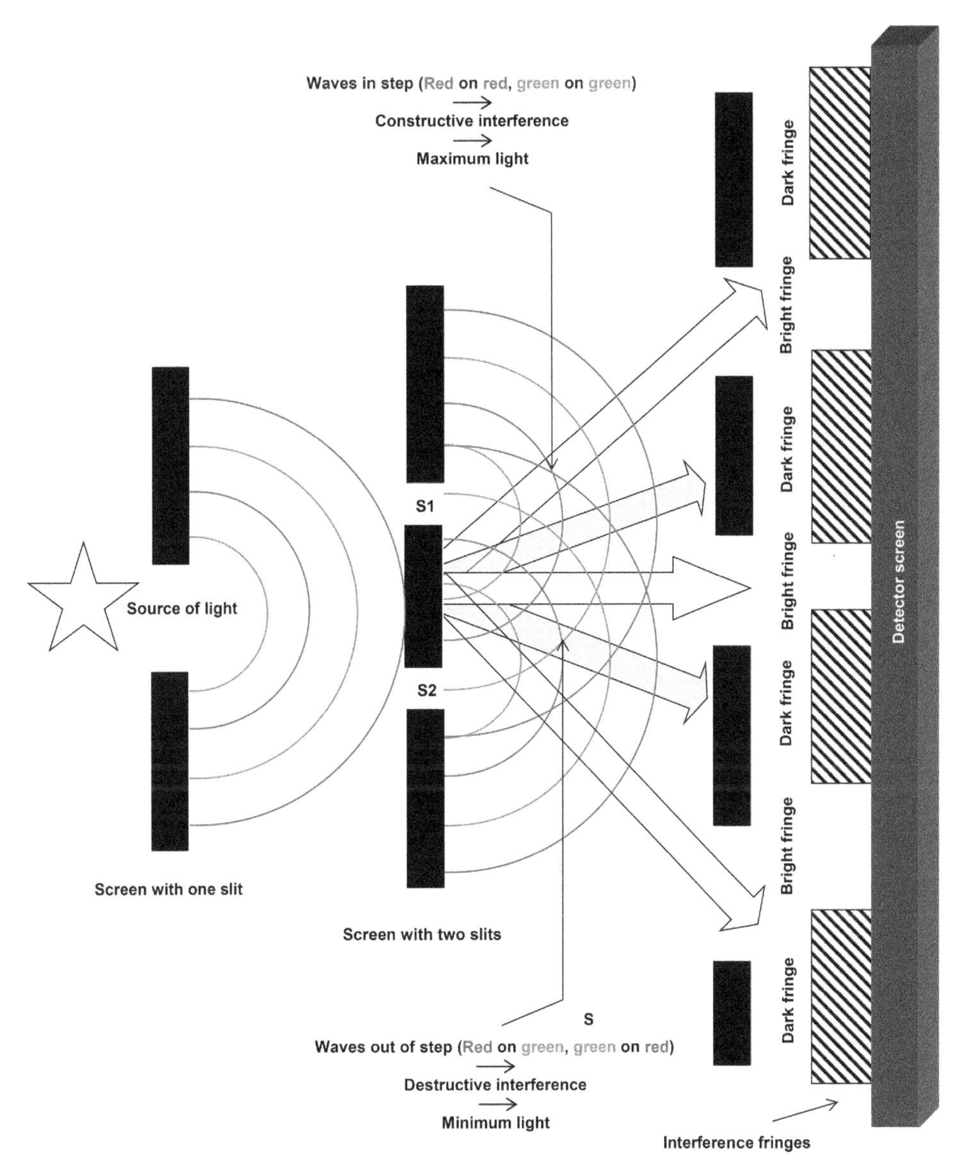

Figure 12.10. Young's double-slit experiment.

experiment, a monochromatic light beam strikes two slits. The light waves emerging from the slits are diffracted. The spread out beams meet in synchronization at some points. Here, they are aligned crest-to-crest and trough-to-trough. At these locations, they interfere constructively. At other points, they fall out of synchronization. At these points, they interfere destructively. Thus a pattern consisting of alternate bright and dark bands called interference fringes is formed on a screen placed behind the slits. Similar to the optical experiment, the critical current intensity of a DC SQUID is modulated in proportion to the magnetic flux. The critical current shows an ideal Fraunhoffer interference pattern. The two JJs behave as the two slits. The interfering waves are the supercurrents flowing through the opposite halves of the ring. The difference in the phases of the supercurrents is produced by the magnetic field.

12.5.2 The AC or RF SQUID

Unlike the DC SQUID, an AC SQUID contains only one JJ. Its sensitivity is less, its fabrication is simpler and it is less costly. The sample is moved through the superconducting pick-up coil. Its movement produces an alternating flux in the pickup coil. The pickup coil is part of a superconducting circuit used for transferring the magnetic flux from the sample to the RF SQUID placed inside a liquid-helium bath. The magnetic flux received by the SQUID is converted into a voltage. The voltage signal is amplified and readout by an electronic circuit.

The reason for the lower sensitivity of the single junction RF SQUID in comparison to the dual-junction DC SQUID is that in the RF SQUID, only the junction is engaged in gathering the flux whereas in the DC SQUID, the total area of the loop takes part in picking up the magnetic flux, not the junctions alone.

The JJs in SQUIDs are fabricated from pure niobium or a lead alloy containing 10% Au or In, because pure Pb is thermally unstable. The tunnel barrier is formed by oxidation of the surface of the niobium base electrode. Over the tunnel barrier, the top electrode of the lead alloy is deposited, forming a JJ of structure: niobium–oxide–lead alloy.

12.6 Rapid single flux quantum logic

12.6.1 Difference from traditional logic

Operation of traditional logic circuits is based on two voltage levels, a high voltage level corresponding to the full power supply voltage and a low voltage level for zero voltage. The high voltage level represents the logic high or logic 1 state. The low voltage level indicates a logic low or logic 0 state. Rather than using two voltage levels, RFSQ logic works with two distinct conditions, one in which the voltage pulse is present and a second in which it is absent (Hutchby *et al* 2002). These voltage pulses are generated from the quantization of magnetic flux. As mentioned in section 12.3.4, magnetic flux threading a superconducting loop does not change continuously. It changes in discrete steps. Each step is known as a quantum of magnetic flux (Φ_0) or fluxon.

12.6.2 Generation of RFSQ voltage pulses

The device, which produces the voltage pulse corresponding to the change of magnetic flux, is the non-hysteric JJ consisting of a JJ shunted by an external resistor R_n. It is said to be non-hysteric in the sense that its current–voltage characteristics are non-hysteric. Suppose at a particular instant of time, the junction is biased in the superconducting state ($V = 0$) by supplying a biasing current of magnitude lower than the critical current I_C. Suppose at this instant, the flux supported by the loop equals one fluxon. An input signal current pulse of magnitude exceeding the critical current (typically 100 μA) switches the junction into the resistive state ($V \neq 0$). For this purpose, a short-duration DC pulse may be used, such as from a semiconductor device. Due to switching of the JJ, a very short voltage pulse is induced across the JJ. This voltage pulse has an area given by

$$\text{area} = \int V(t)\mathrm{d}t = 2.07\ \text{mV ps}. \tag{12.172}$$

If the pulse is 1 ps wide, the amplitude will be 2 mV. If is <1 ps wide, the amplitude is proportionally higher. Generally, the pulse amplitude is given by the equation

$$\text{amplitude} \approx 2I_C R_n. \tag{12.173}$$

Consequent upon temporary closure of the JJ, a quantum of flux is ejected out of the loop comprising the JJ. After the input signal current pulse is over, the JJ again switches back to its original superconducting state ($V = 0$). Such restoration of the circuit to its initial state is analogous to the 2π rotation of a pendulum bob. The pendulum is said to be overdamped and does not continue its oscillations after reaching the starting point.

Logic circuits are constructed around the aforesaid voltage pulses that are produced each time that a fluxon is ejected out or enters a superconducting loop. The voltage pulses propagate along superconducting transmission lines energizing complex circuits implementing NOT, AND, OR, XOR, etc, logic functions.

12.6.3 RFSQ building blocks

By organization of three basic structural units (Brock *et al* 2000), one can perform various binary logic functions:

(i) *Active transmission stage.* This consists of a JJ and an inductor with small inductance.

(ii) *Storage loop.* This loop contains a JJ and an inductor with a large inductance through which a persistent current I_p is maintained around a loop.

(iii) *Decision-making pair.* This consists of two JJs of different sizes and hence different critical currents I_{C1} and I_{C2}.

12.6.4 RFSQ reset–set flip-flop

Looking at the circuit diagram of reset–set (R–S) flip-flop (figure 12.11), it is easily surmised that the circuit is essentially a DC two-JJ SQUID. The two JJs comprising

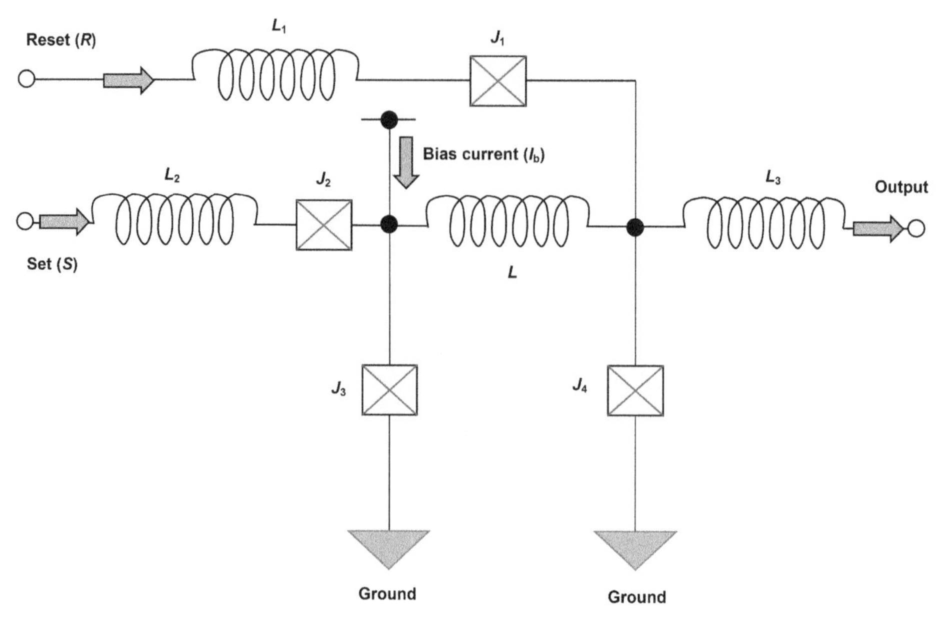

Figure 12.11. An RFSQ R–S flip-flop.

the DC SQUID are the junctions J_3 and J_4. This flip-flop functions as follows (Brock *et al* 2000): in the initial condition, the circuit is in the logic low state. In this logic low state, the persistent current flowing through JJ J_3 has a magnitude I_p and its direction is anticlockwise. From the circuit diagram, it is obvious that the direction of the persistent current I_p is the same as the direction of the biasing current $I_b/2$ in JJ J_3 (Likharev 1991). The biasing current is halved because it is divided into two branches while entering JJ J_3. Hence, the currents $I_b/2$ and I_p add up to give the total current flowing through JJ J_3 as

$$I_3 = I_b/2 + I_p \tag{12.174}$$

For JJ J_4 the persistent current I_p and the biasing current I_b are flowing in opposite directions. Hence the total current I_4 flowing through J_4 is obtained by subtracting I_p from $I_b/2$, so that we can write

$$I_4 = I_b/2 - I_p. \tag{12.175}$$

Clearly $I_4 < I_3$.

On arrival of a set pulse, JJ J_1 transmits it to the storage loop. This set pulse can trigger a 2π jump in JJ J_3 carrying a larger current I_3 and not in JJ J_4, which carries a smaller current I_4. Consequent upon this switching of the JJ by 2π, the direction of current flowing through JJ J_3 reverses from the original anticlockwise direction to the clockwise direction. This current circulating around the storage inductance through JJ J_3 constitutes the logic high state of the flip-flop. Thus in the logic high state, the persistent current flowing through JJ J_3 has the same magnitude I_p as in the logic low state but its direction of flow is reversed from the anticlockwise to clockwise. Due to this reversal of current, the two currents flowing through JJ J_3 are I_p in the clockwise direction and $I_b/2$ in the anticlockwise direction, so that the total current I_3 flowing through JJ J_3 becomes

$$I_3 = I_b/2 - I_p. \quad (12.176)$$

Similarly, the two currents flowing through JJ J_4 are I_p and $I_b/2$. Both these currents are flowing in the clockwise direction. By addition of these two currents, we find the combined current I_4 flowing through JJ J_4 as

$$I_4 = I_b/2 + I_p. \quad (12.177)$$

As soon as a reset pulse arrives, a 2π jump is triggered in JJ J_4 carrying a larger current whereby the current flowing through the junction J_4 starts flowing in the clockwise direction. But current flow through JJ J_4 in the clockwise direction corresponds to current flow in the anticlockwise direction through JJ J_3, which is the logic low state, from which we started. At the same time, a single flux quantum pulse $V(t)$ is produced across the JJ J_2 which serves as an output signal F.

Now consider what happens if the flip-flop is in the logic high state with current I_p flowing in the clockwise direction through JJ J_3, and the set pulse arrives. Then the set pulse cannot cause switching of the JJ carrying a low current

$$I_3 = I_b/2 - I_p. \quad (12.178)$$

Instead, it switches the junction J_2 carrying a higher current $I_b/2$. Due to this switching the pulse falls out of the circuit, i.e. the fluxon escapes from the circuit. Hence, no pulse appears at the output. Similarly, the junction J_1 protects the source of the RFSQ pulse from the reverse reaction of the interferometer on receipt of an erroneous reset pulse.

Thus the flip-flop has two stable states, one in which the current I_p is flowing through JJ J_3 in the anticlockwise direction (logic low state) and the other in which the current I_p is flowing through JJ J_3 in the clockwise direction (logic high state).

12.6.5 RFSQ NOT gate or inverter

The RFSQ NOT gate (figure 12.12) contains one DC SQUID consisting of two parallel JJs J_2 and J_3 (Likharev 1991). An additional JJ J_1 is connected between JJ J_3 and the ground terminal. The circuit contains two parallel branches for the flow of current: J_2–J_1–ground and L_1–J_3–J_1–ground. In the initial state of the circuit, suppose a high current is flowing through JJ J_2 while low current is flowing through JJ J_3. Thus a high current is flowing along the branch J_2–J_1–ground whereas a low current is flowing along the branch L_1–J_3–J_1–ground. The input pulse is zero. Application of a clock pulse switches the JJ J_1 in the high-current branch J_2–J_1–ground. It does not trigger the junction J_3 in the low-current branch L_1–J_3–J_1–ground. The resultant RFSQ pulse produced in JJ J_1 appears at the output of the circuit. Thus in the absence of any input pulse, a pulse appears at the output. This is NOT operation.

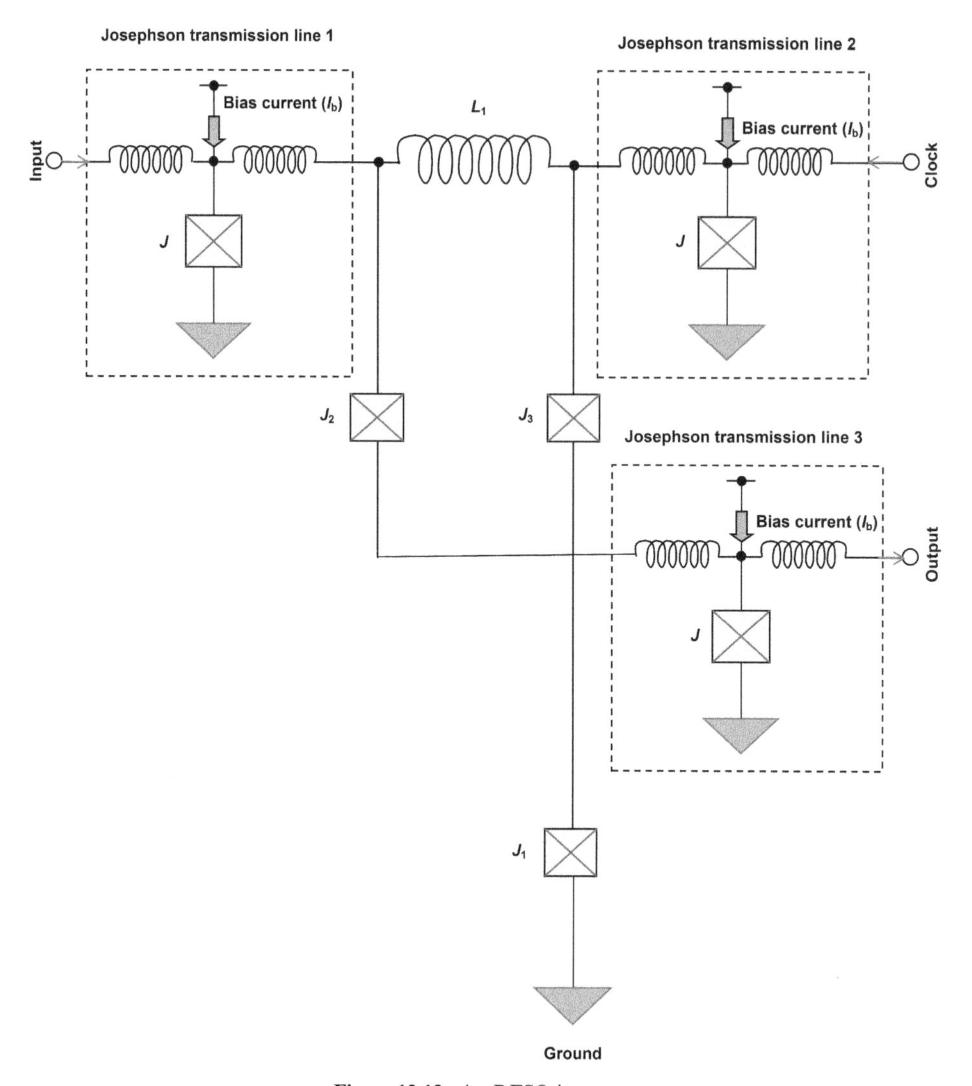

Figure 12.12. An RFSQ inverter.

Now if an input pulse arrives, the operation is similar to that of an R–S flip-flop with a pulse applied to the set terminal. The JJ J_2 switches so that JJ J_2 carries a low current and JJ J_3 carries a high current. At this stage, a high current is flowing along the branch L_1–J_3–J_1–ground whereas a low current is flowing along the branch J_2–J_1–ground. On applying a clock pulse, the junction J_3 in the high-current branch L_1–J_3–J_1–ground is switched whereas the junction J_1 in the low current branch J_2–J_1–ground remains unaffected. Consequently, there is no pulse at the output. Hence, the output pulse is zero when the input pulse is present. Once again, NOT operation is verified. Thus in both cases, the circuit behaves as a NOT gate or inverter.

12.6.6 RFSQ OR gate

In this circuit (figure 12.13), two DC SQUIDs can be identified (Likharev 1991).The first SQUID consists of the JJs (J_1, J_2). The two stable states of this SQUID are: (1) high current flowing through JJ J_1 with low current through JJ J_2 and (2) high current flowing through JJ J_2 with low current through JJ J_1. The second SQUID comprises the JJs (J_5, J_6). Its two stable states are: (i) high current flowing through J_5, low current in J_6 and (ii) high current in J_6, low current in J_5. The following cases arise:

(i) There are no input pulses during the clock period. The next clock pulse produces RFSQ pulses across J_4, J_8, which do not reach output.

(ii) Only input pulse IN1 arrives during the clock period. High-current-carrying junction J_2 switches due to the CLK pulse. The generated RFSQ pulse passes through L_3, triggering an RFSQ pulse across J_9, and

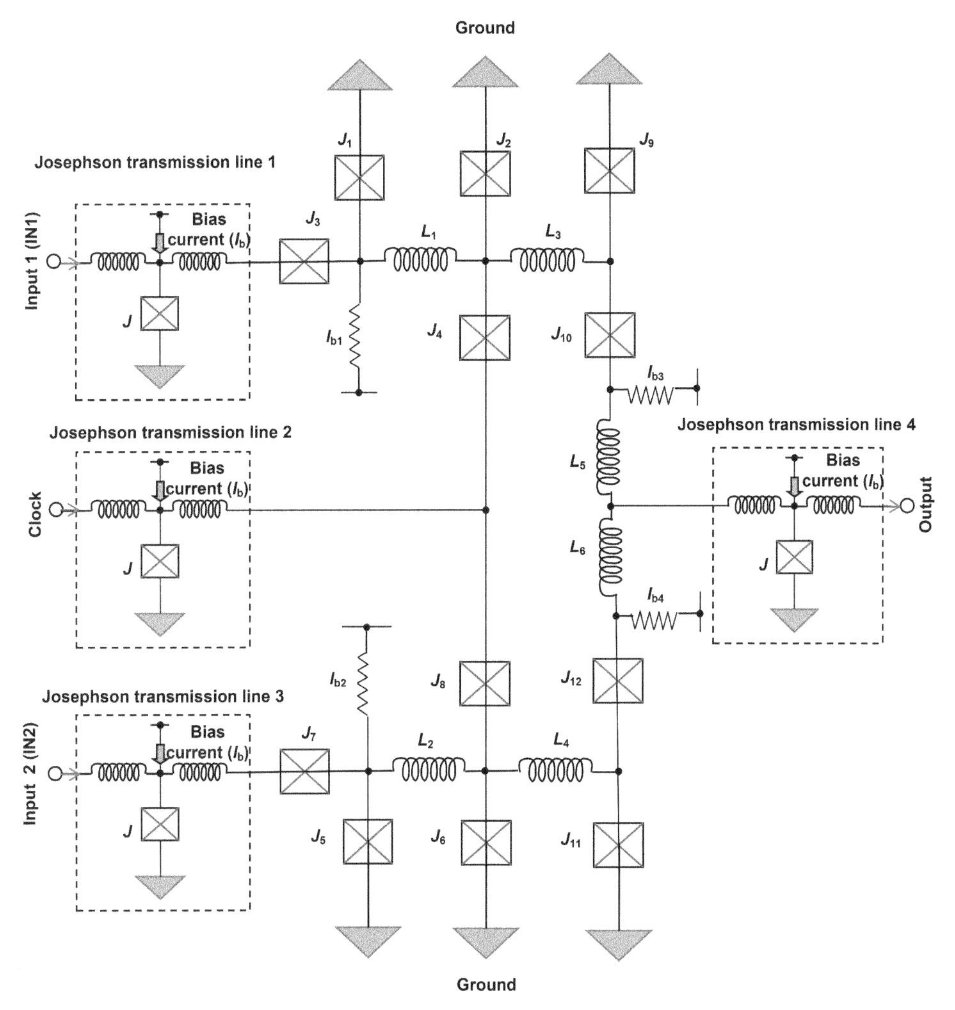

Figure 12.13. An RFSQ OR circuit.

passes to the output through J_{10} and L_5. At the same time, an RFSQ pulse is produced across J_{12}.

(iii) Only input pulse IN2 arrives during the clock period. A similar process to step (ii) occurs in the symmetrically located lower part of the circuit.

(iv) Both input pulses IN1 and IN2 arrive during the clock period. Both J_2, J_6 produce RFSQ pulses at the same time. RFSQ pulses are created across J_9 and J_{11} at the same time. Only one RFSQ pulse occurs at the output because J_9, J_{11} are connected in parallel whereas J_{10} and J_{12} remain in the superconducting state.

All the above operations are in conformity with that of an OR gate. Hence, the circuit works as an OR gate.

12.6.7 Advantages of RFSQ logic

The fabrication technology for RFSQ logic is simple. The commonly used superconductor is niobium or niobium nitride. The tunnel barrier in the JJs is aluminum oxide. The power consumption of RFSQ logic is extremely low. Each time that an RFSQ pulse passes through the junction, the energy consumed = circulating current (100×10^{-6} A) × quantity of flux = 10^{-4} A × 2×10^{-15} Wb = 2×10^{-19} J. The voltage pulses are conveyed to neighboring gates either along microstrip transmission lines or through active lines containing JJs to introduce timing delays necessary for synchronization. The superconducting interconnections are virtually loss-free. The losses are practically zero at DC and very low in comparison to metals up to frequencies around 750 GHz. Together with low power, the RFSQ logic circuits also provide superfast operation at frequencies of several hundred gigahertz.

12.6.8 Disadvantages of RFSQ logic

The main disadvantage is the cooling requirement for cryogenic operation, but the popularity of small-size closed-cycle refrigeration units has considerably eased the situation, making possible the integration of cryo-cooled RFSQ circuits with room-temperature electronics.

12.7 Discussion and conclusions

Superconductive electronics encompasses various fields such as SQUIDs, filters, RFSQ logic, power transmission and current limiting devices, to name a few. Superconductors belong to one of the two categories, low-temperature and high-temperature. A HTS does not work at a high temperature in reality; the term is used in a relative context to distinguish this class from the low-temperature class. Superconducting behavior vanishes when any of the three parameters, either critical temperature, critical current or critical magnetic field, is surpassed. A superconductor does not permit entry of any magnetic field inside it. The prohibition or expulsion of magnetic field by a superconductor is known as the Meissner effect. The correlation of superconductivity with magnetic phenomena leads to the BCS theory of superconductivity. The most important device in superconductive electronics is the JJ. A phenomenon of paramount significance is the Josephson effect, which is of two types: a DC Josephson effect characterized by current flow across a JJ in the absence of an applied voltage and an AC Josephson effect in which a DC voltage produces an alternating electric field. By defining the wave functions of the Cooper pairs in the two superconducting electrodes of the JJ, the theory of the Josephson effect is developed. The theory leads to two simple equations, one concerning the rate of change of Cooper pair density in the two superconducting electrodes, and the other dealing with the rate of change of phase difference between the wavefunctions. These equations are used to interpret the DC and AC Josephson effects. The theory is placed on a rigorous foothold by introducing a gauge-invariant phase difference.

Unlike the BCS theory of superconductivity, the Ginzburg–Landau theory is a macroscopic theory in which the two basic equations are derived by minimizing the free energy function with respect to an order parameter. The Ginzburg–Landau theory does not provide any explanation of the physical origin of superconductivity. On the contrary, it commences on the presumption that superconductivity exists. Then it proceeds towards modeling of the superconducting properties.

London's first equation is derived by writing the equation of motion of an electron in an electric field. For an electron in a superconductor, the retarding force due to electron scattering is set to zero. By taking the curl of London's first equation and applying Maxwell's third equation, London's second equation is obtained. Applying London's second equation and Maxwell's fourth equation, with the help of Maxwell's second equation, shows after mathematical manipulation that a magnetic field can be uniform inside a superconductor only if it is zero everywhere. The differential equation for the spatial dependence of magnetic induction shows that the magnetic induction decays exponentially with distance from the exterior surface of the superconductor, explaining the Meissner effect.

The inverse AC Josephson effect involves the creation of a supercurrent by exciting a JJ with a combination of small amplitude, high-frequency AC signal on a large DC signal. The DC current contains a series of discrete steps called Shapiro steps at multiples of the frequency of the AC signal.

Magnetic fields of very feeble strengths can be accurately measured using quantum interference devices based on superconductivity. These devices are known as SQUIDs. A DC SQUID consists of two or more parallel-connected JJs. On applying a biasing current on the SQUID and superimposing an external magnetic flux on the loop, the supercurrents flowing through the two halves of the loop undergo interference, producing a periodic voltage waveform, which is processed by an electronic circuit. Thus the SQUID transduces the flux variations into voltage changes. An AC SQUID contains only a single JJ. The sample is moved through a pickup coil, which is part of the SQUID circuit. The resulting time-varying flux produced in the pickup coil influences the SQUID converting the flux changes into voltage signals.

The rudimentary principles of RFSQ logic circuits have been introduced. This logic is differentiated from routine voltage-based logic. The key switching elements in this logic are overdamped JJs. Operations of R–S flip-flop, NOT and OR gates are outlined.

Review exercises

12.1. What is a superconductive material? Define superconductive electronics and elaborate its scope mentioning the fields of study which fall under it.
12.2. What is the full form of: (a) SQUID and (b) RFSQ? What are they used for?
12.3. What is a low-temperature superconductor? Can the current flow through such a material for years without decreasing?
12.4. What happens when the temperature of a superconductor is lowered to a value close to absolute zero? How does a normal conductor behave near absolute zero? What is meant by the critical temperature of a superconductor? Give three examples of superconductors. Mention the critical temperature in each case.
12.5. What is the Meissner effect? What is meant by the London penetration depth of a superconductor? Write the equation for penetration depth and perform its calculation.
12.6. What are the three parameters whose value when exceeded leads to loss of the superconducting property? Define these parameters.
12.7. Define the following terms for a superconductor: (a) critical temperature, (b) critical magnetic field and (c) critical current density.
12.8. What are type I and type II superconductors? Give one example of each class. Which type is practically more useful? How do type I and type II superconductors differ with regard to the Meissner effect?
12.9. What behavior of a superconductor in a magnetic field leads to the correlation of superconductivity with the magnetic properties of materials? How are the electron spins arranged when a material is magnetized?
12.10. What is a Cooper pair? What is meant by the coherence length of a superconductor?
12.11. What type of particle is an electron: fermion or boson? What type of particle is a Cooper pair: fermion or boson? Do bosons obey Pauli's exclusion principle?
12.12. Explain the main idea of the BCS theory of superconductivity in terms of mediation of electrons in a Cooper pair via phonon interaction.
12.13. Why are the electrons paired in the form of Cooper pairs at low temperatures but become unpaired at high temperatures?
12.14. Describe the applications of superconductivity in MRI machines and particle accelerators.
12.15. How does a HTS differ from a low-temperature superconductor? Is the high temperature required for a superconductor really high or it is a relative comparison with temperatures near absolute zero required for conventional superconductors?
12.16. What is a JJ? What are the two types of structures used for fabricating JJs?
12.17. What is the DC Josephson effect? How is it interpreted on the basis of tunneling of Cooper pairs? How does the tunneling of Cooper pairs differ from the tunneling of electrons? How is the Josephson current

related with the phase difference between wavefunctions of Cooper pairs on opposite sides of the JJ? How does the flow of Josephson current refute common physics laws?

12.18. What is the AC Josephson effect? Do the oscillations produced across a JJ depend on the size of the superconductors? Does it depend on the voltage applied across the junction?

12.19. What happens when the wavefunctions ψ_1 and ψ_2 of the Cooper pairs in the two superconducting electrodes of a JJ come closer together? Write the time-dependent Schrödinger equations for ψ_1 and ψ_2? What is the opacity constant? Derive the following governing equations for the Josephson effect:

$$\frac{\partial \rho_1}{\partial t} = -\frac{\partial \rho_2}{\partial t}$$
$$\frac{\partial \phi}{\partial t} = \frac{2qV}{\hbar},$$

where ρ_1, ρ_2 are densities of Cooper pairs in superconductors 1 and 2, respectively, $\phi = \phi_2 - \phi_1$, where ϕ_1, ϕ_2 are the phases of ψ_1 and ψ_2, V is the applied voltage, q is the electronic charge and $\hbar = h/2\pi$; h is Planck's constant. Utilizing these basic equations, explain the phenomena of DC and AC Josephson effects.

12.20. Define gauge-invariant phase difference θ. What is the magnetic flux quantum Φ_0? Taking the gauge-invariant phase difference into account, derive the equation relating the modified critical current density J_C when a gauge-invariant phase difference θ is considered, with the critical current density j_C when a phase difference ϕ is taken in place of θ:

$$J_C = Lj_C|\sin(\pi\Phi/\Phi_0)|/|(\pi\Phi/\Phi_0)|,$$

where L is the length across which the JJ extends and Φ is the flux enclosed by a contour passing through the superconducting electrodes and the tunnel barrier.

12.21. How does the Ginzburg–Landau theory of superconductivity differ from the BCS theory? What is the wavefunction ψ proposed in this theory? Expand the free energy function with respect to ψ and derive the first equation of Ginzburg–Landau by equating to zero its derivative with respect to ψ. Equate to zero the derivative of free energy with respect to **A** and derive the second equation of Ginzburg–Landau.

12.22. Derive London's two equations starting from the equation of motion of an electron moving in an electric field. Provide an explanation for the Meissner effect by applying London's second equation.

12.23. What is the inverse AC Josephson effect? Show that application of a (DC + AC) field to a JJ produces a DC supercurrent containing a series of steps at multiples of the AC frequency. What are these steps called?

12.24. What is a SQUID? Mention one important application of this device.
12.25. How many JJs does a DC SQUID contain? How is the quantum interference produced in a DC SQUID? Draw the similarity between the operation of a DC SQUID and Young's two-slit experiment. To which parts of the DC SQUID do the two slits in Young's experiment correspond? What physical entities in a DC SQUID represent the interfering waves in Young's experiment?
12.26. How many JJs does an AC SQUID contain? How does an AC SQUID work? Why is the AC SQUID less sensitive than a DC SQUID?
12.27. Describe the use of flux quantization for performing logic operations. What is this logic called?
12.28. What are the main structural units of RFSQ logic? Why is a shunt resistor included along with the JJ?
12.29. Draw the circuit diagram of an R–S flip-flop in RFSQ logic. Explain its operation. How do you define the two stable states of the R–S flip-flop?
12.30. It is necessary to implement a NOT operation in RFSQ logic? How many SQUIDs are used for this implementation? Explain the operation of the NOT gate with its circuit diagram.
12.31. Explain with the help of a circuit diagram the operation of an OR gate in RFSQ logic.
12.32. Is RFSQ logic inter-operable with CMOS logic? What are its advantages and disadvantages relative to CMOS logic?

References

Brock D K, Track E K and Rowell J M 2000 Superconductor ICs: the 100-GHz second generation *IEEE Spectrum.* December pp 40–6

Grosso G and Parravicini G P 2014 *Solid State Physics* (New York: Academic) pp 842–3

Hutchby A, Bourianoff G I, Zhirnov V V and Brewer J E 2002 Extending the road beyond CMOS *IEEE Circuits Devices Mag.* pp 28–41

Likharev K and Semenov V 1991 RSFQ logic/memory family: A new Josephson-junction technology for sub-terahertz clock frequency digital systems *IEEE Trans. Appl. Supercond.* **1** 3–28 http://www.physics.sunysb.edu/Physics/RSFQ/Projects/WhatIs/rsfqre2m.html

Mehta N 2011 *Applied Physics for Engineers* (New Delhi: PHI Learning Private) p 955

Scientific American 1997 What are Josephson junctions? How do they work? *Sci. Am.* November www.scientificamerican.com/article/what-are-josephson-juncti/

Superconductivite 2015 (March 2) Electronics with superconductors *Superconductivite* www. supraconductivite.fr/en/index.php?p=applications-electronique-more

Tsang T 1997 *Classical Electrodynamics* (Singapore: World Scientific) pp 381–2

Chapter 13

Superconductor-based microwave circuits operating at liquid-nitrogen temperatures

HTS-based compact planar microwave circuits have been designed, fabricated and packaged, and their functioning has been corroborated. Prominent materials used for substrates include sapphire, magnesium oxide and lanthanum aluminate. Practical realizations of circuits largely exploit microstrip designs that are fabricated utilizing thin-film techniques. High-temperature superconducting films of YBCO and TBCCO have been used. These HTS thin-films fulfil the critical insertion loss specifications of cellular communication networks, satellite broadcasting, international navigation and space flights. While filters fabricated by traditional technologies decrease in sensitivity if selectivity is increased, HTS filters can provide maximum sensitivity concomitantly with maximum selectivity. These filters exhibit characteristics closely approximating those of an ideal filter, namely 100% acceptance of frequencies in the passband and 100% rejection of frequencies in the stopband. They assure complete rejection of frequencies in proximity to the allowed band, thereby reducing interference. The combination of steep-skirted filters exhibiting rapid attenuation of transmission outside the desired band with related microwave components enables the realization of a low-noise front end of a receiver. These HTS microwave circuits provide the necessary weight and volume reduction, although the cryogenic cooling equipment for keeping temperatures in the 60–80 K range is an indispensable appendage to these circuits.

13.1 Introduction

A superconductor does not offer any resistance to the flow of DC or zero-frequency current, i.e. the opposition to zero frequency current flow is zero. However, similar remarks are not applicable to a current of any frequency. In the case of an AC, the superconductor acts differently from its DC behavior. In fact, the resistance of a superconductor must be mentioned with reference to frequency, because it changes

doi:10.1088/978-0-7503-1155-7ch13

with the frequency of the current. The higher the frequency of current, the larger is the resistance of a superconductor.

Although during operation at a frequency in the microwave range, the surface resistance of a superconductor might be quite large, it is still appreciably less than that of a conductor at that frequency, e.g. at 1 GHz the surface resistance of a superconductor at liquid-nitrogen temperature (77 K) is 3–4 orders of magnitude lower than that of copper at the same temperature. For this reason, it is advantageous to use a superconductor in a microwave circuit in place of a conductor. Although this has been known for a long time, the application of superconductors in microwave circuits did not gain popularity because of the associated expensive and cumbersome cooling gadgetry required for building such circuits. However, with the advent of HTSs requiring liquid-nitrogen temperatures instead of liquid-helium temperatures necessary for low-temperature superconductors, the circumstances improved and the stage was arranged for acceptance of superconductors in microwave circuits (Vendik *et al* 2000).

Substitution of a normal conductor by a HTS helps to decrease the insertion loss drastically. Additionally, a microwave component can be fabricated at an enormously reduced dimensional scale leading to a tremendously compact form factor. For microwave resonators, an extremely high quality factor is achievable by using HTS. The performance of microwave resonators, filters, phase shifters and antennas has been vastly improved by following this strategy (Willemsen 2001, Faisal 2012).

13.2 Substrates for microwave circuits

The following are desirable properties for substrates to be used for microwave circuits. First, the loss tangent of the substrate material must be very low at the high frequencies. Second, the material must be mechanically strong. Third, the temperature coefficient of the substrate material must match closely with that of the HTS film to be deposited over it. If this matching is not very close, the thickness of the HTS film is restricted by temperature-induced stresses, which causes the films to crack and delaminate. Fourth, the substrate material should not undergo any chemical reaction with the HTS film, in order that its chemical integrity is preserved. Lastly, but most importantly, the material should be economically worthwhile. Only when it is available at an affordable cost will it be widely adopted.

Materials which fulfill these criteria, to some extent, are sapphire (aluminum oxide, Al_2O_3), lanthanum aluminate ($LaAlO_3$ or LAO) and magnesium oxide (MgO). Sapphire has an anisotropic dielectric constant $\varepsilon = 8.9-11$ and its thermal expansion properties match poorly with those of the HTS films. However, it is relatively inexpensive. Its mechanical robustness combined with the availability of large-size sapphire substrates may lead to its extensive adoption. LAO has a relative permittivity of 25. It is expensive and suffers from local permittivity variations when subjected to thermal cycling owing to structural transitions of phase. LAO single crystals are used for epitaxially growing cuprate superconductors. MgO has found widespread use as a substrate material for microwave filters (Ramesh *et al* 1990).

13.3 HTS thin-film materials

13.3.1 Yttrium barium copper oxide

YBCO, also called yttrium barium cuprate or Y123, and represented by the formula $YBa_2Cu_3O_{7-\delta}$ ($0 \leqslant \delta \leqslant 1$) is a black crystalline solid with pervoskite structure. When $\delta = 0.07$, the material becomes superconducting at 92 K. The process known as the solid-state reaction is used to prepare YBCO. The process starts by mixing the precursor powders, yttrium oxide (Y_2O_3), barium carbonate ($BaCO_3$) and copper oxide (CuO), in the stoichiometric ratio 1:2:3 for 2–3 h in an agate mortar. The mixture is calcined at 840 °C for 14 h in an alumina crucible in a programmable furnace:

$$Y_2O_3 + 4BaCO_3 + 6CuO \rightarrow 2YBa_2Cu_3O_{7-\delta} + 4CO_2. \tag{13.1}$$

The temperature is increased to 950 °C to remove CO_2 gas. The calcined powder is pressed into pellets, followed by sintering and annealing in oxygen ambient.

YBCO thin-films are deposited by techniques such as sputtering from a ceramic target or using pulsed laser deposition (PLD) by melting and evaporating the target material. MOCVD is another method applied for HTS thin-film deposition.

13.3.2 Thallium barium calcium copper oxide

TBCCO films in the 2212 phase ($Tl_2Ba_2CaCu_2Oy$) have a transition temperature $T_C > 105$ K, so that a smaller cooling system with low input power suffices. To prepare a TBCCO film, a Ba–Ca–Cu–O precursor film free from thallium is first deposited epitaxially. This film is thallinated by annealing at 800 °C in the presence of thallium.

13.4 Fabrication processes for HTS microwave circuits

The process steps for the fabrication of these circuits are borrowed from standard semiconductor technology. Necessary modifications are performed wherever applicable. The steps followed are photolithography, ion milling, metallization for contact formation, and dicing. Ion milling is used in the absence of a reliable wet or dry etching chemistry. The aim is to provide reliable control of the line width of cuprate superconductors. The usual contact metals are used. Lift-off or chemical etching processes are employed for pattern definition. During dicing, care must be taken to avoid edge damage. The brittleness of MgO or ceramic substrates makes dicing difficult.

13.5 Design and tuning approaches for HTS filters

HTS filters are predominantly based on the popular microstrip technology. The microstrip is a planar transmission line. This transmission line is used for conveying signals at microwave frequencies. It consists of a conducting strip separated from a ground plane by a dielectric layer. This dielectric layer is the insulating substrate.

The filter design is based on four parameters: (i) the size of the filter (and hence the chip area), (ii) the quality factor, (iii) the power handling capability and (iv) the insensitivity of filter response to manufacturing variations (Simon *et al* 2004). For a distributed element resonator, maximization of Q and power handling requires a large line width. The disadvantage of this design is the large size necessary for a half-wavelength resonator at high frequencies. The constraints on wafer size and cost have led to two possibilities of reducing the die size. These are the lumped-element resonator or folded half-wavelength resonator. A lumped-element resonator yields a compact structure as a capacitively loaded inductance (figure 13.1). But the extremely small line widths required make it prone to variations in fabrication.

The folded half-wave type resonators include the clip resonator (figure 13.2) and spiral in/spiral out (SISO) resonator. In filter circuits, the commonly used geometry for distributed element resonators is the single spiral resonator (figure13.3) because of the compactness in size achieved. The main issue with the single spiral structure is the inconvenience in accessing its inner terminal for which either a via hole or an air bridge is necessary. This difficulty is avoided by using the SISO geometry. For further size miniaturization, an on-chip capacitor is used as shown in figure 13.4. The terminals of SISO geometry are easily accessible. Two feedline structures are used. These are: insert-tapped and insert-coupled (figure 13.5).

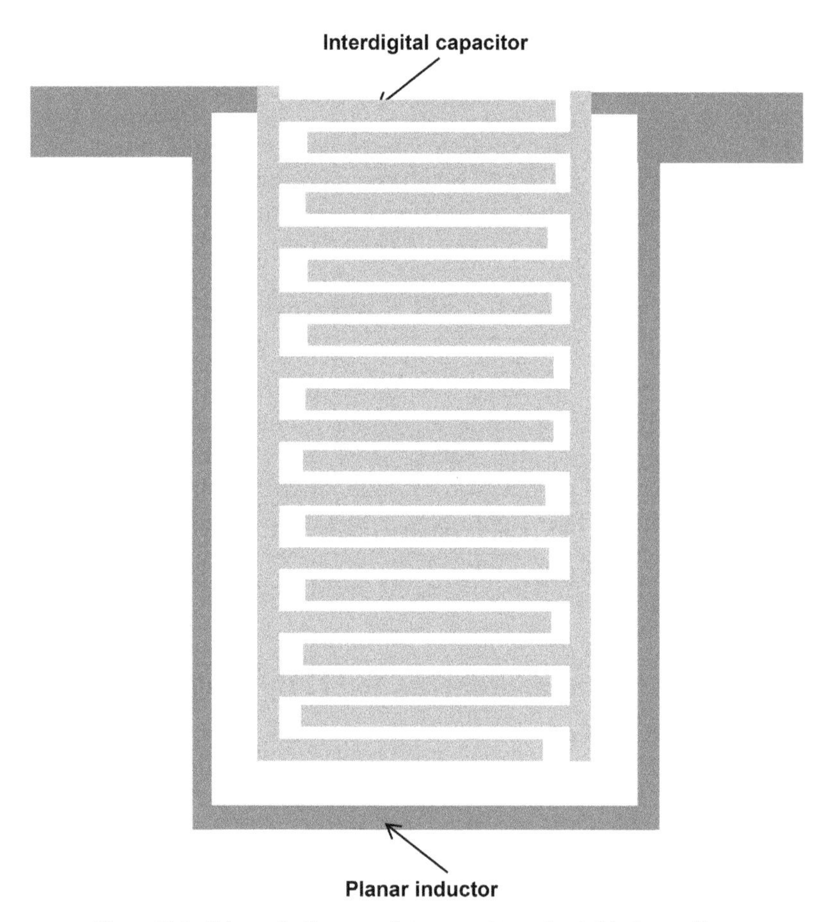

Figure 13.1. Schematic diagram of the capacitance loaded inductor filter.

Figure 13.2. The clip resonator configuration.

Figure 13.3. The single spiral resonator.

The clip resonator serves as a high-performance filter for wireless applications. Quasi-elliptic filters of band-pass and band-reject type are fabricated with the SISO resonator.

In a practical application, accurate adjustment and alignment of resonators constituting an HTS filter is essential. Generally, the frequency response deviates marginally from the designed value due to manufacturing tolerances. The frequency response is adjusted to the desired value by a process called tuning or trimming the filter. Filter tuning involves transferring the passband of the filter to the desired frequency as well as optimization of return loss of the filter. The conventional method using mechanical tuning elements such as metal screws introduces disappointingly high losses. Therefore, a low-loss device, e.g. a sapphire cylinder fixed to a threaded metal element, is employed. As the resonant frequency of an HTS resonator varies with temperature, it can be reset by a temperature control scheme.

Figure 13.4. On-chip capacitor-loaded SISO resonator.

For electronically tuning a filter, either the capacitance or inductance of its resonator is changed.

Often during the design of an HTS thin-film microwave device, a ferroelectric thin-film such as single-crystal strontium titanate (STO) is included (Wooldridge *et al* 1999) for *in situ* tuning of the frequency response based on the electric field dependence of the ferroelectric material. A change in permittivity with electric field allows tunability of capacitance.

13.6 Cryogenic packaging

The package uses a cryo-cooler, a standalone cooler and a Dewar; the Dewar is a vacuum insulated vessel, built to provide long-lasting, reliable operation. The cryo-cooler must be small in size to decrease the overall size of the system and also reduce power consumption. Typically, the cryo-coolers must provide several watts of

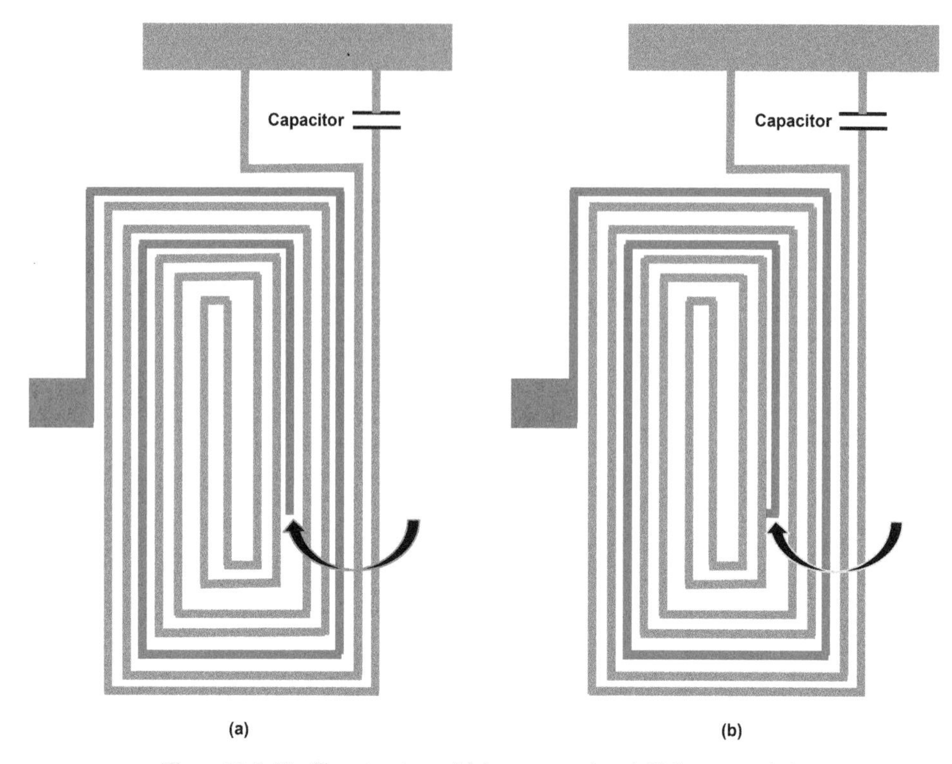

Figure 13.5. Feedline structures: (a) insert-tapped and (b) insert-coupled.

cooling capacity at filter operating temperatures ~60–80 K and surround temperatures up to 50 °C, with minimal maintenance and at affordable costs.

A long-life Dewar may be built by reducing vacuum leaks and controlling the outgassing by materials. The Dewar is permanently welded after evacuation and the DC/RF feedthroughs are sealed for hermeticity at leakage rates $<1 \times 10^{-10}$ cm^3 s^{-1} helium. For outgassing, the Dewar must be maintained in a vacuum for several days to weeks at a temperature ~100 °C prior to sealing. A gettering agent may be incorporated to remove any leftover gas. Although a higher temperature quickens the outgassing, it may degrade the HTS material.

13.7 HTS bandpass filters for mobile telecommunications

Highly selective HTS filters improve the performance of mobile communication networks by increasing the data rates and redeeming frequency spectrum (Shen *et al* 1997).

13.7.1 Filter design methodology

A filter with a passband of 10 MHz is considered (figure 13.6). The response of the filter follows a ten-pole quasi-elliptic function. It is designed using a special topological layout. This geometry is a cascaded quadruplet trisection (CQT) coupling structure (Hong *et al* 2005). To design the filter, a set of design parameters

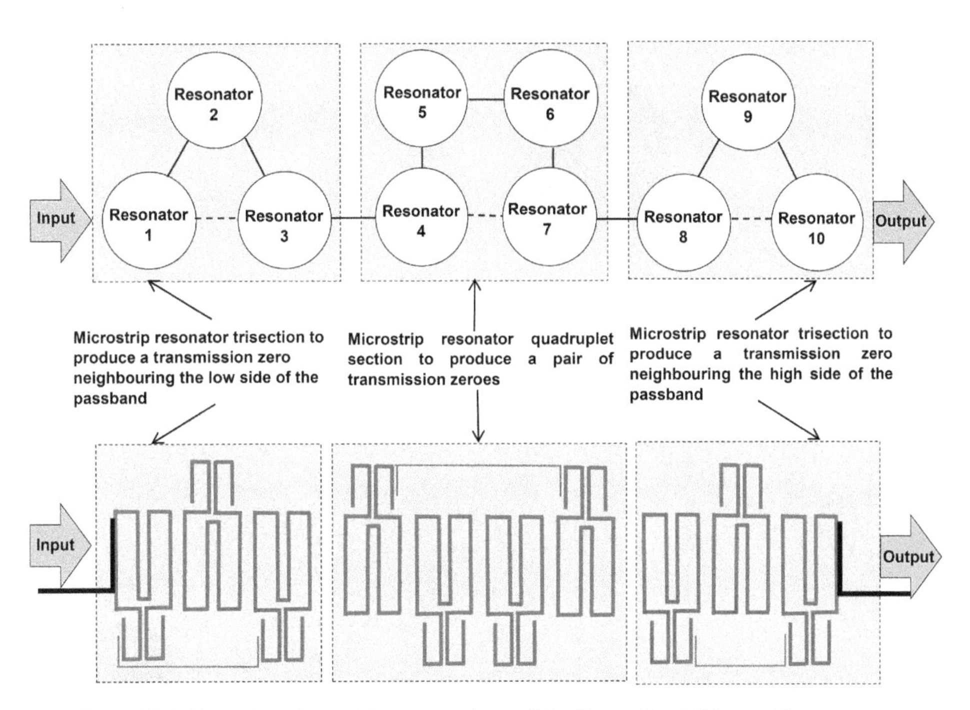

Figure 13.6. Execution plan and layout topology of the filter with a CQT coupling structure.

is extracted from filter synthesis. The frequency response of the synthesized filter is analyzed using a circuit model, which has the required frequency offsets, coupling coefficients and external quality factors. The advantage of CQT topology is that the asymmetrical design leads to a smaller number of resonators and lower insertion loss in the passband. Furthermore, the efforts in tuning are minimized because the cross-couplings in the filter are mutually independent. The subsequent step in filter design consists in implementation of the circuit model with a suitable microstrip structure. The filter is fabricated on an *r*-cut sapphire substrate. Due to anisotropy of the dielectric properties of sapphire, its relative permittivity is represented by a tensor. Arrangement of microstrip resonators on the anisotropic substrate is done in such a manner that all of them experience the same permittivity tensor. The frequency response is confirmed by full-wave electromagnetic simulation. Sensitivity analysis is performed against the variation in the thickness of substrate. Significant effects on the sensitivity of the filter response were noticed for a thickness tolerance of ±5 μm.

13.7.2 Filter fabrication and characterization

The designed filter is fabricated on a sapphire substrate (Hong *et al* 2005). This substrate is 430 μm thick. It is coated with 0.3 μm thick YBCO films on both sides. The characteristic temperature of YBCO films is 87 K. For RF contacts, gold electroplating is carried out on both sides up to a thickness of 0.3 μm. The filter thus fabricated on a substrate of size (47 mm × 17 mm) is mounted on an Au-plated Ti carrier in a brass enclosure placed inside a cryogenic Dewar. The frequencies of all the HTS resonators in the filter are tuned using sapphire tuners. A minimum

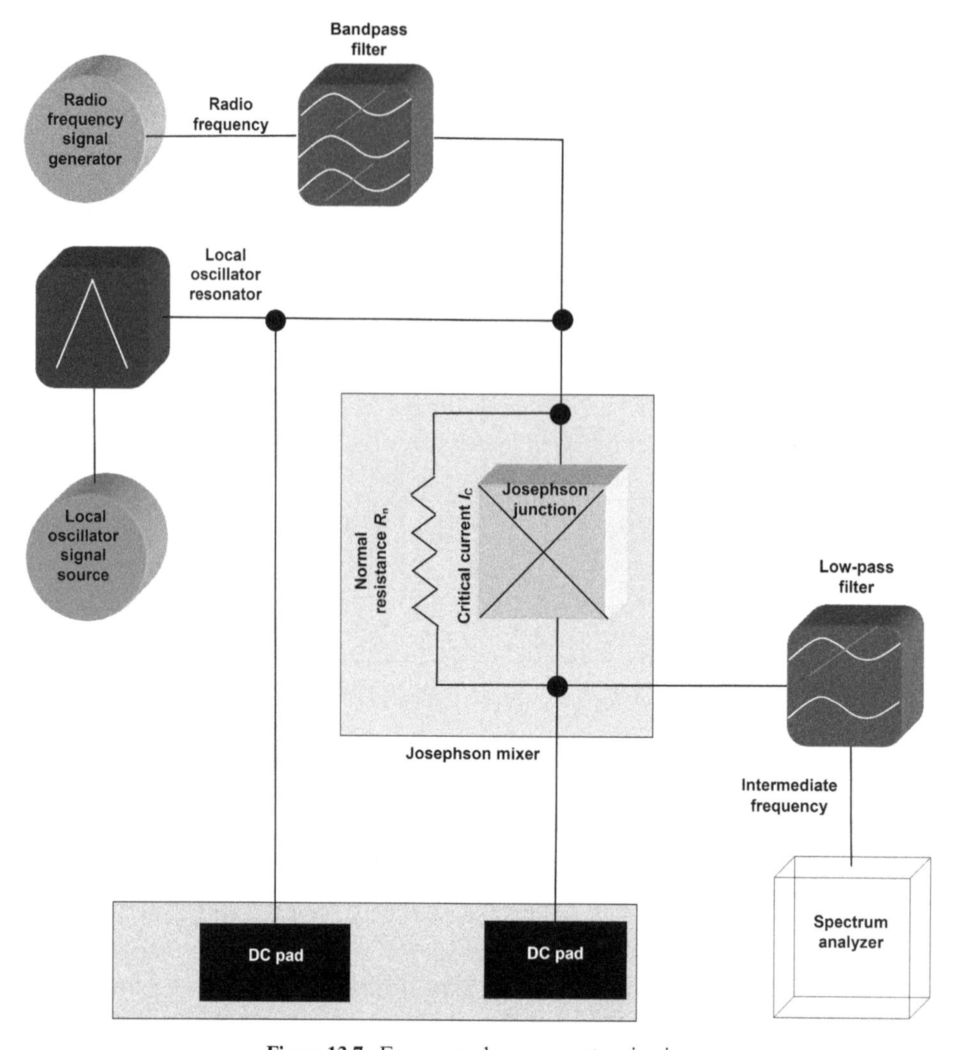

Figure 13.7. Frequency down converter circuit.

insertion loss of 0.2 dB was measured in the passband. The return loss was better than −12 dB over the passband. Rejection better than 85 dB was recorded over a broad range covering the universal Mobile Telecommunication System transmission band extending from 2110 MHz to 2170 MHz.

13.8 HTS JJ-based frequency down-converter

HTS technology can be utilized in wireless telecommunication front-end receivers. An all-HTS monolithic microwave IC (MMIC) JJ frequency down-converter module with nearly zero conversion loss consists of four components integrated on a (10 mm × 20 mm) YBCO film on a single substrate of magnesium oxide (Du *et al* 2012, 2013, 2014, 2015; Bai *et al* 2013). These components are (figure 13.7):

(i) A step-edge JJ mixer circuit.

(ii) A 10–12 GHz bandpass filter (BPF) circuit for RF signal input. This eight-pole filter uses stepped-impedance hairpin resonators. Its best insertion loss is measured as 0.4 dB at 40 K and 0.8 dB at 77 K. The out-of-band rejection is better than 40 dB.
(iii) A low-pass filter (LPF) having a ~4 GHz cut-off frequency for intermediate frequency (IF) output.
(iv) A capacitively coupled HTS microstrip line resonator at 8 GHz to isolate the local oscillator (LO). Effective RF signal isolation is provided by the BPF, LPF and the resonator at the LO port, eliminating the losses through RF leakage and by wire-bonded connections amongst components.

HTS step-edge grain boundary technology is applied for the fabrication of the MMIC chip (Du *et al* 2014). The step-edge pattern is delineated on a 0.5 mm thick, 10 mm wide and 20 mm long MgO substrate. The techniques used are photolithography, e-beam evaporation and argon ion beam etching, giving a step height of 0.4 μm and a step angle of 35°. A *c*-axis oriented YBCO film of thickness 0.22 μm is epitaxially deposited. Gold film of thickness 0.05 μm is deposited by e-beam evaporation. Patterning and etching the Au/YBCO film helps to form the junction. The gold film is removed from most regions, except for the contact pads and the lines feeding the filter. For packaging the MMIC chip, a gold-coated copper enclosure is used. Incorporating an RF blocking network and resistors in the DC biasing lines of the JJ prevents RF interference. The filters are linked to the connectors by silver epoxy. The chip is also grounded to the base plate by Ag-epoxy. The module has a total volume of $25 \times 27 \times 15$ mm^3.

In comparison to a single Josephson mixer, the DC and RF characteristics showed a 7 dB improvement in conversion efficiency because of efficient RF coupling, isolation between the RF, IF and LO ports, and elimination of wire bonding. The MMIC circuit is integrated in a miniature cryo-cooler, and the complete assembly is packed inside a portable (350 mm × 350 mm × 250 mm) box. The circuit performed satisfactorily in an unshielded condition, confirming its usefulness in wireless communications.

13.9 Discussion and conclusions

Lowering of surface resistance of HTS materials by $\geqslant 10^3$-fold relative to cryo-cooled metallic conductors enables the design of resonators with quality factors $\geqslant 10^5$ in the unloaded condition (Simon 2003). A gamut of devices has been demonstrated by utilizing the exceptional properties of HTS thin-films, notably high-Q resonators, filters, oscillators and mixers based on JJs. Representative examples of devices have been described.

HTS microwave microelectronics is an outcome of high-temperature superconductivity, offering widespread use and commercialization. The popularity of HTS technology arises from its ability to provide sensitivity and the rejection of out-of-band interference along with a wide dynamic range. All these virtues are essential for surveillance, reconnaissance and intelligence pursuits. The availability of cryogenic refrigerators for cooling the HTS materials at affordable costs is responsible for easy adoption of this technology in RF applications.

Review exercises

13.1. The resistance of a HTS increases with frequency. Give arguments in favor of using these materials in microwave circuits in place of a normal conductor.
13.2. List the desirable properties of a material that qualify it as a substrate for the fabrication of microwave circuits. How far do the following materials meet these qualifications: MgO and alumina?
13.3. How is YBCO prepared by a solid-state reaction? What techniques are used for depositing thin-films of this material?
13.4. Which material has a higher transition temperature: YBCO or TBCCO? Write the full forms of these acronyms.
13.5. What is the primary technology on which the design of HTS filters is based? Name four important parameters involved in design.
13.6. How is the deviation in the frequency response of a filter corrected? Why is the conventional method of mechanical tuning of filters not applicable to HTS filters?
13.7. How is cryogenic packaging of an HTS filter done? What precautions are taken to build a long-life, reliable Dewar?
13.8. What are the main steps in the design of an HTS filter for mobile telecommunications? How is the filter fabricated and packaged?
13.9. What are the four components of an all-HTS JJ-based frequency down-converter? Describe its fabrication and packaging. Discuss the characteristics of the circuit.
13.10. Name four types of microwave circuits realized using HTS thin-films.

References

Bai D D, Du J, Zhang T and He Y S 2013 A compact high temperature superconducting bandpass filter for integration with a Josephson mixer *J. Appl. Phys.* **114** 133906

Du J, Bai D D, Zhang T, Guo Y J, He Y S and Pegrum C 2014 Optimized conversion efficiency of a HTS MMIC Josephson down-converter *Supercond. Sci. Technol.* **2** 7105002

Du J, Wang J, Zhang T, Bai D, Guo Y J and He Y 2015 Demonstration of a portable HTS MMIC microwave receiver front-end *IEEE Trans. Appl. Supercond.* **25** 1500404

Du J, Zhang T, Guo Y J and Sun X W 2013 A high temperature superconducting monolithic microwave integrated Josephson down-converter with high conversion efficiency *Appl. Phys. Lett.* **102** 212602

Du J, Zhang T, Macfarlane J C, Guo Y J and Sun X W 2012 Monolithic HTS heterodyne Josephson frequency downconverter *Appl. Phys. Lett.* **100** 262604

Faisal W M 2012 High T_c superconducting fabrication of loop antenna *Alexandria Eng. J.* **51** 171–83

Hong J-S, McErlean E P and Karyamapudi B M 2005 A high-temperature superconducting filter for future mobile telecommunication systems *IEEE Trans. Microw. Theory Tech.* **53** 1976–81

Ramesh R, Hwang D M, Barner J B, Nazar L, Ravi T S, Inam A, Dutta B, Wu X D and Venkatesan T 1990 Defect structure of laser deposited Y–Ba–Cu–O thin films on single crystal MgO substrate *J. Mater. Res.* **5** 704–16

Shen Z-Y, Wilker C, Pang P, Face D W, Carter C F and Harrington C M 1997 Power handling capability improvement of high-temperature superconducting microwave circuits *IEEE Trans. Appl. Supercond.* **7** 2446–53

Simon R 2003 High-temperature superconductor filter technology breaks new ground *Mobile Devices Des.* pp 28–37 http://mobiledevdesign.com/site-files/mobiledevdesign.com/files/archive/mobiledevdesign.com/images/archive/308RF_Simon28.pdf

Simon R W, Hammond R B, Berkowitz S J and Willemsen B A 2004 Superconducting microwave filter systems for cellular telephone base stations *Proc. IEEE* **92** 1585–96

Vendik O G, Vendik I B and Kholodniak D V 2000 Applications of high-temperature superconductors in microwave integrated circuits *Mater. Phys. Mech.* **2** 15–24

Willemsen B A 2001 HTS filter subsystems for wireless telecommunications *IEEE Trans. Appl. Supercond.* **11** 60–7

Wooldridge I, Turner C W, Warburton P A and Romans E J 1999 Electrical tuning of passive HTS microwave devices using single crystal strontium titanate *IEEE Trans. Appl. Supercond.* **9** 3220–3

IOP Publishing

Extreme-Temperature and Harsh-Environment Electronics
Physics, technology and applications
Vinod Kumar Khanna

Chapter 14

High-temperature superconductor-based power delivery

As the aging electricity infrastructure of metropolitan cities fails to cope with the burden of increasing power demands from the customers, there is a dire need to face the challenge by introducing novel scientific ways of reducing the transmission losses incurred in sending electricity from generating plants to users. The widespread adoption of a HTS-based power network is envisaged as a potent answer to the power crisis. In this chapter, the methods developed for producing current-carrying wires from brittle superconducting ceramics are described. Salient features of the first and second generation superconducting wires are mentioned. As the HTSs need to be incessantly cooled with liquid nitrogen, different architectures of the cables have evolved. The warm dielectric and cool dielectric designs are discussed. As the superconductors can readily undergo transition from low impedance to a high impedance state, they possess inherent fault limiting capabilities, which are utilized in restive, shielded core and saturable core devices for protecting against surges. Superconducting wire transformers offer several benefits, such as fault limitation and the absence of fire hazards, in addition to the incredibly low power losses.

14.1 Introduction

To cater to escalating electric power loads, attention must be paid to both (i) the generation of power as well as (ii) the transmission and distribution of power. As far as the transmission and distribution of power are concerned, the deployment of high-temperature superconductors provides a convenient way of meeting this requirement. This calls for modernizing and upgrading the present power infrastructure. Let us therefore briefly review the existing power transmission structure to understand its shortcomings and limitations.

doi:10.1088/978-0-7503-1155-7ch14

14.2 Conventional electrical power transmission

14.2.1 Transmission materials

Two elements constitute the backbone of power delivery. These are copper (electrical resistivity = 1.68×10^{-8} Ωm) and aluminum (resistivity = 2.65×10^{-8} Ωm). Copper is superior to aluminum due to its lower resistivity. Even though these resistivity values are small, they are finite and sufficient to introduce appreciable losses over long distances.

14.2.2 High-voltage transmission

Power transmission loss = I^2R (I = current and R = resistance of wire). To reduce the transmission loss, power is conveyed over long distances in the form of AC by stepping up its voltage and reducing the current. At the utility point, the power is stepped down to the needed lower voltage. Thus power transmission is essentially based on the principle of circumventing the resistive losses using step-up and step-down transformers. Large, high-voltage transformers are oil filled. The transformer coil is a hydrocarbon mineral coil composed of aromatics, paraffins, napthenes and olefins. It serves as an insulating medium between the internal parts and a coolant. In the case of short circuits, transformer fires causing oil vaporization and spreading are fearsome events, often accompanied by an explosion.

In addition to transformers, capacitive and inductive elements are used in power transmission to keep the waveform in synchronization. Further, it must be noted that the higher the voltage to which the current is boosted to, the larger is the electromagnetic field surrounding the conducting wire, and therefore the larger are the losses incurred due to this field.

14.2.3 Overhead versus underground power delivery

Both modes have their advantages and drawbacks. The initial installation costs of overhead lines are lower than for underground lines. But overhead lines have to withstand the impact of weather attacks such as rains, storms, ice, etc. Overhead lines are easily repaired and tapped. In underground lines, fault-finding becomes difficult because it is necessary to dig the Earth and trace the fault location by trenching. Due to this reason, the duration of line outage may sometimes be longer for underground lines than for overhead lines.

14.3 HTS wires

The high-temperature ceramic materials are not only brittle, they are also fragile. They can provide flexibility at very small length scales only. Therefore, it is impossible to draw them out into long wires. The obvious solution is to take the help of either a metallic matrix or a thin bendable metallic substrate. These two approaches form the bases, respectively, for the first and second generation of HTS wires. The former employs a metallic matrix while the latter utilizes a thin metallic substrate as the supporting medium for superconducting particles. It must also be emphasized that the current conduction through the superconductor is influenced by the orientation of the crystalline structure. The rated currents are restricted by the grain boundaries of the crystalline microstructure. This microstructure also plays a decisive role in determining the magnetic field strength up to which the HTS wires remain superconducting. The value of this field strength is very critical in energy technology applications. In view of the above constraints, the process of making long HTS wires is riddled with technological challenges, which have been faced by different companies and solved through various methods to achieve the goals.

14.3.1 First generation (1G) HTS wire

This wire has a bismuth-based multi-filamentary composition embedded in a silver matrix (figure 14.1). It is represented by the acronym BSCCO (bismuth–strontium–calcium–copper oxide $(Bi, Pb)_2Sr_2Ca_2Cu_3O_{10}$) or Bi-2223 (BINE 2010). It has a transition temperature T_C = 110 K. This kind of wire is commercially

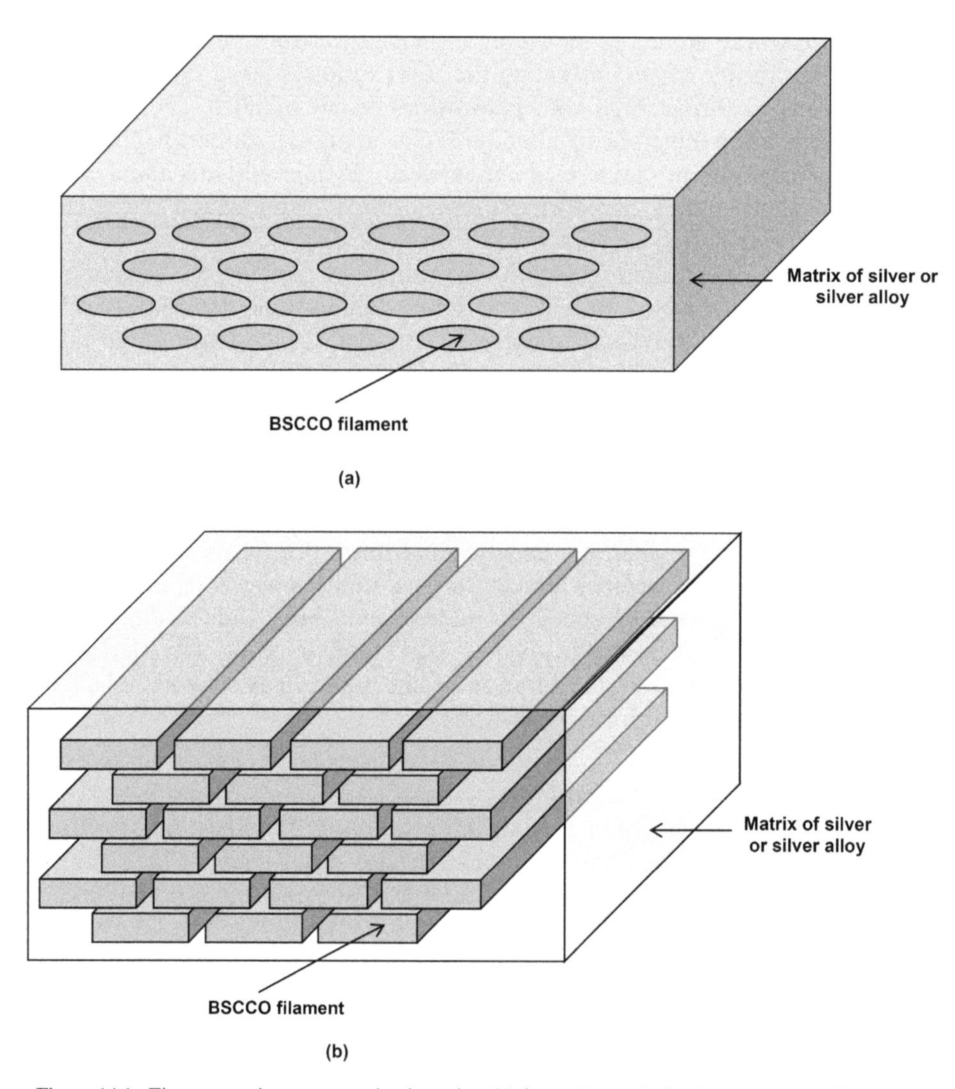

Figure 14.1. First generation superconducting wire: (a) front view and (b) view along the wire length.

available in kilometer lengths. The oxide-powder-in-tube process can be used to produce reliable 1G HTS wires. In this process, the superconductive material BSCCO is crushed into a fine powder and filled into a hollow, malleable tube of a silver alloy. The tube is extruded to a thin silver wire from 35 mm to 2 mm, whereby the particles in the powder adjust to the smaller cross-section wire. In this way, several silver wires are formed. A large number of these silver wires containing the fine ceramic powder are combined together in a silver tube. The resulting silver tube is shaped into a multi-filament wire. This multi-filament wire is annealed in several stages at temperatures up to 950 °C. After undergoing annealing, the superconductive phase is arranged as a fine crystallite in parallel filaments. The metallic sleeves separate the individual filaments. By multiple rolling treatments intervening the annealing processes, the surfaces of the crystallites and the conducting CuO are aligned parallel to each other. The conductivity decreases if all the crystallite axes are not perfectly aligned.

The 1G HTS wire can carry 100 times more current than a copper wire of identical size but also costs 100 times the price of the copper wire because of its high silver content, ⩾60%. Hence, the extraordinary current-carrying advantage is offset by the high cost disadvantage. The research focus was therefore diverted towards the second generation HTS wire. Nonetheless, a superior version of BSCCO wire exhibiting high critical current and displaying more strength has been demonstrated with larger yield on a commercial scale by Sumitomo Electric Industries (SEI), than conventional BSCCO (Hirose *et al* 2006). This version is called dynamically innovative BSCCO (DI-BSCCO). It is produced by a specially developed controlled over pressure (CO-OP) process.

14.3.2 Second generation (2G) HTS wire

The 2G HTS wire is a thin-film rare earth barium copper oxide (ReBCO) tape where Re = yttrium, samarium, neodymium, gadolinium, etc. The 2G wire has the shape of a band while the 1G HTS wire was shaped like a wire. The 2G wire is less costly and can achieve higher current densities. It is more suitable for magnetic field applications.

There are several layers in this tape (figure 14.2): metal substrate–buffer layer–ReBCO HTS layer–stabilizing Ag layer–stabilizing Cu layer. In the SuperPower process, a thin Ni alloy substrate (50–100 μm) is electrochemically polished to a surface roughness <2 nm (Hazelton *et al* 2009). It is sufficiently smooth for the formation of a 150 nm thick textured MgO-based buffer stack by ion beam-assisted deposition (IBAD)/sputtering. Then 1–5 μm thick ReBCO HTS film is formed by the MOCVD process. The deposition rate is 0.7 μm min^{-1}. A capping silver layer (2 μm) is sputtered for good electrical contact. The full structure is electroplated with a 2 × 20 μm surround copper layer for conductor stabilization. The thickness of the copper film is variable and is determined by the application.

The peak critical current at an angle ~5° between the magnetic field (1 T) and the surface of the film is 100 A cm^{-1} in width for 0.7 μm ZrGdYBCO and 158 A cm^{-1} in

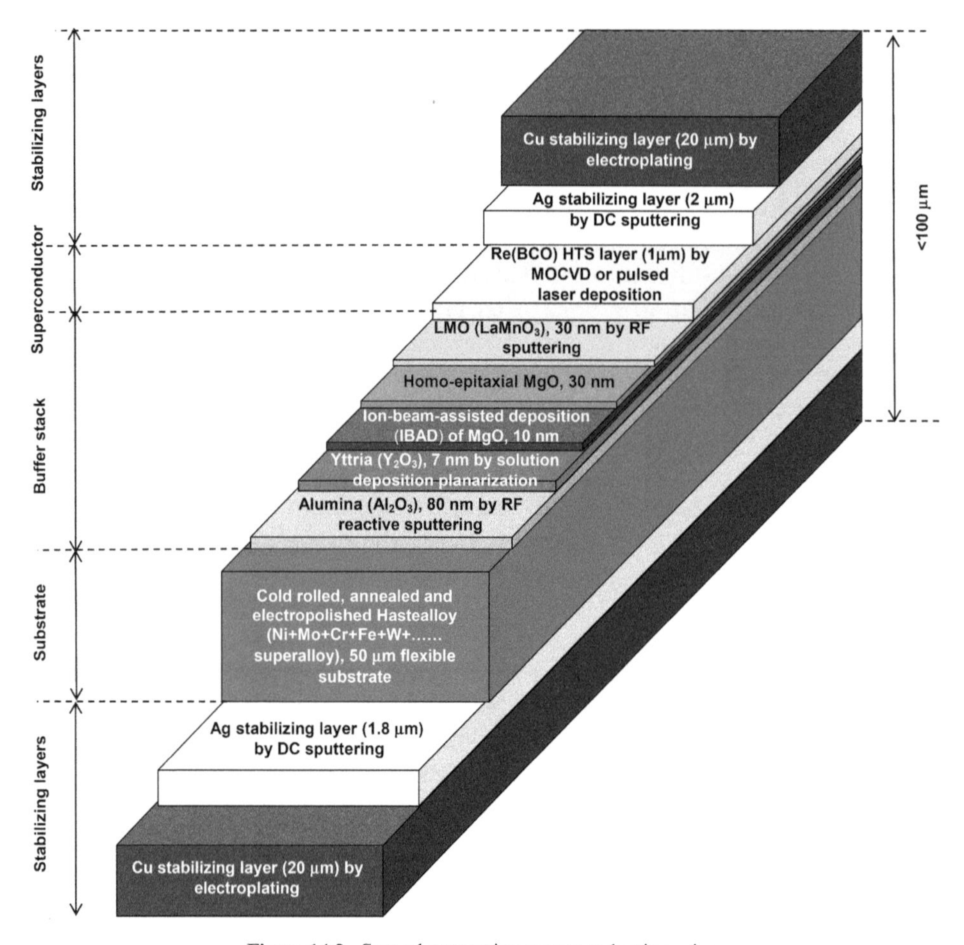

Figure 14.2. Second generation superconducting wire.

width for 0.7 μm SmYBCO (Hazelton *et al* 2009). The wires can tolerate 700 MPa of stress before the critical current is irreversibly degraded.

In the process of Sumitomo Electric Industries, the substrate is a long metal tape of high mechanical strength and low magnetism to keep AC loss small (Yamaguchi *et al* 2014). To prepare the textured metal substrate, a high-strength non-magnetic metal (100 μm thick) is laminated with Cu (20–50 μm thick) and the Cu film is plated with nickel (3 μm thick). The Cu film can be textured by thermal treatment while nickel does not allow copper to oxidize. RF sputtering of three films produces the intermediate layer on the textured metal covered substrate: CeO_2, cerium (IV) oxide, 0.1–0.2 μm thick as a seed layer; yttria stabilized zirconia (YSZ), 0.2–0.4 μm thick for preventing dispersion; and CeO_2 0.1–0.2 μm thick for matching the lattice structure with the ensuing superconducting layer. The superconducting film $GdBa_2Cu_3O_y$ (1–4 μm thick) for power transmission is coated by pulse laser deposition. The stabilizing layer consists of a DC sputtered Ag layer (2–8 μm thick) followed by electroplated Cu film (20 μm thick). It also prevents the superconducting layer from suffering any mutilation. The critical current I_C is 200 A cm^{-1} $width^{-1}$ over the 30 mm conductor with a maximum value of 500 A cm^{-1} $width^{-1}$ in its central region (Yamaguchi *et al* 2014).

In the process of American Superconductor (Li *et al* 2009), 2G YBCO HTS wire is manufactured in continuous lengths by a full-scale reel-to-reel process. The process is based on an economical wide-strip technology. The wide-strip process using 4 cm wide strips commences with the deformation-texturing of a metal alloy substrate. The alloy used is Ni 5 at%W. The substrate is recrystallized in a reel-to-reel system. A cube-textured alloy substrate is thereby formed. A fast reactive sputtering technique is employed to deposit the buffer layers. These layers include: an Y_2O_3 seed layer (75 nm thick), an YSZ barrier layer (75 nm thick) and a CeO_2 capping layer (75 nm thick). After deposition of the buffer layers, an YBCO precursor film doped by rare earth is slot-die coated. The loading is 4800 mg $Y(Dy_{0.5})Ba_2Cu_3O_{7-\delta}$ m^{-2} of template corresponding to a calculated thickness of 0.8 μm $YBa_2Cu_3O_{7-\delta}$. It is pyrolyzed at a temperature below 600 °C. In a reel-to-reel system, it is transformed into an epitaxial $YBa_2Cu_3O_{7-\delta}$ film at 750 °C to 800 °C. Capping with a silver layer is followed by oxygenation at 500 °C. The produced strip of width 4 cm is divided into several thin wires and laminated to metallic stabilizers. As a result, a three-ply wire known as 344 superconductors is produced. If the metallic stabilizer is made of steel, the HTS wire is represented by 344S; if brass is used, it is 344B. In a pre-pilot device, the critical current achieved is $I_C > 100$ A (250 A cm^{-1} $width^{-1}$).

14.4 HTS cable designs

14.4.1 Single-phase warm dielectric HTS cable

At the center is a hollow flexible core through which liquid nitrogen flows (figure 14.3). The HTS wires are wound around this liquid nitrogen-cooled core and therefore are cooled by the liquid nitrogen. A vacuum jacket surrounds the HTS wires to thermally insulate them from the environment. Still outward is the dielectric layer for electrical insulation. Two HTS wires are necessary to complete the circuit: one for current inflow into the circuit and the other for current return from the circuit. In the region surrounding the warm dielectric HTS wire, a magnetic field exists. This magnetic field issue is absent in the cool dielectric HTS wire. Furthermore, the warm dielectric design has a shorter life span.

14.4.2 Single-phase cool dielectric HTS cable

Recall the constructional features of a coaxial cable or coax used for electrical connections. It consists of a central core conductor surrounded by an insulating sheath. Over this sheath is the metallic conducting coating. A final dielectric sheath may enclose the above layers.

Like the coaxial cable, the cool dielectric HTS cable consists of a complex assembly comprising several tubular parts and sheaths (Demko *et al* 2000). It consists of an innermost flexible metal former through which liquid nitrogen is supplied (figure 14.4). The main HTS surrounds the metal former. It is the conductor

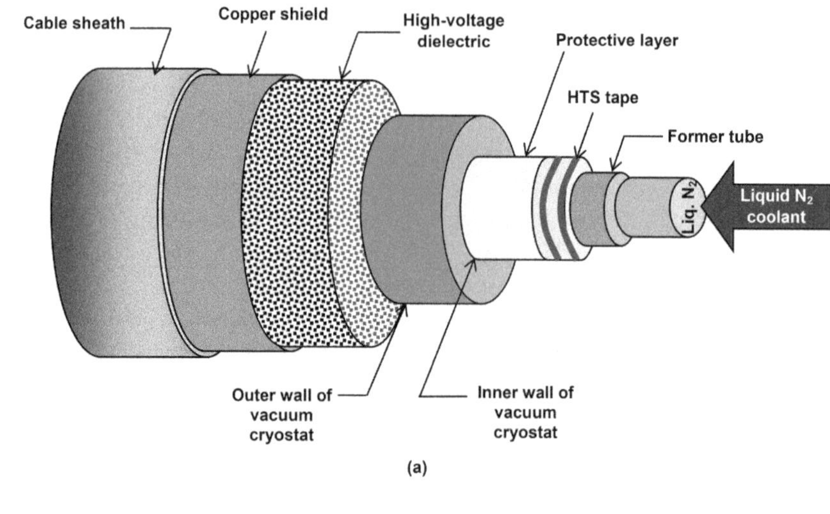

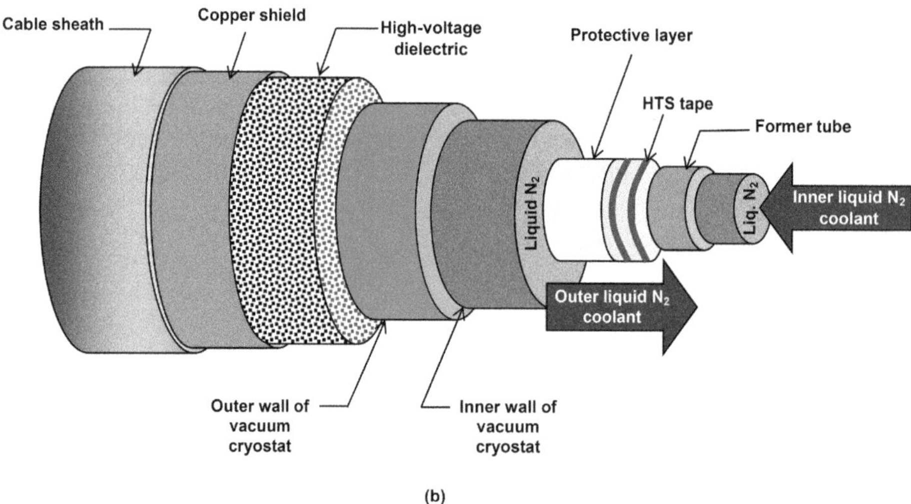

Figure 14.3. Two designs of a warm dielectric superconductor cable with: (a) inner liquid nitrogen coolant flow and (ii) inner and outer liquid nitrogen coolant flows.

carrying the supercurrent. The layer covering the main HTS contains the dielectric and shielding layers for the low-temperature operation. The shielding HTS envelops these layers. This shielding HTS, which is separated from the main HTS by the dielectric material, carries the return current sent by the main HTS. Therefore, a single cool dielectric HTS cable is adequate for both forward and return supercurrents. The outermost layer is the flexible cryostat, the apparatus used to maintain low temperature for the enclosed components. Liquid nitrogen flows in the annular space between the shielding HTS and the flexible cryostat. The innermost flexible metal former and the outermost flexible cryostat are made of corrugated stainless steel tubing. At one extremity of the cooling arrangement is a refrigeration unit. This unit provides a constant supply of liquid nitrogen.

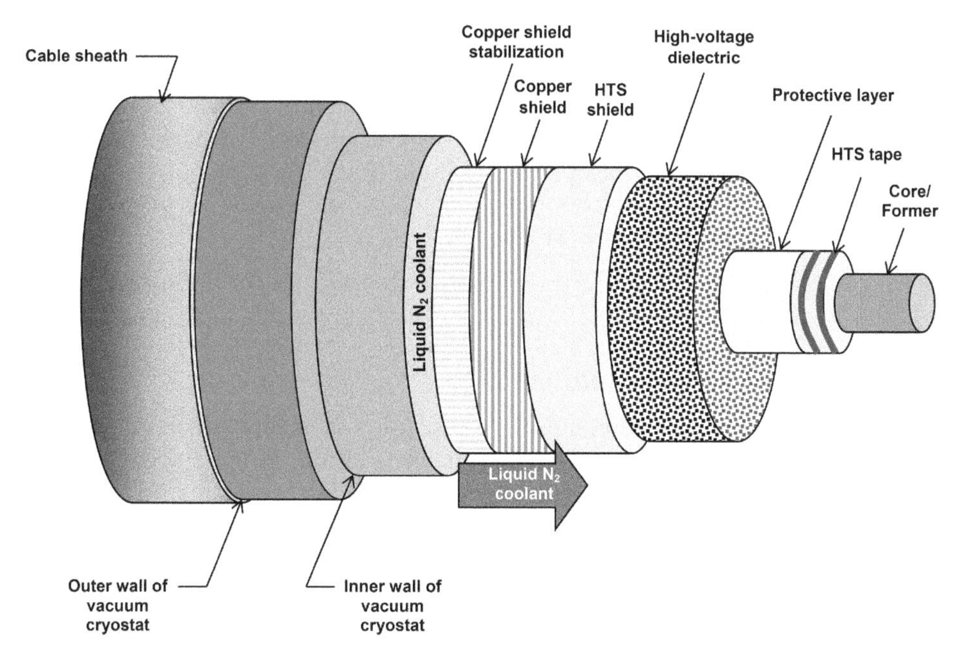

Figure 14.4. Cold dielectric superconductor cable.

The cool dielectric architecture is an enhancement over the warm dielectric design because only one cool dielectric cable is necessary instead of two warm dielectric cables. Moreover, it can carry higher current. Additionally, the external magnetic field produced by it is nearly zero. The life expectancy of the cool dielectric design is longer.

The parallel-flow cooling arrangement

Liquid nitrogen flows in the same direction in the innermost flexible metal former and annular region between the shielding HTS and the flexible cryostat. The path for returning of liquid nitrogen is a separate duct jacketed inside a vacuum.

The counter-flow cooling arrangement

The directions of flow of liquid nitrogen in the metal former and the annulus are opposite. Liquid nitrogen enters through the metal former. It cools the terminations as well as the HTS. Then, it returns through the annulus.

14.4.3 Flow rate, pressure drop and HTS cable temperatures

In both cooling arrangements, as the temperature of the HTS decreases, AC losses decrease due to the superconductivity effect while thermal losses increase due to the larger temperature difference between the superconductor and the surroundings according to Newton's law of cooling. Further, in both cooling arrangements, lower flow rates result in a higher temperature of the HTS. At the higher temperature, the critical current decreases and the AC loss increases.

As compared to the parallel flow arrangement, the counter-flow arrangement has a lower heat load. In addition, for the same length of the HT superconductor and flow rate, the counter-flow arrangement yields a higher temperature of the HTS and larger pressure drops. The larger pressure drop experienced in this arrangement imposes a limitation on the length of the counter-flow arrangement.

14.4.4 Three-phase cold dielectric HTS cable

For three-phase current supply, three phases are enclosed in a single cryogenic envelope, as shown in figure 14.5. They can also be put inside separate envelopes.

14.5 HTS fault current limiters

Sometimes under heavy load condition or during a lightning strike, a high-value fault current suddenly appears, damaging the installation. In these situations, fault current limiters (FCLs) come to the rescue. They are devices which limit the current to safe value fast enough to prevent damage. Superconducting FCLs (SFCLs) utilize the non-linear properties of superconducting materials, such as non-linearity in response towards temperature, current and magnetic field changes for restraining the current magnitudes. At a constant temperature and under a constant magnetic field, when the fault current becomes high, the increase in fault current makes a section of the superconductor so resistive that the heat produced cannot be removed locally. Through transference of excess heat to the neighboring regions of the superconductor, the temperature of these regions is raised. The combined effect of current and temperature effects is that the superconductor starts behaving as a normal conductor. This propagation of the normal conductor zone through a superconductor is known as quenching. Effectively, the material undergoes a transition from a low impedance superconducting state to a normal high-impedance conducting state in a fault episode.

14.5.1 Resistive SFCL

In a resistive SFCL (figure 14.6), the main current carrying conductor under normal operating conditions of the transmission system is the superconducting material (Electric Power Research Institute 2009). When a fault occurs, the current increases, and the superconductor undergoes quenching, accompanied by an exponential rise in resistance. Due to the speedy increase in resistance of the superconductor, a voltage is produced across the superconductor. As a result, the current is transferred to a shunt consisting of a resistor and an inductor. Further increase in voltage across the superconductor is thereby restricted. Thus the superconductor acts as a switch responding in milliseconds. This switch causes a changeover of the load current to the shunt impedance. The uneven heating of the superconductor during quenching leads to hot spot generation, damaging the HTS material. To some extent, this has been mitigated by modification of the superconductor material process and innovative device designs.

The heat produced in the superconductor during quenching is carried away by the cryogenic cooling system. Thus a finite time elapses before the cooling system

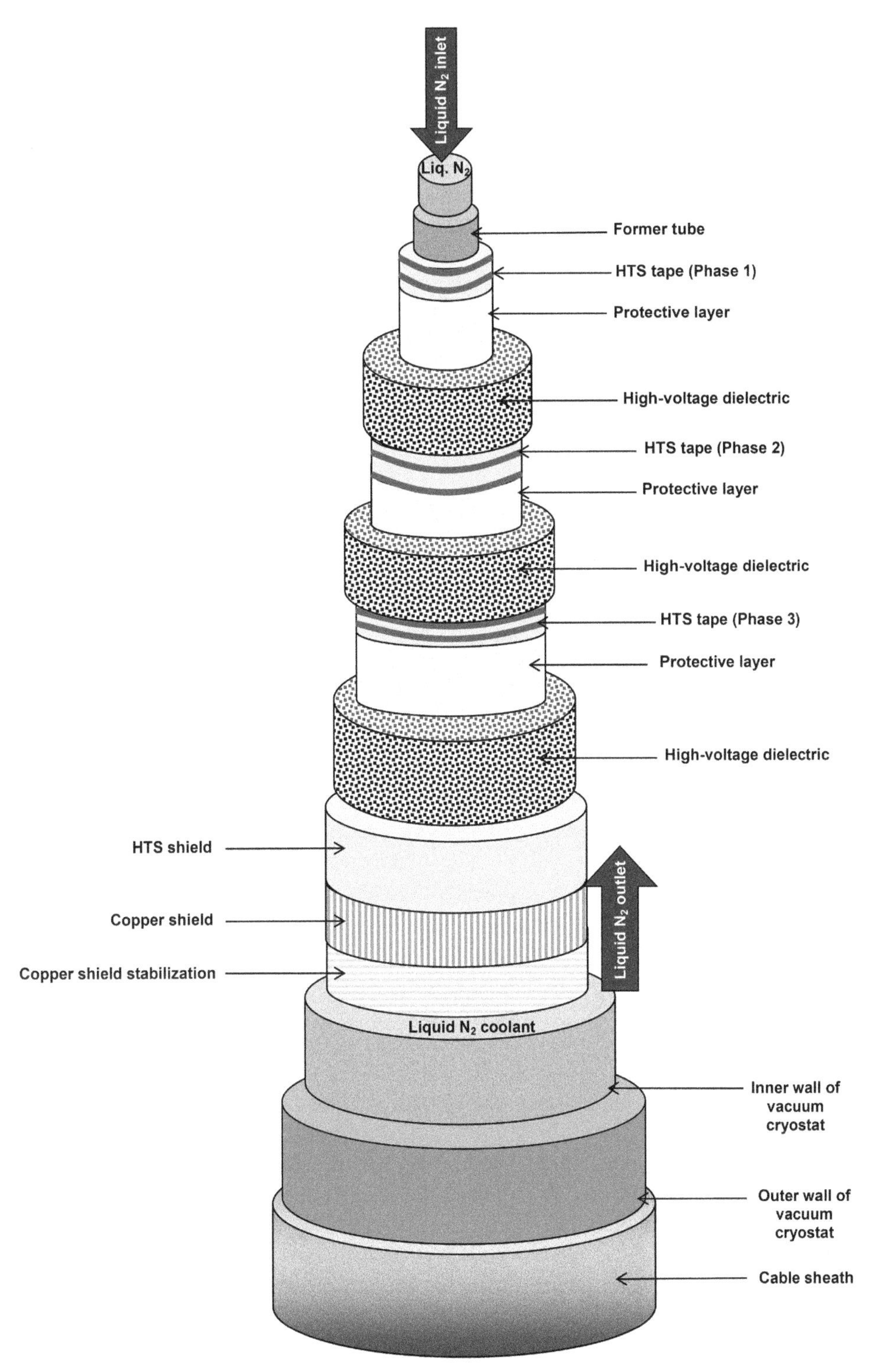

Figure 14.5. Triaxial cable with a cold dielectric arrangement.

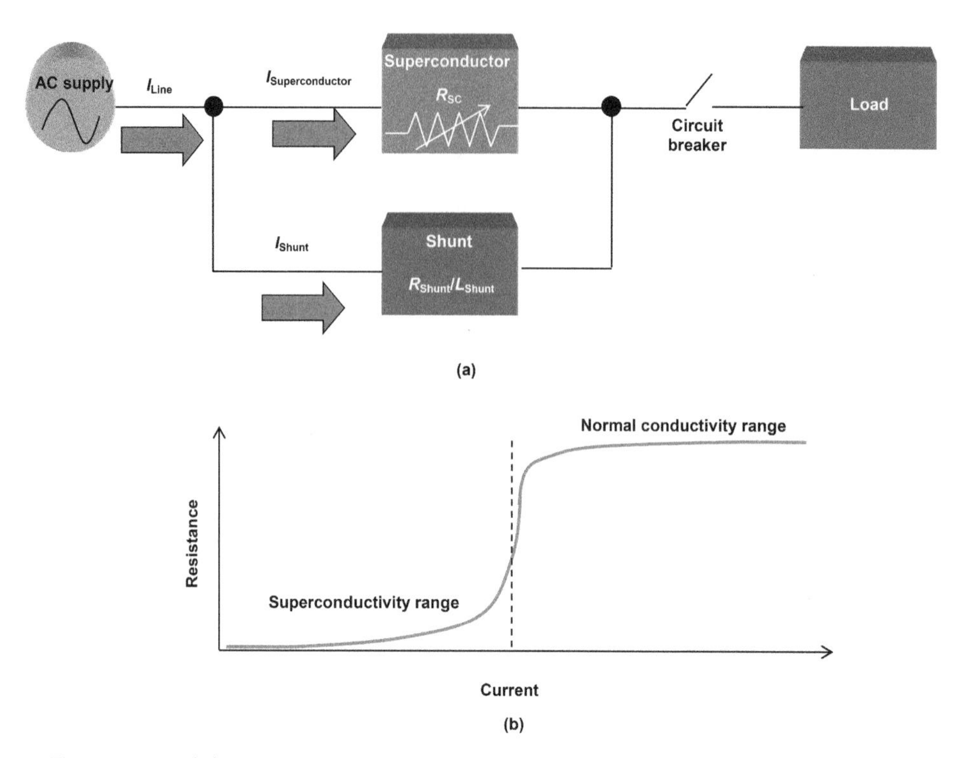

Figure 14.6. Resistive type SFCL: (a) circuit diagram and (b) its resistance–current characteristics.

restores the material back to the superconducting state and this time duration is referred to as the recovery time of the system. In some SFCLs, a fast switching element is connected in series with the superconductor to separate the superconductor after transitioning of a large proportion of the current to the shunt element. This switch decreases the maximum temperature in the superconductor, hastening the recovery, and thereby reducing the recovery time.

14.5.2 Shielded-core SFCL

The shielded-core SFCL (figure 14.7) is a variant of the resistive SFCL in which the cryogenic cooling system is mechanically separated from the remaining circuit. The power line and the HTS element are not directly connected. They are indirectly connected through mutually coupled AC coils via a magnetic field. The SFCL device is similar to a transformer in construction. In this transformer, the secondary side is shunted by an HTS element. Whenever a fault current is produced, the current on the secondary side increases. Consequently, the HTS element quenches. The associated rising voltage across the primary coil is in such a direction that it opposes the fault current. By fiddling with the ratio of the number of turns in the transformer coils, the hot spot problem caused by non-uniform heating of the superconductor is avoided. However, like a resistive SFCL, the shielded-core SFCL must cool down after a current limiting action has taken place. A major disadvantage of the

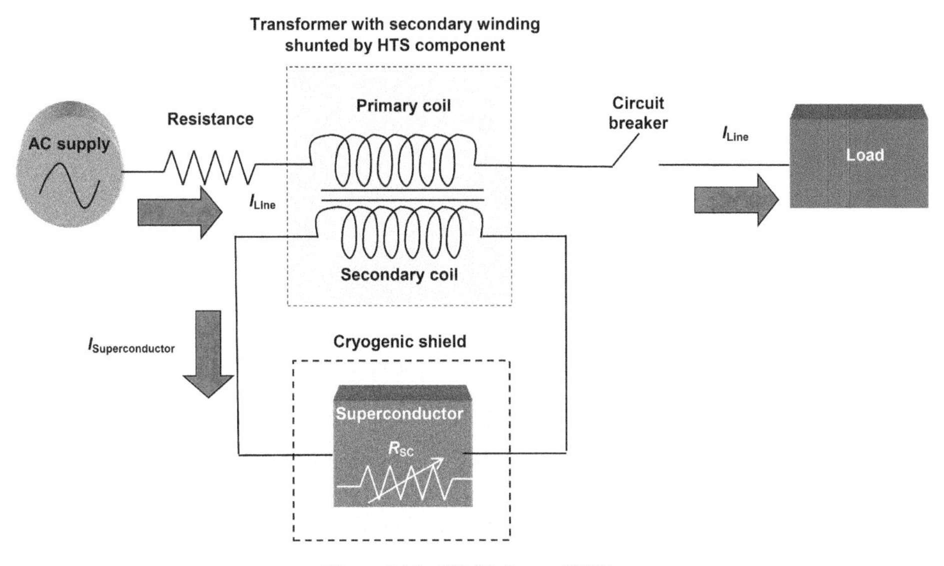

Figure 14.7. Shielded-core SFCL.

shielded-core SFCL is that it occupies four times the volume of a resistive SFCL. Also, it has four times the weight of the resistive SFCL.

14.5.3 Saturable-core SFCL

The saturable-core SFCL (figure 14.8) uses two iron cores and two AC windings for each phase. The AC windings comprise conventional conductors. These conductors are wound around the iron core forming an inductance in series with the AC line. A HTS winding carrying a constant current is also wound around the iron core. This winding provides a magnetic field. When the grid current is within normal limits, the HTS winding completely saturates the iron. Under this condition, its relative permeability is unity. For the coils carrying AC, the iron core acts like air. The AC impedance or inductive reactance resembles an air-core reactance component. But when the fault current is produced, the core is pulled out of saturation by the positive and negative peaks of current. Consequently, the line impedance increases during part of each half cycle. Hence, the peak fault current is appreciably reduced. Thus the saturable-core SFCL is an iron-core reactive component having variable inductance. Normally, it is an air-core reactive component. Under a fault current condition, its impedance becomes very high. The saturable-core SFCL differs from the resistive SFCL. It can control a series of faulty events successively without needing to cool down like a resistive SFCL. This happens because it does not quench. However, the volume and weight of the heavy iron core required to build a saturable core SFCL is a serious impediment to its pervasive utilization.

14.6 HTS transformers

A HTS transformer is similar to an ordinary transformer except that the primary and secondary windings of the HTS transformer are made from zero-resistance HTS

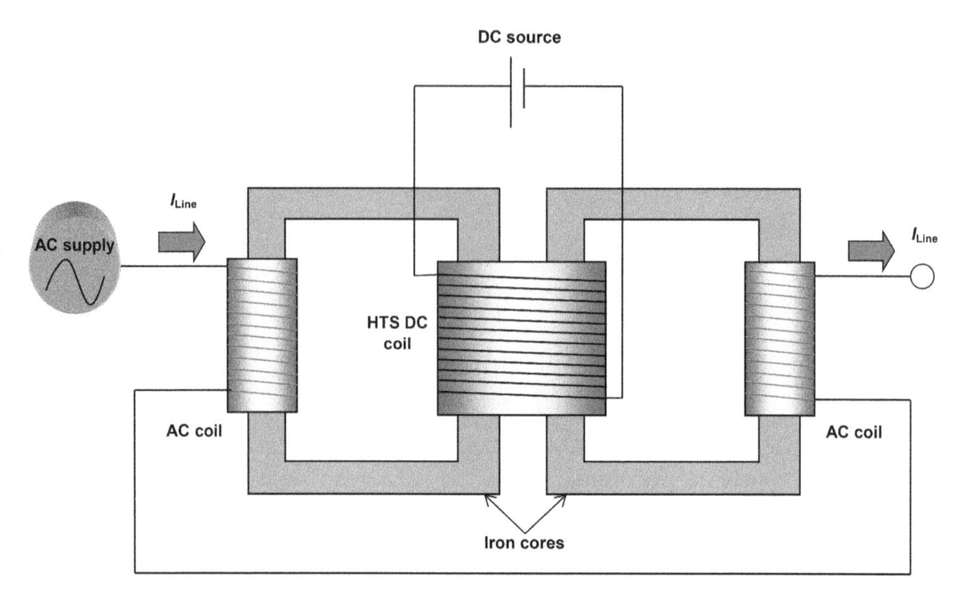

Figure 14.8. Saturable iron-core SFCL.

wire in place of the copper or aluminum windings used in ordinary transformers (US Department of Energy 2000). All the advantages of superconductivity are therefore derived in a HTS transformer as compared to an ordinary transformer. The HTS transformer is cooled down to 77 K in order that it operates in the superconductive mode, providing lower losses by approximately 30% at a 10–30 times higher current density than a regular transformer and with a 45% smaller weight than the same. A HTS transformer offers unmatched fault current limiting ability. But the biggest advantage of a HTS transformer is that it does not require any insulating oil coolant. The fire and environmental hazards of the oil coolant related to a conventional transformer are therefore automatically eliminated.

14.7 Discussion and conclusions

HTSs have changed the fundamental component in electrical and electronic circuits, namely, the 'conducting electrical wire'. Since the wire is a key component without which no circuit implementation is possible, the impact of superconductivity on electricity and electronics is expected to be profound and revolutionary. This chapter provided a glance at some of the ways that superconductors can reshape the electric power infrastructure and change the ways in which power is distributed and controlled. All these efforts pave the way towards maximizing the capabilities and thereby the available facilities while minimizing electricity bills and hence the financial liability of consumers.

Review exercises

14.1. Name the two elements used in making wires for transmitting electricity? Which one has a higher conductivity?

14.2. Why is power transmitted at high voltages and low currents? What is the role of transformers in power transmission? What types of hazards are possible due to transformer faults?

14.3. In what ways is overhead power transmission better than underground delivery? In what ways is it worse?

14.4. Why is it difficult to draw high-temperature superconducting materials into long wires? What are the two approaches followed in first and second-generation wire making from superconductors?

14.5. What is the acronym for the first generation HTS wire? What do the letters of the acronym represent?

14.6. What are the main steps in the oxide-powder-in-tube process of making 1G HTS wire? What is the main drawback of 1G HTS wire? What is a DI-BSCCO wire?

14.7. What is the acronym representing a 2G high-temperature superconducting wire? Mention three advantages of the 2G wire over the 1G wire.

14.8. How many layers of different materials are present in a 2G HTS wire? What are the names of these materials? What is the role of each material?

14.9. What are the main steps involved in the full scale reel-to-reel process of making a 2G HTS wire? Describe the function of each step.

14.10. What is the substrate material used in a 2G HTS wire? What is the buffer layer and what is its function? What are the HTS materials used? What elements are used in the top stabilizing layers?

14.11. Explain with a diagram the construction of a warm dielectric HTS cable? What are its disadvantages?

14.12. Draw a labelled diagram and explain the constructional features of a cool dielectric HTS cable. Mention its advantages.

14.13. Explain the parallel-flow and counter-flow cooling arrangements used in a cool dielectric HTS cable. What is the length limitation of the counter-flow arrangement?

14.14. What is the function of the FCL in a power transmission circuit? What property of a superconductor helps in incorporating fault limitation in the circuit? What is quenching?

14.15. Draw the circuit diagram of a resistive SFCL. Explain its operating mechanism with reference to the circuit diagram. How is the recovery of the SFCL speeded up?

14.16. How does a shielded-core SFCL operate? Explain with a diagram.

14.17. What is the operating principle of a saturable-core SFCL? How does it differ from a resistive SFCL?

14.18. What is an HTS transformer? Point out the advantages offered by it over an ordinary transformer.

References

BINE 2010 High temperature superconductors *Projektinfo* 06/10 www.bine.info/fileadmin/content/Publikationen/Englische_Infos/projekt_0610_engl_internetx.pdf

Demko J A, Lue J W, Gouge M J, Stovall J P, Butterworth Z, Sinha U and Hughey R L 2000 Practical AC loss and thermal considerations for HTS power transmission cable systems *Appl. Supercond. Conf.* (*Virginia Beach, VA 17–22 September*) pp 1–4

Electric Power Research Institute 2009 Superconducting fault current limiters *Technology Watch* 1017793 www.suptech.com/pdf_products/faultcurrentlimiters.pdf

Hazelton D W, Selvamanickam V, Duval J M, Larbalestier D C, Markiewicz W D, Weijers H W and Holtz R L 2009 Recent developments in 2G HTS coil technology *IEEE Trans. Appl. Supercond.* **19** 2218–22

Hirose M, Yamada Y, Masuda T, Sato K–i and Hata R 2006 Study on commercialization of high-temperature superconductor *SEI Technical Review* No. 62, June, pp 15–23

Li X, Rupich M W, Thieme C L H, Teplitsky M and Sathyamurthy S *et al* 2009 The development of second generation HTS wire at American Superconductor *IEEE Trans. Appl. Supercond.* **19** 3231–5

US Department of Energy 2000 Selected high-temperature superconducting electric power products, January, www.nrel.gov/docs/gen/fy00/supercon_products.pdf

Yamaguchi T, Shingai Y, Konishi M, Ohya M, Ashibe Y and Yumura H 2014 Large current and low AC loss high temperature superconducting power cable using REBCO wires *SEI Technical Review* No. 78, April, pp 79–85

Part II

Harsh-environment electronics

IOP Publishing

Extreme-Temperature and Harsh-Environment Electronics
Physics, technology and applications
Vinod Kumar Khanna

Chapter 15

Humidity and contamination effects on electronics

High humidity is a primary initiator and booster of devastating effects on electronics. Great damage to electronics can be anticipated if moisture is allowed to seep into packages and reach the circuits. Contaminants introduced either during manufacturing or later by contact with handling workers or from the environment, partner with moisture to intensify the destruction. Mechanisms for destruction by corrosion are many and originate from different causes, e.g. electrochemical, anodic, cathodic, galvanic, etc. All these mechanisms, whereby high humidity along with dust, dirt and organic matter pose a hazard to the reliability of electronics, are dealt with in this chapter. The conditions favoring these mechanisms are mentioned. An understanding of these mechanisms is helpful to create conditions and devise suitable counteracting measures to obliterate their adversative consequences.

15.1 Introduction

Electronic devices and circuits have proliferated worldwide into newer applications, both indoors and outdoors. They have permeated into everyday activities. All this has led to mounting concerns on the effects of humidity on electronic equipment. The primary cause of this worry is the invariable association of a high-humidity environment with corrosion of electronic components or parts thereof. Ideally, the optimum humidity level prescribed for electronics is the 40%–60% RH range.

Additionally, progressive miniaturization of components in computers, telecommunications and control systems has made them more susceptible to humidity-induced failures, even in gentler environments. In nanoelectronics, a flawed patch measuring a few nanometers is more detrimental than a several millimeter wide blotch in large steel structures. An automobile can function with the loss of several grams of material, but

doi:10.1088/978-0-7503-1155-7ch15

for an IC even a few picograms of material make the difference. To deal with this serious problem, the root mechanisms underpinning such failures must be elucidated.

15.2 Absolute and relative humidity

Moisture-induced failure modes exhibit a strong correlation with RH instead of absolute humidity (AH; Tencer and Moss 2002). There exists a threshold RH level at which a certain thickness of adsorbed water is formed on the surface of the component, leading to its non-performance. AH is the mass of water vapor (m) present in a given volume (V) of air, expressed in units of mg L^{-1}:

$$AH = m/V. \tag{15.1}$$

The temperature of air is not taken into consideration in this definition. The RH at a given ambient temperature is a percentage representing the ratio of AH at that temperature to the maximum possible AH at the same temperature under saturation conditions. It is defined as a ratio of partial pressures: the partial pressure of water vapor p_{H_2O} present in the air at a given temperature to the equilibrium vapor pressure $p^*_{H_2O}$ at that temperature:

$$RH = p_{H_2O}/p^*_{H_2O}. \tag{15.2}$$

15.3 Relation between humidity, contamination and corrosion

Corrosion is the deterioration of the properties of a material, e.g. a metal, inorganic or organic compound, through chemical reactions of the material with environmental gases such as oxygen and water vapor. Adsorbed moisture is a key enabler and driver of corrosion. The rate of corrosion increases rapidly at RH > 60% and in the presence of contaminants such as dust and microbes. The high RH causes condensation of large quantities of water vapor on the surface of the material, e.g. in a humid environment, water droplets condense on circuit boards (figure 15.1). Ions can flow through the condensed water film, accelerating corrosion. Together with dust and dirt, they can trigger corrosion, which would not certainly take place if the circuit boards were clean. Very small quantities of surface contaminants suffice to initiate corrosion. The sources of these contaminants are: (i) the manufacturing process steps in which the boards are exposed to gases, etching and plating solutions, soldering fluxes, etc; (ii) contact with workers handling the boards and customers mounting them. Handling contamination comes from oils and grease on fingers, salts in sweat, skin flakes, cosmetics and creams on the skin, etc.

During processing, if any ion-releasing compounds are left over, they can promote corrosion. If not properly cleaned, chloride ions, salts and sulfur compounds can play havoc in the presence of water. Careful cleaning of the boards ensures avoidance of corrosion via ionic contents.

Corrosion looks like a natural outcome of the presence of water, heat, dust, microbes and oxygen on the surfaces of materials. The corrosion rate increases at high temperatures and becomes slow at low temperatures.

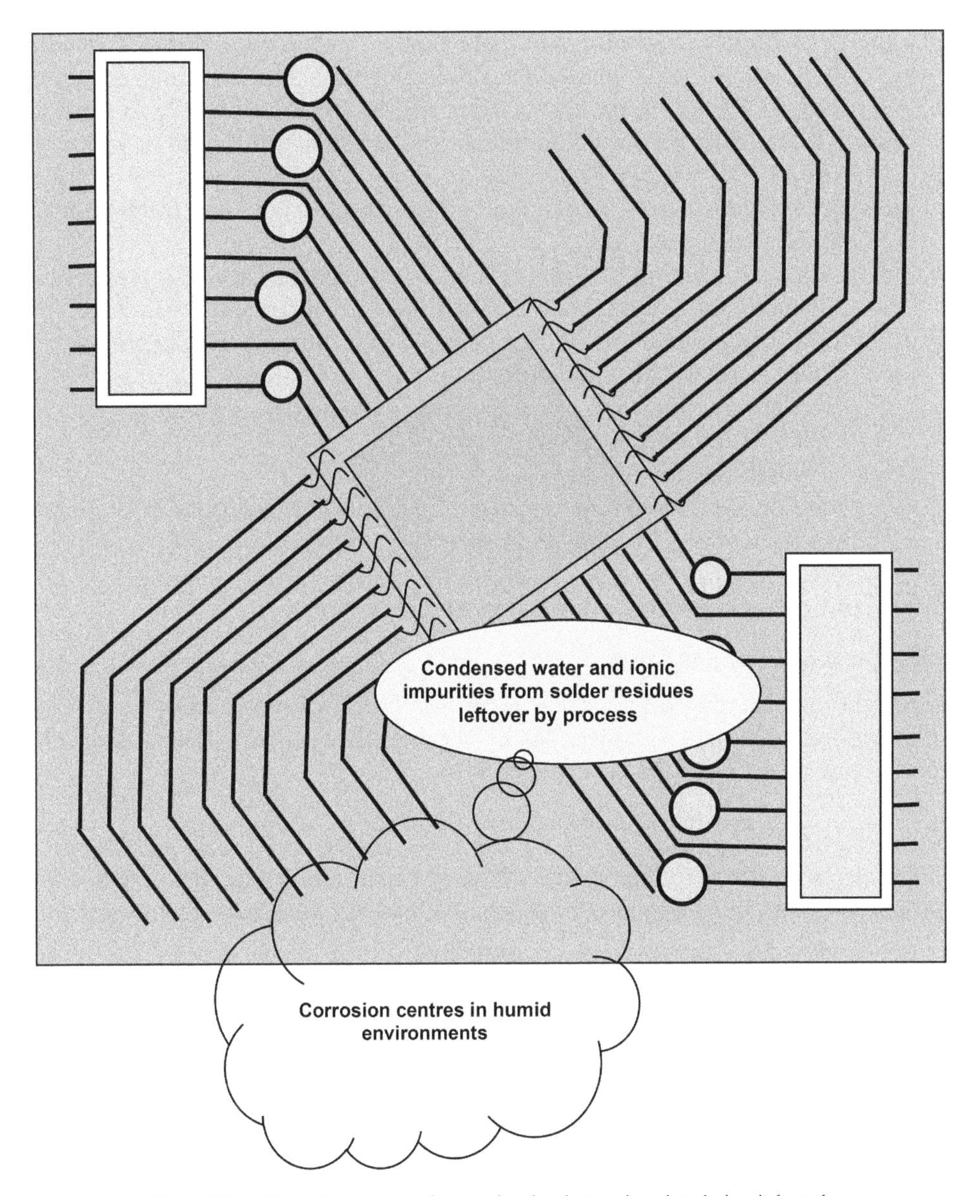

Figure 15.1. Humidity-promoted corrosion in electronic printed circuit boards.

15.4 Metals and alloys used in electronics

It is the metallic parts of electronic equipment, which have to suffer the maximum onslaught of corrosion. In electronic chips, PCBs and packages, a multiplicity of metals and alloys are used, some of which are:

(i) Chips of discrete semiconductor devices and ICs: aluminum, gold, silver, copper in addition to silicon (a semi-metal).

(ii) Lead frames connecting the gold bumps on the chip with surrounding PCB: gold or silver-treated Cu/Zn37, CuFe2, FeNi42 or CuNiZn (on the chip side) and a solderable coating such as Ni–Au on the PCB side.
(iii) Wire bonding between gold bumps and the lead frame: gold or aluminum wire.
(iv) Wire bonding metal aluminum pad on the chip and lead frame: Au/Au, Au/Ni, Au/Ag, Al/Al.
(v) Electronic chip packages: galvanized/chromated steel or magnesium. Polymer packages are coated with Cu/Ni or Al for EMI shielding.
(vi) Connecting lines on PCB: copper with solderable electroless Ni-immersion Au (IM Au) coating.
(vii) Electronic connectors: gold electroplated copper and its alloys, e.g. CuSn6, CuBe2, CuNi10Sn2.
(viii) Lead-free solders: Sn–Ag–Cu.
(ix) Hard discs of computers: a platter made of Al electroplated with Ni over which there is a Co alloy with carbon coating.

15.5 Humidity-triggered corrosion mechanisms

15.5.1 Electrochemical corrosion

This corrosion, exemplified by the rusting of iron, is a two-step process. (i) An atom at the surface of a metal dissolves in the condensed water film. The metal is left negatively charged. For a bivalent metal M,

$$\text{anodic reaction: } M(s) \rightarrow M^{2+}(aq) + 2e^{-}. \tag{15.3}$$

(ii) The electrons travel to the outside of the water droplet where they interact with an electron acceptor called the depolarizer, such as oxygen from the atmosphere:

$$\text{cathodic reaction: } O_2(g) + 2H_2O(l) + 4e^{-} \rightarrow 4OH^{-}(aq). \tag{15.4}$$

Inside the water droplet, the hydroxyl ions move inwards to react with metal ions:

$$M^{2+}(aq) + 2OH^{-}(aq) \rightarrow M(OH)_2(s). \tag{15.5}$$

The metal hydroxide quickly oxidizes:

$$4M(OH)_2(s) + O_2(g) \rightarrow 2M_2O_3 \cdot H_2O(s) + 2H_2O(l). \tag{15.6}$$

In such atmospheric corrosion, changing weather conditions play a major role.

15.5.2 Anodic corrosion

This corrosion (figure 15.2) occurs when there is a potential difference between two conductors, such as between two soldering points on a PCB, connected by a thin film of liquid water (Lighting Global 2013). The metal ions dissolve from the anode. Depending on their stability in the aqueous solution, they either deposit on the cathode (e.g. Sn, Pb, Cu, Ag, etc) or may dissolve in water forming hydroxide, such as for aluminum, which is a non-migrating metal. The Sn, Pb, Cu or Ag ionic

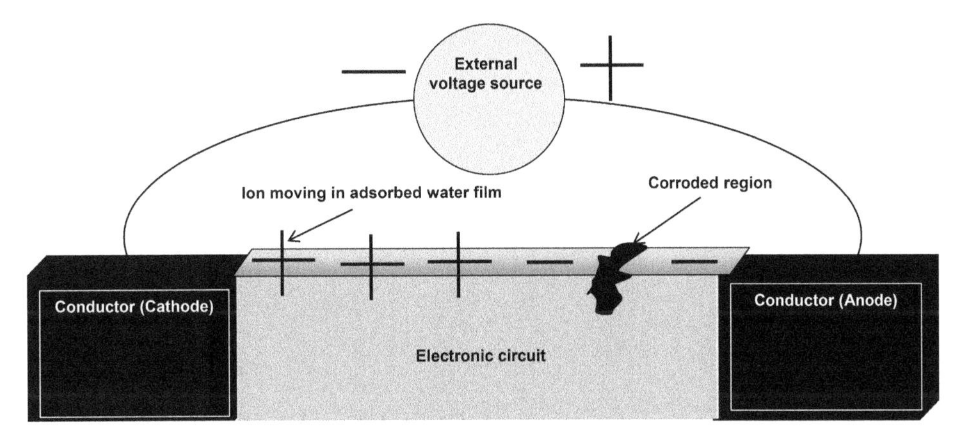

Figure 15.2. Anodic corrosion.

migration causes dendrite growth from the cathode metal towards the anode metal, bridging the gap between them and thus leading to a short-circuiting effect. Failures due to dendrite growth can take place between copper lines on a PCB.

Looking at the scenario of relentlessly shrinking dimensions, the spacing between the conductor lines is becoming very small, increasing the electric field between them. Higher electric fields hasten ionic migration. Micro-soldering processes carried out using low-residue, no-clean fluxes in inert gas ambient are gaining popularity. The long-term stability of circuits with narrow spacing between lines assembled using micro-soldering processes is not yet proven.

15.5.3 Galvanic corrosion

This corrosion, also called 'dissimilar metal corrosion' is a type of preferential corrosion of one metal relative to another when two dissimilar metals with distinctly different electrochemical potentials are brought in contact and are exposed to an electrolytic solution. The two different metals produce a potential difference driving the corrosion process without any help from an externally applied voltage. Water condensation from the environment or immersion of the object in water provides the electrolyte. The relative nobility of a metal is known from its corrosion potential, and can be obtained from the galvanic series (figure 15.3). The corrosive effect is utilized in primary cells (figure 15.4). Many possibilities of galvanic corrosion are shown in figures 15.5 and 15.6.

The gold/aluminum wire bonding in ICs is vulnerable to this type of corrosion and so also is the lead frame, which is coated with nobler coatings than base metals, if there are any cracks or mechanical defects in the coating metal. For touch connectors, graphite can substitute gold because of its slow anodic/cathodic reaction kinetics.

In a mobile phone keypad, the keys are made of IM Au and electroless nickel (EL Ni). If there is any damage to the IM Au layer, the EL Ni layer will be corroded with IM Au, acting as the cathode. Pitting of the EL Ni will expose Cu and cause further damage.

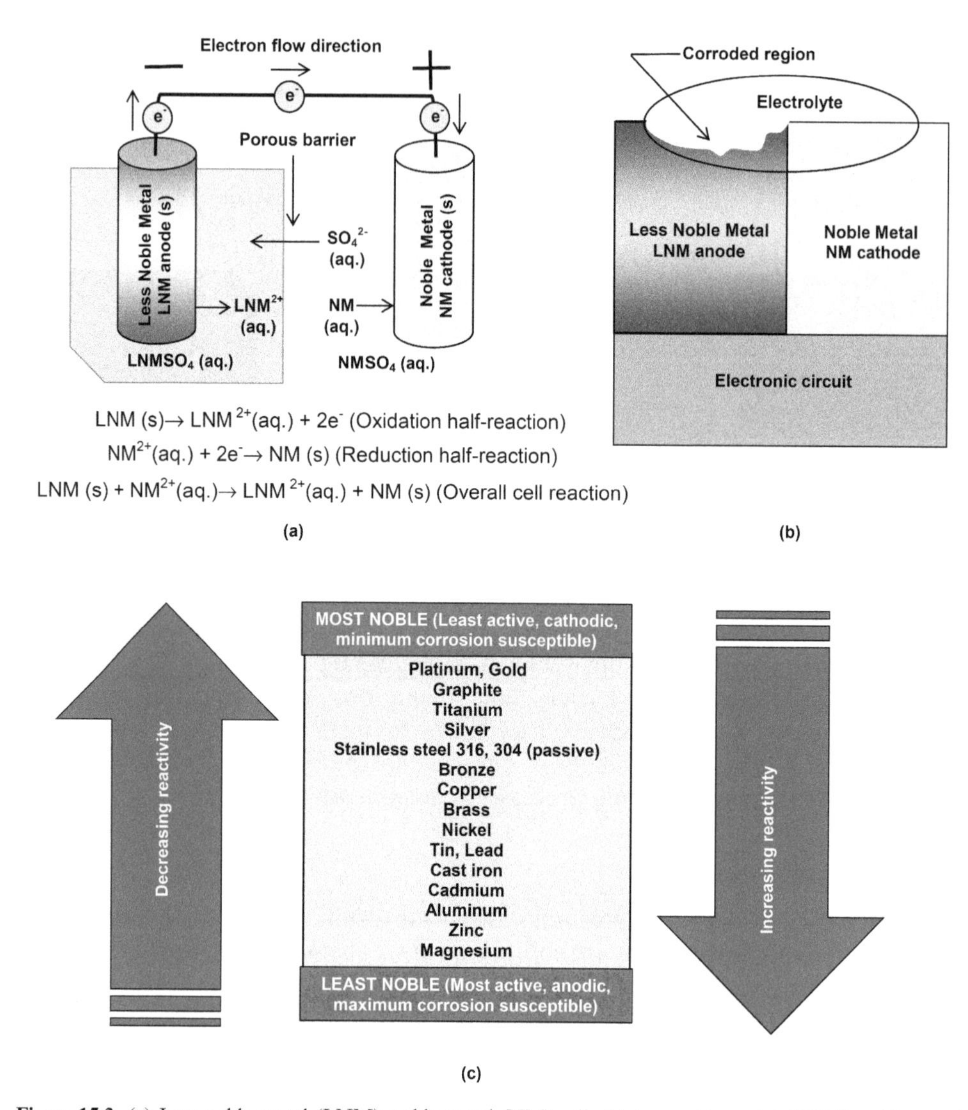

Figure 15.3. (a) Less noble metal (LNM)–noble metal (NM) cell, (b) less noble metal (LNM)–noble metal (NM) corrosion cell; both metals are taken as bivalent and (c) partial galvanic series.

15.5.4 Cathodic corrosion

A few metals used in electronics dissolve in acidic and alkaline solutions over a broad range of pH and potentials, e.g. aluminum and zinc are soluble in these solutions above the equilibrium dissolution potential. In an electrochemical cell deriving energy from redox reactions, called the galvanic cell (see above), OH^- ions are liberated by oxygen reduction at the cathode in accordance with equation (15.4). The resultant shift of pH towards the alkaline side enables dissolution of aluminum. In IC chips, the aluminum conducting lines fall prey to this type of cathodic corrosion.

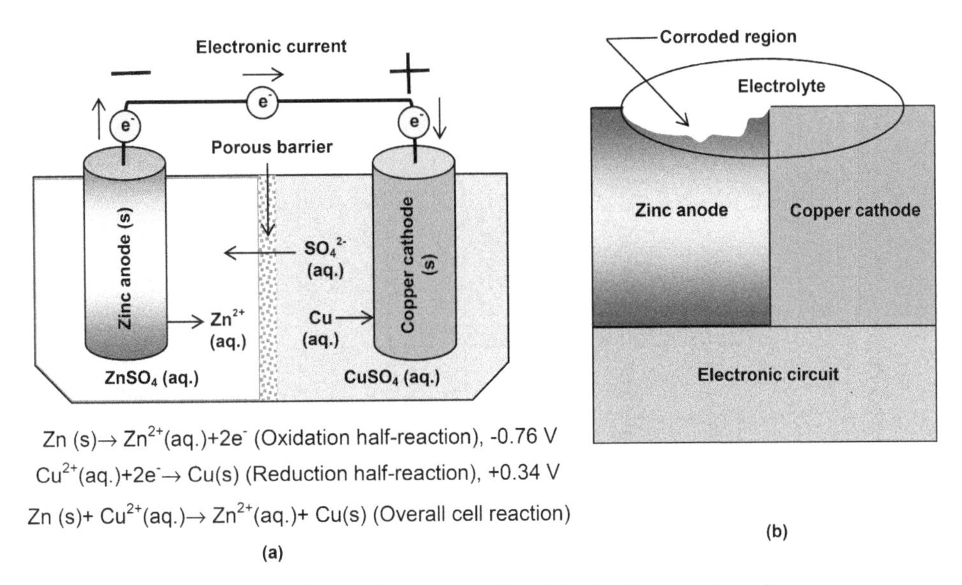

Figure 15.4. (a) Zn–Cu galvanic cell and (b) Zn–Cu corrosion cell.

15.5.5 Creep corrosion

Immersion silver as the surface finish of PCB on copper is prone to creep corrosion in the presence of sulfur and moisture due to Ag_2S and Cu_2S formation (Savolainen and Schueller 2012). Sulphur is present in the elemental form or as H_2S in environments adjacent to paper mills, as well as the cement and rubber manufacturing industries. Sulfur is also available near mining industries and waste-water treatment plants. Creep corrosion begins by dendritic growth and takes place equally in all directions.

Creep corrosion does not require any external potential difference to be applied to the PCB. The mechanism seems to be driven by a galvanic process because of the involvement of two metals, silver and copper, with copper behaving anodically with respect to silver and subject to more vigorous attack. The area of the anodic copper layer is much smaller than that of the cathodic silver covering layer. As a result, the attack on the anode metal is profoundly aggressive, compelling stronger corrosion reaction of copper with sulfur and moisture. The creep corrosion is described as a three step process (Chen *et al* 2012): (i) dissolution of the corrosion products in adsorbed water layers on the surface, (ii) their movement by diffusion under the concentration gradient and (iii) their re-deposition at another site.

High-temperature organic solderability preservative (OSP) or anti-tarnish and lead-free hot air solder leveling (HASL) have been found to be good protection for mitigating creep corrosion. Proper surface cleanliness is of paramount importance. If silver finish is to be retained, several design modifications are necessary such as exclusion of metal features demarcated by solder mask (polymer coating on copper traces in PCB), thorough coverage of non-test vias with a solder mask, keeping smoothed rounded corners for component pads, etc.

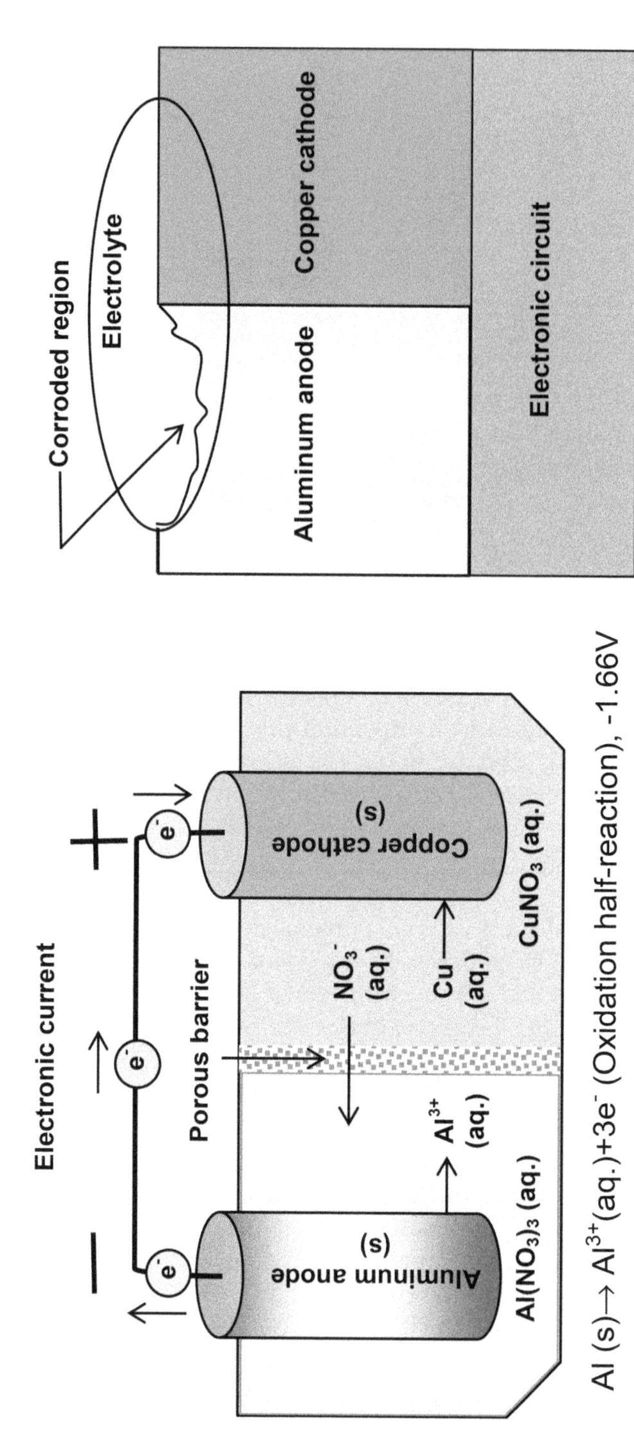

Figure 15.5. (a) Al–Cu galvanic cell and (b) Al–Cu corrosion cell.

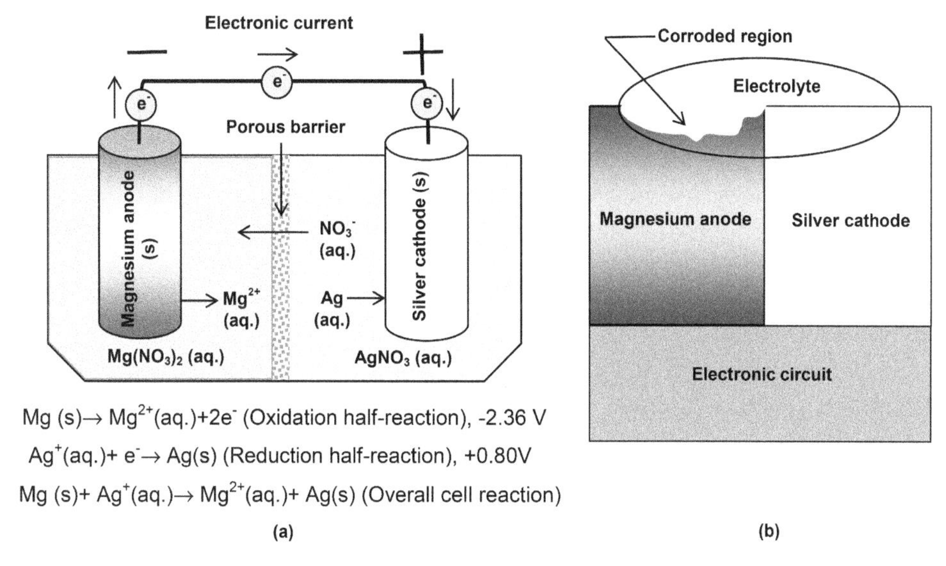

Figure 15.6. (a) Mg–Ag galvanic cell and (b) Mg–Ag corrosion cell.

15.5.6 Stray current corrosion

Exposing conductors to high electric and magnetic fields can cause high-magnitude stray current flow. Microwaves can induce turbulent liquid flow on a wet aluminum surface. The eddy currents flowing in the liquid film can aggravate corrosion problems.

15.5.7 The pop-corning effect

Sometimes, if a plastic encapsulated IC is not moisture baked before reflow soldering, the absorbed moisture pockets trapped inside may vaporize and expand by heating during the soldering process, thereby spoiling the substrate, die and the wire bonds (Chen and Li 2011). These are caused by built-up internal stresses, which may rupture the package and produce delamination between the molding compound and the lead frame. This phenomenon is known as the pop-corning effect from its resemblance to pop-corn, which dilates as the pressure builds up within the kernel accompanied by a small explosion. The pop-corning effect is easily avoided by moisture baking the plastic packaged microelectronic device or circuit before assembly and performing reflow within the stipulated time after moisture baking so that moisture has not gained re-entry into the IC.

15.6 Discussion and conclusions

High humidity leads to the formation of a liquid water film on the surfaces of electronic circuits. This water film has the same properties as bulk water in which electrolytic conduction takes place with ions as the charge carriers. Corrosion-causing ionic conduction on chip surfaces does not always need an externally applied potential difference. A class of corrosion reactions works on the electrochemical principle wherein the metal dissolves as positive ions and a depolarizer like oxygen from the atmosphere neutralizes the negative electronic charge on the surface. This class does not need that the circuit be powered. Another class of corrosion reactions is based on the galvanic cell principle. Such potential differences may be produced by galvanic cell formation between different metals on the chip. No external potentials are necessary. A third category is electrolytic ion migration, which happens under an applied voltage supply. Thus corrosion phenomena in electronics are driven by multiple factors.

Review exercises

15.1. Comment on the statement, 'the miniaturization of electronic components has increased the chance of corrosion-assisted failure'.
15.2. How does AH differ from RH?
15.3. Elucidate the interrelationship between ambient humidity, contamination and corrosion of electronics.
15.4. What metals are used in the following: (i) electronic chips, (ii) lead frames, (iii) bonding wires, (iv) electronic packages and (v) connecting lines on a PCB?
15.5. What metals are used in a computer hard disc?
15.6. Write down the chemical equations for the reactions taking place in an electrochemical corrosion process taking a bivalent metal as an example. Do these reactions require the existence of an externally applied voltage?
15.7. Explain the mechanism of PCB failure due to short-circuiting by anodic corrosion taking place by migration of Cu, Ag, Pb or Sn ions.
15.8. What is galvanic corrosion? Which parts of an IC are susceptible to galvanic corrosion and why?
15.9. How do the aluminum lines in an IC succumb to cathodic corrosion?
15.10. What is creep corrosion? Does it require the PCB to be connected to a power supply?
15.11. Describe the mechanism of creep corrosion. Why is it considered to be a three-step process? What are the preventive measures?
15.12. What is the pop-corning effect? With what type of IC packages does this effect take place? Why? How can it be avoided?

References

Chen C, Lee J C B, Chang G, Lin J, Hsieh C, Liao and Huang J 2012 The surface finish effect on the creep corrosion in PCB www.smtnet.com/library/files/upload/creep-corrosion.pdf

Chen Y and Li P 2011 The 'pop-corn effect' of plastic encapsulated microelectronic devices and the typical case study *Int. Conf. Quality, Reliability, Risk, Maintenance, and Safety Engineering* (*Xi'an, June*) pp 482–5

Lighting Global 2013 Protection from the elements. Part III: Corrosion of electronics *Technical Notes* Issue 14 September www.lightingglobal.org/wp-content/uploads/bsk-pdf-manager/63_issue14_part-iii_corrosion_technote_final.pdf

Savolainen P and Schueller R 2012 Creep corrosion of electronic assemblies in harsh environments *DfR Solutions* In: *IMAPS Nordic Ann. Conf. Proc.* pp 54–58 www.dfrsolutions.com/publications/creep-corrosion-of-electronic-assemblies-in-harsh-environments/

Tencer M and Moss J S 2002 Humidity management of outdoor electronic equipment: methods, pitfalls, and recommendations *IEEE Trans. Compon. Packag. Technol.* **25** 66–72

Chapter 16

Moisture and waterproof electronics

The exposure of electronic products to humidity and water from rain, sweat, washing, accidental dipping, etc, is a prominent cause of failure. Amongst the design approaches to protect from moisture are evolving a fault-tolerant chip layout, enclosing heat-generating parts of the circuit in an air circulating chamber with the cooler parts confined in a water-tight compartment, using the heat produced by the circuit to drive away moisture and by careful choice of materials for boundary surfaces to avoid galvanic corrosion. From the practical viewpoint, various coating materials such as parylene, superhydrophobic coatings, volatile corrosion inhibitors and silicones are used to thwart moisture attacks. This chapter presents both design and technological methods to achieve water imperviousness of electronics.

16.1 Introduction

Many terms are interchangeably used with the word 'waterproof', e.g. watertight, water repellant, submersible, sealed, etc. Corrosion prevention is achievable at the design stage of a product by adroit and innovative design artifices. Depending on the intended application, the obvious solutions to waterproof electronics after chip fabrication include either encapsulation of the chip inside hermetic packages, using moisture-absorbent dessicants, covering the relevant surfaces with noble metals or protective films, or using volatile corrosion inhibitor coatings. This chapter will present a step-by-step description of both strategies.

16.2 Corrosion prevention by design

16.2.1 The fault-tolerant design approach

If the magnitudes of the current–voltage signals, leakage currents and impedances are fixed by design at values just sufficient for correct operation of the circuit, the design will be intolerant to any deviations. It will be easily offset by the smallest changes caused by moisture and therefore will be highly corrosion-sensitive. An

doi:10.1088/978-0-7503-1155-7ch16

expert designer will include the likely changes in series resistances of joints, whether they be present in switch contacts, relays, or other electromechanical connectors or at soldering points, in circuit boards and displays, that are expected to take place in the circuit in a humid environment. He/she shall do so by properly calculating the numerical tolerances using information from corrosion specialists. Corrosion risk is effectively reduced by minimization of the number of electromechanical connections. Notably, the connections consisting of male-ended plugs/pins and female-ended receptacles should be minimally used.

16.2.2 Air–gas contact minimization

Contact of polluted air containing dust and other undesirable particles with the circuit should be avoided. A simple mechanical cover helps in many instances. Wiring can be configured in a two-layer structure wherein the dense wiring containing small components is isolated from external airflow whereas the wiring containing the power devices and heat-generating components receives strong airflow for efficient cooling. Thus a major part of the circuit is confined in an inert mass of air in which moisture cannot enter.

16.2.3 The tight dry encasing design

Electronics should always be maintained in a dry condition. Sometimes, the heat produced by the circuit itself may be used for this purpose. It aids in keeping the casing dry. Obviously, a circuit warmer than its surroundings is easy to keep dry. Moisture is swept away by the cooling air circulating over the circuit. If water condenses on the outer cover and drips into circuit boards, the circuit is at great risk. The designer should ensure that such dripping never happens.

16.2.4 A judicious choice of materials for boundary surfaces

In electronics, various boundary surfaces or interfaces exist in the microcircuits, switches, connectors, wiring and PCBs. In all probability, galvanic corrosion takes place at a spot where two metals are in contact if the atmospheric humidity is high. A high temperature further worsens the situation. Corrosion is minimum when two surfaces of the same metal come in contact. It is maximum if the contact metal surfaces are made from metals with widely different electric surface potentials, as seen from a glance at the electrochemical and galvanic series of metals. The use of different metals cannot be avoided in electronics, but the designer can exercise care in metal pair selection to choose metals which are the least likely to create trouble.

16.3 Parylene coatings

16.3.1 Parylene and its advantages

Parylene is a generic name. It stands for poly-para-xylylenes (Parylene Coatings and Applications: PCI 2017). Parylene coatings offer several advantages. The first is hydrophobicity. Next comes its chemical resistance to acids, caustic solutions and gases. The coating is pinhole free at a thickness >0.2 μm, serving as thin, smooth, chemically inert layer. Besides providing biostability and biocompatibility, it also gives thermal stability from −200 °C to +125 °C. Resistant to fungi and bacteria, it is sterilized by ethylene oxide (ETO) and gamma rays. It can be used for coating implantable medical electronics devices. It has high electrical impedance, providing excellent electrical isolation. It has a low dielectric constant. Last but not least, it can be deposited at room temperature in a vacuum system. It conforms to any surface clinging to it molecule by molecule with good adhesion. Bridging or pooling issues with liquid conformal coatings are avoided. Avoidance of a curing step helps to prevent curing stresses. No catalysts, solvents or plasticizers are required.

16.3.2 Types of parylene

There are several kinds of parylene, which differ in their moisture permeability, dielectric strength, biocompatibility and temperature-withstanding capability. The choice of parylene is decided by the requirements of the application at hand. Parylene N or (natural) exhibits a high dielectric strength with a low loss tangent and a frequency-invariant dielectric constant. It has a linear structure. Each molecule is formed by the combination of carbon and hydrogen. Its penetrating power is very high. Substituting one aromatic hydrogen atom with a chlorine atom in parylene N produces parylene C. It is characterized by low permeability to water vapor and corrosive gases. It provides a pinhole free conformal insulating coating widely used for critical electronic devices. Parylene D is obtained by substituting two aromatic carbon atoms by two chlorine atoms in parylene N. It can withstand higher temperature than parylene C up to 125 °C. But it is not biocompatible. In parylene HT, a fluorine atom replaces the primary hydrogen atom of parylene N, extending the temperature limit up to 350 °C and making it UV-stable. Due to the small size of its molecule, it can reach every nook of the object to be coated.

16.3.3 The vapor deposition polymerization process for parylene coatings

A process resembling vacuum metallization is used to form parylene coatings (figure 16.1). But in contrast with vacuum metallization which is performed at a vacuum $\leqslant 10^{-5}$ Torr, parylene deposition is carried out at a vacuum level ~0.1 Torr, giving a mean free path ~0.1 cm at which all sides of the object are uniformly covered to produce a conformal coating (Parylene Engineering 2016). The process consists of three main steps:

(i) Vaporization of the solid dimer di-para-xylylene at 150 °C, 1 Torr.
(ii) Pyrolysis of the dimer at 680 °C, 0.5 Torr, giving a stable para-xylylene monomer by cleavage at the two methylene–methylene bonds.
(iii) Parylene deposition at 25 °C, 0.1 Torr, by adsorption and polymerization on the object. The deposition chamber is followed by a cold trap at −70 °C and a mechanical vacuum pump.

Two parameters determine the parylene thickness: (i) the quantity of dimer vaporized and (ii) the dwelling time in the deposition chamber. The achievable thickness accuracy is within ±5% of the targeted value. The parylene deposition rate

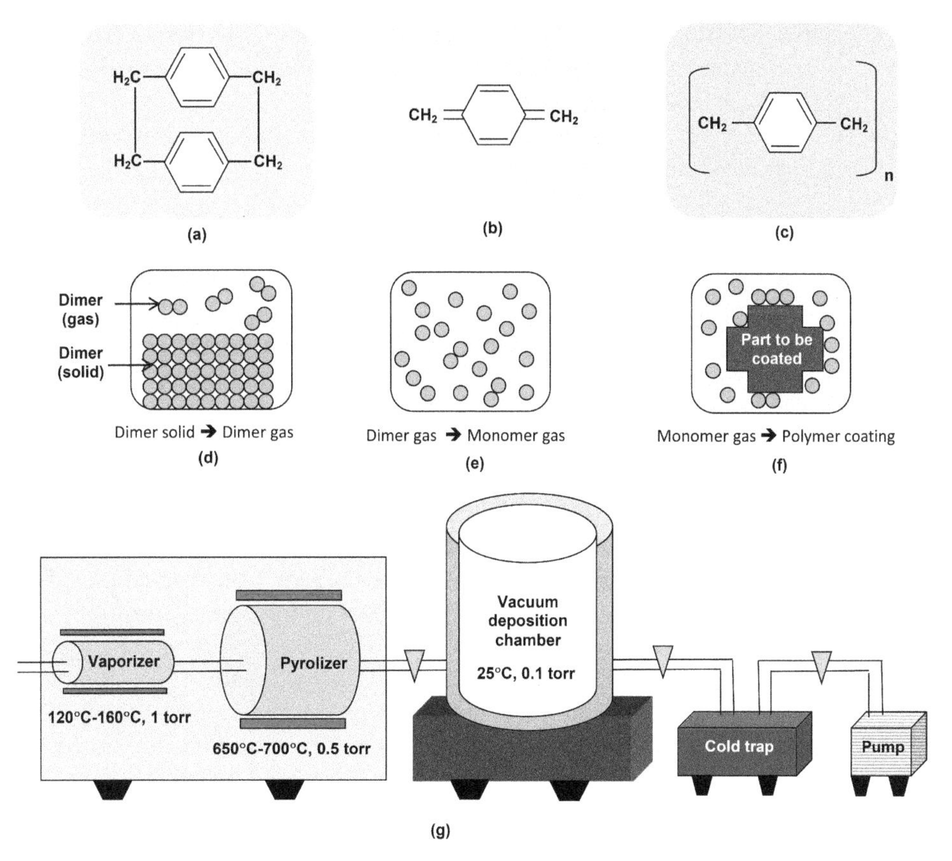

Figure 16.1. Vapor deposition polymerization: (a) dimer, [2, 2]paracyclophane, (b) monomer, paraxylylene, (c) polymer, poly[paraxylylene], (d) solid dimer to gaseous dimer conversion, (e) monomer gas, (f) parylene deposition on the workpiece and (g) the device.

is 0.2 ml h^{-1}. The deposition rate is directly proportional to the square of the monomer concentration. It varies inversely with the absolute temperature.

16.3.4 Typical electrical properties

For parylene N, the dielectric strength is 7 000 V mil^{-1}, volume resistivity is 1.4×10^{17} Ωcm, the dielectric constant at 60 Hz is 2.65 and the dissipation factor at the same frequency is 0.0002. Corresponding values for parylene C are 5 600 V mil^{-1}, 8.8×10^{16} Ωcm, 3.15 and 0.020.

16.3.5 Applications for corrosion prevention

Applications are as follows:

(i) As a protective coating for PCBs: 0.001″ thick parylene C intrudes and covers a 0.002″ space between the wire and the board.

(ii) Parylene thicknesses up to 0.4 μm are free from pinholes and can coat bonded wires and the conductor metal lines at the same thickness as the semiconductor surface. On the performance level, parylene coated ICs with transfer molded epoxide encapsulation are equivalent to those with hermetically sealed packaging.

(iii) As an electrically insulating coating for biomedical devices, protecting against the biological environment.

16.4 Superhydrophobic coatings

16.4.1 Concept of superhydrophobicity

A superhydrophobic or ultrahydrophobic coating is a nanoscopic coating, which forms a surface on the electronic component, that can be wetted with great difficulty. Examples include zinc oxide polystyrene nanocomposite or manganese dioxide polystyrene nanocomposite. Silica-based gels are also used.

An idea about the wettability of a solid surface by a liquid is obtained from the contact angle created by the liquid with the solid surface (Yuan and Lee 2013). Contact angle θ is defined as the angle subtended by the liquid–vapor interface with the liquid–solid interface, obtained by drawing a tangent from the contact point along the liquid–vapor interface in the profile of the liquid droplet (figure 16.2). The contact angle is an inverse measure of wettability. The smaller the contact angle, the larger is the tendency of the liquid to wet the solid. If $\theta = 0°$, the liquid completely spreads on the surface of the solid like a flat puddle at a rate limited by the viscosity of the liquid and the roughness of the solid. If $\theta < 90°$, the liquid is said to be a wetting liquid, the wetting is favored and the liquid will spread over a large area. The surface is called a hydrophilic surface. But when $\theta > 90°$, the liquid is non-wetting, the wetting is disfavored and the liquid tries to minimize its contact with the solid. It forms a compact liquid drop. The surface is termed hydrophobic. For contact angles $150° < \theta < 180°$, there is almost no contact between the liquid and the solid. The liquid forms a sphere or bead with minimal surface area and rolls off the surface without any adsorption. Such a surface is known as a superhydrophobic surface. The leaves of a lotus plant exhibit such an extreme degree of repellence of water. Hence, this effect is called the lotus effect. A superhydrophobic coating can create a moisture barrier far more effective than normal coatings.

16.4.2 Standard deposition techniques versus plasma processes

The standard coating techniques are spray coating, screen or spread coating, and dip or immersion coating in liquid. These techniques provide poor adhesion to substrates, involve the use of a large quantity of coating material, and often need a high-temperature curing step to improve the quality of layers. The thickness of liquid coatings depends on the viscosity and working temperature/humidity, and are controllable with an accuracy of ±50% of the desired value. High-quality conformable coatings can be formed on objects of different shapes by plasma processes. Further, these coatings can be of nanoscale thickness so that they are virtually transparent and invisible. The plasma process is performed at room temperature. It is an environmentally benign process. In this process, ions, excited molecules

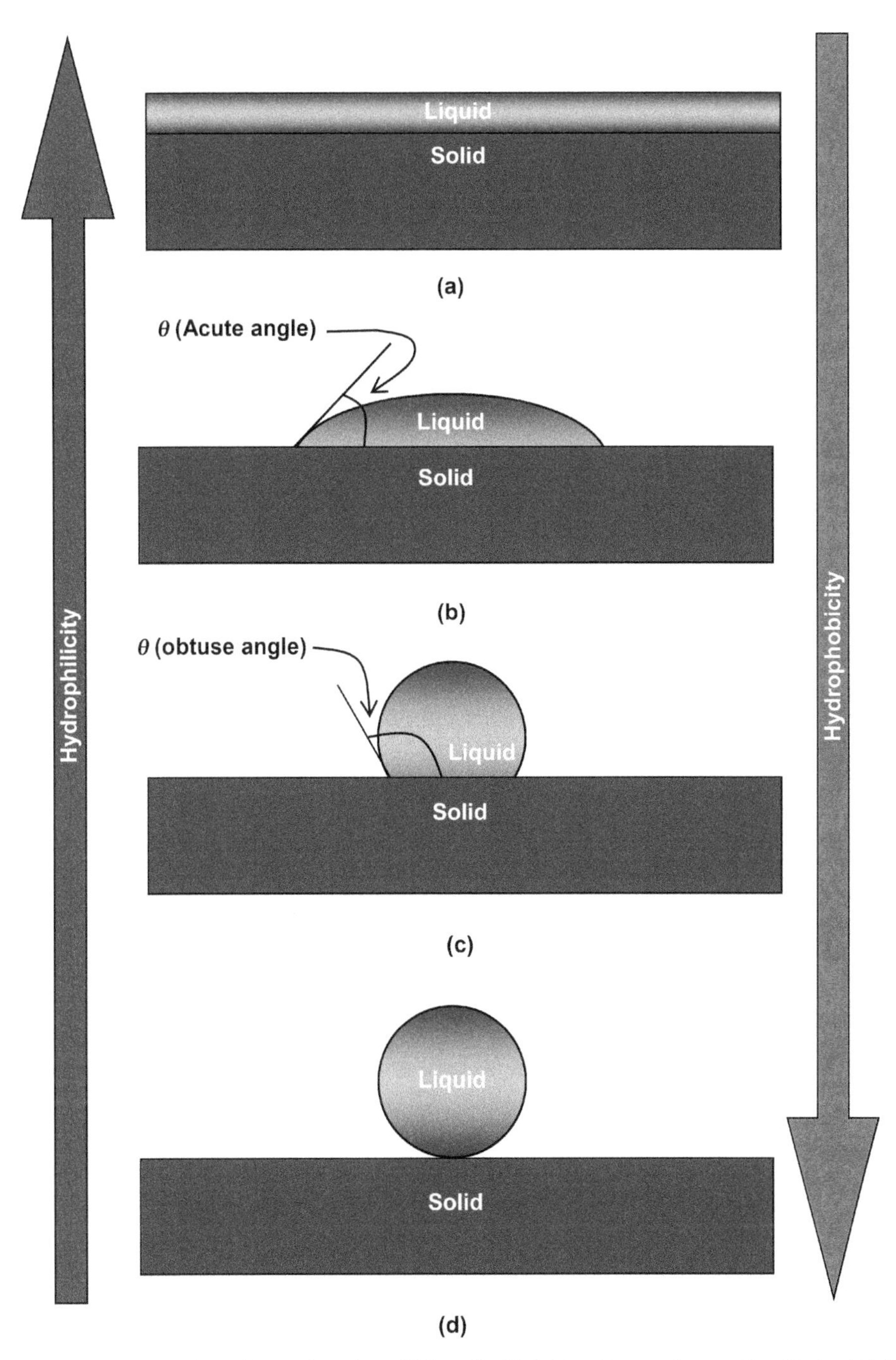

Figure 16.2. Contact angles θ for wetting of a solid by a liquid: (a) $\theta = 0°$, complete wetting; (b) $\theta < 90°$, high wetting and spreading; (c) $\theta > 90°$, low wetting and repelling; and (d) $\theta = 180°$, no wetting.

and radicals produced by a RF electric field bombard the object to be coated, modifying its properties.

16.4.3 The main technologies

Three companies are intensively engrossed in research on nanodeposition processes, namely P2i, GVD and Semblant (Tulkoff and Hillman 2013). The key technologies are:

- P2i: pulsed plasma and halogenated polymer, specifically fluorocarbon.
- GVD: initiated CVD using PTFE and silicone.
- Semblant: plasma deposition and halogenated hydrocarbon.

Hydrophobicity is related to the number and length of fluorocarbon groups, and their concentration on the surface. Whereas traditional plasma fragments the monomer, in the pulsed plasma approach, the monomer structure is retained. Retention of this structure provides a high liquid repellency. The process is applied industrially using roll-to-roll plasma processing equipment (Rimmer 2014).

16.4.4 Applications

The main application is the reliability enhancement of electronic gadgetry, such as tablets and smartphones as well as headphones, headsets and hearing aids (P2i 2015).

16.5 Volatile corrosion inhibitor coatings

VCIs are organic compounds (figure 16.3) such as dicyclohexylammonium nitrite (DICHAN), $C_{12}H_{24}N_2O_2$; tolytriazole ($C_7H_7N_3$), etc. They come in solid form for easy handling (VCI; VCI2000 2016). They have sufficient vapor pressure ~10^{-3}–10^{-5} mm Hg at 21 °C, favoring easy vaporization (Vimala *et al* 2009). In an enclosed space, their vapors disseminate until equilibrium is attained, as determined by their partial pressure. Vaporization is followed by condensation and adsorption of vapors on a metal surface. Adsorbed VCI is in the form of microcrystals. Even very small quantities of water vapor cause dissolution of these crystals. As a result, protective ions are formed. These ions are also adsorbed on the surface of the metal. Consequently, an ultra-thin monomolecular passivating film is produced, covering the metal surface and breaking down the contact between the metal and electrolyte. This film prevents electron transport between anodic and cathodic areas on the metal, retarding the corrosion process. Thus the VCIs chemically condition the environment near the metal surface to make it less corrosive to the metal. It must be emphasized that the shielding layer formed by VCIs produces no interference with the resistivity of the underlying metal layer, which is crucial for an electronics application, and is in many respects superior to the indirect method of passivation by a coating. The VCI films are self-healing. In the case that the film is scratched from a portion of the metal, the VCI film from the adjoining region evaporates and deposits on the scratched portion (Prenosil 2001).

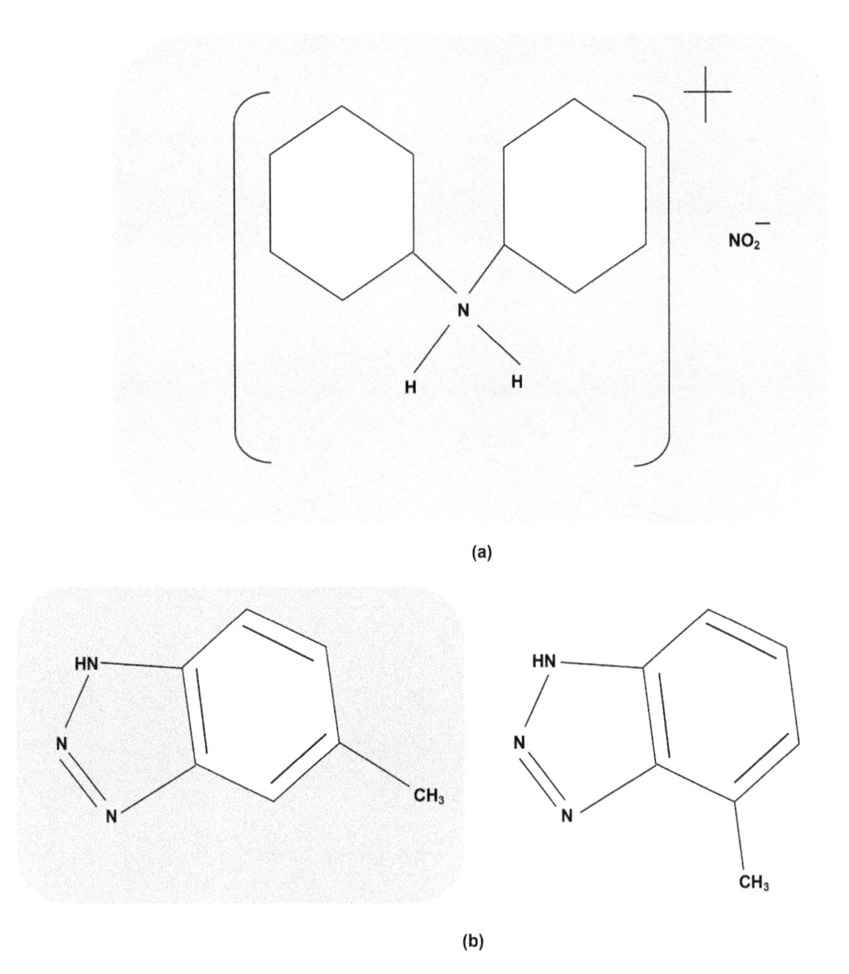

Figure 16.3. (a) DICHAN and (b) tolyltriazole.

16.6 Silicones

The silicones, e.g. silicone rubber (figure 16.4), silicone resin, silicone oil, silicone grease, etc, are synthetic polymers containing long chains of silicon and oxygen atoms alternately, such as $(\text{–Si–O–Si–O–})_n$. Organic groups, e.g. the methyl group $-CH_3$, attached to silicon atoms determine their properties. Their structural unit is R_2SiO where R is an organic group. Silicones contain carbon, hydrogen, oxygen and occasionally other elements as well. Silicones are known for their water repellence, chemical inertness, thermal stability (−100 to 250 °C), resistance to degradation by ozone and UV, and fluidic, resinous, lubricating, rubbery, electrical insulating, adhesive, stress-relieving, and shock and vibration absorbing properties. Silicones are prepared by decomposing halides of organic silicon compounds.

Silicone adhesives are of three types (DOW CORNING ©2000–2006 Dow Corning Corporation): (i) moisture cure for room-temperature processing;

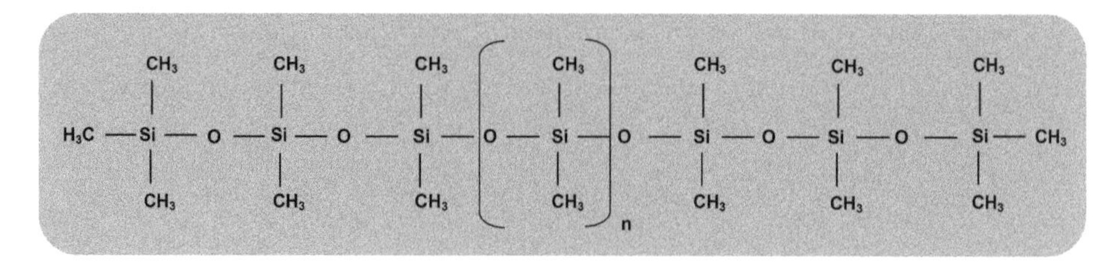

Figure 16.4. Silicone rubber.

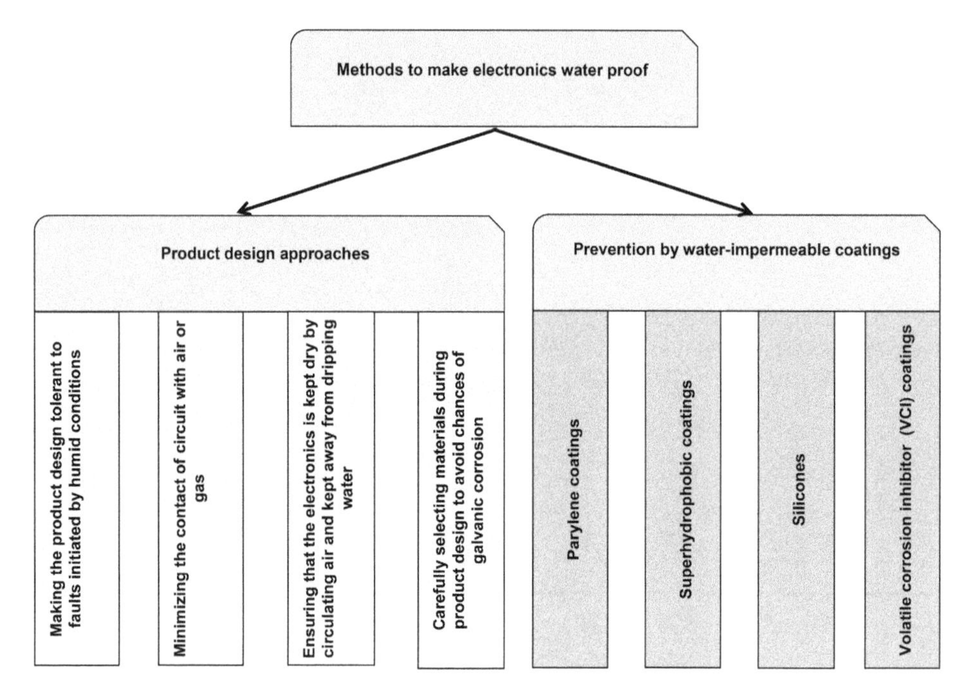

Figure 16.5. Schemes for moisture-free electronics.

(ii) condensation cure offering room-temperature and deep-section curing; and (iii) heat cure for fast processing. All the three varieties provide primerless adhesion with ceramics, reactive metals and plastics.

The one-part moisture-cure room temperature vulcanization (RTV) silicone is cured at room temperature in 30%–80% RH. With 24–72 h curing, >90% of physical properties are obtained. For condensation cure silicone in the two-component format, curing occurs at room temperature, resulting in good strength within an hour, but full properties are only attained after several days. The heat cure silicon is cured at ⩾100 °C. Thorough cleaning and/or degreasing of the surface is mandatory before silicone application.

RTV silicone is used for sealing electronic equipment and modules. It is used for fixing parts on PCBs of power supplies and cathode ray tubes (CRTs). Liquid crystal display (LCD)/LED modules are also assembled without affecting their optical properties. Condensation cure silicone is mainly used for gasketing and housing seals. Heat cure silicone is used for gasketing electronic modules. It is also used for sealing capacitors and electronic components; flyback transformers too are fixed with this silicone.

16.7 Discussion and conclusions

The necessary effort to make an electronic product water-impermeable commence right at the outset from its inception, and span from product designing up to the complete fabrication and assembly level. Security against the harmful effects of dampness, humidity and water spillage can be provided by perfectly covering the product with water-repellant nanotechnological films, such as those possessing superhydrophobic properties. In less critical conditions, various other alternative coatings are applicable. Figure 16.5 summarizes the strategies to be adopted for making electronics safe in damp and wet conditions.

Review exercises

16.1. What is meant by a fault-tolerant electronic design? Explain with examples.

16.2. How do the precautions taken regarding the choice of materials for boundary surfaces in an electronic product influence the chances of galvanic corrosion and impact its reliability?

16.3. What does parylene stand for? State five advantages of parylene as a coating material for protecting electronics from the harmful effects of moisture.

16.4. What are the salient features of the four types of parylene: parylene N, parylene C, parylene D and parylene HT?

16.5. How does the vapor deposition polymerization process for coating parylene differ from vacuum metallization? What are the three main steps in this process? How is parylene thickness controlled?

16.6. Which of the two parylenes, parylene N or parylene C, has a higher dielectric constant? Which one has the lower loss tangent?

16.7. Describe two corrosion prevention applications of parylene.

16.8. Define the contact angle of a liquid with a solid surface. How is it related to the wettability of the solid surface by the liquid?

16.9. Explain the following statements: (a) the contact angle of a liquid with a solid is $<90°$, (b) the contact angle is $>90°$ and (c) the contact angle is $>150°$ but $<180°$.

16.10. How does a hydrophobic surface differ from a hydrophilic surface in terms of: (a) contact angle and (b) wetting properties?

16.11. When is a surface said to be superhydrophobic? What is the lotus effect?

16.12. In what respects are the plasma processes for deposition of conformable coatings superior to conventional coating processes such as spraying and dip coating?

16.13. Describe three industrial approaches for the nanodeposition of super-hydrophobic coatings, indicating the materials and process technologies used.
16.14. Give two examples of VCI coatings. What is the mechanism of corrosion prevention by these coatings?
16.15. Do VCI coatings alter the resistivity of the underlying metal? How does the self-healing property of a VCI help in repairing any abrasion of the coating from any portion?
16.16. What is a silicone? What is the basic structural unit in a silicone? What are the special properties of silicones which make them attractive as a coating material for electronic use?
16.17. How do the three types silicones differ? What are the typical applications of the three types?

References

DOW CORNING ©2000–2006 Dow Corning Corporation Information about Dow Corning® brand adhesives/sealants—silicones and electronics *Product Information* http://www.ellsworthadhesives.co.uk/media/wysiwyg/files/dowcorning/DC-adhesives-sealants-info.pdf

P2i 2015 Electronics specialists: P2i's focus is on the consumer electronics industry, www.p2i.com/electronics/

Parylene Coatings and Applications: PCI 2017 http://www.pcimag.com/articles/87074-parylene-coatings-and-applications

Parylene Engineering 2016 Deposition process: parylene *Parylene Engineering* www.paryleneengineering.com/parylene_deposition_process.htm

Prenosil M 2001 Volatile corrosion inhibitor coatings *CORTEC Corp. ORP Supplement to Materials Performance* January pp 14–17

Rimmer N 2014 Vacuum plasma deposition of water and oil repellent nano-coatings *P2i* www.aimcal.org/uploads/4/6/6/9/46695933/rimmer_presentation.pdf AIMCAL Web Coating & Handling Conference 2014 Europe, 8 pages

Tulkoff C and Hillman C 2013 Understanding nanocoating technology *Electronic System Technologies Conference and Exhibition (May)* www.dfrsolutions.com/wp-content/uploads/2013/07/ESTC-Hillman-Understanding-New-Coating-Tech.pdf

VCI2000 2016 Anticorrosive VCI Packaging *VCI* http://vci2000.com/technology/

Vimala J S, Natesan M and Rajendran S 2009 Corrosion and protection of electronic components in different environmental conditions—an overview *Open Corros. J.* **2** 105–13

Yuan Y and Lee T R 2013 Contact angle and wetting properties *Surface Science Techniques* ed G Bracco and B Holst (Berlin: Springer) pp 1–34

IOP Publishing

Extreme-Temperature and Harsh-Environment Electronics
Physics, technology and applications
Vinod Kumar Khanna

Chapter 17

Preventing chemical corrosion in electronics

The corrosion of electronic circuits has two principal sources: the leftover corrosive materials in the fabrication of microelectronic circuits and PCBs, and the interaction of corrosive gases in the atmosphere with the surfaces of metal layers of the circuits. Therefore, corrosion-prevention must follow a two-pronged strategy, by identifying the possible reasons from both aspects of the problem and planning corrosion inhibition measures methodically. In this chapter, both the processing and environment-based likely causes of corrosion are indicated and remedial measures are described. Common corrosion prevention techniques, such as the application of conformal coatings, potting, plastic molding, hermetic packaging by ceramic and metal packages as well as by glass passivation, are elaborated.

17.1 Introduction

Microelectronic chip fabrication and packaging employs several metals and alloys, such as aluminum, copper, silver, tin, nickel and steel, to name a few (Salas *et al* 2012). They are used in conductive metal patterns, die attach and solder pastes, support frames, heat sinks, etc. All these metals/alloys have to bear the vagaries of atmospheric pollutants and climatic conditions. The escalation of cost-effective plastic packages replacing hermetically sealed types has provided easier access for corrosive gases to inner components because they can penetrate through most polymers. The main gases implicated in the corrosion of electronics are hydrogen sulfide, carbonyl sulfide, sulfur dioxide, nitrogen dioxide and chlorides.

17.2 Sulfidic and oxidation corrosion from environmental gases

The corrosion products formed on copper and silver surfaces are copper sulfides (Cu_2S and CuS) and silver sulfide (Ag_2S). The common reactions are:

$$2Cu + H_2S + \tfrac{1}{2}O_2 \rightarrow Cu_2S + H_2O \tag{17.1}$$

$$Cu + H_2S + \tfrac{1}{2}O_2 \rightarrow CuS + H_2O \tag{17.2}$$

$$2Ag + H_2S + \tfrac{1}{2}O_2 \rightarrow Ag_2S + H_2O. \tag{17.3}$$

In the presence of oxygen of the atmosphere, H_2S forms sulfuric acid, a strong corrosive agent:

$$H_2S + 2O_2 \rightarrow H_2SO_4. \tag{17.4}$$

The tarnish film on silver becomes darker with increasing thickness. The morphology of the grown film is non-uniform due to the formation of fern-shaped dendrites or filamentary whiskers.

Silver spontaneously oxidizes in air through the reaction:

$$4Ag + O_2 \rightarrow 2Ag_2O. \tag{17.5}$$

In an oxygen environment, copper oxidizes as follows:

$$4Cu + O_2 \rightarrow 2Cu_2O; 2Cu_2O + O_2 \rightarrow 4CuO. \tag{17.6}$$

The gold film covering a copper or silver layer can wear out. If the environment contains sulfur or chlorine, the corrosion material formed at the tattered regions will increase the contact resistance.

Due to the reactivity of lead, the lead component of lead–tin solder is readily oxidized in the atmosphere. When the weight percentage of tin is below 2%, the unstable lead oxides undergo further reaction with chloride, sulfate and phosphate ions leading to more wear.

17.3 Electrolytic ion migration and galvanic coupling

Aluminum metallization can corrode by ionic migration in an electric field using atmospheric moisture and halide contamination. In an applied electric potential gradient in a humid environment, silver ions migrate to the negatively charged surface forming silver dendrites, which can grow to sufficient lengths to close the gap between contacts, creating a short-circuiting condition.

In magnetic and magneto-optical memory devices, corrosion occurs at the sites where the carbon coating is damaged, exposing the Co-based layer and Ni–P substrate. Under an applied voltage between carbon and the metallic substrate, a galvanic couple is formed, dissolving the magnetic material.

17.4 Internal corrosion of integrated and printed circuit board circuits

The metals used in device fabrication may interact mutually. One example is the Al–Cu metallization in which intermetallic compounds like Al_2Cu are formed along the grain boundaries of aluminum. The intermetallic compound Al_2Cu acts as a cathodic site with respect to aluminum. Consequently, aluminum is dissolved and micropits are formed during rinsing after chemical etching.

The liquid and vapor-phase halogenated solvents used in IC and PCB manufacturing react with aluminum-containing devices meddling with their performance. If diluted with an aromatic alcohol, the halogenated solvent breaks down forming chlorine ions, which react with Al.

17.5 Fretting corrosion

Fretting is a kind of wearing away which takes place by rubbing (figure 17.1). Fretting corrosion impairs the functioning of switches using non-noble metals as contact materials. Gold does not form an oxide layer under exposure to atmosphere. But it is very expensive. So, one cannot afford to make switches from gold. Instead, tin-plated copper is used for making switches. When these switches experience vibration such as due to floor vibration or thermal stresses, mechanical rupturing of oxide layers occurs. It breaks down the oxide layers at the contact points. Fresh oxide layers are therefore formed. These are also destroyed in subsequent rupturing.

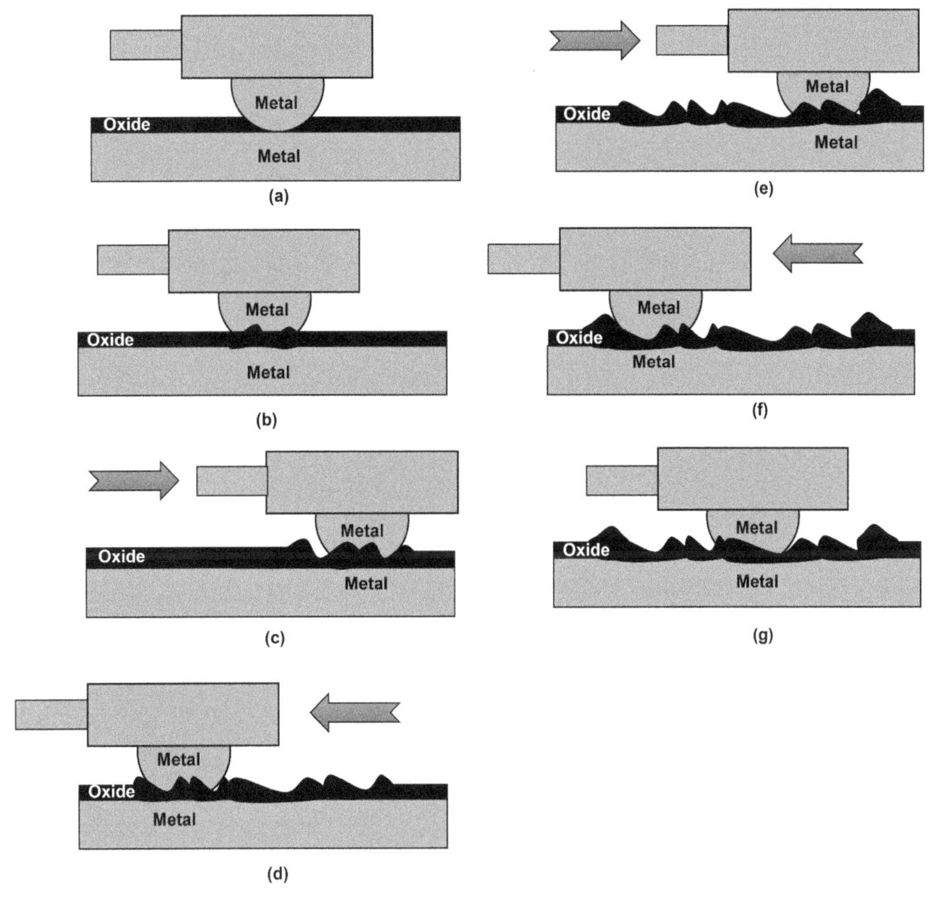

Figure 17.1. Fretting corrosion: (a) initial metal-to-metal contact, (b) interfacial native oxidation, (c) rightward micromotion, (d) leftward micromotion, (e) oxide debris pileup on the right, (f) oxide debris pileup on the left and (g) oxide film to oxide film contact after N cycles.

Thus oxide debris gathers at the contact points. This debris increases the contact resistance. Demolition of the oxide film is caused by the wiping action of plugging/ unplugging of the contact. Internal wire-to-wire contacts are not prone to this kind of threat. Notwithstanding, the internal wire-to-board contacts are confronted with a similar problem. They face the risk of fretting corrosion.

17.6 Tin whisker growth

Adoption of lead-free processes has led to tin whisker issues, which are not forms of corrosion but behave similarly. Thin, hard whiskers grow on tin-plated leads of the electronic components in reaction to surface compression stress in the tin coating. These whiskers short-circuit adjoining leads.

17.7 Minimizing corrosion risks

17.7.1 Using non-corrosive chemicals in device application and assembly

Flux residues have been pinpointed as a reliability threat to printed circuit board assemblies (PCBAs; Woolley 2013). In the early days, rosin-activated fluxes were used. These fluxes contained chlorine or bromine. These reactive halides were used primarily as a cleaning agent to remove the copper oxide layer, which grows on copper conductors in air, and inhibits the formation of an intimate contact between the solder and the copper layer. The removal of the oxide layer was essential for establishing a secure solder connection.

Several post-soldering cleaning methods evolved. But if the PCBA surface is not properly cleaned after soldering, remnant chlorine or bromine traces left on the surface lead to ionic contamination, which is a dominant cause of corrosion. Needless to say, the compounds used for cleaning the flux residues should be free from halides and phosphates.

To overcome the problem of leftover halide traces, restriction of hazardous substances (RoHS)-compliant low-solid 'no-clean fluxes' were introduced. These fluxes were not chlorinated. In place of the halogen element, they contained organic acids such as adipic acid, $C_6H_{10}O_4$, and citric acid, $C_6H_8O_7$, for cleaning purposes. They did not require any cleaning after the soldering step because the organic acids underwent decomposition at the soldering temperature. Cleaning was a non-issue during reflow soldering. However, during wave soldering, the flux wedged between the wave soldering pallet and the PCB is not bared to the molten solder. This happens because the pallet serves as a thermal impediment between the solder and the wave. The lingering flux, being acidic and hygroscopic in nature, participates in corrosion.

As a precaution, to avoid contamination of future products, the wave solder pallets should be always cleaned to remove the built-up residues acquired during fluxing. This may be done using a fiber-free cloth with the same solvent, which was used to dilute the flux.

17.7.2 Device protection with conformal coatings

A conformal coating is a thin polymeric film preserving the angles and topological features of the surface of a PCB. Its thickness is typically 25–100 μm. Generally, it is electrically insulating. The coating serves as a barrier separating the electrical circuit from the harmful effects of the hostile atmospheric gases, inimical chemical vapors and moisture. It also protects the circuit from organic attack by microbial action. It is thus a valuable booster of longevity of the PCB under actual operating conditions.

Types of conformal coatings
Choice exists between several conformal coatings (HumiSeal 2016). Ideally, a coating should be easily applicable and easily removal for repair work, allowing a simple process both for initial coating and for later repair of the PCB for replacement of any damaged components. But in practice, the coatings differ widely in terms of their removability.

(i) *Acrylic coatings.* These are plastic materials containing the acryloyl group obtained from acrylic acid. These coatings are very easily removed using specific stripping chemistries, either locally or completely. The PCB can then be repaired and its operation restored.
(ii) *Polyurethane coatings.* These are thermosetting/thermoplastic polymers containing organic units interlinked by carbamate or urethane –NH–(C=O)–O– groups. The higher chemical resistance of the silicone coatings than acrylic coatings makes their removal a little more difficult than the acrylics. Nonetheless, they are removable, permitting reworking of the PCB.
(iii) *Silicone coatings.* These are polymers constructed from recurring units of siloxane containing alternate S and O atoms, in combination with C, O and H atoms. These coatings are the most difficult to remove. By mechanical abrasion, they can be removed from small areas. So, at the most local repairing is possible.
(iv) *UV cured coatings.* These are the most chemically resistant and mechanically rugged of all coatings. Hence, repairing is extremely difficult. Removal by powder abrasion or by local melting in a small area can be attempted. Following repair, the board is re-coated.

Glob top
This is a variant of conformal coating (EPOXY Technology 2001). In appearance, it looks like a black spot on the circuit board. It is widely used in chip-on-board (COB) assembly as a protective dome over a fragile semiconductor die and associated wire bonding, thereby reinforcing it mechanically and making it tamper-proof, in addition to isolating it from the environmental effects and eliminating chances of static discharge.

Glob tops are of two types (figure 17.2): single part hemispherical and two-part dam-and-fill. In the former type of glob-tops, a thixotropic material is used to erect a protective dome over the electronic component and wiring. The thixotropic property

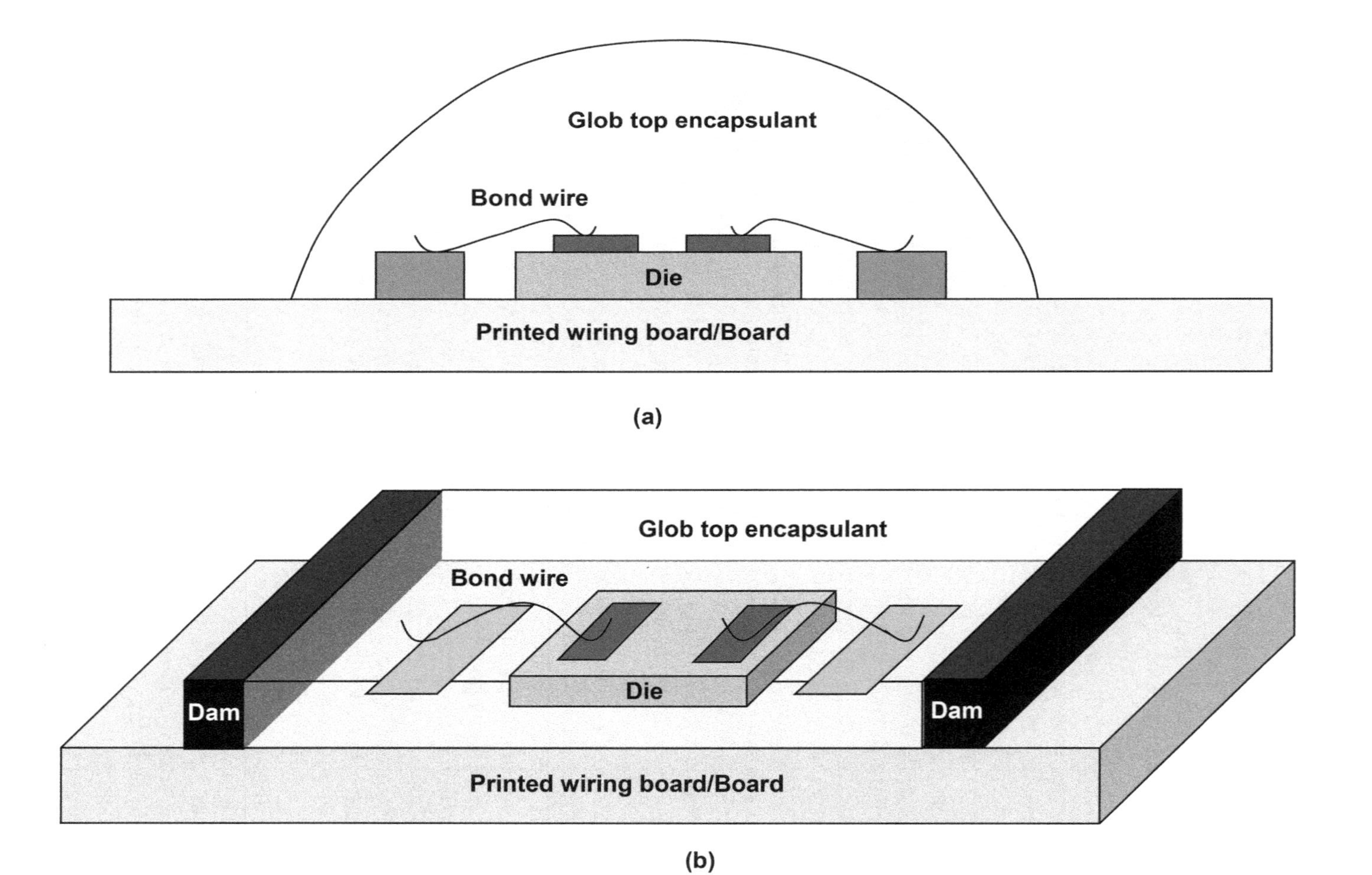

Figure 17.2. Types of glob tops: (a) hemispherical and (b) dam and fill.

is a property exhibited by certain materials of liquid-like flowing (thin and less viscous) upon stirring, agitation or shaking, and thick and viscous behavior under static conditions.

In the latter type of glob tops, a dam is formed surrounding the electronic component using a thixotropic epoxy. Subsequently, the cavity is filled up using a low-viscosity solution. Thus it is a two-step process involving dam building and cavity filling. Both types of glob tops were found to sustain operation at 85 °C and 85% RH for 2000 h without any degradation (EPOXY Technology 2001), promising acceptable reliability assurance.

Application methods and inspection

The coatings are applied by dip coating, spraying and by robotic methods at defined locations. Voids or bubbles left in the film thus formed serve as a path for the ingress of contaminants. By careful inspection, any defects in coating should be identified and immediately repaired. Neither any solder joints nor any sharp edges of components or their leads should be left bare. Otherwise the aggressive chemicals will intrude to the electronic devices and the whole purpose will be defeated.

17.8 Further protection methods

17.8.1 Potting or overmolding with a plastic

In applications demanding exceptional protection from environmental corrosion, the conformal coating is covered with a potting compound, which is a gelatinous substance for sealing and encapsulating the whole assembly (figure 17.3). The available potting compounds are one- or two-part formulations. The formulations are epoxies, silicones, polyurethanes and UV-curable substances. The liquid is poured over the circuit board and allowed to cure. It solidifies providing good mechanical strength to the electronic components. Immersing the assembly in a liquid resin is another way of potting. Taking it out and curing it completes the process. The electronic item is sometimes kept in a pre-fabricated potting shell. The shell is filled with the potting compound and the compound is cured. Then the item is taken out of the shell.

Void-free structures can be formed by a method called vacuum potting. To accomplish vacuum potting, the electronic item is placed in a vacuum chamber when the resin is in the liquid state, i.e. prior to its solidification. In this way, air is pumped out of the internal cavities. Then the vacuum is withdrawn and the item is exposed to atmospheric pressure. Under the influence of atmospheric pressure, the internal cavities collapse, forcing the liquid resin to fill any vacant spaces where it has not so far penetrated and thus leaving no voids. Vacuum potting is suitable for resins that polymerize, rather than those which work by the solvent evaporation mechanism.

The potting of the PCB can prevent it even from immersion in salt water. Resistance against shock, vibration, thermal cycling and fungal attack are additional benefits. In addition to solenoids and transformers, potting is extensively used in electronic and electrical components such as capacitors, rectifiers, sensors, switches, ignition coils, motors, etc.

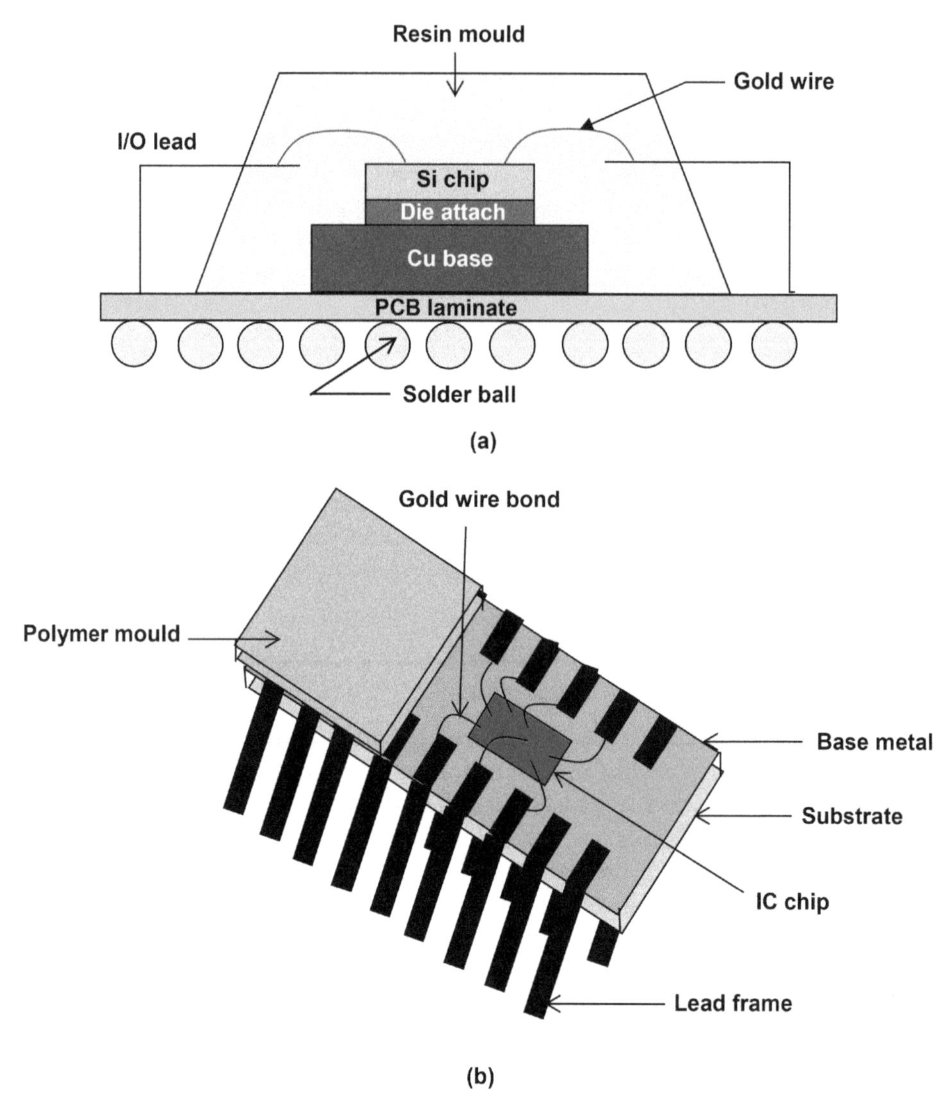

Figure 17.3. Plastic packages: (a) two leads and (b) multiple leads.

17.8.2 Porosity sealing or vacuum impregnation

In porosity sealing or vacuum impregnation (figure 17.4), the electronic item is submerged in a plastic resin followed by lowering the ambient pressure to vacuum level. On releasing the vacuum, i.e. raising the ambient pressure to atmospheric level, the liquid solution flows into the most inaccessible cavities. The electronic item is taken out, the excess resin is drained out and the item is cured. The polymeric resin hardens. In the case of a solvent-based resin, the solvent evaporates. The method is applied to solenoids, transformers and other high-voltage components to ensure good dielectric strength. It prevents the components from ionization-induced failures.

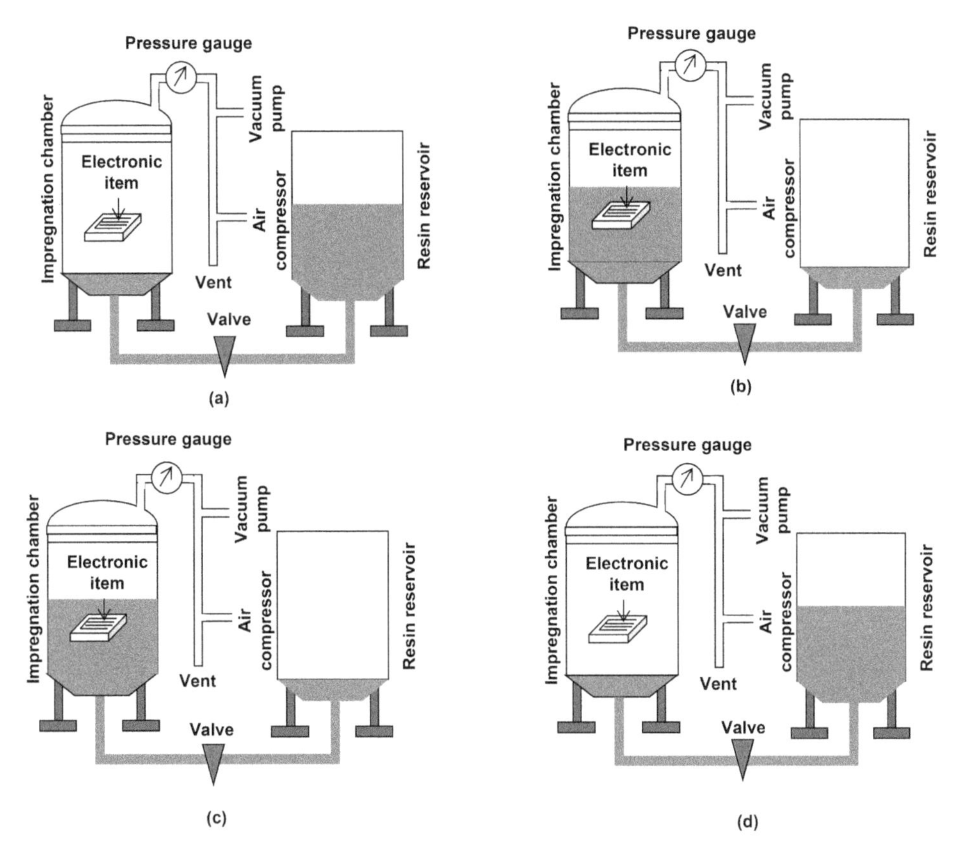

Figure 17.4. Steps of the vacuum impregnation process: (a) preheating the equipment, loading the electronic item into impregnation chamber, creating and maintaining a vacuum for 30 min; (b) flooding the impregnation chamber with resin; (c) releasing the vacuum and applying a positive pressure for 30 min and (d) releasing the pressure and discharging the resin back to the reservoir.

17.9 Hermetic packaging

'Hermetic sealing' refers to the quality of a container being airtight, completely prohibiting any entry or exit of air, oxygen, moisture and other gases. Perfect hermetic sealing is not possible because every material is permeable to gases, even though the permeability may be negligibly small for practical applications. A hermetic package can withstand higher operating temperatures than a plastic package, primarily due to the temperature-withstanding properties of the materials used in hermetic packaging. Different types of hermetic packages are described below.

17.9.1 Multilayer ceramic packages

Metallization, lamination and firing of ceramic tape layers constitute the three main steps to form the body of the package (Texas Instruments 1999), see figure 17.5.

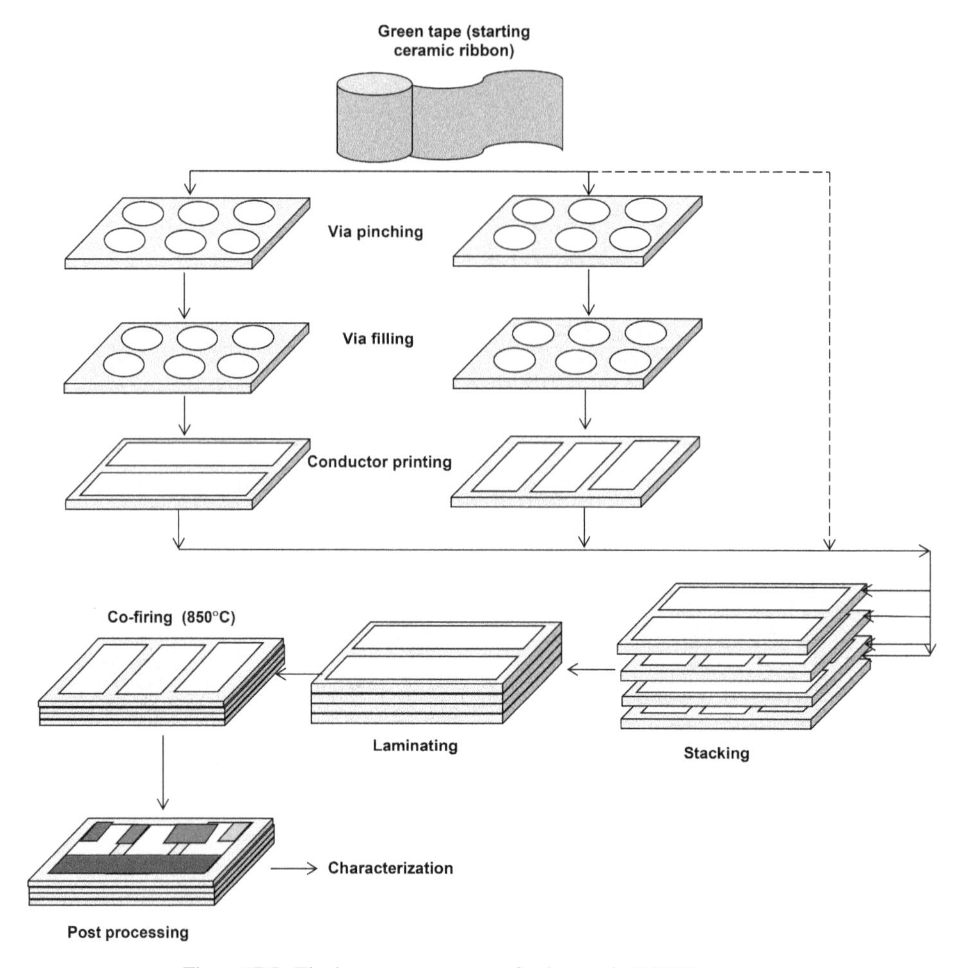

Figure 17.5. The low-temperature co-fired ceramic (LTCC) process.

The multilayer ceramic tape construction helps the package designer to introduce useful electrical features into the package. Notable features include the power and ground planes, shield planes and controlled characteristic impedance of signal lines. The power and ground planes decrease inductance while the shield planes minimize crosstalk.

Nickel–gold electroplating is done on the metallized regions of the package. By soldering a metal ring into the metallized and plated seal ring, hermetic sealing is accomplished. These packages are therefore designated solder-seal packages (figure 17.6).

Brazing is the process used for insertion of leads into the package body (figure 17.7).

17.9.2 Pressed ceramic packages

These packages consist of three parts: the base, lid and lead frame. By pressing ceramic powder into the required shape and firing, the base and the lid are made.

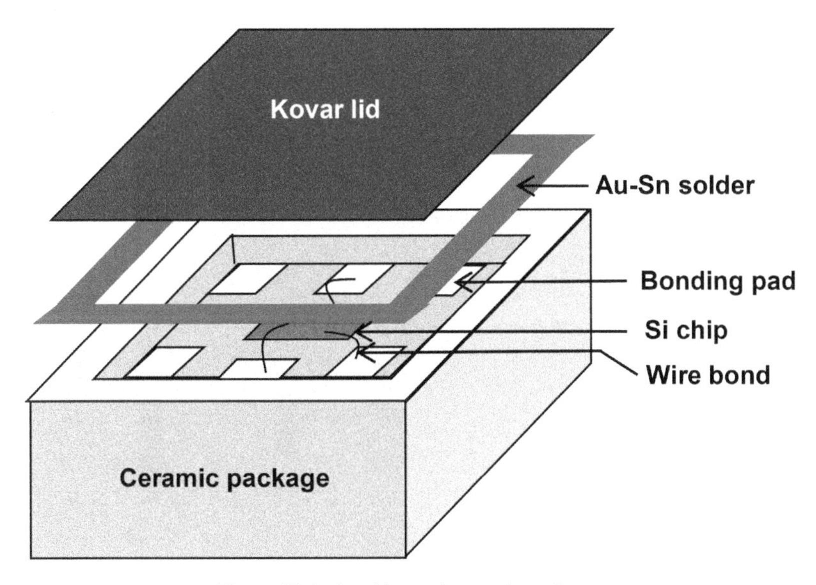

Figure 17.6. A solder-seal ceramic package.

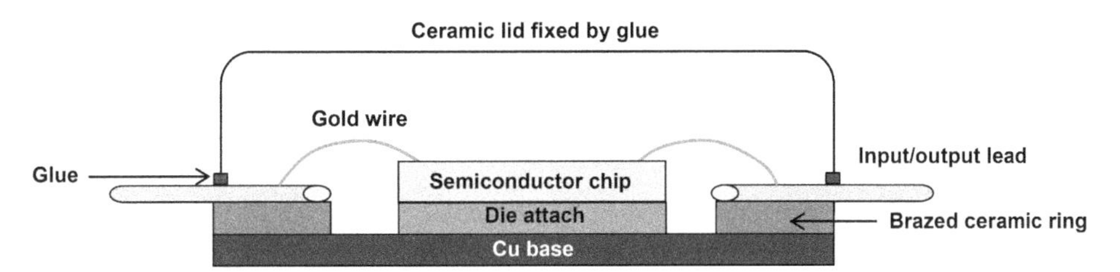

Figure 17.7. Insertion of leads into the package by brazing a ceramic ring with a copper base plate.

Upon the fired base and lid, glass paste is screened and fired. A separate lead frame is entrenched inside the base glass. The glass is melted over the base and the lead frame for hermetic sealing. This sealing method is known as glass frit sealing and therefore, the package is called a glass frit seal package.

From the cost consideration, pressed ceramic packages are less expensive than multilayer ceramic packages. However, their structural simplicity does not permit incorporation of additional electrical features.

17.9.3 Metal can packages

These packages have a metallic base (figure 17.8). The leads come out from a glass seal, which is either a compression seal or a matched seal. Resistance welding is used to attach a metal can or lid to the metallic base. Hermetic sealing is thereby achieved. The lead count in metal can packages is usually low, <24. Cost wise, they are preferable to other packages.

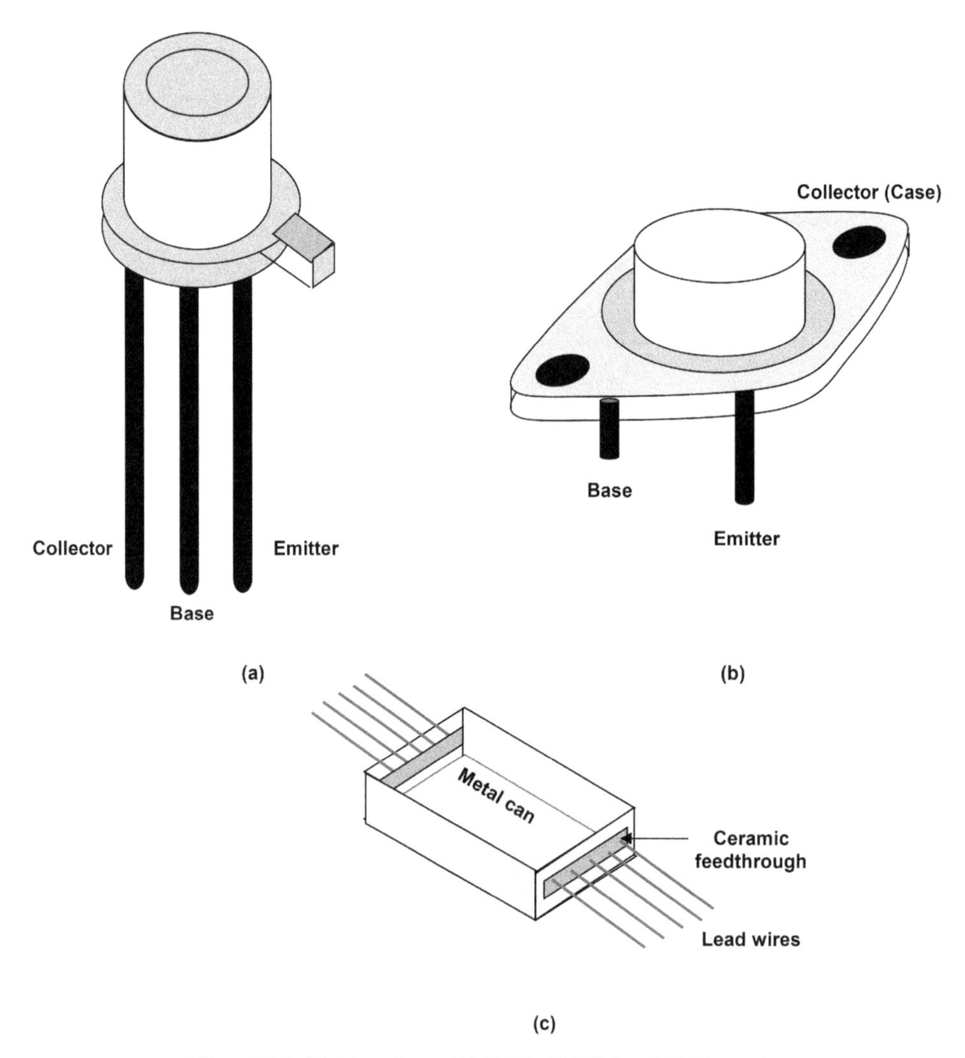

Figure 17.8. Metal packages: (a) TO-3, (b) TO-5 and (c) flat pack.

17.10 Hermetic glass passivation of discrete high-voltage diodes, transistors and thyristors

Low alkali content, particularly Na^+ ion glasses provide high insulation resistance on p–n junction surfaces (figure 17.9). Thermal stress-induced cracking is minimized by matching the CTE of the semiconductor material with that of the glass used. A slurry of the glass powder in deionized water—suspended ceramic particles along with binders, dispersants or plasticizers—is applied to the surface. A relative motion between a blade and the substrate spreads the slurry in a thin uniform layer. A gel-type hermetic layer is formed on drying. The method is called doctor blading or tape casting (Berni *et al* 2004).

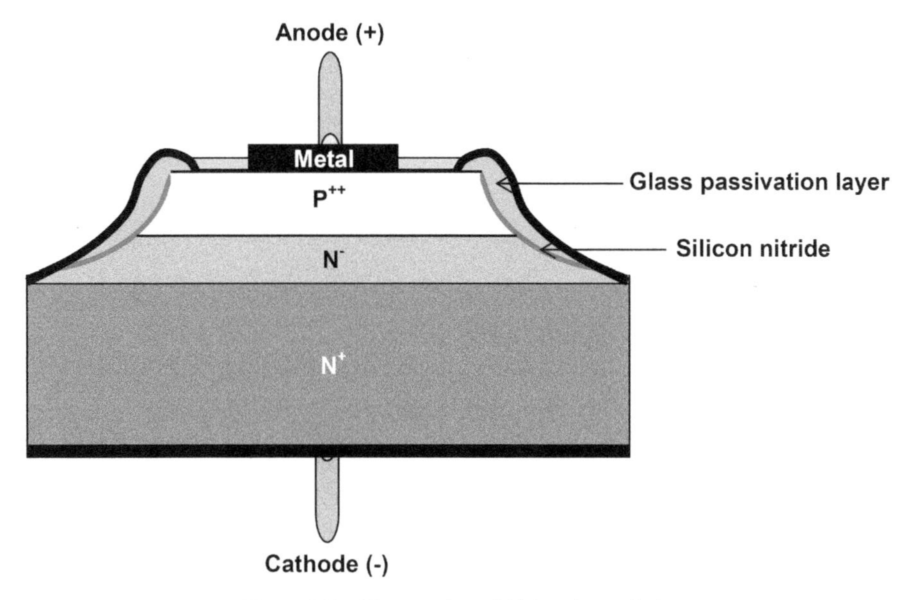

Figure 17.9. Glass-passivated high-voltage diode.

In another laser-based technique of forming a glass passivation layer, the part of the semiconductor substrate on which the layer is to be deposited is simultaneously irradiated with two laser beams: a laser beam which is absorbed in glass and another laser beam which is absorbed in the substrate (Iesaka and Yakushizi 1984).

A CMOS-compatible process for hermetic packaging using low-temperature BSG involves plasma-enhanced e-beam deposition of the glass layer at a temperature <100 °C followed by patterning using a photoresist lift-off technique (Leib *et al* 2009).

17.11 Discussion and conclusions

The package represents a barrier between the semiconductor chip/circuit/PCB and the environment, protecting the electronics from the detrimental effects of corrosive chemicals. Various options are available from the packaging side. One can select from a diversity of packages according to convenience. A wide range of coating materials is also available. These materials give more strength to packaging by providing additional protection. Taking care to reduce the remaining chemicals left on the chip and PCB surfaces is equally important to fabricating a reliable electronic product, because these residues act as corrosion-triggering centers during the operational life of the circuit.

Review exercises

17.1. Name a few metals used in the fabrication of microelectronic circuits. Write the names of three gases which are responsible for the corrosion of electronic circuits.

17.2. How does hydrogen sulfide react with copper and silver? Is the morphology of film growth on silver uniform?
17.3. How does corrosion of lead–tin solder take place in the atmosphere? What happens if the percentage of tin in lead–tin solder is less than 2% by weight?
17.4. How do silver ions cause short-circuiting problems in a humid climate? How are magnetic storage devices corroded when the carbon coating is abraded?
17.5. How do aluminum and copper interact mutually in Al–Cu metallization? How are micropits formed?
17.6. What is the role of chlorine in rosin fluxes? What happens if the residual flux left after soldering is not properly cleaned from the PCB surface?
17.7. What organic acids are used as cleaning agents in no-clean fluxes? Write their molecular formulae. Do they contain any halogens?
17.8. Why does the flux trapped between the wave soldering pallet and PCB trigger corrosion?
17.9. What is a conformal coating? How does it increase the reliability and lifespan of a PCB?
17.10. What are acrylics, polyurethane and silicone compounds? Discuss their relative merits/demerits in terms of the removability for PCB repair.
17.11. What is a glob top? Name the two types of glob top. How are they formed?
17.12. How do the voids or defects in conformal coatings lead to corrosion?
17.13. What is meant by potting an electronic component? How is this done?
17.14. How is vacuum potting done? How does it help in producing void-free structures?
17.15. What is vacuum impregnation? Name a few applications of vacuum impregnation.
17.16. What is meant by hermetic sealing? Can hermetic packages be used at higher temperatures than plastic packages?
17.17. How is the body of a multilayer ceramic package formed? What process is used for insertion of leads in the package? Why is this package called a solder-seal package?
17.18. What utility-enhancement electrical features can be introduced by multilayer ceramic tape construction of packages?
17.19. Why is the pressed ceramic package called a glass frit seal package? How does it compare with the multilayer ceramic package in terms of cost and provision of enhancement features?
17.20. How is hermetic sealing performed in a metal can package? How are the leads taken out?
17.21. Describe the doctor blading method of glass passivation?
17.22. Can glass passivation be performed in a CMOS-compatible process? How is this done?

References

Berni A, Mennig M and Schmidt H 2004 Doctor blade *Sol–Gel Technologies for Glass Producers and Users* ed M A Aegerter and M Mennig (Berlin: Springer) pp 89–92

EPOXY Technology 2001 Glob-top *Selected Application* www.epotek.com/site/files/brochures/pdfs/Glob_Top_SAS(1).pdf

HumiSeal 2016 What is conformal coating *Chase Electronic Coatings* www.humiseal.com/conformal-coating/

Iesaka S and Yakushizi S 1984 Method of manufacturing a glass passivation semiconductor device *US Patent* US4476154 A

Leib J, Gyenge O, Hansen U, Maus S, Fischer T, Zoschke K and Toepper M 2009 Thin hermetic passivation of semiconductors using low temperature borosilicate glass—benchmark of a new wafer-level packaging technology *59th Electronic Components and Technology Conf. (San Diego, CA 26–29 May* pp 886–91

Salas B V, Wiener M S, Badilla G L, Beltran M C, Zlatev R, Stoycheva M, de Dios Ocampo Diaz J, Osuna L V and Gaynor J T 2012 *H_2S Pollution and its Effect on Corrosion of Electronic Components* (Rijeka: InTech) chapter 13 pp 263–86

Texas Instruments 1999 Hermetic packages *Texas Instruments Literature Number* SNOA280 www.ti.com/lit/an/snoa280/snoa280.pdf

Woolley M 2013 Flux residues can cause corrosion on PCB assemblies *EDN Network* October, www.edn.com/electronics-blogs/all-aboard-/4423581/Flux-residues-can-cause-corrosion-on-PCB-assemblies

IOP Publishing

Extreme-Temperature and Harsh-Environment Electronics
Physics, technology and applications
Vinod Kumar Khanna

Chapter 18

Radiation effects on electronics

Radiation is part and parcel of our environment. It is present in the skies, on the Earth and within our bodies. It is found naturally, but it is also a man-made hazard. One important effect of radiation on semiconductors is the TID effect, which either causes ionization of lattice atoms or creates lattice defects by dislodgement of atoms from their regular lattice positions. EHP production by ionization impacts the performance of MOS devices through a shift in threshold voltage, transconductance degradation and leakage current enhancement. Lattice defect creation influences bipolar device operation via minority-carrier lifetime killing. Another significant class of radiation effects on semiconductor circuits is represented by SEEs, which are either temporary or permanent in nature. Temporary effects include single event upset (SEU) and single event transient (SET). The permanent effects are single event latchup (SEL), single event burnout (SEB) and single event snapback (SES).

18.1 Introduction

Radiation is energy that is emitted from a source and travels through space or a material medium, either in the form of particles (the particle form of radiation) or waves (the wave form of radiation). The particle form of radiation consists of charged or uncharged particles such as electrons, protons, alpha particles, neutrons, etc. The wave form of radiation consists of electromagnetic waves such as visible light, x-rays, gamma rays, and so on. The former form of radiation may be either uniform or non-uniform depending on the distribution of particles in the beam, so that while part of the target electronic item is affected, the rest may not be influenced at all. The latter form is uniform in its impact on the target.

Whether on Earth or in space, it is impossible to find a place without radiation. Radiation is present in every nook and corner of the Universe. Only its degree and energy vary. On Earth, we are protected to a large extent by the blanket of atmosphere enveloping our planet.

doi:10.1088/978-0-7503-1155-7ch18

18.2 Sources of radiation

18.2.1 Natural radiation sources

Human beings living on Earth and so also the electronic equipment they use can never escape from natural radiation exposure. A primary source of radiation is the cosmic radiation or cosmic rays coming from the Sun and stars (USNRC 2014a). Also important are the van Allen radiation belts around the Earth containing trapped charged particles. Another source is terrestrial radiation, from radioactive materials such as uranium, thorium and radium inside the Earth's crust; radioactive carbon and potassium from plants and animals; and radon and thoron gases inhaled from the atmosphere. The third source is internal radiation due to the carbon-14, potassium-40 and lead-210 isotopes present within our bodies.

Cosmic rays

Noticeably impacting the functionality of electronic devices, these rays (figure 18.1) comprise high-energy radiation propagating at approximately the velocity of light with some fraction having energies up to 10^{14}–10^{15} eV. Their intensity increases with altitude above the Earth indicating their origin from outer space, beyond the Solar System. It also varies with latitude, suggesting presence of charged particles which are affected by the Earth's magnetic field. Compositionally, the primary cosmic rays, accelerated by the energy of detonation waves of supernova vestiges and striking the Earth's atmosphere, are composed of 99% nuclei of elements ranging from the lightest to the heaviest, and 1% electrons similar to β-particles. Of the nuclei, 90% are hydrogen nuclei or protons, 9% are helium nuclei or α-particles and the remaining 1% are nuclei of heavy elements. Antimatter particles such as anti-protons constitute a very small percentage (NASA Goddard Flight Center 2014). Particles with energies in the range 10^8–2×10^{10} eV cause the most damage. As the atmosphere acts as an armor against cosmic rays, the electronic gadgetry of aircraft flying at high altitudes and space vehicles is most susceptible to the detrimental effects of cosmic ray exposure.

Van Allen radiation belts

These are two belts or layers (figure 18.2), discovered by James Van Allen, extending at an altitude of 1 000–6 000 km above the Earth's surface and containing collections of energetic particles, mainly electrons with energies up to 10^7 eV and protons having energies of hundreds of mega-electronvolts, from cosmic rays and solar wind, confined by the magnetic field of the Earth. Satellites spending a sizeable time in this region must take precautions against radiation (Fox 2014).

18.2.2 Man-made or artificial radiation sources

Major sources of artificial radiation are medical diagnostic x-rays such as radiography, mammography, computerized tomography and fluoroscopy; airport x-ray machines; nuclear medicine and radiation therapy using isotopes such as I-131,

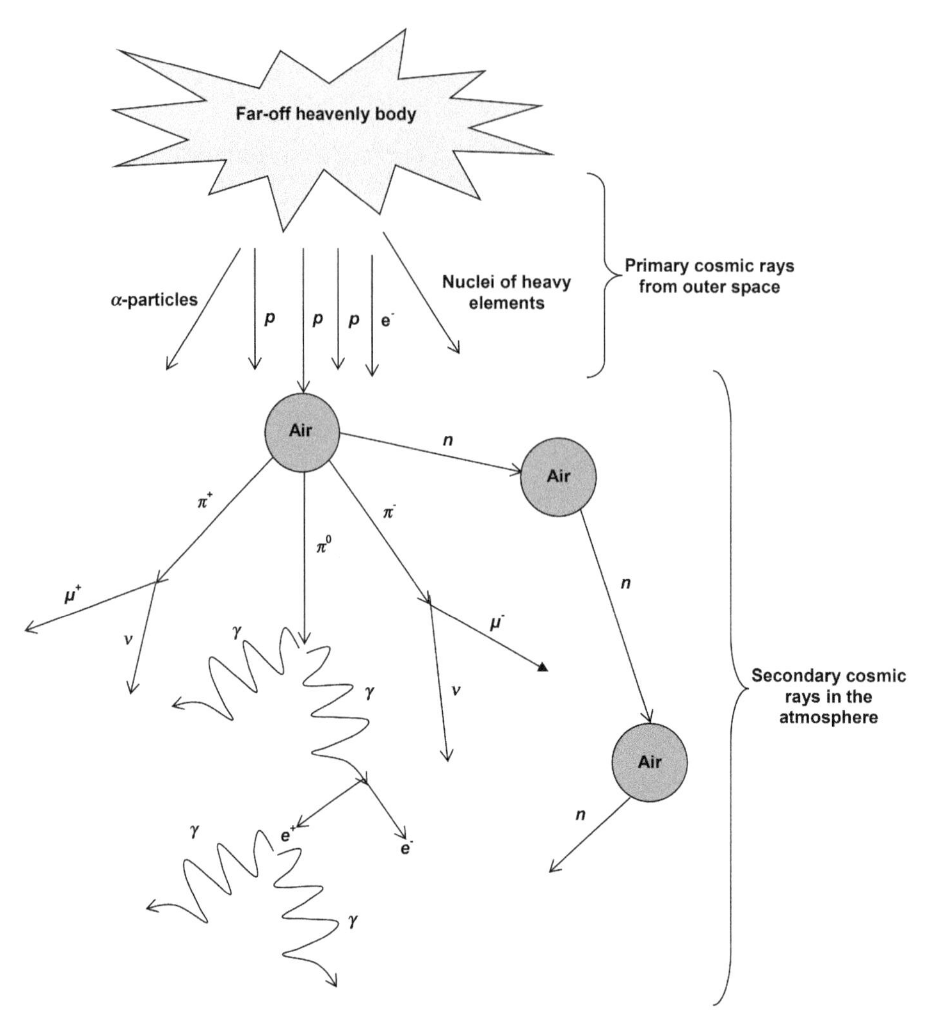

Figure 18.1. Primary and secondary cosmic rays.

Tc-99m, Co-60, Ir-192, Cs-137, etc; nuclear reactors in power plants; nuclear accidents; nuclear bombs or weapons used in warfare; particle accelerators; in addition to tobacco (thorium), lantern mantles (thorium), luminous dials in watches (tritium), smoke detectors (americium), and so forth (USNRC 2014b).

The initial radiation released from a nuclear explosion during the first minute consists of mostly gamma and neutron radiation. Residual radiation over longer periods of time comes from the radioactive fallout, including the debris, fission products and soil.

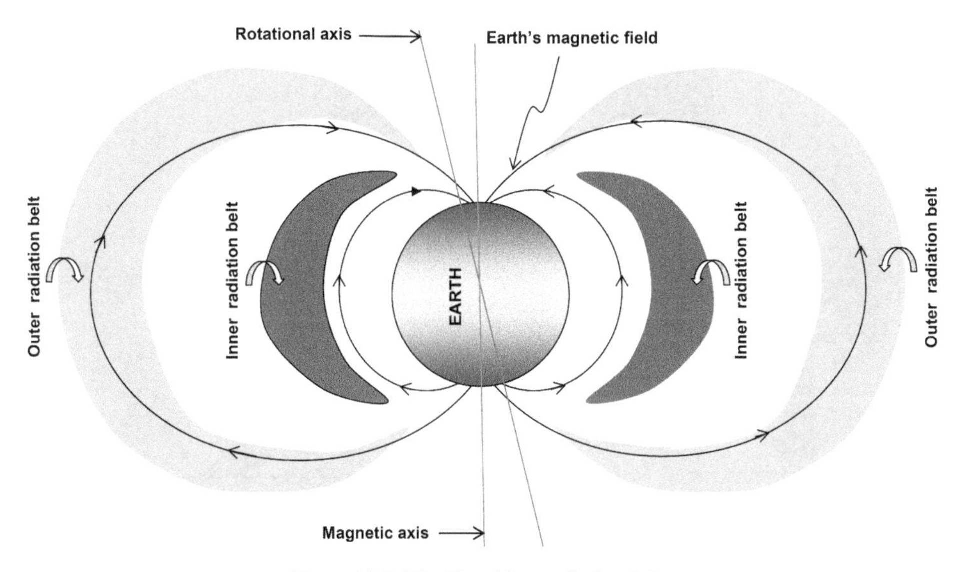

Figure 18.2. The Van Allen radiation belts.

18.3 Types of radiation effects

Based on extensive experimental studies, the effects of radiation on electronic devices and circuits are broadly classified into the three categories described below.

18.3.1 Total ionizing dose (TID) effect

The unit for measurement of TID is rad (radiation absorbed dose). The SI unit of TID is the Gray (Gy), 1 Gy = 100 rads. TID is a long-term failure mechanism. It accumulates over an extended period of time. To find the TID over this period, the total or cumulative radiation dose administered during the period is measured. The TID effect is observed as trapping of charges in oxides and the creation of defects by DD, which occurs when the atoms are pushed out from their regular lattice positions. A change of the threshold voltage of a MOSFET, a decrease in its transconductance and increase of leakage current take place. In the case of a bipolar transistor, current gain is degraded.

18.3.2 Single event effect

An SEE is an instantaneous temporary or permanent damage inflicted by particle strike. Soft errors, latchup or burnout follow. If SSE is like puncturing a football bladder, the TID effect bears resemblance to its aging over several years.

18.3.3 Dose-rate effect

The dose-rate effect is caused by the generation of photocurrent over the full circuit upon exposure to an extremely high dose rate over a very short time span. Latchup or burnout occurs accompanied by data loss.

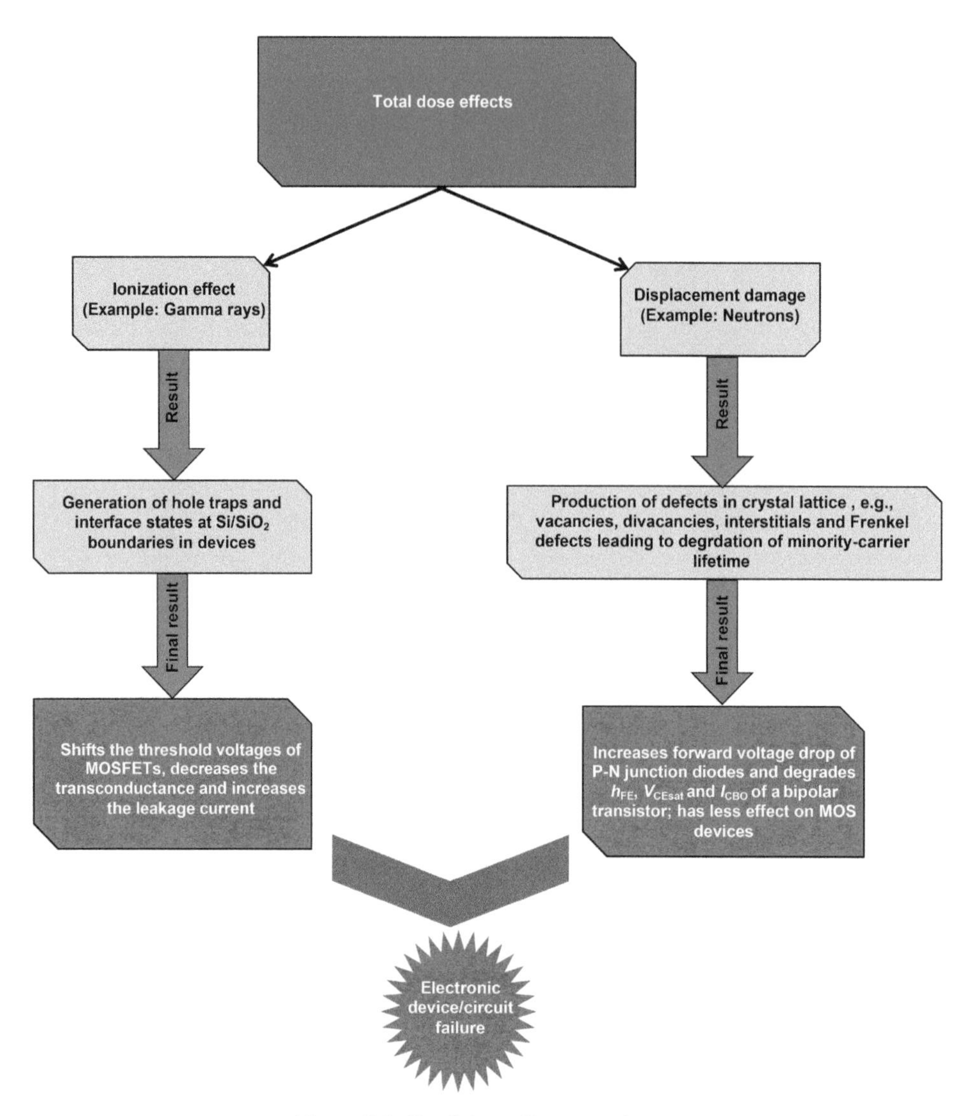

Figure 18.3. Total dose effects at a glance.

18.4 Total dose effects

Particles and photons in the radiation interact with the crystal in two fundamental ways (figure 18.3):

(i) The ionization effect, by removal of the outer shell electrons, and creation of EHPs through the impact of charged particles or photons on the crystal.

(ii) DD, e.g. neutrons bumping against atoms in the crystal lattice dislodge them from their positions and move them into vacancy or interstitial sites. As a result, Schottky or Frenkel defects are produced in the crystal.

Effects (i) and (ii) can co-exist, e.g. a heavy particle can bombard the nucleus producing charged particles having sufficient energy to cause ionization by collision.

18.4.1 Gamma-ray effects

Gamma rays predominantly cause the ionization effect (figure 18.4). EHPs are produced in semiconductors such as silicon, as well as insulators such as silicon dioxide. In silicon dioxide, the mobility of electrons is much higher than that of holes by four orders of magnitude. So, electrons move fast under an applied electric field leaving holes behind. These remnant holes create a charge buildup in SiO_2 and at the Si–SiO_2 interface, which has two effects:

(i) *Production of hole traps.* The subgroup of holes accumulated in SiO_2 serve as traps. Some of these positive traps recombine while others anneal out with time, but the remaining traps persist as electrically active defect centers for years.

(ii) *Interface state generation at the Si/SiO_2 interface.* The subgroup of holes lying near the Si–SiO_2 interface causes re-arrangement of atomic bonds at the Si/SiO_2 interface. Consequently, new interface states are produced.

Threshold voltage shifts in NMOS and PMOS transistors

The threshold voltage of an NMOS transistor is positive. The build-up positive hole charges discussed above resemble a type of persistent positive biasing of the gate. They produce the same effect as the positive gate voltage. Therefore, less gate voltage has to be applied to produce the same drain–source current. Hence, the threshold voltage of an NMOS transistor decreases, e.g. the change in threshold voltage ΔV_{Th} is negative.

A PMOS transistor has a negative threshold voltage. To counteract the positive hole charge, more negative voltage must be applied. Only then is it possible to obtain the same drain–source current as without positive charge. Therefore, threshold voltage increases. This means that the change in threshold voltage ΔV_{Th} is positive.

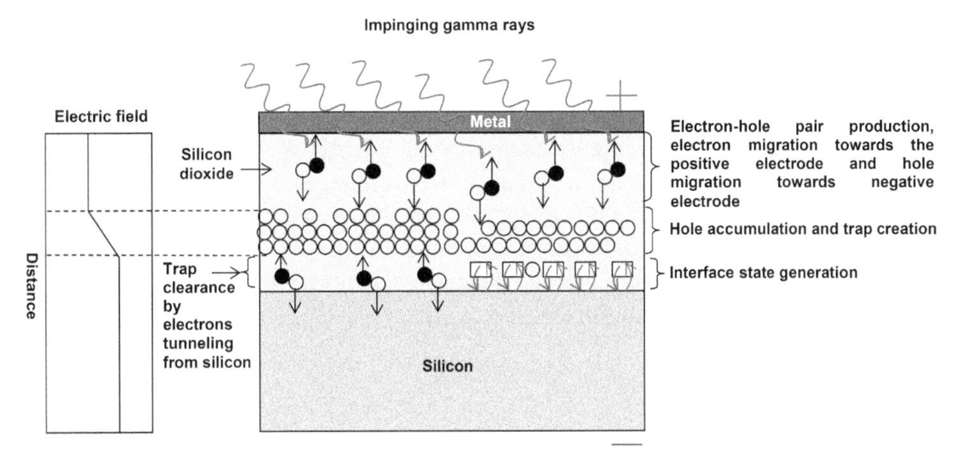

Figure 18.4. The effect of gamma rays on a metal–oxide–semiconductor structure. The electric field disturbance caused by the sheet of hole charge is shown on the left.

Thus the operations of NMOS and PMOS transistors are affected in the opposite directions.

Superior radiation hardness of the PMOS over the NMOS transistor
The position of the built-up sheet of charge depends on the sign of the gate voltage. The smaller the distance between the gate terminal and charge sheet, the smaller is the influence of the charge sheet on silicon for inversion layer formation, because the charge sheet is located far away from the silicon. Thus the additional electric field due to the charge sheet is smaller as is the shift in threshold voltage. In an NMOS transistor, the charge sheet is repelled a larger distance away from the positive gate terminal towards silicon. Hence, the additional electric field is also larger and a substantial threshold voltage shift results. For a similar reason, a comparatively insignificant threshold voltage shift is observed in a PMOS transistor. The obvious implication is that a PMOS transistor is more radiation-hard than an NMOS transistor. PMOS transistors are used as dosimeters in radiotherapy for accurate measurement of the gamma radiation doses delivered to patients.

Decrease of transconductance g_m and increase of leakage current
The charges trapped in the silicon dioxide and at the silicon–silicon dioxide interface decrease the carrier mobility in accordance with equation (Makowski 2006)

$$\mu_\mathrm{f} = \mu_\mathrm{i}/(1 + \alpha_\mathrm{it}\Delta N_\mathrm{it} + \alpha_\mathrm{ot}\Delta N_\mathrm{ot}), \qquad (18.1)$$

where μ_f is the final mobility, μ_i is the initial mobility, ΔN_it is the interface trap density, ΔN_ot is the oxide trapped charge density, α_it is the coefficient for the interface states and α_ot is the coefficient for oxide-trapped charges.

The mobility reduction brings down the transconductance of the device.

The aforementioned positive space charges accumulate n-type surface layers. They also invert p-type surface layers. The creation of such parasitic conduction paths or channels between the source and drain, which are not under gate control, increases the MOSFET leakage currents. The leakage current flowing between a MOS transistor and other circuit components is also increased. The resultant breakdown voltage degradation calls for incorporation of guard ring structures in the design. Overall, the functioning of a MOSFET is adversely affected through a shift in its threshold voltage, a decrease in its transconductance, increase of the drain–source leakage current and a fall in breakdown voltage. The n-channel MOSFET is turned on easily at a lower gate voltage while the p-channel MOSFET becomes difficult to turn on, requiring a higher gate voltage. Eventually, this leads to a situation in which an NMOSFET in a circuit may be permanently on while the PMOSFET may be permanently off, causing failure of the circuit.

18.4.2 Neutron effects

Neutrons are heavy particles. They are also charge-less. When a high-energy neutron impinges on a semiconductor device, it imparts energy to the atom struck through an elastic scattering process. An elastic scattering is like a collision between two hard elastic spheres in which the kinetic energy of the neutron is conserved with modification of its direction of motion. Consequent, upon this impact, the target atom may be kicked off from its lattice position and may occupy another position, either at a lattice site or between atoms. The recoiling atom of the semiconductor lattice may collide with other lattice atoms and trigger further collisions, initiating a cascade of collisions, depending on its energy, or it may participate in an EHP production process. On the whole, the neutron generates an assortment of lattice defects, including vacancies, divacancies, interstitials, Frenkel defects, etc (figure 18.5). The creation of defects is tantamount to the introduction of additional energy levels deep within the bandgap of the semiconductor; hence called 'deep-level defects'. The defects facilitate recombination of charge carriers by acting as stepping-stones for carrier recombination. In this way, they decrease the minority-carrier lifetime of the semiconductor. Operation of bipolar junction transistors (BJTs) is impaired because the carrier lifetime reduction leads to degradation of common-emitter current gain h_{FE} via enhanced recombination in the base. For this reason, narrow base width BJTs, particularly those for high-frequency operation, tend to be more radiation resistant. Furthermore, due to the loss of free charge carriers through increased recombination favored by a low carrier lifetime, the silicon resistivity increases, raising the saturation voltage V_{CEsat} of the transistor.

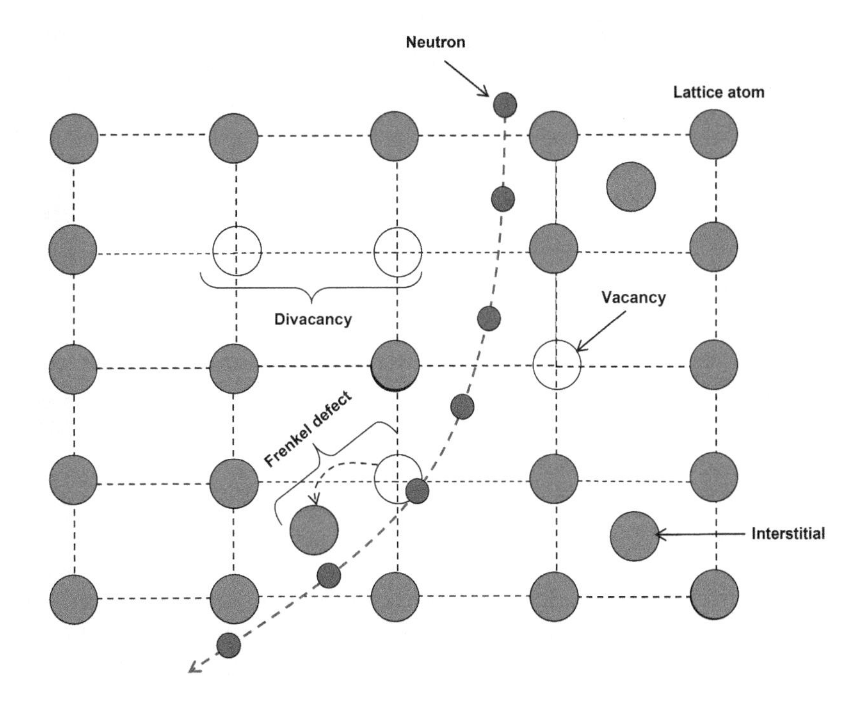

Figure 18.5. Atomic displacement defects such as vacancies, divacancies, interstitials, and Frenkel defects produced in a semiconductor crystal by neutron bombardment.

Additionally, the collector-base leakage current I_{CBO} increases due to charge build-up in the SiO_2 layer adjoining the junction. Thus the h_{FE}, V_{CEsat} and I_{CBO} parameters of a BJT deteriorate upon neutron irradiation.

The decrease in carrier lifetime from its initial value τ_i prior to neutron bombardment, to the final value τ_f after bombardment with neutrons of particle fluence ϕ, is expressed as (Makowski 2006)

$$\Delta(1/\tau) = 1/\tau_f - 1/\tau_i = K_\tau\phi, \tag{18.2}$$

where K_τ is the damage coefficient associated with carrier lifetime. The consequential decrease in current gain h_{FE} of a p–n–p transistor is written as (Makowski 2006)

$$\Delta(1/h_{FE}) = (1/2)\left(W_B^2/D_{pB}\right) \times K_\tau\phi = K_{FE}\phi, \tag{18.3}$$

where W_B is the base width, D_{pB} is the diffusion coefficient of holes in the base and K_{FE} is the damage coefficient related to current gain. The larger the neutron fluence ϕ, the greater is the degradation in current gain.

Contrary to the high-energy neutron, a low-energy neutron may be captured by the nucleus of a lattice atom resulting in an excited atomic nucleus. This is the neutron capture mechanism. During de-activation of the excited nucleus, gamma rays are produced which produce EHPs by ionizing the atoms on their way.

Fortunately, MOSFETs are majority-carrier devices. So, their operation is little influenced by carrier lifetime killing. Whereas bipolar devices start showing pronounced changes in electrical parameters at neutron fluxes of 10^{10}–10^{11} neutrons cm^{-2}, CMOS transistors are imperceptibly affected up to 10^{15} neutrons cm^{-2}. Like the BJTs, other minority-carrier devices such as solar cells have to cope with the problems of generation–recombination centers, which are trivial issues for MOSFETs.

18.5 Single event effects

As suggested by the name, SSEs are ushered by a single, energetic particle, and appear as momentary pulses in logic circuits or as flipping of bits in memory cells (Gaillard 2011). They are of two types (figure 18.6):

(i) Non-destructive, appearing as soft or temporary errors, recoverable by re-starting the circuit or re-processing the data.

(ii) Destructive, appearing as firm, hard or permanent errors and damaging the device or circuit.

With technological advancement towards smaller feature sizes, a smaller quantity of charge is necessary to store a logic high state than in primitive methods. Thus the logic states of ultra-small dimension technologies are more easily upset than those of coarse ones.

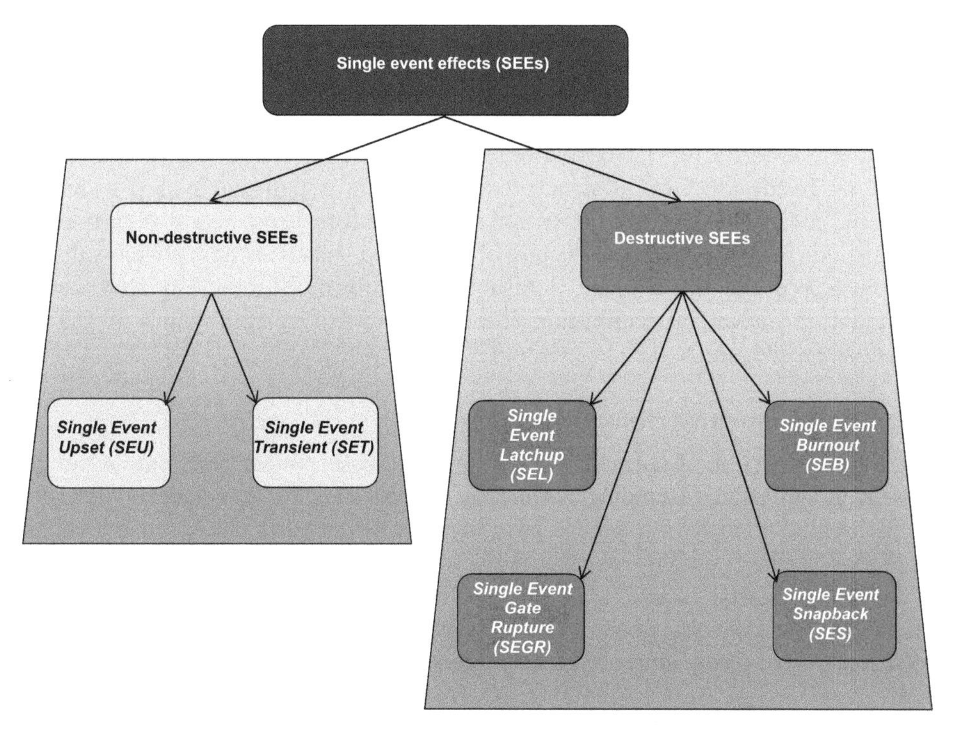

Figure 18.6. Classification of SEEs on semiconductor electronics.

18.5.1 Non-destructive SEEs

Single event upset

An SEU is a change of state induced in a microprocessor, field programmable gate array (FPGA), memory chip or power transistor by passage of an energetic particle striking a sensitive node of the circuit. A charged particle may cause ionization, creating EHPs inside or in the neighborhood of the sensitive node. An uncharged particle may also cause ionization through its ionizing fragments. An SEU is a transient error, which is corrected by resetting the device. In a multiple-bit SEU, two or more bits are affected concomitantly. An SEU of more serious concern is the single event functional interrupt (SEFI). The SEFI stops device operation. A power reset is required to restore its functionality (Holbert 2006).

Single event transient

An SET is a spurious voltage or current pulse in a voltage comparator, OP-AMP or a digital circuit, which is triggered by the charge collected during the ionization event. This pulse travels through the analog or digital circuit, causing a perturbation in the output signal.

18.5.2 Destructive SSEs

Single event latchup

SEL is a condition damaging the device, which is provoked by a single event current state. It is harmful for CMOS circuitry, which contains NMOS and PMOS transistors. Inside the CMOS structure are parasitic n–p–n and p–n–p transistors (figure 18.7). These transistors constitute a parasitic four-layer n–p–n–p or p–n–p–n thyristor. Normally, the reverse biasing of the n-well–substrate and p-well–substrate junctions keeps the thyristor in a disabled condition. But the thyristor can be triggered into action by irradiation if the product of current gains of the two transistors exceeds unity, i.e.

$$\beta_{\mathrm{npn}} \times \beta_{\mathrm{pnp}} > 1, \tag{18.4}$$

together with injection of adequate currents into the base–emitter junctions of the transistors, and provision of holding current by the power supply. Whenever the above conditions are complied with, regenerative thyristor action takes place, causing latchup. The latchup can be cleared by resetting through power on–off to disconnect the supply. If not done immediately, excessive overheating due to high current flow may lead to catastrophic failure and irreversible damage to metallization or bond wires. Latchup is a strong function of temperature. Its susceptibility increases as temperature increases.

Single event burnout

SEB is a damaging condition encountered due to high current flow in high-power MOSFETs, such as vertical double-diffused MOS (DMOS) transistors, caused by a heavy-ion component of cosmic rays and solar flares. Suppose a vertical DMOS is in the off-state. It is blocking a high drain–source voltage. Let a heavy ion pass through the transistor structure across the source region. As it leaves behind a tail of EHPs

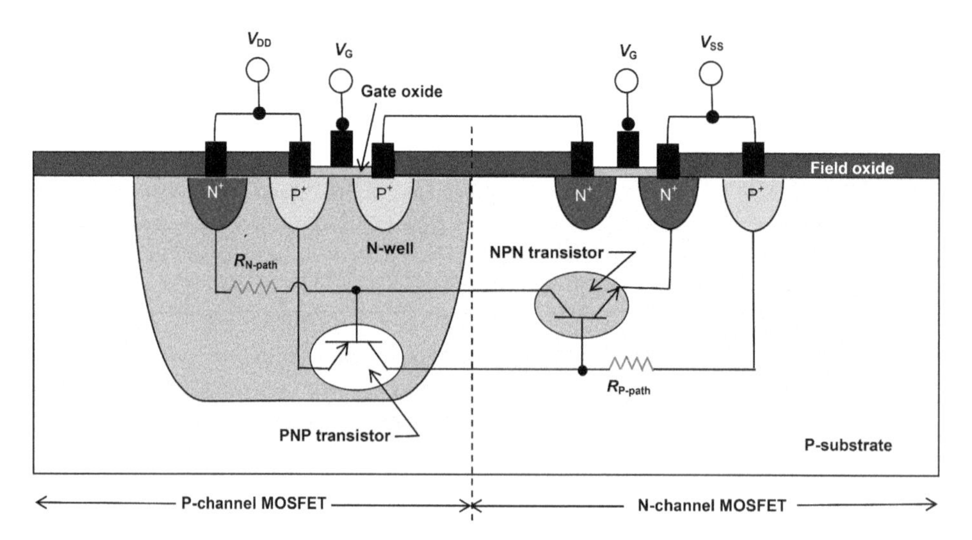

Figure 18.7. Parasitic p–n–p and n–p–n bipolar transistors constituting the thyristor, which is responsible for regenerative action causing latchup in an n-well CMOS device.

along its path, and because a large drain–source voltage is applied, a high current density ~10^4 A cm $^{-2}$ is generated in the parasitic bipolar transistor below the source. As soon as the voltage drop across the base–emitter junction of the parasitic BJT is >0.7 V, it is turned on by avalanche multiplication of collector current. Forward-bias second breakdown is initiated. The localized junction heating leads to burning of the device, provided sufficient short-circuit energy is available.

Power MOSFETs are highly vulnerable to SEB at $V_{DSS} \geqslant 100$ V. Nonetheless, SEBs have also occurred at $V_{DSS} < 30$ V. In the beginning, SEBs were considered to be problems afflicting MOS devices only. But SEBs are not confined to MOSFETs alone and have also taken place in BJTs at voltages >400 V and thyristors at voltages >1000 V (Foutz 2005).

Single event gate rupture

A single event gate rupture (SEGR) is a destructive burnout of the gate dielectric consequent upon the formation of a conducting path through it by the passage of a heavy ion traversing across the gate insulator through the neck region of a MOSFET cell, evading the p-regions and moving through the n-epilayer, applying the drain potential to the gate. A plasma filament of EHPs is produced in the n-epilayer. The electrons in the plasma quickly diffuse away. But the slower holes pile up and this hole accumulation generates a transient electric field of magnitude exceeding the breakdown voltage of the gate insulator. If this lasts for a short duration, the device may recover, but if it continues too long, it destroys the gate oxide either via increased gate leakage current or ruptures it through an increase of temperature in the vicinity, completely damaging the same.

SEGR is witnessed in non-volatile memories such as EEPROMs during write or erase operations. In these operations, high voltages are applied to the gate electrodes. SEGR also takes place in BJTs.

Single event snapback

An SES is similar to an SEL, except that no p–n–p–n thyristor is involved. The ionization induced by the energetic particle turns on the parasitic bipolar transistor of the MOSFET structure. Recovery is possible by properly sequencing the signals without reduction of the power supply. SES may lead to circuit destruction if the local current density is very high.

18.6 Discussion and conclusions

Both MOS and bipolar devices and associated circuits are prone to the radiation effects of either short-term or long-term nature. Changes in MOSFET parameters upon radiation exposure are manifested in CMOS circuits as variations in low and high digital logic levels, a decrease in output current, lengthening of propagation delay and upswing of quiescent current values. Digital circuits are vulnerable to SEEs of short-lived or long-lasting nature, mainly SEU and SEL.

Review exercises

18.1. What are the two principal forms of radiation? Give examples.
18.2. Name and describe the three natural sources of radiation.
18.3. What leads us to believe that cosmic rays come from outer space?
18.4. What argument leads to the inference that cosmic rays contain charged particles?
18.5. What protects us from the harmful effects of cosmic radiation on Earth's surface?
18.6. What is the typical composition of primary cosmic rays?
18.7. What are the Van Allen radiation belts? What particles are they composed of?
18.8. Give five examples of man-made radiation sources?
18.9. What is the composition of initial radiation from a nuclear explosion?
18.10. What is the TID effect of radiation on semiconductor devices and circuits? What are the different forms in which it is generally observed? What electrical parameters of MOS and bipolar devices undergo degradation by TID?
18.11. What are SEEs and dose-rate effects of radiation on semiconductors?
18.12. What happens when a charged particle or photon strikes a semiconductor crystal?
18.13. What happens when a semiconductor crystal is subjected to neutron bombardment?
18.14. Gamma rays are striking a layer of silicon dioxide. What effects are observed within the silicon dioxide layer and at the silicon–silicon dioxide interface?
18.15. What are the effects of positive charge build-up by gamma-ray irradiation of a silicon dioxide gate insulator on the threshold voltage of: (a) an n-channel MOSFET and (b) a p-channel MOSFET? Why is the PMOS transistor more radiation-hard than the NMOS transistor?
18.16. What is the effect of gamma-ray irradiation on the transconductance and leakage current of a MOSFET device?
18.17. How are deep-level effects created in a semiconductor crystal by bombardment with high-energy neutrons? How does the neutron-created damage affect the minority-carrier lifetime in the semiconductor?
18.18. How does a low-energy neutron interact with a semiconductor crystal? Does this interaction affect the carrier lifetime or produce a different result?
18.19. What electrical parameters of a bipolar junction transistor are degraded by a neutron strike and how?
18.20. Why are MOSFETs more immune to the damaging effects of neutrons than bipolar devices?
18.21. What are SEEs of radiation on semiconductor circuits? What are the two types of SEEs? Why are small geometry devices more easily prey to SEEs than the large geometry ones?

18.22. Differentiate between an SEU and an SET.
18.23. How does SEL take place in a CMOS structure? Explain with a diagram.
18.24. What types of MOSFETs are susceptible to SEB? How does this burnout take place?
18.25. How is the gate dielectric of a MOSFET ruptured in an SEE? Give examples of MOSFET-based structures prone to this type of damage.
18.26. In what way is SES different from SEL? Is it recoverable? How?

References

Foutz J 2005 Power transistor single event burnout, www.smpstech.com/power-mosfet-single-event-burnout.htm

Fox K C 2014 NASA's Van Allen probes spot an impenetrable barrier in space *NASA* www.nasa.gov/content/goddard/van-allen-probes-spot-impenetrable-barrier-in-space

Gaillard R 2011 Single event effects: mechanisms and classification *Soft Errors in Modern Electronic Systems* ed M Nicolaidis (Berlin: Springer) pp 27–54

Holbert K E 2006 Single event effects http://holbert.faculty.asu.edu/eee560/see.html

Makowski D 2006 The impact of radiation on electronic devices with the special consideration of neutron and gamma radiation monitoring *PhD Thesis* Technical University of Lodz pp 13, 14, 20, 21

NASA Goddard Space Flight Center 2014 What are cosmic rays? http://imagine.gsfc.nasa.gov/science/objects/cosmic_rays1.html

USNRC 2014a Natural background sources *USNCR* www.nrc.gov/about-nrc/radiation/around-us/sources/nat-bg-sources.html

USNRC 2014b Natural and man-made radiation sources: reactor concepts manual *USNRC Technical Training Center* 0703, pp 6-1–12 www.nrc.gov/reading-rm/basic-ref/students/for-educators/06.pdf

Chapter 19

Radiation-hardened electronics

Various techniques used to immunize electronic devices and circuits against detrimental radiation effects are described. These techniques are subdivided into two categories: process-based and design-based. The principal process-based techniques include the implementation of triple-well and SOI CMOS technologies, in addition to impurity diffusion profile tailoring and carrier lifetime killing. Amongst the design-based schemes, mention may be made of using annular MOSFET geometry, channel stoppers and guard ring structures. The channel aspect ratio may be increased to control the dissipation of charge. Spatial and temporal redundancy concepts may be applied to make devices suitable for operation in radiation-contaminated environments. SEU-resistant flip-flops and RAM structures are designed using DICE memory cells. As the radiation-hardened electronics constitutes a small segment of the total chip market, the fabrication of radiation-hardened designed structures by commercial CMOS processes has received more acceptance than the adoption of high-cost technologies.

19.1 The meaning of 'radiation hardening'

To radiation harden a device or circuit, or to make it 'rad-hard', means that it was designed or fabricated to withstand a certain amount of radiation without malfunctioning. The necessary procedures are executed on the device/circuit during the design or fabrication stage to ensure proper functioning. Sometimes, the term 'radiation hardening' is used in a different context. It implies that the device/circuit has undergone a series of post-fabrication tests to check its survivability and correct operation in radiation-rich environments. Thus the term may have different connotations depending on the contextual setting.

In this chapter, we shall consider radiation hardening as a set of techniques used to mitigate radiation effects in semiconductor devices by developing immunity against radiation onslaught from the very beginning of device conception to the last

doi:10.1088/978-0-7503-1155-7ch19

phase of their delivery to the user. Radiation-hardened electronics is thus the circuitry or equipment in which these immunization steps are fully taken care of.

19.2 Radiation hardening by process (RHBP)

19.2.1 Reduction of space charge formation in silicon dioxide layers

To minimize the radiation-induced positive space charge build up in oxide layers, one strategy is to grow thin oxide layers so that a smaller hole population is created and trapped. In reference to the contemporary downscaling of MOSFET dimensions, this oxide thinning is a boon because further miniaturization is aided by reduced gate oxide thicknesses. So, with unabated shrinkage of transistor sizes, radiation hardening of gate oxides is in harmony with technological trends, and requires no major effort. The major bottleneck is the prevention of charging of the thicker field oxides of MOSFETs. This is achieved by decreasing the mechanical stress in the oxides as well as ion-implantation induced damage, by using a smaller quantity of hydrogen in the process and by using lower process temperatures after gate formation (Myers 1998).

19.2.2 Impurity profile tailoring and carrier lifetime control

Increasing the substrate doping and adjusting the impurity doping in MOS-well regions helps to decrease the SEU and SET sensitivity. SEL immunity is enhanced by preventing positive feedback action of parasitic transistors in a CMOS structure by choosing an appropriate doping profile and decreasing the minority carrier lifetime in the base region.

19.2.3 Triple-well CMOS technology

There are four CMOS processes: n-well, p-well, twin-well and triple-well (or deep n-well), as depicted in figure 19.1. Some of the electrons produced by radiation exposure are swept away into the deep n-well. Many of these electrons fail to reach the drain of the n-channel MOSFET. As the drain of the NMOS device either does not collect them at all or collects a very small number, these electrons are unable to disturb its operation. The likelihood of an SEU is thereby reduced. In other words, the upset is prevented in most cases by strengthening the isolation of the n-channel MOSFET in the p-substrate.

19.2.4 Adoption of SOI technology

Two major shortcomings of the junction isolation technique are readily overcome by the adoption of SOI technology. These shortcomings are leakage current and latchup. In the SOI structure, there are neither any leakage paths along device edges nor any p–n–p and n–p–n transistors. SOI technology guarantees freedom from both these problems. Susceptibility to SEEs is greatly reduced in devices fabricated using SOI wafers, because these devices have a much smaller volume for collection of charges and energy due to irradiation, in comparison to those made on bulk silicon wafers (figure 19.2). By confining the transistor in a thin layer of silicon over a thin

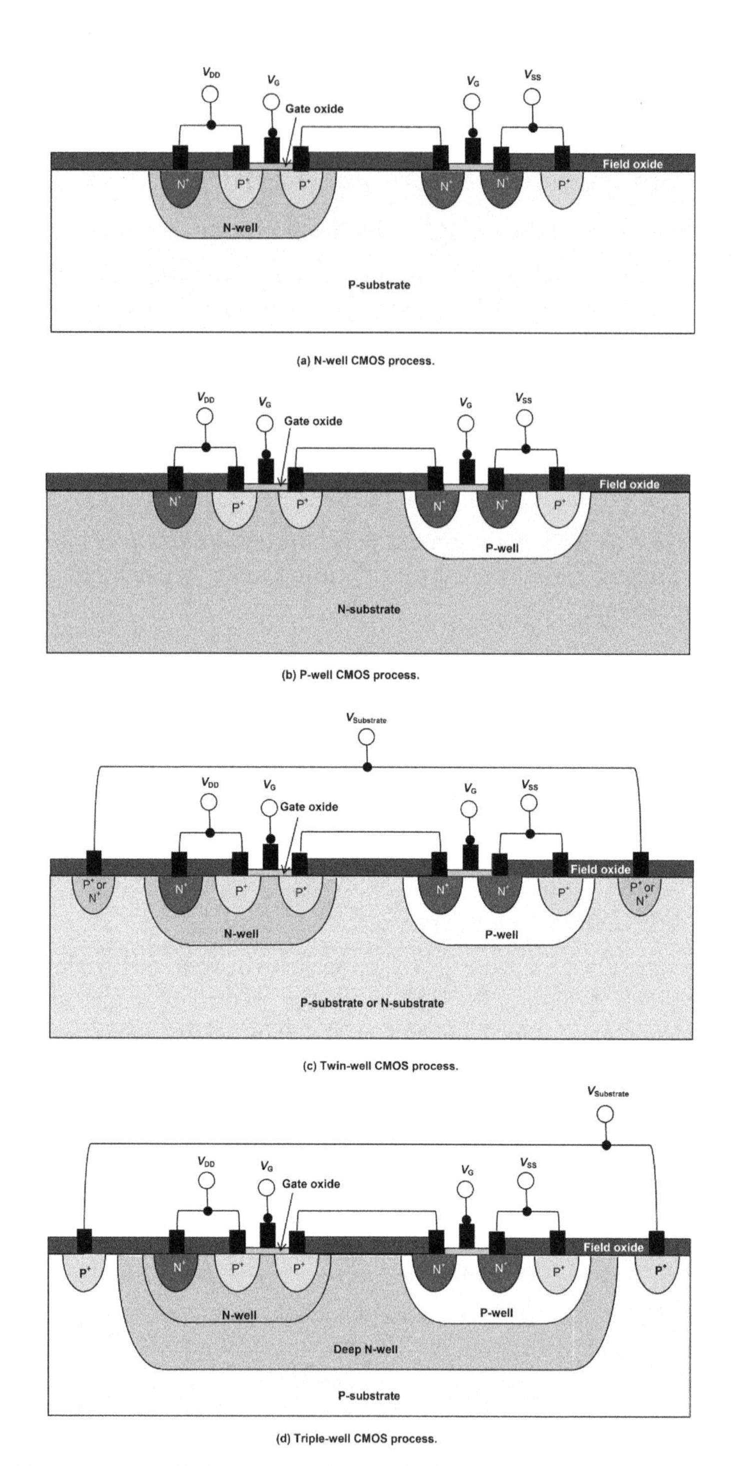

Figure 19.1. Four kinds of CMOS processes: (a) n-well, (b) p-well, (c) twin-well and (d) triple well. V_{DD} = drain supply, V_{SS} = source supply, V_G = gate supply and $V_{Substrate}$ = substrate supply.

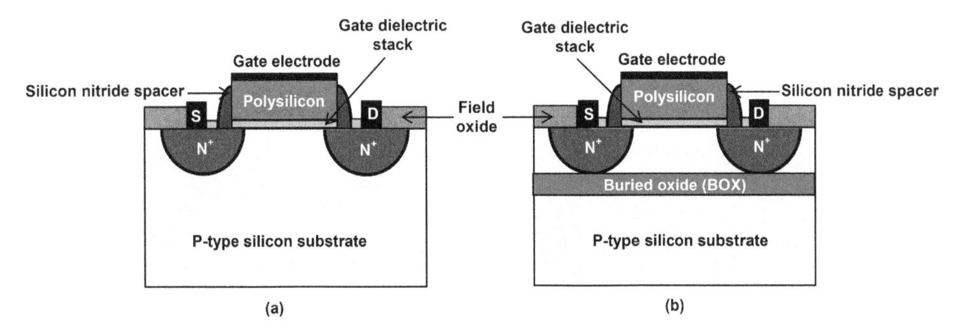

Figure 19.2. Planar MOSFETs on (a) a bulk silicon wafer and (b) an SOI wafer.

insulator, a much smaller sensitive volume is available for charge pile up than in a bulk-silicon transistor having a larger sensitive volume in a thick substrate, offering a long trajectory of particle movement in which undesired positive charges are produced and stored. The smaller this volume, the smaller is the hole charge storage and the more robust is the device against radiation effects. In addition to improving radiation hardness, SOI technology also reduces capacitive loading. The capacitance between the source and substrate, and also that between the drain and substrate, is decreased, offering high-speed operation. Power consumption too is less. The noise level is low and so is the cross-talk.

There are two offshoots of SOI-MOSFETs. In the partially depleted SOI-MOSFET (figure 19.3(a)), only part of the region between the gate and BOX is depleted. In the fully depleted SOI-MOSFET (figure 19.3(b)), this region is completely depleted. Therefore, the fully depleted version represents the smallest volume for charge generation and has a minimal risk for radiation-induced malfunction.

The drawback of SOI technology is its prohibitive cost and the accumulation of positive charge in the BOX layer due to radiation impact. The threshold voltage shift caused by this positive charge in a partially depleted transistor leads to inversion of the bottom part of the channel (back gate side), increasing the drain–source leakage current. The unwarranted leakage current causes failure of the circuit.

19.3 Radiation hardening by design

The radiation-hardening by design (RHBD) methodology combines the advantages offered by the introduction of manufacturing processes mitigating radiation effects in commercial state-of-the-art foundries, together with novel device and circuit layouts, geometries and topologies to enhance the radiation tolerance. Very often, a price has to be paid in terms of increased power consumption, larger chip area and lower operating frequency.

19.3.1 Edgeless or annular MOSFETs

Traditional MOSFET design geometry (figure 19.4) has the drawback that holes trapped in the bird's beak region at the edge of the MOSFET, where the gate overlaps the field oxide, cause a threshold voltage shift and contribute to leakage current flow along the gate edges. To overcome these issues, an edgeless MOSFET

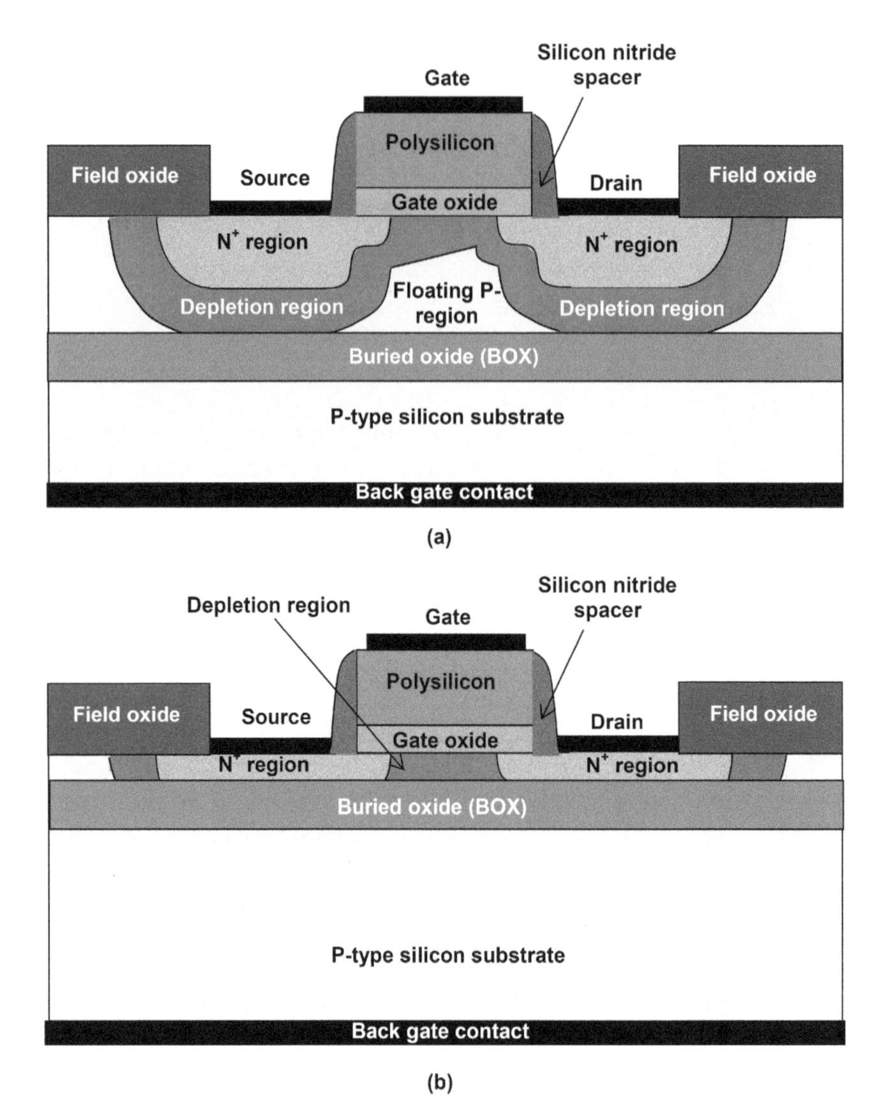

Figure 19.3. SOI-MOSFET structures: (a) a partially depleted MOSFET (PD-MOSFET) and (b) a fully depleted MOSFET (FD-MOSFET).

design is used in which either the source region completely surrounds the drain region or the drain region completely surrounds the source region (figure 19.5). In either case, the gate region completely surrounds the source/drain region. For circular geometry, the gate region circumscribes the source/drain region.

The enclosed geometry transistor eliminates the leakage current flowing between the drain and source along the edges of the gate in conventional geometry, due to the ionization-induced charge generation by irradiation. On the downside, the annular transistor is always larger in area than the traditional one, thereby increasing the intrinsic capacitance and decreasing the packing density. The larger input

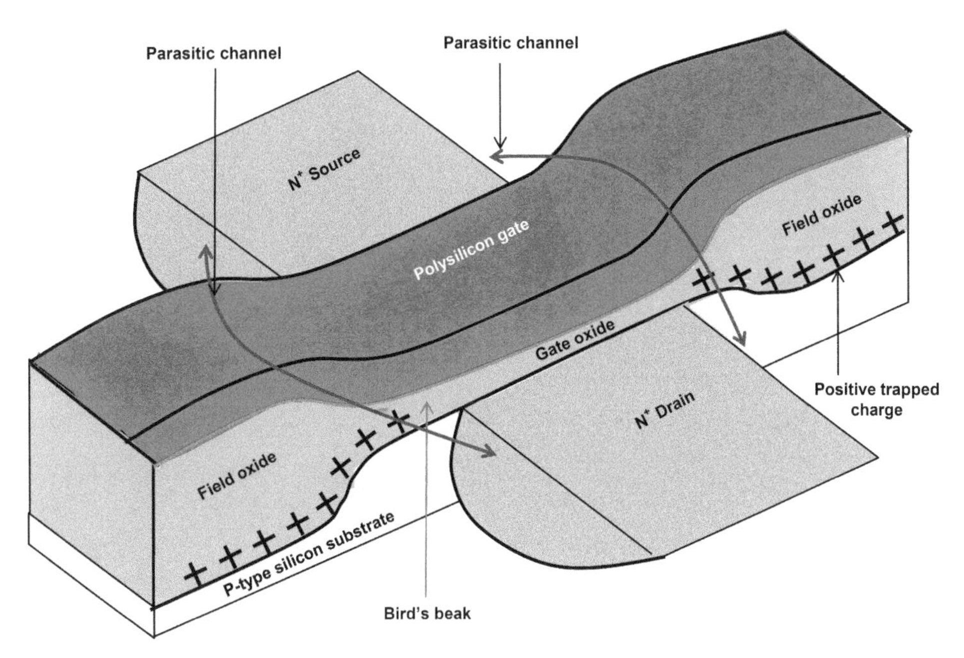

Figure 19.4. Bird's beak formation in the MOSFET isolation region and the holes trapped inside.

capacitance prolongs the switching time, slowing down the device. Therefore, annular transistors mainly find use in radiation environments.

19.3.2 Channel stoppers and guard rings

Leakage currents flowing between neighboring n-channel MOSFETs are stopped by forming p-type channel stoppers and guard rings. A channel stopper (figure 19.6(a)) is a heavily doped region, which cannot be easily inverted by surface charges. The guard ring (figure 19.6(b)) is a deeper diffusion of the same polarity and larger radius of curvature, and therefore higher breakdown voltage, surrounding a shallow diffused junction of smaller radius of curvature and hence lower breakdown voltage. The deep diffused guard ring improves the breakdown voltage of the shallow junction.

19.3.3 Controlling the charge dissipation by increasing the channel width to the channel length ratio

The charge Δq collected at a circuit node due to a particle strike is related to the effective capacitance C of the node and the change in node voltage ΔV caused by the strike, through the familiar equation

$$\Delta q = C\Delta V. \tag{19.1}$$

The impact of MOSFET downscaling is that the minimum value of C has decreased with the square of feature size. However, Δq has remained the same. The implication

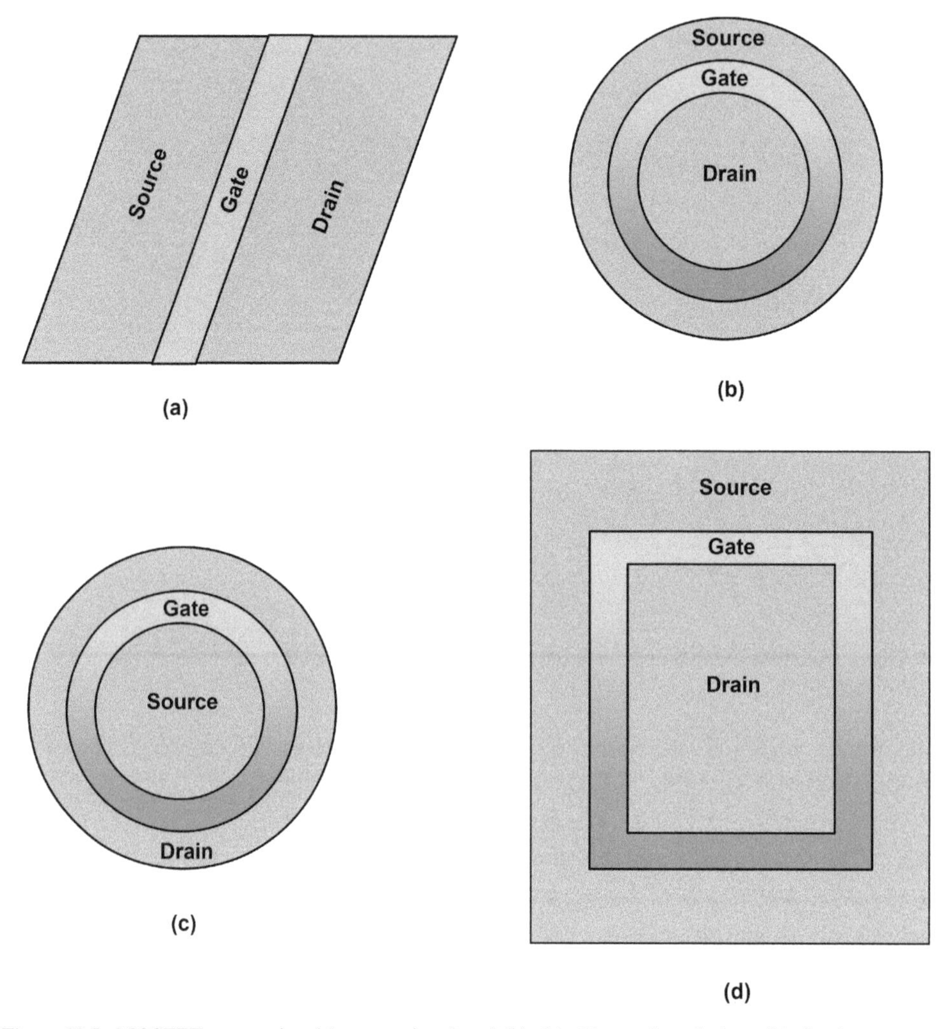

Figure 19.5. MOSFET geometries: (a) conventional and (b), (c), (d) annular edgeless; (b) circular geometry with an external source and internal drain; (c) circular geometry with an external drain and internal source; and (d) rectangular geometry.

is that the disturbance ΔV in node voltage has drastically increased. Additionally, IC power supply voltages have continuously decreased. The decrease has made the proportionate consequence of the ΔV upsurge even more pronounced by comparison.

Insofar as the influence of the operating frequency value on SSEs is concerned, we note that

$$\Delta q = I\Delta t, \tag{19.2}$$

where I is the maximum charging/discharging current at the node and Δt is the time taken by the charge to dissipate from the node. The present trend is that the frequency f of the clock signal of ICs is continuously increasing. So, the periodic time

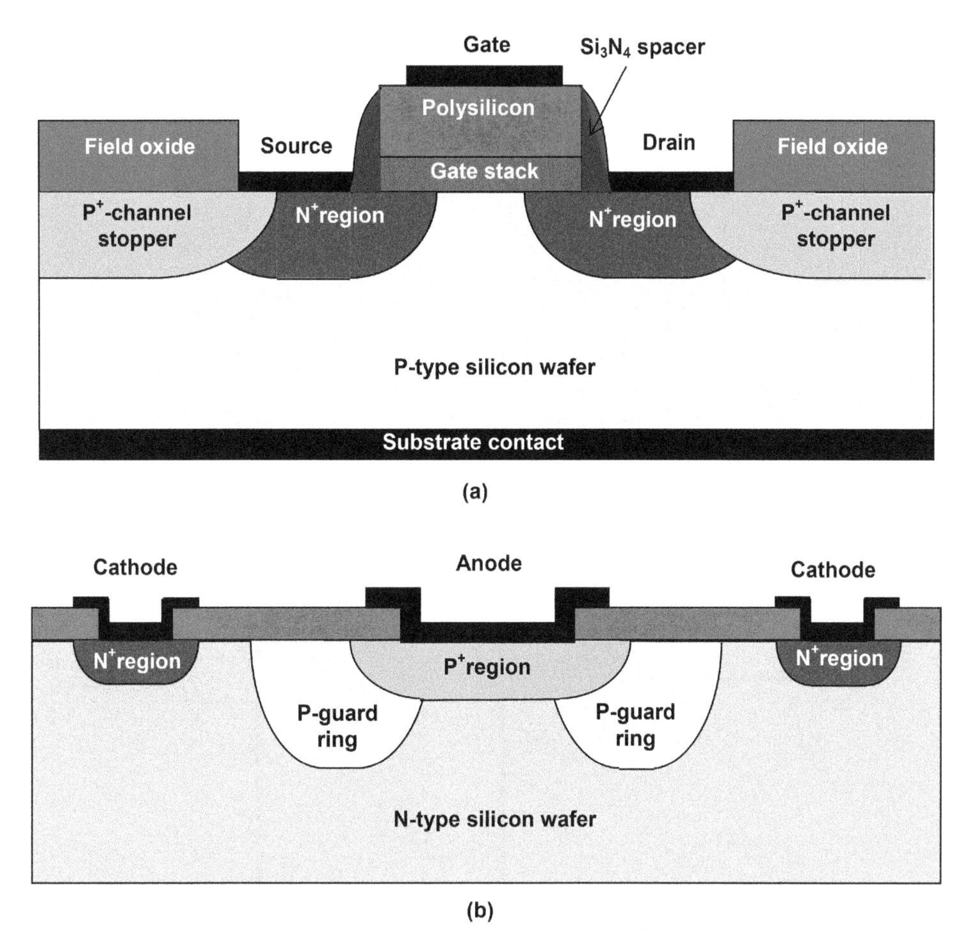

Figure 19.6. Channel stopper and guard ring structures: (a) a channel stopper in an n-channel MOSFET diode and (b) a guard ring in a p–n junction.

T of the clock signal, being the reciprocal of frequency, is decreasing. As $T \rightarrow \Delta t$ and T becomes $<\Delta t$, the chances of an SEU moving through the remaining circuit increase. Thus there is a greater danger of SEU in higher operating frequency circuits, causing more uncertainty.

Reduction of Δt appears to be the obvious remedy to obviate SEU. This is accomplished by applying the principle of charge dissipation (Holman 2004). For this purpose, the maximum charging/discharging current I at a critical node is increased. In the inverter circuit shown in figure 19.7, the logic high signal turns on the n-channel MOSFET while the p-channel MOSFET is turned off. Any charge deposited on the output capacitance C_{out} is discharged through the n-channel MOSFET. The strategy followed is to increase the ratio of channel width/channel length to raise the maximum drain–source current I_{DS}. This enables the dissipation of total change in capacitor voltage due to a particle strike before the succeeding stage can disseminate the resulting transitory voltage as a bit error. It should be

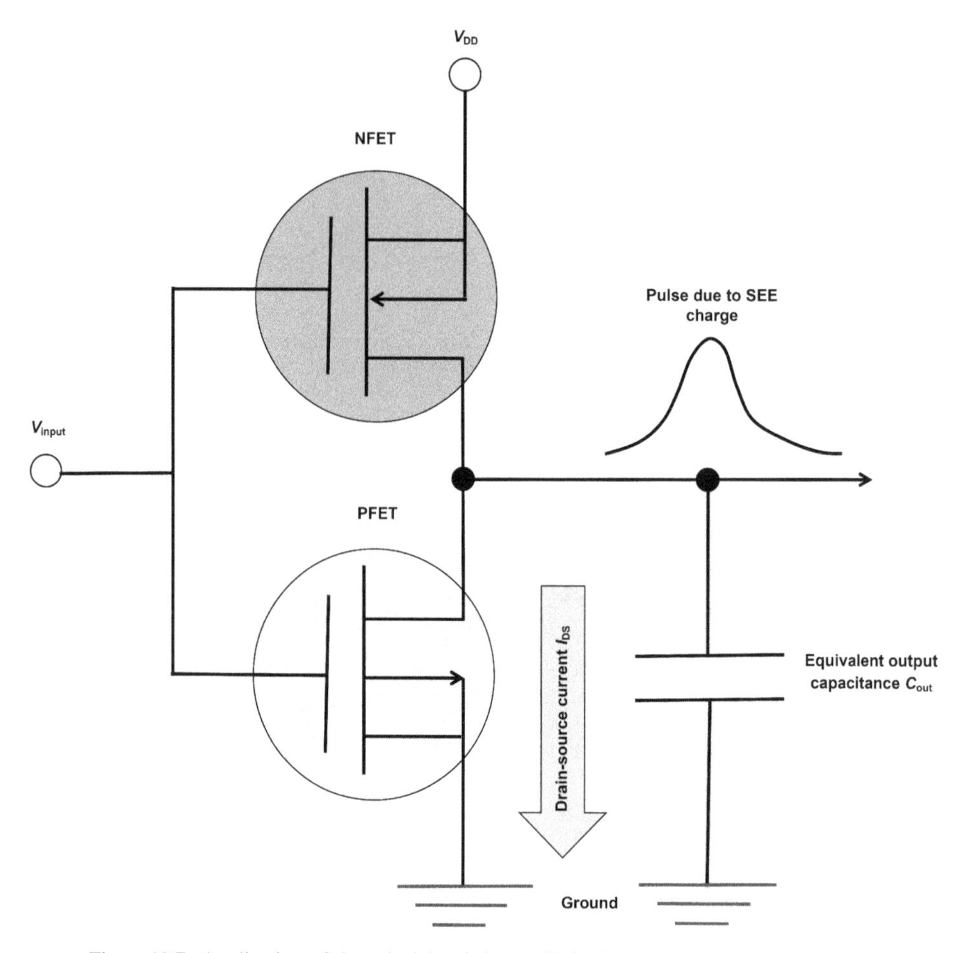

Figure 19.7. Application of the principle of charge dissipation to a digital inverter circuit.

noted that the above strategy for increasing I has the disadvantage of a larger area and more power consumption of the circuit.

19.3.4 Temporal filtering

This involves insertion of resistive and/or capacitive components at critical nodes of the circuit (figure 19.8). The circuit modification provides low-pass filtering of the fast voltage pulses generated by SSEs. Principally due to the area and speed disadvantages of this method, it is rarely used in high-performance circuits.

19.3.5 Spatial redundancy

A combinational logic circuit with triple modular redundancy (TMR) is implemented using three physically separated data paths to feed logic circuits X, Y and Z (figure 19.9). The aim is to ensure that any single particle strike cannot disturb more than one logic circuit (Holman 2004). A voting circuit hardened against errors produces a valid result only when a minimum of two calculations is in agreement.

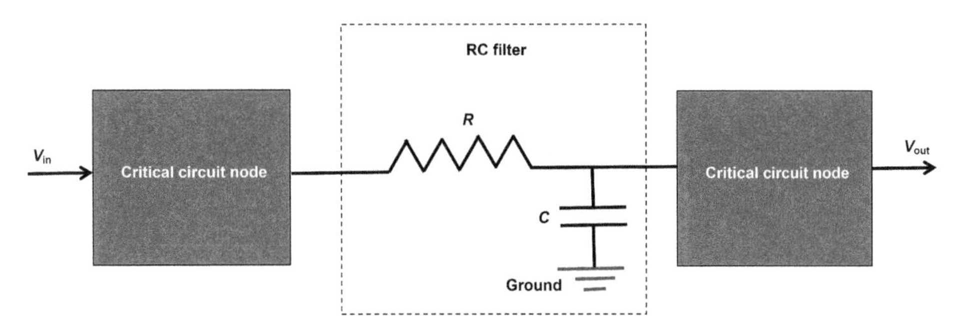

Figure 19.8. Temporal (low-pass) filtering.

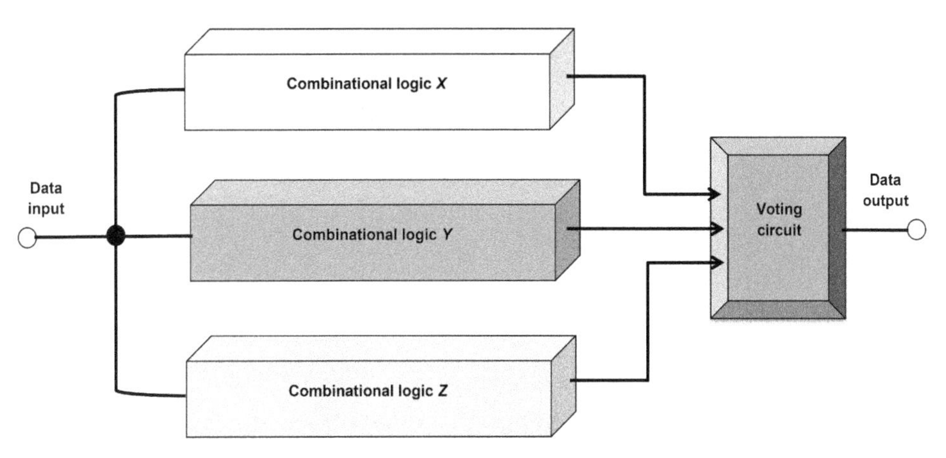

Figure 19.9. Spatial TMR.

The circuit provides protection against SEEs with penalty of larger area consumption.

TMR techniques involving storage cell replication and majority voting are also applicable to sequential logic circuits, such as flip-flops and registers, at the cost of system level overhead and power dissipation.

In TMR-based circuits, SEU tolerance may be lost due to error latency, i.e. a delay between the stimulus receipt and the response to it. A first upset is tolerated. But thereafter a second upset may take place before recovery from the effect of the first upset. This leaves the system prone to interconnected dual errors.

19.3.6 Temporal redundancy

Similar to the protection provided by TMR with respect to SEU by separation of data signals in space, temporal redundancy does so by separating the data signals on a time scale (Holman 2004). Figure 19.10 illustrates triple-path temporal redundancy implementation in an SEU-resistant data latch. The circuit consists of a dual input multiplexor connected in a feedback loop, delay circuits and a voting circuit. There are three separate data paths to the voting circuit. Two of these paths insert

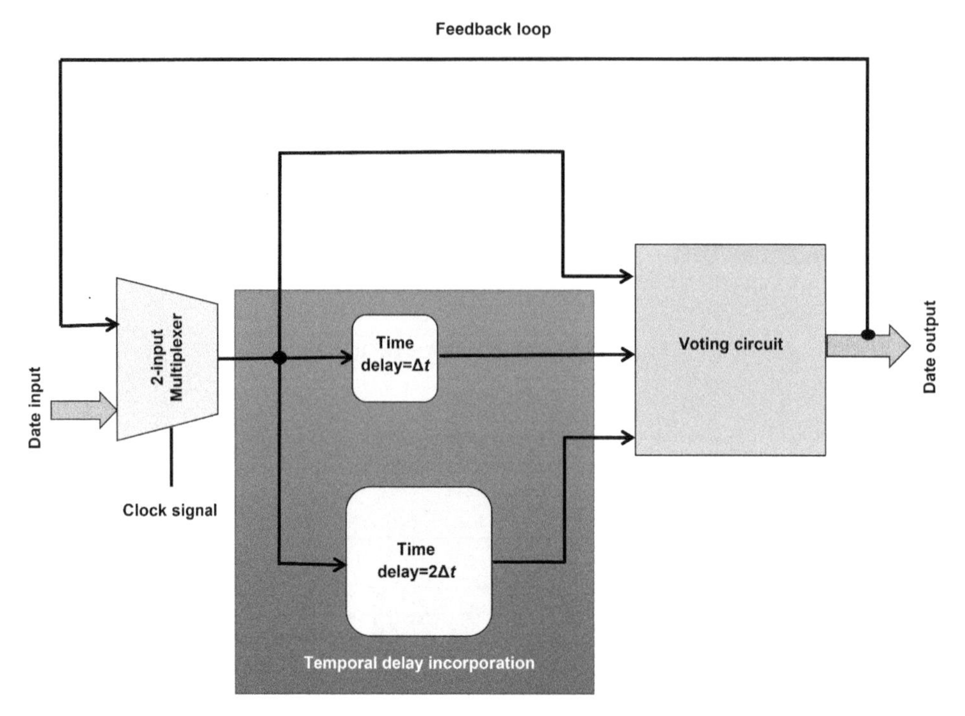

Figure 19.10. Three-path temporal redundancy; Δt = time taken to dissipate the maximum anticipated charge due to ion strike.

time delays, one time delay is Δt and the other time delay is $2\Delta t$, where Δt is the time during which the maximum charge of a particle strike can be dissipated. The penalization of the circuit is that a waiting period = $2\Delta t$ seconds must be allowed to execute a calculation. At a frequency of 50 MHz this latency is small, but at higher frequencies it becomes unacceptable.

19.3.7 Dual interlocked storage cell

The dual interlocked storage cell (DICE) memory cell is used to make CMOS flip-flops and static RAMs resistant to SEU (Calin *et al* 1996, Mukherjee 2008). It is a redundant structure containing four internal nodes V_0, V_1, V_2, V_3. Data are stored on these nodes. Alternate logic levels are always present on the nodes. Logic 0 is represented by 0101, i.e. $V_0V_1V_2V_3 = 0101$. Logic 1 is represented by 1010, i.e. $V_0V_1V_2V_3 = 1010$. To perform read and write operations, the enable input C is asserted and data accession is performed through bidirectional lines D, $\overline{D}$.

The DICE structure consists (figure 19.11) of two horizontal cross-coupled inverter latch structures (N_0, P_1) and (N_2, P_3). These horizontal latch structures are connected by vertical bidirectional feedback inverters (N_1, P_2) and (N_3, P_0). The inverters are p-channel or n-channel MOS transistors, as indicated by the corresponding p and n symbols. These transistors are arranged in feedback loops. The PMOS transistors (P_0, P_1, P_2, P_3) constitute a clockwise loop. Likewise, the NMOS

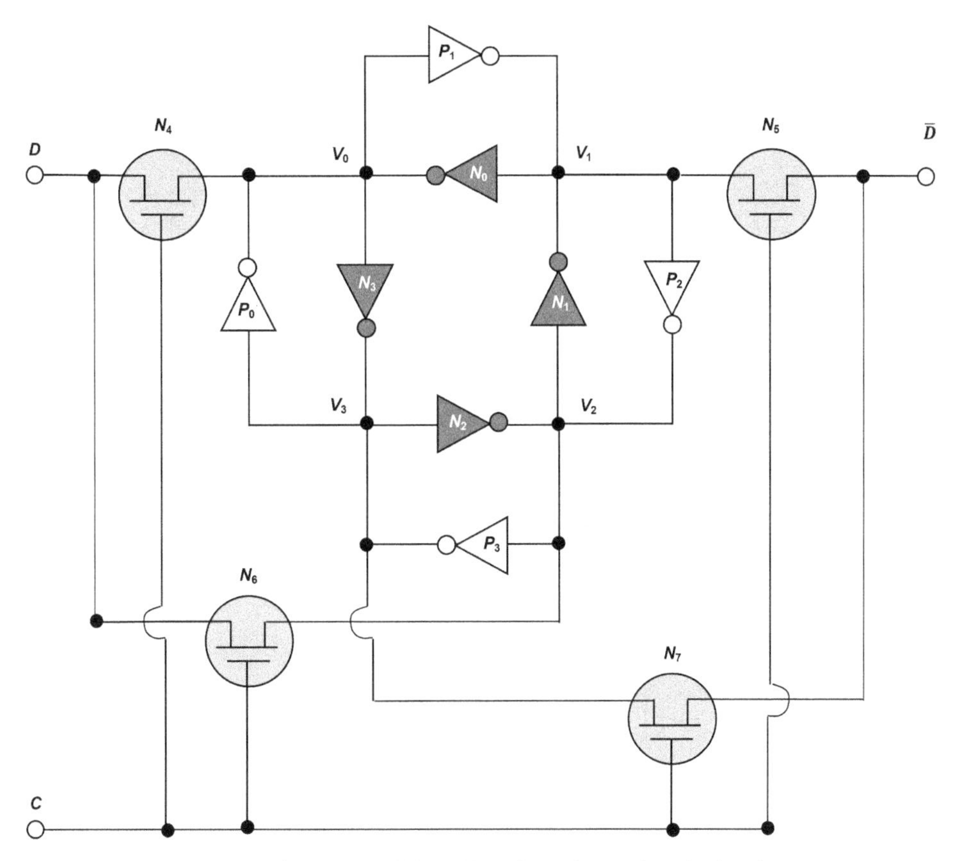

Figure 19.11. The concept of the DICE using a four-node redundant layout.

transistors represent a counterclockwise loop. The DICE circuit showing the PMOS and NMOS transistors is given in figure 19.12.

The DICE is based on the redundancy approach. By using two latch sections that store the same data, redundancy in the circuit maintains a source of uncorrupted data after an SEU. The data in the uncorrupted section furnish feedback for the restoring state for the recovery of corrupted data. Immunity to SEU is obtained through dual node feedback control. This implies that two adjoining nodes situated on the opposite diagonal govern the logic states of each of the four nodes of the cell. There is no direct interdependence of two nodes on each diagonal. Two nodes of the other diagonal control the states of these nodes.

Let us consider a node V_i ($i = 0,1,2,3$). Here the subscript i is a modulo 4 number, e.g. $0 \equiv 0$ modulo 4, $1 \equiv 1$ modulo 4, $2 \equiv 2$ modulo 4, $3 \equiv 3$ modulo 4, $4 \equiv 4$ modulo 4, but $-1 \equiv -1 \times 4 + 3 = 3$ modulo 4. With the help of single transistor complementary feedback connections through transistors N_{i-1} and P_{i+1}, any node V_i controls two complementary nodes on the opposite diagonal V_{i-1} and V_{i+1}. For $i = 0$, using transistors N_3 and P_1, the node V_0 controls the nodes V_3 and V_1.

In the logic state 0, i.e. $V_0V_1V_2V_3 = 0101$, the horizontal inverter loops (N_0, P_1) and (N_2, P_3) are in the conduction state. They represent two horizontal latches.

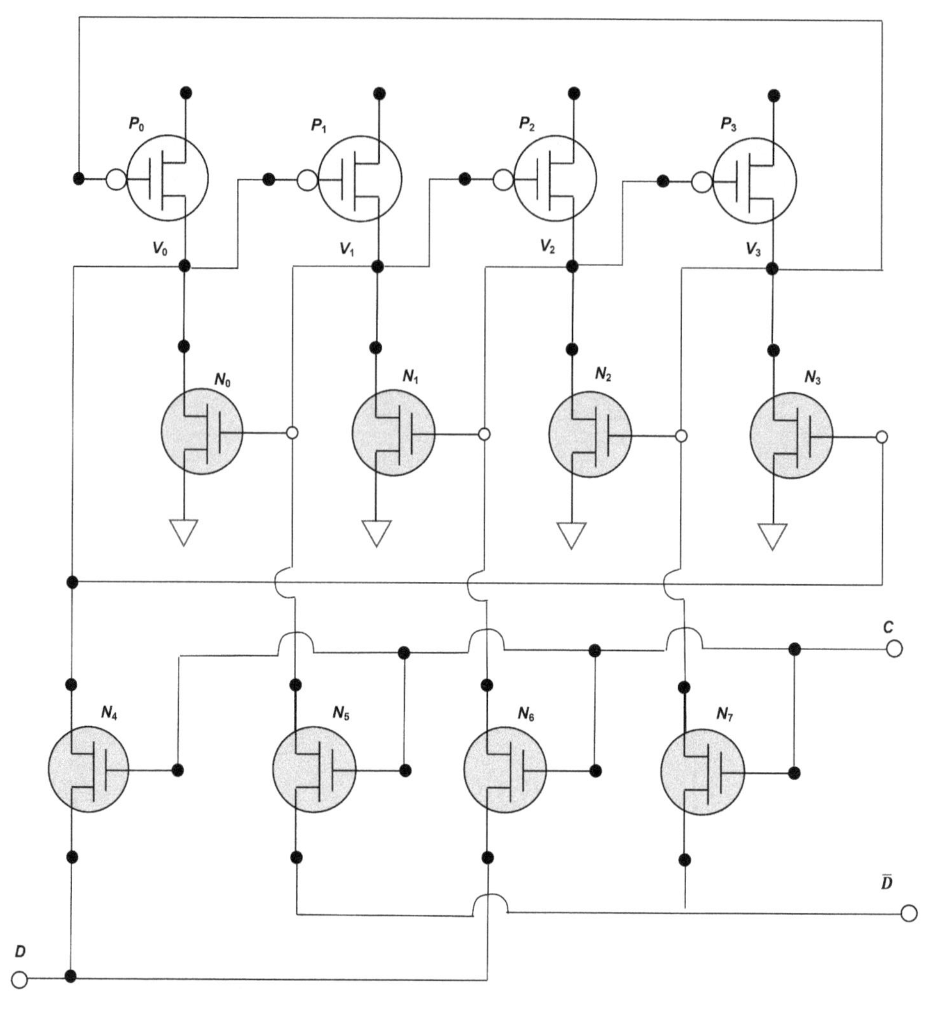

Figure 19.12. Transistor-level diagram of the DICE.

These latches store the same data at their nodes (V_0, V_1) and (V_2, V_3). Under this condition, the pairs of transistors (N_1, P_2) and (N_3, P_0) of the vertical inverters are non-conducting. They perform the role of an interlocking function by isolating the two horizontal latches from each other.

Similarly, in the logic state 1, i.e. $V_0V_1V_2V_3 = 1010$, the vertical inverter loops (N_1, P_2) and (N_3, P_0) are in the conduction state. They represent two vertical latches. These latches store the same data at their nodes (V_0, V_1) and (V_2, V_3). Under this condition, the pairs of transistors (N_0, P_1) and (N_2, P_3) of the horizontal inverters are non-conducting. They perform the role of interlocking function by isolating the two vertical latches from each other.

As the four nodes V_0, V_1, V_2, V_3 are sensitive to particle strikes, let us examine their immunity against radiation by taking the node V_i as an example.

- *Case I.* Suppose a negative upset pulse is produced at the node V_i. This negative pulse will induce a positive pulse disturbance at node V_{i+1} by feedback via PMOS transistor P_{i+1}. But it will be unable to exert any influence on the same logic state stored at the node V_{i-1}. This happens due to blockage of the feedback transistor N_{i-1} by the negative upset pulse at node V_i. Furthermore, the positive pulse disturbance induced at node V_{i+1} will not be onward transmitted through PMOS transistor P_{i+2}. Isolation of the nodes V_{i-1} and V_{i+2} helps in preservation of their logic states. Thus logic state changes take place only at the two nodes V_{i-1} and V_{i+1}. These changes are temporary. After the upset transient has departed, state-reinforcing feedback occurs by the nodes V_{i-1} and V_{i+2} via transistors P_i and N_{i+1}.
- *Case II.* Suppose a positive upset pulse is produced at the node V_i. The perturbation can be analyzed in a similar manner to case I. The positive pulse will disturb the node V_{i-1} via NMOS transistor N_{i-1}. After the upset transient has disappeared, the nodes V_{i-1} and V_{i+2} will capacitively maintain their states. They will restore the correct logic states of the two perturbed nodes via transistors N_i and P_{i+1}.
- *Case III.* If a single particle flips two sensitive nodes of the cell together at the same time, which store the same logic state, e.g. either nodes (V_0, V_2) or (V_1, V_3), recovery of the cell is not possible. Spacing apart the drain regions of the node pairs that are simultaneously sensitive on the cell layout can diminish the probability of occurrence of such a failure. This will minimize the chances of collection of the threshold charge that is adequate for upsetting the working of the cell.

It may be noted that ensuing latches may capture any glitch that occurs at the output of the cell. Although the cell itself can recover, such a glitch cannot be prevented from proceeding further.

A DICE may be 1.7–2.0 times larger than original cell, but this factor can be reduced to 1.5 by careful design.

19.4 Discussion and Conclusions

The process and design techniques for realizing radiation-tolerant ICs have been discussed. The radiation-hardened IC market constitutes only a small segment of the total IC outlay, which is dominated by the commercial IC market. Therefore, the trend of the semiconductor industry is towards reinforcing RHBD techniques using commercial IC processes instead of spending more capital on radiation-hardening IC processes.

Review exercises

19.1. Elucidate the meaning of the term ‘radiation-hardened electronics’. Distinguish between RHBP and RHBD.

19.2. How is the technological trend of downscaling MOS devices to smaller dimensions in synchrony with the requirement of a small volume of gate oxide for radiation hardening? Explain. How will you protect the thicker field oxides from radiation effects?

19.3. How do the adjustments of the impurity doping profiles and minority carrier lifetime help in offsetting radiation effects?
19.4. How do the n-well, p-well and triple-well CMOS processes differ? In what ways is the triple-well process helpful in avoiding SEU?
19.5. How does SOI technology overcome leakage current and latchup problems in semiconductor devices caused by radiation exposure?
19.6. What is the basic reason which makes SOI-based MOSFETs more immune to SEEs in comparison to planar bulk silicon MOSFETs? Besides radiation hardness, what are the additional advantages of adopting SOI technology?
19.7. What are the two types of SOI MOSFETs? Which type is more radiation hard? Why?
19.8. How does the positive charge in the BOX layer of a partially depleted SOI-MOSFET increase its leakage current?
19.9. What is the main idea behind RHBD? What sacrifices have to be made if RHBD is done?
19.10. How do the holes trapped in the bird's beak region of a conventional MOSFET degrade its performance? How does the edgeless MOSFET design avoid this degradation? What is the penalty paid?
19.11. What is a channel stopper? How does it decrease the leakage current flow?
19.12. What is guard ring? How does it increase the breakdown voltage of a shallow diffused p–n junction?
19.13. How does increasing the channel width to channel length ratio of a MOSFET aid in obviating SEU?
19.14. What is the temporal filtering technique for SEE mitigation? Why is it unpopular?
19.15. Explain the concept of triple modular redundancy? How does it decrease proneness to SEEs?
19.16. How is the likelihood of SEUs decreased by triple-path temporal redundancy implementation? Explain with reference to a data latch circuit.
19.17. Draw the circuit diagram of a DICE memory cell. Explain its operation when a negative/positive upset pulse is produced at an internal node and if a single particle flips two sensitive nodes of the cell simultaneously.

References

Calin T, Nicolaidis M and Velazco R 1996 Upset hardened memory design for submicron CMOS technology *IEEE Trans. Nucl. Sci.* **43** 2874–8

Holman W T 2004 Radiation-tolerant design for high performance mixed-signal circuits *Radiation Effects and Soft Errors in Integrated Circuits and Electronic Devices* ed R D Schrimpf and D M Fleetwood (Singapore: World Scientific) pp 76–7

Mukherjee S 2008 *Architecture Design for Soft Errors* (Amsterdam: Elsevier) pp 72–3

Myers D R 1998 A brief discussion of radiation hardening of CMOS microelectronics *SciTech* www.osti.gov/scitech/servlets/purl/2453

Chapter 20

Vibration-tolerant electronics

Since there is a limit to the ultimate strength of materials used in the fabrication of electronic devices and circuits, the logical way to protect an electronic assembly from the deleterious effects of shock and vibration is to isolate the device/assembly from the vibrating platform supporting it. This has led to the development of passive and active techniques of vibration isolation. This chapter attempts to describe some of the opportunities at hand under both categories of vibration isolators, and provides glimpses of the applications where they are deployable. The mathematical theory of passive vibration isolation is developed, starting from free undamped vibrations and moving to forced damped vibrations. It is shown how at frequencies above the natural frequency of the system, the vibration isolator acts as a low-pass mechanical filter. The clear-cut superiority of active vibration isolators at very low frequencies is emphasized, where they are used in a plethora of sophisticated high-precision instrumentation and machines, notably nanoscale microscopies, lithography equipment for semiconductor device fabrication, and so forth.

20.1 Vibration is omnipresent

The moment an electronic device leaves the factory on its way to the showroom or the customer, its survivability is time and again put to test against two adversaries: shock and vibration. Whether it is transported by road or through aerial or water routes, vibration is always a serious threat to it, although to different degrees. The applications of electronic equipment are diverse, covering domestic, office, industrial, military and space deployment. In a building, everything vibrates in the presence of heavy machinery, high-voltage AC (HVAC) transmission, nearby traffic and weather effects. In the industrial sector, shocks of 5–10 g may be encountered, and in military and space applications, shocks >100 g are expected. Even if handled with the utmost care, the chances of an accidental fall anywhere are never ruled out. Therefore, vibration-withstanding capability has to be built in the equipment to

doi:10.1088/978-0-7503-1155-7ch20

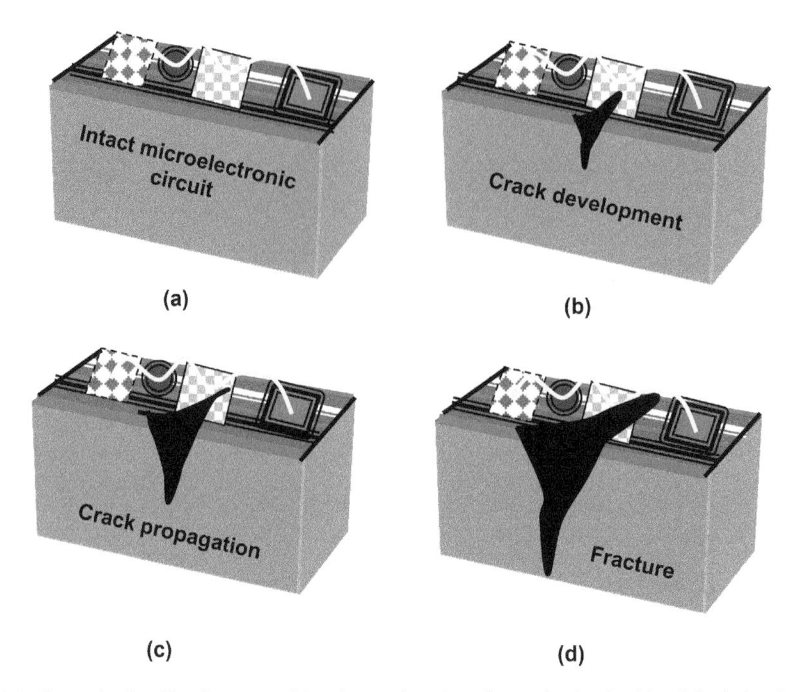

Figure 20.1. Stages in the vibration-caused breakage of a microelectronic circuit: (a) original circuit, (b) crack initiation, (c) crack spreading and (d) complete damage.

enable it to face real world situations. Figure 20.1 illustrates the progress of vibration-induced deterioration in microelectronic circuits.

20.2 Random and sinusoidal vibrations

Practically, vibrations are of random nature consisting of several frequencies. Electronic equipment loaded in a truck moving over a highway route experiences a random vibration according to the roughness and undulations of the surface of the road, whereas equipment placed near a rotating electric pump motor feels sinusoidal vibrations. A robust vibration-enduring chassis is required for protection against both types of vibration.

20.3 Countering vibration effects

There are two ways to counter vibration effects. One way is to make the devices and connections sufficiently sturdy and durable, so that they do not crumble under vibration attacks. But there is a limit to the ultimate strength achievable. The other solution is to prevent the vibration from harming the fragile parts of the equipment. Isolating the delicate parts from vibration effects can help to protect them, and this line of thought leads us to the interesting field of vibration isolators. A plethora of vibration isolators exist and these can be applied, as per convenience.

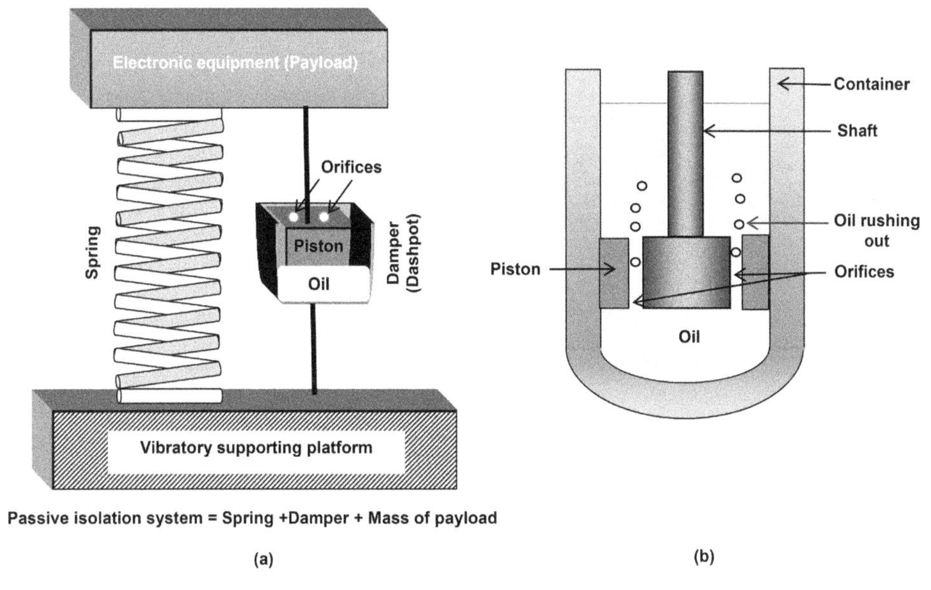

Figure 20.2. (a) Passive isolator and (b) dashpot.

20.4 Passive and active vibration isolators

Vibration isolators can be subdivided broadly into two classes. Passive isolators are conventional systems using springs and dampers (figure 20.2), which do not need any electrical power. Active vibration isolators use an electronic circuit and work on the feedback/feedforward principle to apply the calculated cancellation force of correct magnitude to impede the vibration. Let us begin with the passive vibration isolators. Figure 20.3 provides a glimpse of different vibration isolation systems.

20.5 Theory of passive vibration isolation

The primary role of a vibration isolator is to decouple or separate the vibrating supporting platform from the electronic circuit or device to be isolated. By providing a resilient connection between the vibrating supporting platform and the electronic circuit/device, it dissipates the energy of vibrations within itself so that this energy is not transmitted to the circuit/device, and the vibration fails to cause any performance perturbation.

An elementary vibration isolator consists of a mass, a spring and a dashpot or damper. A dashpot is a mechanical device to constrain vibration through viscous friction. A common example of a dashpot is the one used in a door closer to prevent it from slamming loudly. A linear dashpot consists of a piston moving in a hydraulic cylinder containing pressurized hydraulic fluid. Typically, oil is used as the fluid.

For simplification, let us assume the damping to be zero. Further, let us assume that the vibrations take place in the vertical direction x only. Displacement in the downward direction is taken as positive. In the upward direction it is taken as negative. Let us analyze the free vibrations of this system without damping.

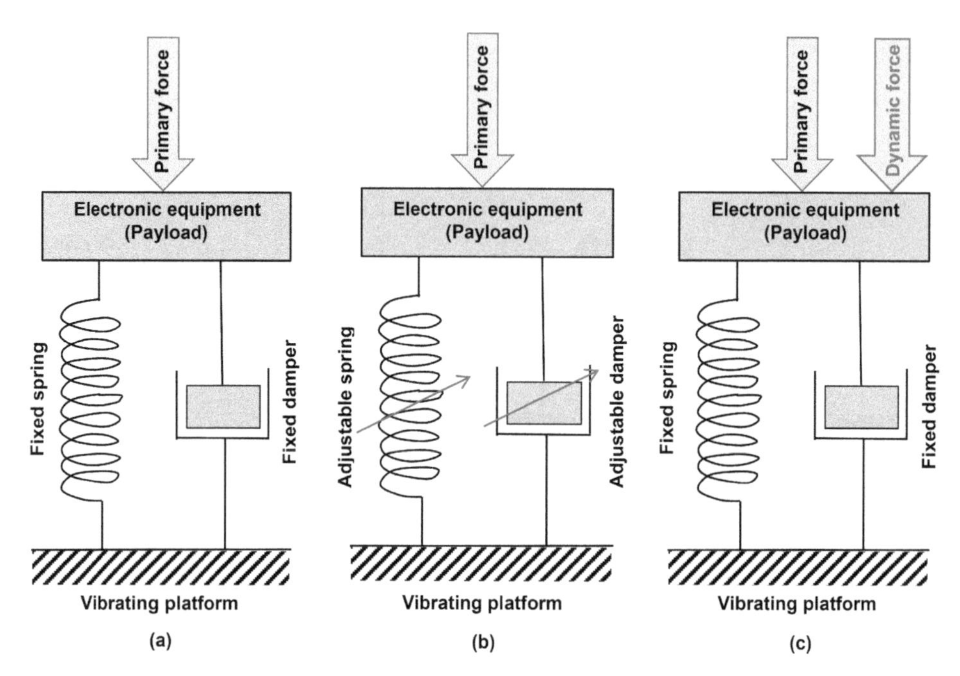

Figure 20.3. Vibration isolation systems: (a) fully passive with fixed values of spring and damper; (b) adaptive passive or semi-active with a variable spring stiffness and damping property to minimize the vibrations; and (c) fully active with dynamic forces applied to mitigate vibrations.

20.5.1 Case I: free undamped vibrations

For an undamped system (figure 20.4), if the payload (electronic equipment) has a mass m kg and the spring has stiffness k N m^{-1}, and if the deformation of the spring in static equilibrium is d, the force F acting on the spring is

$$F = kd \tag{20.1}$$

according to Hooke's law.

If g is the acceleration due to gravity, applying Newton's second law of motion, the gravitational force acting on mass m is

$$F = mg. \tag{20.2}$$

This force is also represented by F because the spring force = gravitational force.

From equations (20.1) and (20.2), we have

$$F = kd = mg. \tag{20.3}$$

Applying Newton's second law of motion to the mass m, we have

$$m\frac{\mathrm{d}^2x}{\mathrm{d}t^2} = \sum F = mg - k(d + x) = mg - kd - kx = mg - mg - kx = -kx$$

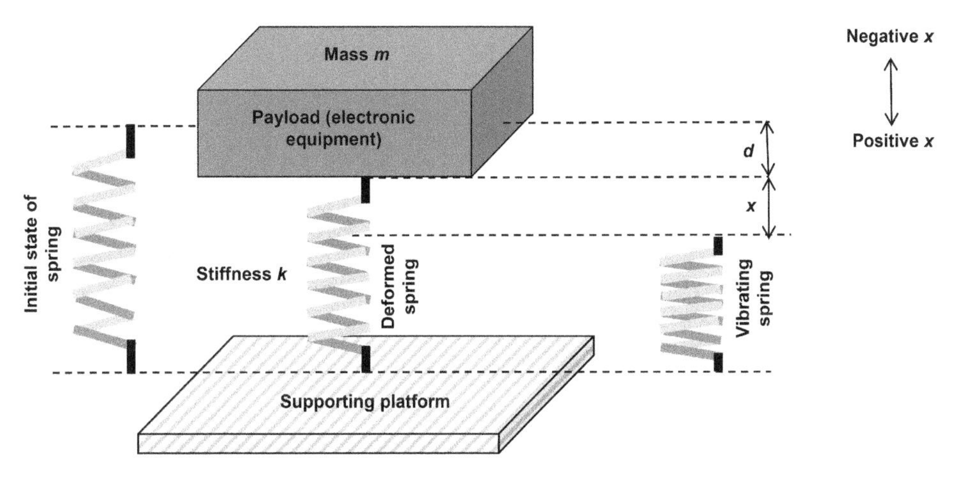

Figure 20.4. Free undamped vibrations of an electronic device.

or

$$m\frac{d^2x}{dt^2} + kx = 0. \tag{20.4}$$

This is a homogeneous second order linear differential equation with the solution

$$x(t) = \exp(pt). \tag{20.5}$$

Since

$$\begin{aligned}\frac{d^2x}{dt^2} &= \frac{d^2\{\exp(pt)\}}{dt^2} = \frac{d}{dt}\left[\frac{d\{\exp(pt)\}}{dt}\right] = \frac{d}{dt}\{p\exp(pt)\} \\ &= p\frac{d}{dt}\{\exp(pt)\} = p \times p\exp(pt) = p^2\exp(pt) \\ &\therefore mp^2\exp(pt) + k\exp(pt) = 0\end{aligned} \tag{20.6}$$

or

$$mp^2 + k = 0$$

or

$$p^2 = -k/m$$

or

$$p = \pm i\sqrt{k/m} = \pm i\omega_n \tag{20.7}$$

by putting

$$\omega_n = \sqrt{k/m}. \tag{20.8}$$

Hence the solution is

$$\begin{aligned} x(t) &= C\exp(+i\omega_n t) + D\exp(-i\omega_n t) \\ &= C\{\cos(\omega_n t)+i\sin(\omega_n t)\} + D\{\cos(\omega_n t)-i\sin(\omega_n t)\} \\ &= i(C-D)\sin(\omega_n t) + (C+D)\cos(\omega_n t) = A\sin(\omega_n t) + B\cos(\omega_n t), \end{aligned} \tag{20.9}$$

where the arbitrary constants A and B are obtained from the initial conditions.

$$\begin{aligned} &\text{At } t = 0,\ x(t) = x(0);\ \text{hence } x(0) = A\sin(0)+B\cos(0) = 0 + B = B \\ &\quad \text{or, } B = x(0). \end{aligned} \tag{20.10}$$

$$\text{At } t = 0,\ \dot{x}(t) = \dot{x}(0). \tag{20.11}$$

Since,

$$\dot{x}(t) = A\omega_n\cos(\omega_n t) - B\omega_n\sin(\omega_n t) \tag{20.12}$$

$$\dot{x}(0) = A\omega_n\cos(0) - B\omega_n\sin(0) = A\omega_n - 0 = A\omega_n \tag{20.13}$$

$$\therefore A = \dot{x}(0)/\omega_n. \tag{20.14}$$

Substituting the values of A and B into equation (20.9), we obtain

$$x(t) = \{\dot{x}(0)/\omega_n\}\sin(\omega_n t) + x(0)\cos(\omega_n t) \tag{20.15}$$

where

$$\begin{aligned} \omega_n &= \sqrt{k/m} = \sqrt{\text{stiffness/mass}} \\ &= \text{natural angular frequency of the system.} \end{aligned} \tag{20.16}$$

Hence, the natural frequency f_n is given by

$$f_n = \frac{\omega_n}{2\pi} = \frac{1}{2\pi}\sqrt{\frac{k}{m}}. \tag{20.17}$$

This frequency is the frequency at which the system will vibrate if it is pulled from its equilibrium static position and released.

20.5.2 Case II: forced undamped vibrations

Suppose the supporting platform undergoes sinusoidal vibrations under an excitation frequency Ω_s, and a sinusoidal applied force

$$F = F_0\sin(\Omega_s t), \tag{20.18}$$

where F_0 is the peak value of the force waveform following a trigonometric sine curve. The vibrating supporting platform is shown in figure 20.5.

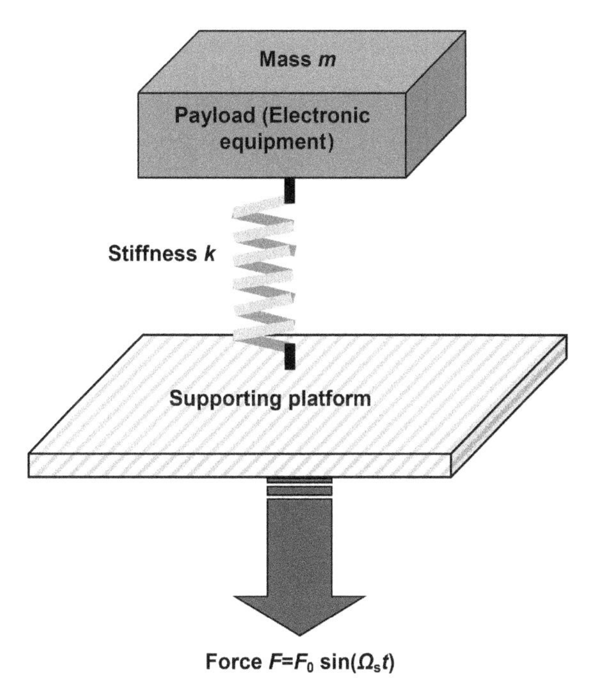

Figure 20.5. Forced vibrations of an electronic equipment without any damping.

Then equation (20.4) is modified to

$$\begin{aligned} m\frac{\mathrm{d}^2x}{\mathrm{d}t^2} &= \sum F = mg + F_0 \sin(\Omega_s t) - k(d + x) \\ &= mg + F_0 \sin(\Omega_s t) - kd - kx \\ &= mg + F_0 \sin(\Omega_s t) - mg - kx = F_0 \sin(\Omega_s t) - kx \end{aligned} \tag{20.19}$$

$$\begin{aligned} &\text{or } m\frac{\mathrm{d}^2x}{\mathrm{d}t^2} + kx = F_0 \sin(\Omega_s t) \\ &\text{or } \frac{\mathrm{d}^2x}{\mathrm{d}t^2} + \frac{k}{m}x = \frac{F_0}{m} \sin(\Omega_s t) \\ &\text{or } \frac{\mathrm{d}^2x}{\mathrm{d}t^2} + \omega_n^2 x = \frac{F_0}{m} \sin(\Omega_s t), \end{aligned} \tag{20.20}$$

from equation (20.16).

The solution of this second order differential equation comprises two parts: the complementary solution and the particular solution. To find the complementary solution, the right-hand side of the equation is set to zero obtaining

$$\frac{\mathrm{d}^2x}{\mathrm{d}t^2} + \omega_n^2 x = 0, \tag{20.21}$$

which turns out to be the same as equation (20.4) with $\omega_n = \sqrt{(k/m)}$. Therefore, it has the same solution as for the case of free vibration.

To find the particular solution, let us try the following value as a solution

$$x = X_0 \sin(\Omega_s t). \tag{20.22}$$

Differentiating twice with respect to t, we obtain

$$\frac{dx}{dt} = \Omega_s X_0 \cos(\Omega_s t) \tag{20.23}$$

$$\frac{d^2x}{dt^2} = -\Omega_s^2 X_0 \sin(\Omega_s t). \tag{20.24}$$

Therefore, from equation (20.20),

$$-\Omega_s^2 X_0 \sin(\Omega_s t) + \omega_n^2 X_0 \sin(\Omega_s t) = \frac{F_0}{m} \sin(\Omega_s t). \tag{20.25}$$

Dividing both sides by $\sin(\Omega_s t) \neq 0$, we obtain

$$\begin{aligned} &\text{or } -\Omega_s^2 X_0 + \omega_n^2 X_0 = \frac{F_0}{m} \\ &\text{or } \left(\omega_n^2 - \Omega_s^2\right) X_0 = \frac{F_0}{m}. \end{aligned} \tag{20.26}$$

Dividing throughout by ω_n^2,

$$\frac{\left(\omega_n^2 - \Omega_s^2\right) X_0}{\omega_n^2} = \frac{1}{\omega_n^2} \times \frac{F_0}{m} = \frac{1}{k/m} \times \frac{F_0}{m} = \frac{m}{k} \times \frac{F_0}{m} = \frac{F_0}{k} \tag{20.27}$$

$$\therefore X_0 = \frac{\omega_n^2}{\omega_n^2 - \Omega_s^2} \times \frac{F_0}{k} = \frac{F_0/k}{1 - \Omega_s^2/\omega_n^2} = \frac{F_0/k}{1 - r^2}, \tag{20.28}$$

where we have put

$$r = \Omega_s/\omega_n. \tag{20.29}$$

The force transmitted through the spring is written as

$$P = kx = kX_0 \sin(\Omega_s t) = P_0 \sin(\Omega_s t), \tag{20.30}$$

where

$$P_0 = kX_0 \tag{20.31}$$

is the amplitude of the transmitted force.

Transmissibility T is defined as the ratio of two amplitudes, namely, the amplitude of vibration of the electronic circuit/device to the amplitude of vibration of the supporting platform. It is expressed as

$$\begin{aligned} T &= \frac{\text{amplitude of the transmitted force}}{\text{amplitude of the applied force}} = \left|\frac{P_0}{F_0}\right| = \left|\frac{kX_0}{F_0}\right| \\ &= \left|\frac{k}{F_0} \times \frac{F_0/k}{1 - r^2}\right| = \left|\frac{1}{1 - r^2}\right|. \end{aligned} \tag{20.32}$$

The effectiveness of the vibration isolator is given by the equation

$$\% \text{ isolation} = (1 - T) \times 100. \tag{20.33}$$

A high value of T signifies poor isolation. On the opposite side, a low value of T indicates higher percentage isolation. The transmissibility curve showing the plot of transmissibility against frequency is sketched in figure 20.6. From the diagram, it may be noted that at frequencies Ω_s of the supporting platform which are lower than the natural frequency ω_n of the system, transmissibility is approximately 1. This means that the amplitude of vibration of the electronic circuit/device is the same as that of the supporting platform. At the above frequency values, the incoming vibration energy of the supporting platform is neither amplified nor decreased. The vibrating supporting platform and the electronic circuit/device vibrate in phase.

When the frequency Ω_s of vibration of the supporting platform is equal to the natural frequency ω_n of the system, the transmissibility reaches a peak value. This is the resonance condition under which the electronic circuit/device vibrates with maximum amplitude. The vibration energy of the supporting platform has been amplified. A transmissibility value >1 indicates amplification. The phase of the electronic circuit/device is shifted by 90° with respect to that of the supporting platform.

Now if the frequency Ω_s of vibration of the supporting platform exceeds the natural frequency ω_n of the system, the transmissibility starts to decrease and falls considerably at high frequencies. A transmissibility value <1 suggests attenuation. The electronic circuit/device vibrates in antiphase with the vibration source.

Thus lower frequencies Ω_s of vibration of the supporting platform are conveyed to the electronic circuit/device as such, producing vibrations of the same displacement in it as in the supporting platform. But frequencies Ω_s higher than the natural frequency ω_n of the system are weakened, so that the electronic circuit/device vibrates with progressively smaller amplitudes. This essentially means that higher frequencies of vibration of the supporting platform are prevented from affecting the electronic circuit/device; hence the electronic circuit/device is successfully separated from the fast vibrating supporting platform. Another noteworthy inference is that the higher frequencies have been filtered. So, the behavior of the vibration isolator is analogous to that of a low-pass mechanical filter. Obviously, the best vibration isolation system will be one with a very low natural frequency because all higher frequencies will be sieved off.

For effective vibration isolation, the ratio (Ω_s/ω_n) should be >1.4 and ideally, it should be >2–3.

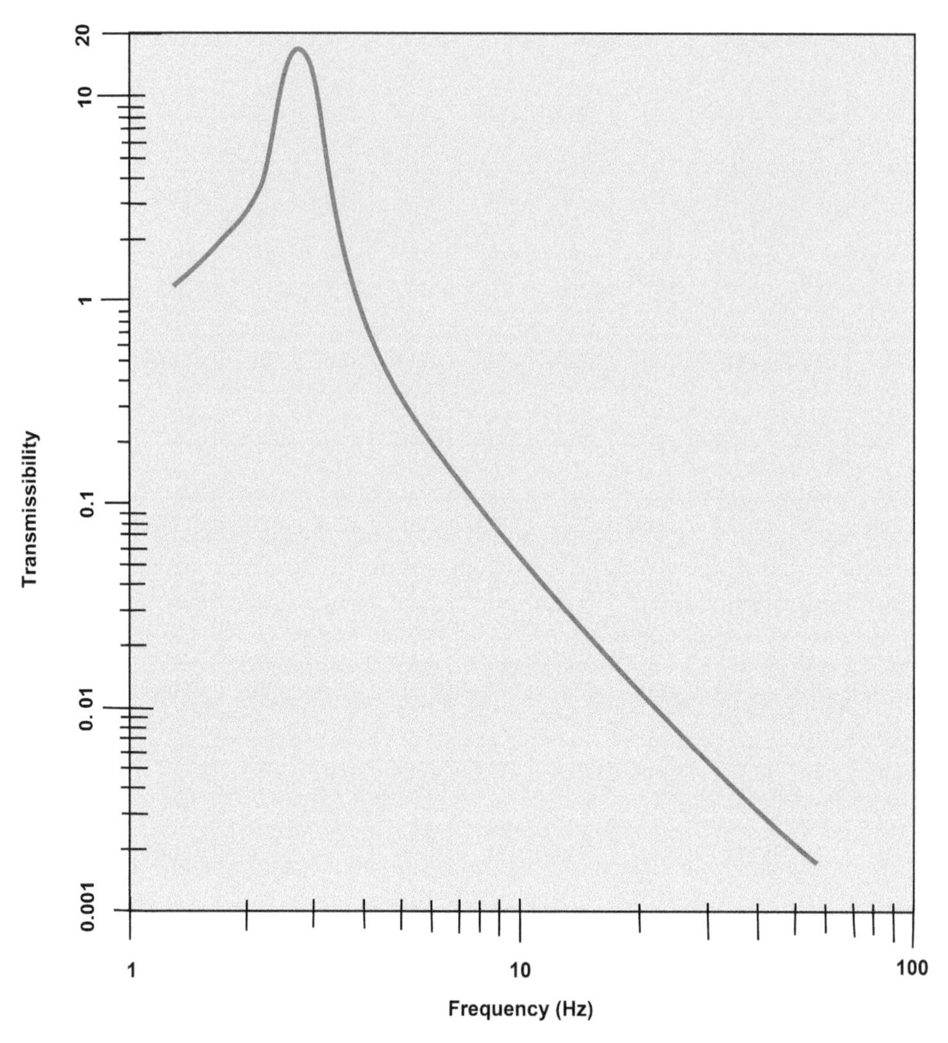

Figure 20.6. Transmissibility curve.

20.5.3 Case III: forced vibrations with viscous damping

For this situation, the equation of motion for undamped forced vibrations is modified as follows by including the damping effect through the term $c(\mathrm{d}x/\mathrm{d}t)$

$$m\frac{\mathrm{d}^2x}{\mathrm{d}t^2} + c\frac{\mathrm{d}x}{\mathrm{d}t} + kx = F_0 \sin(\Omega_s t), \tag{20.34}$$

where c is the damping coefficient (figure 20.7).

By dividing both sides by m, this equation is rewritten in the form

$$\frac{\mathrm{d}^2x}{\mathrm{d}t^2} + \frac{c}{m}\frac{\mathrm{d}x}{\mathrm{d}t} + \frac{k}{m}x = \frac{F_0}{m}\sin(\Omega_s t) \tag{20.35}$$

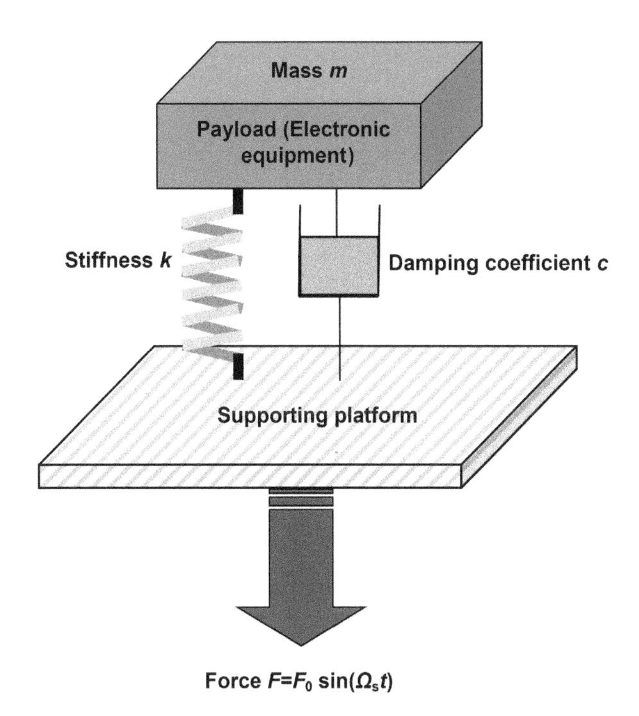

Figure 20.7. Forced vibrations of an electronic device with viscous damping.

$$\text{or } \frac{d^2x}{dt^2} + \frac{2c}{2\sqrt{km}} \times \sqrt{\frac{k}{m}}\frac{dx}{dt} + \frac{k}{m}x = \frac{F_0}{m}\sin(\Omega_s t)$$
$$\text{or } \frac{d^2x}{dt^2} + 2\zeta\omega_n\frac{dx}{dt} + \omega_n^2 x = \frac{F_0}{m}\sin(\Omega_s t), \tag{20.36}$$

where we have put

$$\frac{c}{2\sqrt{km}} = \zeta \tag{20.37}$$

and

$$\sqrt{\frac{k}{m}} = \omega_n. \tag{20.38}$$

Let us try a particular solution to this equation of the form

$$x = X_0 \sin(\Omega_s t - \phi). \tag{20.39}$$

Then,

$$\frac{dx}{dt} = \Omega_s X_0 \cos(\Omega_s t - \phi) \tag{20.40}$$

and

$$\frac{d^2x}{dt^2} = -\Omega_s^2 X_0 \sin(\Omega_s t - \phi). \tag{20.41}$$

Hence, equation (20.36) becomes

$$\begin{aligned} &-\Omega_s^2 X_0 \sin(\Omega_s t - \phi) + 2\zeta\omega_n\Omega_s X_0 \cos(\Omega_s t - \phi) \\ &+\omega_n^2 X_0 \sin(\Omega_s t - \phi) = \frac{F_0}{m}\sin(\Omega_s t) \end{aligned} \tag{20.42}$$

$$\begin{aligned} &\text{or } \left(\omega_n^2 - \Omega_s^2\right)X_0 \sin(\Omega_s t - \phi) + 2\zeta\omega_n\Omega_s X_0 \cos(\Omega_s t - \phi) = \frac{F_0}{m}\sin(\Omega_s t) \\ &\text{or } \left(\omega_n^2 - \Omega_s^2\right)X_0\left(\frac{m}{F_0}\right)\sin(\Omega_s t - \phi) + 2\zeta\omega_n\Omega_s X_0\left(\frac{m}{F_0}\right)\cos(\Omega_s t - \phi) = \sin(\Omega_s t). \end{aligned} \tag{20.43}$$

$$\text{Let us put } \left(\omega_n^2 - \Omega_s^2\right)X_0\left(\frac{m}{F_0}\right) = \cos\phi \tag{20.44}$$

and

$$2\zeta\omega_n\Omega_s X_0\left(\frac{m}{F_0}\right) = \sin\phi. \tag{20.45}$$

Then

$$\cos\phi\sin(\Omega_s t - \phi) + \sin\phi\cos(\Omega_s t - \phi) = \sin(\Omega_s t) \tag{20.46}$$

$$\text{or } \sin\left(\phi + \Omega_s t - \phi\right) = \sin(\Omega_s t), \tag{20.47}$$

which means that the equation is satisfied. Now,

$$\begin{aligned} \cos^2\phi + \sin^2\phi = 1 &= \left\{\left(\omega_n^2 - \Omega_s^2\right)X_0\left(\frac{m}{F_0}\right)\right\}^2 + \left\{2\zeta\omega_n\Omega_s X_0\left(\frac{m}{F_0}\right)\right\}^2 \\ &= X_0^2\left(\frac{m}{F_0}\right)^2\left\{\left(\omega_n^2 - \Omega_s^2\right)^2 + (2\zeta\omega_n\Omega_s)^2\right\} \end{aligned} \tag{20.48}$$

$$\begin{aligned} \therefore X_0^2 &= \frac{1}{\left(\frac{m}{F_0}\right)^2\left\{\left(\omega_n^2 - \Omega_s^2\right)^2 + (2\zeta\omega_n\Omega_s)^2\right\}} = \frac{\left(\frac{F_0}{m}\right)^2}{\left\{\left(\omega_n^2 - \Omega_s^2\right)^2 + (2\zeta\omega_n\Omega_s)^2\right\}} \\ &= \frac{F_0^2(k/m)^2 \times (1/k)^2}{\left\{\left(\omega_n^2 - \Omega_s^2\right)^2 + (2\zeta\omega_n\Omega_s)^2\right\}} = \frac{F_0^2\omega_n^4 \times (1/k)^2}{\left\{\left(\omega_n^2 - \Omega_s^2\right)^2 + (2\zeta\omega_n\Omega_s)^2\right\}} \end{aligned} \tag{20.49}$$

giving

$$\begin{aligned} X_0 &= \frac{F_0\omega_n^2/k}{\sqrt{\left(\omega_n^2-\Omega_s^2\right)^2+(2\zeta\omega_n\Omega_s)^2}} = \frac{F_0/k}{\sqrt{\left\{\left(\omega_n^2-\Omega_s^2\right)^2+(2\zeta\omega_n\Omega_s)^2\right\}/\omega_n{}^4}} \\ &= \frac{F_0/k}{\sqrt{\left(1-\Omega_s^2/\omega_n^2\right)^2+(2\zeta\Omega_s/\omega_n)^2}} = \frac{F_0/k}{\sqrt{(1-r^2)^2+(2\zeta r)^2}}. \end{aligned} \tag{20.50}$$

Also,

$$\begin{aligned} \frac{\sin\phi}{\cos\phi} &= \tan\phi = \frac{2\zeta\omega_n\Omega_s X_0\left(\frac{m}{F_0}\right)}{\left(\omega_n^2-\Omega_s^2\right)X_0\left(\frac{m}{F_0}\right)} = \frac{2\zeta\omega_n\Omega_s}{\omega_n^2-\Omega_s^2} = \frac{(2\zeta\omega_n\Omega_s)/\omega_n^2}{\left(\omega_n^2-\Omega_s^2\right)/\omega_n^2} \\ &= \frac{2\zeta(\Omega_s/\omega_n)}{1-\Omega_s^2/\omega_n^2} = \frac{2\zeta r}{1-r^2}. \end{aligned} \tag{20.51}$$

Force transmitted to the electronic circuit/device is

$$\begin{aligned} P &= c\frac{dx}{dt} + kx = c\Omega_s X_0\cos(\Omega_s t-\phi) + kX_0\sin(\Omega_s t-\phi) \\ &= \frac{c\Omega_s(F_0/k)}{\sqrt{(1-r^2)^2+(2\zeta r)^2}}\cos(\Omega_s t-\phi) + \frac{k(F_0/k)}{\sqrt{(1-r^2)^2+(2\zeta r)^2}}\sin(\Omega_s t-\phi) \\ &= \frac{F_0}{\sqrt{(1-r^2)^2+(2\zeta r)^2}}\sin(\Omega_s t-\phi) + \frac{c\Omega_s(F_0/k)}{\sqrt{(1-r^2)^2+(2\zeta r)^2}}\cos(\Omega_s t-\phi). \end{aligned} \tag{20.52}$$

This is of the form

$$\begin{aligned} P &= P_0\sin(\Omega_s t-\phi+\delta) = P_0\sin(\Omega_s t-\phi)\cos\delta + P_0\cos(\Omega_s t-\phi)\sin\delta \\ &= P_0\cos\delta\sin(\Omega_s t-\phi) + P_0\sin\delta\cos(\Omega_s t-\phi). \end{aligned} \tag{20.53}$$

Comparing equations (20.52) and (20.53), we can write

$$\frac{F_0}{\sqrt{(1-r^2)^2+(2\zeta r)^2}} = P_0\cos\delta \tag{20.54}$$

and

$$\frac{c\Omega_s(F_0/k)}{\sqrt{(1-r^2)^2+(2\zeta r)^2}} = P_0\sin\delta \tag{20.55}$$

$$\therefore \frac{F_0{}^2+\{c\Omega_s(F_0/k)\}^2}{(1-r^2)^2+(2\zeta r)^2} = P_0{}^2(\cos^2\delta+\sin^2\delta) = P_0{}^2(1) \tag{20.56}$$

or

$$P_0 = \sqrt{\frac{F_0{}^2 + \{c\Omega_s(F_0/k)\}^2}{(1-r^2)^2 + (2\zeta r)^2}} = F_0\sqrt{\frac{1 + \{c\Omega_s(1/k)\}^2}{(1-r^2)^2 + (2\zeta r)^2}} = F_0\sqrt{\frac{1 + \left\{\frac{2c\Omega_s}{2k}\right\}^2}{(1-r^2)^2 + (2\zeta r)^2}}$$

$$= F_0\sqrt{\frac{1 + \left\{\frac{2c\Omega_s}{2\sqrt{km}\sqrt{\frac{k}{m}}}\right\}^2}{(1-r^2)^2 + (2\zeta r)^2}} = F_0\sqrt{\frac{1 + \left\{\frac{2\zeta\Omega_s}{\omega_n}\right\}^2}{(1-r^2)^2 + (2\zeta r)^2}} = F_0\sqrt{\frac{1 + (2\zeta r)^2}{(1-r^2)^2 + (2\zeta r)^2}}, \quad (20.57)$$

where equations (20.16), (20.29) and (20.37) have been applied.

Further,

$$\frac{P_0 \sin\delta}{P_0 \cos\delta} = \tan\delta = \frac{\frac{c\Omega_s(F_0/k)}{\sqrt{\left(1-r^2\right)^2 + (2\zeta r)^2}}}{\frac{F_0}{\sqrt{\left(1-r^2\right)^2 + (2\zeta r)^2}}} = \frac{c\Omega_s(F_0/k)}{F_0} = c\Omega_s/k = \frac{2c\Omega_s}{2\sqrt{km}\sqrt{\frac{k}{m}}} \quad (20.58)$$

$$= 2\zeta\frac{\Omega_s}{\omega_n} = 2\zeta r$$

by applying equations (20.16), (20.29) and (20.37).

From equation (20.57), the transmissibility T is

$$\boldsymbol{T} = \frac{P_0}{F_0} = F_0\frac{\sqrt{\frac{1 + (2\zeta r)^2}{\left(1-r^2\right)^2 + (2\zeta r)^2}}}{F_0} = \sqrt{\frac{1 + (2\zeta r)^2}{(1-r^2)^2 + (2\zeta r)^2}}. \quad (20.59)$$

The phenomenon of damping is correlated with the resonance condition. The reason for this statement is that damping has a pronounced effect on the transmissibility value at resonance or in its neighborhood. Hence, damping may be deemed essentially to be an effect associated with resonance. At resonance, the value of $r = 1$. Then the transmissibility attains its maximum magnitude. So, under the resonance condition, equation (20.59) reduces to the simplified form

$$T_{\text{maximum}} = \sqrt{\frac{1 + (2\zeta)^2}{(1-1)^2 + (2\zeta)^2}} = \sqrt{\frac{1 + (2\zeta)^2}{0 + (2\zeta)^2}} = \sqrt{1 + \frac{1}{(2\zeta)^2}} \approx \sqrt{\frac{1}{(2\zeta)^2}} = \frac{1}{2\zeta}. \quad (20.60)$$

At resonance, the height of the transmissibility peak is determined by damping. In the absence of damping, the peak value of transmissibility at resonance is infinity. Moreover, without damping the vibrations will perpetuate forever and will not cease after withdrawal of the exciting force. In practice, systems cannot be made free from damping. Damping is always present in all systems. Only the degree of damping varies. The value of damping may range from very small to very large.

20.6 Mechanical spring vibration isolators

This is the kind of vibration isolators which immediately come to mind, because we are accustomed to their widespread deployment in motor vehicles such as cars, trucks and buses. They absorb vibration and provide us a comfortable journey on rough and bumpy roads.

20.7 Air-spring vibration isolators

A tube filled with air serves as a cushion. This cushioning effect is utilized to protect the overlying equipment from damage. Air springs are manually filled air bladders that can be filled with a hand-operated pump. A compressed air tank can also be used to replenish the air.

Air spring vibration isolators are of two types:

(i) *Bellows air bags.* These are configured in single, double, or triple convoluted chambers. The chambers are made of durable reinforced rubber with the capability to load and carry large weights. Bellow air bags find heavyweight applications where sufficient space is obtainable to accommodate them.

(ii) *Rolling sleeve air bags.* They contain an internally mounted sleeve with an internally molded bead and covered with an air bag, which can be inflated. The assembly provides a smaller diameter than the bellows design. They are useful to support lighter weights. They are suitable when vacant space is scarce.

20.8 Wire-rope isolators

These isolators are based on the spring action of a rope coil over which the crucial electronic part is suspended.

20.9 Elastomeric isolators

Elastomer is a generic term applied to all types of rubber. These rubbers may be natural or synthetic. The elastomers are elastic polymers, e.g. India rubber, silicone rubber, fluorosilicone rubber, butyl rubber, etc. Elastomeric isolators utilize the suppleness and bounciness provided by a rubbery compound to recover its size and shape when deformed. The homogeneous nature of elastomers allows them to be shaped into compact forms such as planar, laminate, cylindrical, etc.

Natural rubber serves as a baseline for comparing the relative performance of different elastomers. It has high strength, excellent fatigue properties and provides low-to-medium damping. But it suffers from the drawback of a narrow operating temperature range from −18 °C to +82 °C. It tends to stiffen at lower temperatures. Several silicone elastomers have been developed with broader temperature ranges.

20.10 Negative stiffness isolators

Hold a flexible plastic scale upright on the table (figure 20.8). Press it downwards by hand by applying a force at its top end. It bends (May 2001). As soon as the force is released, the plastic ruler regains its original straight shape. The recovery force acts

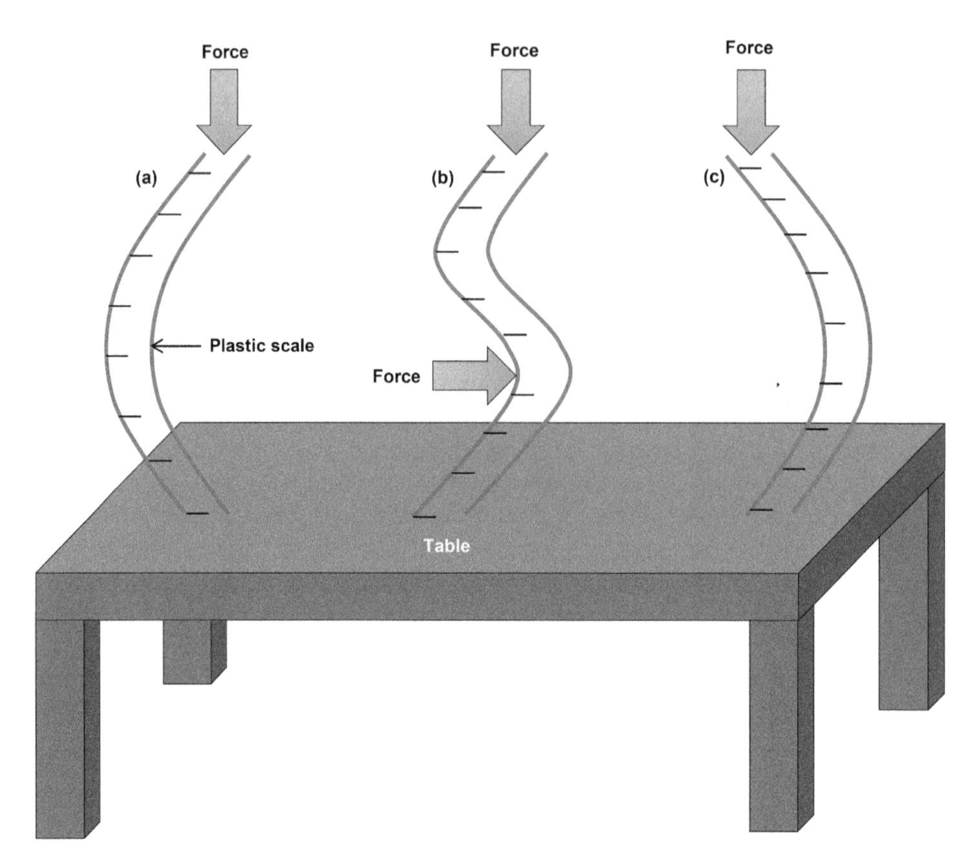

Figure 20.8. Negative thickness effect in the deformation of a plastic scale: (a) a ruler pressed from the top, (b) a ruler pressed from the top and with a finger applied at the middle and (c) subsequent bending of the ruler towards the opposite side.

opposite to the deformation force. This is positive stiffness. Now suppose when the ruler is deformed, a finger is pressed in the middle of the ruler against its bulging side. Then the ruler acquires the *S*-shape shown in the diagram. If the finger pressure is maintained the ruler bulges out on the opposite side. This is a case of negative stiffness. A force acts on the ruler in the same direction as the deforming force and aids the buckling.

To utilize the negative stiffness effect, generally three isolators are organized in series (Platus 1991). A tilt-motion isolator is perched above a horizontal motion isolator placed over a vertical motion isolator.

The vertical motion isolator (figure 20.9) comprises a customary spring squeezed by the weight W to the working point. This spring is connected to a structure made of two bars. It is these bars which provide the negative thickness effect. The bars are hinged at the center. At the extremities, they rest on pivots. Compression forces F, F are applied to the bars. The stiffness k_{Vertical} of the isolator is expressed as

$$k_{\text{Vertical}} = k_{\text{Spring}} - k_{\text{Bar effect}}. \tag{20.61}$$

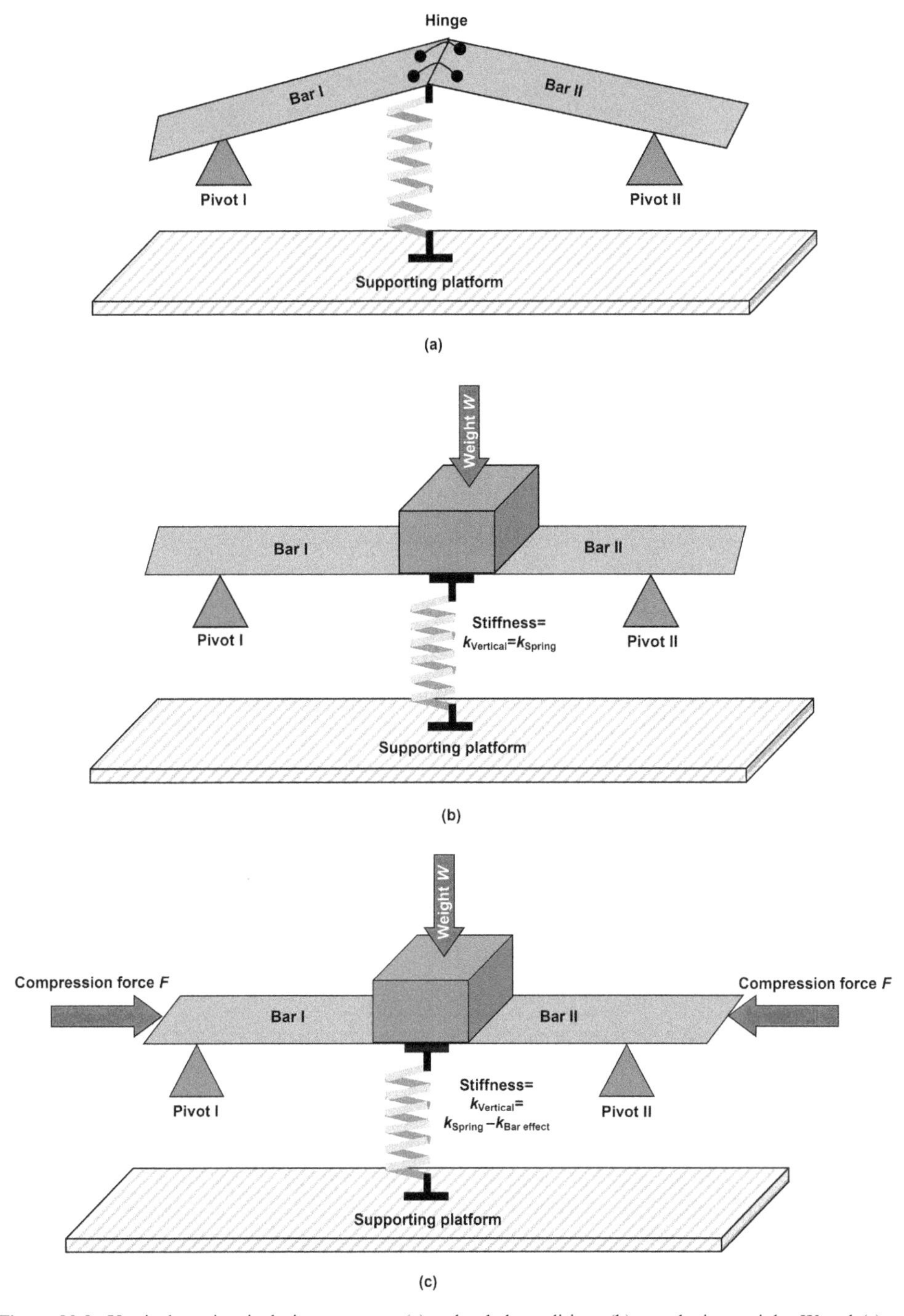

Figure 20.9. Vertical motion isolation concept: (a) unloaded condition, (b) on placing weight W and (c) on applying compressive forces F, F.

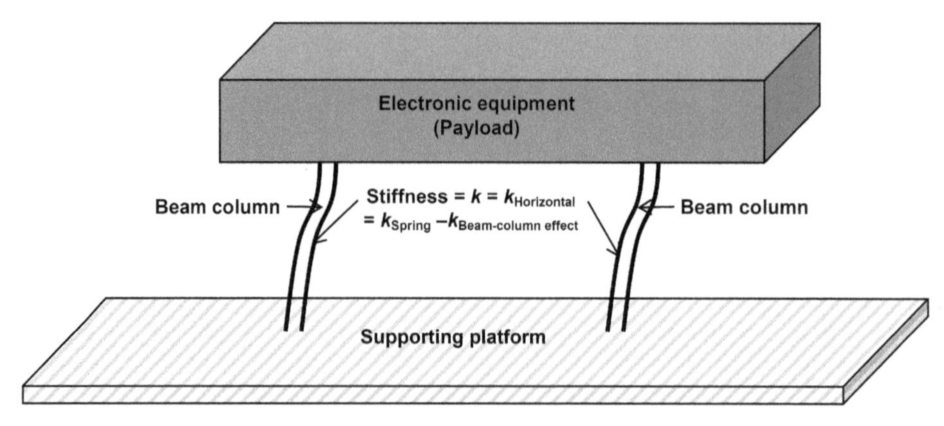

Figure 20.10. Principle of horizontal motion isolation.

$k_{\text{Bar effect}}$ depends on the length of the bars and the force F. The resulting stiffness of the isolator can be reduced to zero while the weight W is supported by the spring.

The horizontal motion isolator (figure 20.10) consists of two beam-column isolators. Considering single isolators, each isolator works in the same manner as two fixed-free beam columns. These beam columns are loaded in the axial direction by the weight W. The horizontal stiffness of the beam columns in the absence of weight load is k_{Spring}. Application of the weight load decreases the lateral bending stiffness by the beam–column effect. The arrangement behaves as a horizontal spring combined with a negative stiffness mechanism. The horizontal stiffness $k_{\text{Horizontal}}$ is given by

$$k_{\text{Horizontal}} = k_{\text{Spring}} - k_{\text{Beam-column effect}}. \tag{20.62}$$

By loading the beam columns in such a way that they attain their critical buckling load, the horizontal stiffness can be made nearly zero.

20.11 Active vibration isolators

All the above isolators belong to the category of passive isolators, which do not use any power supply. We now look at active vibration isolators (Yoshioka and Murai 2002).

20.11.1 Working

Feedback active vibration isolators are reactive in nature. They work by modifying the condition of the vibrating platform on the basis of data retrieved (Shen *et al* 2013). They use a piezoelectric sensor such as an accelerometer to measure the vibration. Along with this sensor, they employ a force actuator such as a loudspeaker voice coil to deliver a counterforce or an anti-phase signal on the vibrating object to oppose the vibration (figure 20.11). In this kind of isolator, a continuous vigil is kept on the platform to be controlled and the required counterforce is applied in accordance with the instantaneous vibration level. In contrast, feedforward active vibration isolators are anticipative in nature. They are deployed in applications after

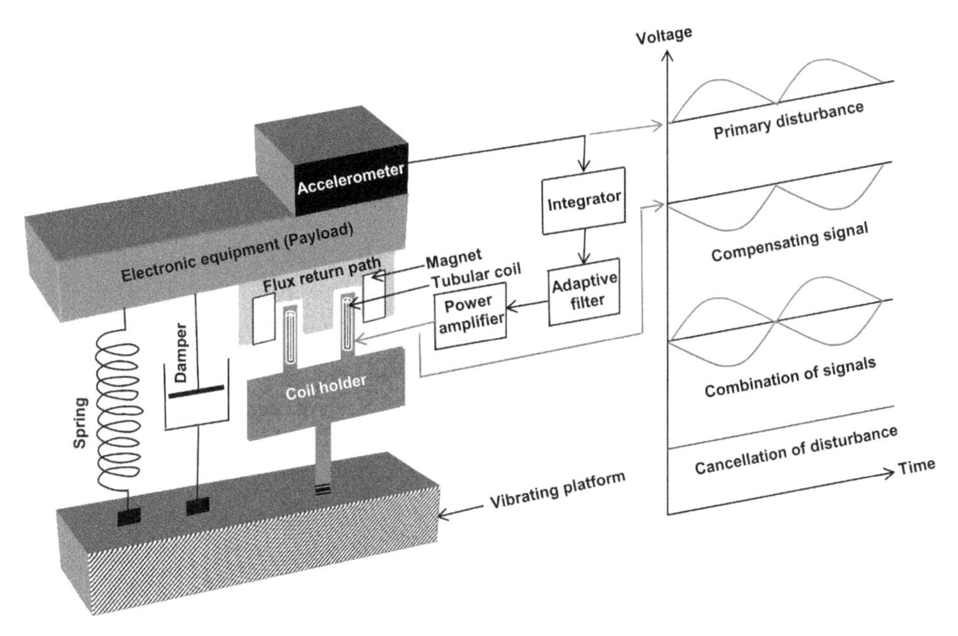

Figure 20.11. Main parts of an active vibration isolator.

detailed information on the vibrating characteristics of the platform has been acquired. Only then it is possible to send appropriate pre-decided signals to control the vibrations.

An active air isolator has a compressor with servo valves (Shaidani 2008). These valves either feed air into the bladder or bleed air from it to maintain the payload at zero vibration level.

20.11.2 Advantages

A major advantage in favor of active isolation systems over passive systems concerns their ability to isolate very low frequencies <10 Hz and even sub-hertz. Such vibrations are common in buildings near subways, and can lead to damage through high-amplitude vibrations via resonances. These low-frequency resonances are easily terminated through closed-loop design by feeding the necessary compensation. Active isolation systems perform better at frequencies <100 Hz; the isolation range is >0.7 Hz. They have neither any resonance nor amplify any vibrations at any frequency. Above 100 Hz, passive systems are useful; the isolation range is >5 Hz. Passive pneumatic systems are known to amplify vibrations instead of suppressing them in the frequency range 1–8 Hz. At frequencies >30 Hz, 98%–99% isolation is achievable with both active and passive systems. Another favorable feature is the increased stiffness, >100 times, of these systems, providing superior directional and positional stability for a precision instrument. Softer systems take a longer time to settle down once disturbed and are also influenced by air currents. Active systems settle down in 10–20 ms whereas passive ones have a longer settling time ~2–10 s. Further, it is possible to adjust the active system robotically, without any human intervention. This adjustment may be performed for a varying load. It can also be performed in the case in which the load is spread out on the platform. However, active systems are more expensive than passive systems.

20.11.3 Applications

Active vibration isolation systems come in a variety of form factors and sizes. They can debilitate vibrations in various translational and rotational modes. They can do so in real time in all six degrees of freedom (x, y, z: roll, pitch and yaw). They are widely used in atomic force and scanning tunneling microscopy (Lan *et al* 2004) where atomic scale resolution is at stake; space telescopes, laser communication systems, as well as in interferometry and metrology. An example of the use of an active system is in lithography process in the semiconductor industry. The silicon wafer is placed on a heavy stage and positioned with great accuracy. The resolution levels are reaching 1 nm and even the slightest vibration can undermine the process. An active system is immensely useful for such critical processes. Similar remarks apply to all manufacturing processes and scientific experiments where such low-dimensional structures are involved.

20.12 Discussion and conclusions

To prevent damage to the delicate electronic equipment from all-prevailing and unavoidable vibration conditions, it is necessary to design ingenious vibration isolation mechanisms. The techniques vary with the intended field of operation of the device, cost considerations and the accuracy expectations of the customer. Obviously, instruments aiming at nanometer accuracy cannot tolerate the least disturbing vibrations, whereas those for crude measurements have a much higher level of endurance. Therefore, each application has a unique solution and no versatile design philosophy of universally applicability can be laid out.

Review exercises

20.1. Why is it necessary to build vibration tolerance capability into electronic equipment? How does sinusoidal vibration differ from random vibration?

20.2. What are the two approaches followed to counter the effects of vibration on electronics? However strong an electronic device may be built to withstand vibration, it is necessary to isolate it from vibration. Why? What are the two categories of vibration isolators? How do they differ?

20.3. What is the main function of a vibration isolator? What are the chief components of a vibration isolator? What is a dashpot?

20.4. Frame and solve the differential equation for free undamped vibrations of a payload of mass m supported by a spring of stiffness k. What are the vital parameters which decide the natural frequency of the system?

20.5. Write and solve the differential equation for forced vibrations of a vibration isolation system consisting of a spring and a damper. How does it differ from that of free vibration system?

20.6. What is meant by transmissibility of the vibration isolator? A vibration isolator has a transmissibility of 0.2. What does this low value of transmissibility indicate? What is the meaning of a transmissibility value = 1?
20.7. Derive an equation for the transmissibility of a vibration isolator in terms of the natural frequency of the payload and the exciting frequency of the platform. How is transmissibility affected by the relative magnitudes of these frequencies?
20.8. The ratio of magnitudes of the natural frequency of the payload to the exciting frequency of the platform is 0.2. Is the vibration isolation effective? Why?
20.9. A vibration isolation system has a damping value of zero. What is the transmissibility of vibrations at resonance? Hence, explain the role of damping in determining transmissibility.
20.10. What is an air-spring vibration isolator? Name the two types of air-spring isolators and describe their operation.
20.11. What is an elastomer? What is the advantage of synthetic silicone elastomers over natural rubber?
20.12. What is meant by positive stiffness and negative stiffness? Explain with an example.
20.13. Explain with a diagram the working of a vertical motion isolator. What are the parameters on which vertical stiffness depends?
20.14. Draw the diagram of a horizontal motion isolator and describe its operation. Indicate the parameters deciding the horizontal stiffness.
20.15. How does a feedback active vibration isolator differ from a feedforward vibration isolator? In what situations are the feedback and feedforward isolation concepts applied?
20.16. What ranges of the vibration frequencies are accommodated by active and passive vibration isolator systems? What will happen if a passive system is used for controlling very low frequency vibrations?
20.17. How do passive and active vibration isolators differ regarding the stiffness parameter? How does the difference in stiffness translate into a difference in settling time after a disturbance from mean position?
20.18. Give examples illustrating the desperate necessity of an active vibration isolator? Justify the criticality of the situations.

References

Lan K J, Yen J-Y and Kramar J A 2004 Active vibration isolation for a long range scanning tunneling microscope *Asian J. Control* **6** 179–86

May M 2001 Getting more stiffness with less *Am. Sci* November-December www.american-scientist.org/issues/pub/getting-more-stiffness-with-less

Platus D L 1991 Negative-stiffness-mechanism isolation systems *Proc. SPIE* **1619** 44–54

Shaidani H 2008 Vibration isolation in cleanrooms: a system for virtually every application *Control. Environ. Mag.* January

Shen H, Wang C, Li L and Chen L 2013 Prototyping a compact system for active vibration isolation using piezoelectric sensors and actuators *Rev. Sci. Instrum.* **84** 055002
Yoshioka H and Murai N 2002 An active microvibration isolation system *Proc. 7th Int. Workshop Accelerator Alignment* pp 388–401

Lightning Source UK Ltd.
Milton Keynes UK
UKHW051607111121
393782UK00003B/116